Glycomicrobiology

Glycomicrobiology

Edited by

Ron J. Doyle
University of Louisville
Louisville, Kentucky

Kluwer Academic/Plenum Publishers
New York, Boston, Dordrecht, London, Moscow

Library of Congress Cataloging-in-Publication Data

Glycomicrobiology / edited by Ron J. Doyle.
    p. cm.
    Includes bibliographical references and index.
    ISBN 0-306-46239-7
    1. Glycoconjugates. 2. Microbial metabolism. 3. Glycoproteins. I. Doyle, Ronald J.

QR92.C3 G59 1999
572'.567293--dc21

99-044858

ISBN 0-306-46239-7

©2000 Kluwer Academic / Plenum Publishers, New York
233 Spring Street, New York, N.Y. 10013

http://www.wkap.nl

10 9 8 7 6 5 4 3 2 1

A C.I.P. record for this book is available from the Library of Congress

Printed in the United States of America

Contributors

Soman N. Abraham • Departments of Pathology and Microbiology, Duke University Medical Center, Durham, North Carolina 27710

Kevin L. Anderson • Department of Biological Sciences, Mississippi State University, Mississippi State, Mississippi 39762

Garabed Antranikian • Technical University Hamburg-Harburg, Department of Biotechnology, Institute for Technical Microbiology, 21071 Hamburg, Germany

Edward A. Bayer • Department of Biological Chemistry, The Weizmann Institute of Science, Rehovot 76100, Israel

Neil T. Blackburn • Department of Microbiology, University of Guelph, Guelph, Ontario, N1G 2W1, Canada

Lori L. Burrows • Canadian Bacterial Diseases Network, Department of Microbiology, University of Guelph, Guelph, Ontario, N1G 2W1, Canada

Anthony J. Clarke • Department of Microbiology, University of Guelph, Guelph, Ontario, N1G 2W1, Canada

Erika Crouch • Department of Pathology, Washington University School of Medicine, St. Louis, Missouri 63110

Mamadou Daffé • Institut de Pharmacologie et de Biologie Structurale du Centre National de la Recherche Scientifique, 31077 Toulouse Cedex, France

Kumari S. Devulapalle • School of Dentistry, Department of Basic Sciences, University of Southern California, Los Angeles, California 90089-0641

Jolyne Drummelsmith • Department of Microbiology, University of Guelph, Guelph, Ontario, N1G 2W1, Canada

Roman Dziarski • Northwest Center for Medical Education, Indiana University School of Medicine, Gary, Indiana 46408

Alvin Fox • Department of Microbiology and Immunology, University of South Carolina, School of Medicine, Columbia, South Carolina 29208

Zhimin Gao • Departments of Pathology and Microbiology, Duke University Medical Center, Durham, North Carolina 27710

Dipika Gupta • Northwest Center for Medical Education, Indiana University School of Medicine, Gary, Indiana 46408

David L. Hasty • Department of Anatomy and Neurobiology, University of Tennessee Memphis, Memphis, Tennessee 38163, and the Veterans' Affairs Medical Center, Memphis, Tennessee 38104

Howard F. Jenkinson • Department of Oral and Dental Science, University of Bristol Dental Hospital and School, Bristol BS1 2LY, England

Joseph S. Lam • Canadian Bacterial Diseases Network, Department of Microbiology, University of Guelph, Guelph, Ontario, N1G 2W1, Canada

Raphael Lamed • Department of Molecular Microbiology and Biotechnology, George S. Wise Faculty of Life Sciences, Tel Aviv University, Ramat Aviv 69978, Israel

Anne Lemassu • Université Paul Sabatier, 31062 Toulouse Cedex, France

Paul Messner • Zentrum für Ultrastrukturforschung und Ludwig Boltzmann-Institut für Molekulare Nanotechnologie, Universität für Bodenkultur Wien, A-1180 Wien, Austria

Sara Moens • Laboratory of Genetics, Faculty of Agricultural and Applied Biological Sciences, Catholic University of Leuven, 3001 Heverlee, Belgium

Gregory Mooser • School of Dentistry, Department of Basic Sciences, University of Southern California, Los Angeles, California 90089-0641

Itzhak Ofek • Department of Human Microbiology, Sackler Faculty of Medicine, Tel Aviv University, Ramat Aviv 69978, Israel

Andrea Rahn • Department of Microbiology, University of Guelph, Guelph, Ontario, N1G 2W1, Canada

Ian S. Roberts • School of Biological Sciences, University of Manchester, Manchester M13 9PT, England

Heather L. Rocchetta • Canadian Bacterial Diseases Network, Department of Microbiology, University of Guelph, Guelph, Ontario, N1G 2W1, Canada

Christina Schäffer • Zentrum für Ultrastrukturforschung und Ludwig Boltzmann-Institut für Molekulare Nanotechnologie, Universität für Bodenkultur Wien, A-1180 Wien, Austria

Jeoung-Sook Shin • Departments of Pathology and Microbiology, Duke University Medical Center, Durham, North Carolina 27710

Yuval Shoham • Department of Food Engineering and Biotechnology, Technion—Israel Institute of Technology, Haifa 32000 Israel

Evgeni V. Sokurenko • Department of Anatomy and Neurobiology, University of Tennessee Memphis, Memphis, Tennessee 38163

Hendrik Strating • Department of Microbiology, University of Guelph, Guelph, Ontario, N1G 2W1, Canada

Artur J. Ulmer • Divison of Cellular Immunology, Research Center Borstel, 23845 Borstel, Germany

Mumtaz Virji • Department of Pathology and Microbiology, School of Medical Sciences, University of Bristol, Bristol BS8 1TD, England

Constantinos E. Vorgias • National and Kapodistrian University of Athens, Faculty of Biology, Department of Biochemistry-Molecular Biology, Panepistimiopolis-Kouponia, 157 01 Athens, Greece

Chris Whitfield • Department of Microbiology, University of Guelph, Guelph, Ontario, N1G 2W1, Canada

Thomas Wugeditsch • Department of Microbiology, University of Guelph, Guelph, Ontario, N1G 2W1, Canada

Preface

At one time, it was thought that only eukaryotic cells could glycosylate proteins. Two major kinds of glycoproteins were recognized in animal cells, consisting of O- or N-linked saccharides. The O-linked saccharides were derived from glycosylation of threonine or serine, whereas N-linked saccharides were derived from asparagine. In recent years, numerous glycoproteins have been found in bacteria, many of which do not possess typical animal cell glycosylation patterns.

One of the purposes of this volume is to provide a thorough discussion of carbohydrate–peptide linkages in bacteria. Interestingly, though Braun's lipoprotein of some gram-negative bacteria was discovered nearly three decades ago, it was never considered to be a glycoprotein, even though it possessed a peptide–carbohydrate linkage. In the past few years, it is clear that even gram-positive cells can covalently bind proteins to their cell walls. The role of carbohydrates in the recognition of animal cells by bacteria is now well established. Saccharide-specific adhesins have been sequenced, cloned and employed as potential vaccines. Furthermore, carbohydrate receptors on animal cells for bacterial adhesins have been identified and characterized.

Another purpose for this volume is to provide a forum for new information on adhesin–receptor complexes involving bacterial pathogens. There now is a burgeoning literature on carbohydrate structure, function, and molecular biology in bacteria. The role of carbohydrates in biotechnology and biomass utilization has become important in the past decade due to new methods for carbohydrate detection and the cloning of biosynthetic and degradative enzymes.

A third purpose of the volume is to provide a modern outlook on the role of microbial glycoconjugates in the emerging field of biotechnology.

The editor recruited acknowledged world experts for this book on glycomicrobiology. All the chapters combine to create the world's first volume in this dynamic area. It is agreed that other chapters could have been added, resulting in an expanded volume. A judgment was made to include only the fastest moving areas of glycomicrobiology. The editor of this volume made the decision to exclude oth-

er areas of glycomicrobiology, some of which are important in disease, biotechnology, and industry.

Ron J. Doyle
University of Louisville

Contents

Chapter 1

Non-S-Layer Glycoproteins: A Review
Sara Moens

Chapter 2

**Glycans in Meningococcal Pathogenesis and the Enigma
of the Molecular Decorations of Neisserial Pili**
Mumtaz Virji

Chapter 3

Polypeptide Linkage to Bacterial Cell Envelope Glycopolymers
Howard F. Jenkinson

Chapter 4

Surface Layer Glycoproteins of Bacteria and Archaea
Paul Messner and Christina Schäffer

Chapter 5

Assembly Pathways for Biosynthesis of A-Band and B-Band Lipopolysaccharide in *Pseudomonas aeruginosa*
Lori L. Burrows, Heather L. Rocchetta, and Joseph S. Lam

Chapter 6

Interactions of Bacterial Lipopolysaccharide and Peptidoglycan with Mammalian CD14
Roman Dziarski, Artur J. Ulmer, and Dipika Gupta

Chapter 7

Pathways for the *O*-Acetylation of Bacterial Cell Wall Polysaccharides
Anthony J. Clarke, Hendrik Strating, and Neil T. Blackburn

Chapter 8

Glycobiology of the Mycobacterial Surface: Structures and Biological Activities of the Cell Envelope Glycoconjugates
Mamadou Daffé and Anne Lemassu

Chapter 9

Biosynthesis and Regulation of Expression of Group 1 Capsules in *Escherichia coli* and Related Extracellular Polysaccharides in Other Bacteria
Chris Whitfield, Jolyne Drummelsmith, Andrea Rahn, and Thomas Wugeditsch

Chapter 10

Mutans Streptococci Glucosyltransferases
Gregory Mooser and Kumari S. Devulapalle

Chapter 11

Glycosyl Hydrolases from Extremophiles
Constantinos E. Vorgias and Garabed Antranikian

Chapter 12

Profiling and Trace Detection of Bacterial Cellular Carbohydrates
Alvin Fox

Chapter 13

Degradation of Cellulose and Starch by Anaerobic Bacteria
Kevin L. Anderson

Chapter 14

**The Cellulosome: An Exocellular Organelle for Degrading
Plant Cell Wall Polysaccharides**
Edward A. Bayer, Yuval Shoham, and Raphael Lamed

Chapter 15

The Expression of Polysaccharide Capsules in *Escherichia coli*:
A Molecular Genetic Perspective
Ian S. Roberts

Chapter 16

Bacterial Entry and Subsequent Mast Cell Expulsion
of Intracellular Bacteria Mediated by Cellular
Cholesterol–Glycolipid-Enriched Microdomains
Jeoung-Sook Shin, Zhimin Gao, and Soman N. Abraham

Chapter 17

The Fim H Lectin of *Escherichia coli* Type 1 Fimbriae:
An Adaptive Adhesin
David L. Hasty and Evgeni V. Sokurenko

Chapter 18

Interactions of Microbial Glycoconjugates with Collectins:
Implications for Pulmonary Host Defense
Itzhak Ofek and Erika Crouch

1

Non-S-Layer Glycoproteins

A Review

Sara Moens

1. INTRODUCTION

For a long time, reports on glycoproteins in prokaryotic organisms (archaea and bacteria) were restricted to a few cases, and therefore the presence of glycoproteins in prokaryotes was rather a controversial matter. As more reports appeared, first mainly about archaebacterial S-layer (surface layer) glycoproteins, the occurrence of glycosylated proteins in prokaryotes generally became accepted. Also, the number of reports on glycosylated proteins in bacteria increased, first concerning mainly S-layer glycoproteins. The term "non–S-layer glycoproteins" was introduced by Sandercock *et al.* (1994), who were the first to review these bacterial glycoproteins. This review illustrated that the once controversial matter should now be viewed as common.

The first fully described non–S-layer glycoproteins were isolated from archaea. One of them is the halobacterial flagellin, of which the complete glycan structure and its linkage to the protein have been determined. Presently, for about 15 bacterial and archaeal glycoproteins the linkage unit has been determined, confirming the covalent attachment of sugars. The evidence for the glycosylation of a protein in many cases, however, is still restricted to lectin-binding and/or sugar-

Sara Moens • Laboratory of Genetics, Faculty of Agricultural and Applied Biological Sciences, Catholic University of Leuven, 3001 Heverlee, Belgium.
Glycomicrobiology, edited by Doyle.
Kluwer Academic/Plenum Publishers, New York, 2000.

staining techniques (see also Table 1 in Moens and Vanderleyden, 1997). The presence of a posttranslational modification such as glycosylation is often postulated from a discrepancy between the molecular weight derived from the gene and that seen for the protein after gel electrophoresis. Detailed studies on the structure of the glycoproteins did not follow the initial reports on glycosylation in all cases.

This chapter reviews reports on prokaryotic (archaeal and bacterial) glycoproteins. It sometimes happened that proteins that were first reported to be glycosylated were shown not to be in later additional studies (Kimura and Stadtman, 1995; Fairchild et al., 1991). It is not unlikely that this also will be the case for some of the glycoproteins here described. The abundance of sugar-containing molecules in many bacterial cell envelopes, such as lipopolysaccharides, makes a thorough purification of the putative glycoprotein essential.

Non-S-layer glycoproteins cover a wide range of locations and functions, both in archaea and bacteria. They are mainly extracytoplasmic proteins, either associated with the cell envelope–surface or secreted in the medium, but also intracellular glycoproteins have been described. In this chapter, the glycoproteins are described in groups according to their nature, including glycoproteins occurring in flagella and pili, glycosylated enzymes, and other (miscellaneous) glycoproteins. In the case of flagella, a clear distinction must be made between archaea and bacteria.

2. FLAGELLAR GLYCOPROTEINS

2.1. Bacteria

2.1.1. BACTERIAL FLAGELLA: STRUCTURAL ASPECTS AND ANTIGENICITY

Bacterial flagella (Winstanley and Morgan, 1997; Macnab, 1996) are ingenious structures that enable bacteria to move to their favorite environment. The global structure of a flagellum can be described as a rotary device anchored in the bacterial cell wall (cell membrane and surrounding layers) and a helical filament protruding in the environment. The rotary device is called *basal body*. Filament and basal body are joined by a structure called *hook*. Every different structural part of the flagellum is built up from one specific flagellar protein. The constituent of the filament is called *flagellin*. Flagellin molecules, ordered in a helical way, form the flagellar filament. In some cases, flagellar filaments are built up from multiple flagellins. Filament growth occurs at its distal end. Flagellin molecules are supposed to travel through a central pore in basal body, hook, and growing filament until they reach the place of their assembly. It is hypothesized that at the cytoplasmic side of the basal body, a selection apparatus is present that distinguishes between flagellar proteins, which can travel out, and other cellular proteins.

Flagellin amino acid sequences have been characterized for a large number of bacterial species (Wilson and Beveridge, 1993). This revealed a strong sequence conservation at both the N-terminal and the C-terminal ends. The central domain is more variable. Native flagellin molecules form a hairpinlike structure. In the filament, the N- and C-termini of the flagellins determine the inter- and intramolecular interactions, defining basic filament structure. The central variable domain of the flagellin is exposed to the medium, defining the filament surface.

Posttranslational modifications of bacterial flagellin molecules have been reported and include phosphorylation (Kelly-Wintenberg et al., 1990, 1993; Logan et al., 1989), methylation (Baker et al., 1983; Joys and Kim, 1979; Glazer et al., 1969; Ambler and Rees, 1959), and glycosylation. The number of glycosylated flagellins is quite small in comparison with the total number of bacterial flagellins that have already been characterized.

Glycosylated flagellins have been reported for *Azospirillum brasilense* (Moens et al., 1995), several *Campylobacter* species (Doig et al., 1996), *Clostridium tyrobutyricum* (Arnold et al., 1998; Bedouet et al., 1998), *Pseudomonas aeruginosa* (Brimer and Montie, 1998), and *Spirochaeta aurantia* (Brahamsha and Greenberg, 1988). A glycosylated flagellar sheath protein has been reported for the spirochete *Serpulina hyodysenteriae* (Li et al., 1993). Spirochete flagella are "endoflagella," located in the periplasmic space, that are surrounded by a proteinaceous sheath.

Flagella are in many pathogenic infections highly immunogenic (e.g., Nachamkin and Yang, 1989). This has even been reported for the endoflagella of spirochetes (Craft et al., 1986; Penn et al., 1985). Interestingly, a similar phenomenon has been reported for plant pathogens, where a *Pseudomonas* flagellin-derived peptide has been shown to elicit a plant defense response (Boller and Felix, 1996).

Some bacteria try to evade their host's immune response by alternating between a flagellated and an aflagellated phase (phase variation) or by expressing different types of flagella (antigenic variation). Both phenomena are well documented for *Campylobacter* sp. (Taylor, 1992). They might be the result of recombination events in the flagellar genes (Wassenaar et al., 1995) and might result in changing surface exposed epitopes of the filament, possibly posttranslationally modified.

2.1.2. REPORTED GLYCOSYLATED BACTERIAL FLAGELLINS

The list of glycosylated bacterial flagellins is not extensive but growing, mainly in the last years. Structural studies of the glycans have not yet been performed, nor is it known to which amino acid the sugar chain is linked. Consequently, it is not known to which domain of the flagellin the glycans are attached. It seems most obvious that the sugars will occur in the central variable domain, which defines the filament surface. In several cases there is some experimental evidence for this.

Azospirillum brasilense is a plant growth-promoting rhizosphere bacterium. Glycosylation of the flagellin of the polar flagellum was shown in different ways (Moens *et al.,* 1995). This includes positive sugar staining, reduced molecular weight after chemical deglycosylation, and recognition by a sugar-specific monoclonal antibody. This antibody clearly binds the filament surface when performing transmission electron microscopy after immunogold labeling. Also, the occurrence of partially glycosylated mutants was reported. One of these mutants was shown to be affected in a gene encoding a phosphomannomutase. This mutant is also affected in the structure of its exopolysaccharides.

In *Campylobacter coli,* two genes that are required for posttranslational modifications of the flagellin were identified (Guerry *et al.,* 1996). The deduced amino acid sequence of one gene, *ptmB,* displays similarity with CMP (cytosine monophosphate)-*N*-acetyl-neuraminic acid synthetases. This prompted the investigators to look for glycosylation of the flagellin. Using mild periodate treatment and biotin hydrazide labeling, the presence of a glycosyl moiety was proven for the flagellins of two antigenic variants of *C. coli* (Doig *et al.,* 1996). This glycosyl moiety was shown to be responsible for the antigenic differences of the flagellins. Glycosylation was also present on the flagellins of other *Campylobacter* strains. Several experimental observations supported the presence of a terminal sialic acid residue in the glycosyl moiety. The amino acids that are glycosylated are not yet identified, but several observations support their location in the surface-exposed part of the flagellin (Doig *et al.,* 1996; Power *et al.,* 1994). The *ptmB* did not correlate with the presence of sialic acid in the lipopolysaccharides, suggesting that its sole function may be the biosynthesis of flagellin posttranslational modifications.

Sugar staining, mild periodate oxidation, and β-elimination were used to prove the glycosylation of the *Clostridium tyrobutyricum* flagellin (Arnold *et al.,* 1998). The sugar moiety contains both Glc and GlcNAc (Bedouet *et al.,* 1998). A monoclonal antibody used for detection of this organism was shown to have its epitope located in the sugar moiety. Immunoelectron microscopy demonstrated that the monoclonal antibody recognized surface-exposed epitopes on the flagellar filament (Arnold *et al.,* 1998).

Recently, glycosylation of the flagellin was shown to occur in a-type *Pseudomonas aeruginosa* but not in b-type (Brimer and Montie, 1998). Evidence was given by a positive biotin–hydrazide glycosylation assay and a shift in molecular weight after chemical deglycosylation. An anti–a-type flagellar monoclonal antibody did still recognize the deglycosylated flagellin, indicating that it has no sugar epitopes.

Lectin binding showed that at least two of the three polypeptides in the flagellar filaments of *Spirochaeta aurantia* are glycosylated (Brahamsha and Greenberg, 1988). They are antigenically cross-reactive with flagellar polypeptides that are more abundantly present, and they may be glycosylated forms of the latter.

2.1.3. GLYCOSYLATION OF THE BACTERIAL FLAGELLAR PROTEINS: WHERE?

As already mentioned, flagellar proteins travel through a central pore in the flagellar structure to their place of assembly, without crossing a membrane (Macnab, 1996). Glycosylation is considered always to occur in a membrane-bound process. A possible location for the glycosylation machinery is the place where an as yet unidentified export apparatus resides at the bottom of the basal body.

2.2. Archaeal Flagella

2.2.1. DIFFERENT FROM BACTERIAL FLAGELLA

Archaeal flagellar structure is not so well established as bacterial flagellar structure, but the global structure seems to be similar (Jarrell *et al.,* 1996). The filament, however, is more narrow than bacterial flagellar filaments, and archaea with relatively thin cell envelopes may stabilize the insertion of flagella with subcytoplasmic membrane layers. Most archaebacterial filaments are composed of multiple flagellins.

Well-documented differences with the bacterial flagella include mainly features of the flagellin molecule itself. Glycosylation of the archaeal flagellins seems to be widespread (see Section 2.2.2). Also, sulfation occurs as posttranslational modification of archaeal flagellins (Wieland *et al.,* 1985).

The archaeal flagellin amino acid sequences are similar to each other but different from bacterial flagellin sequences. They are related, however, to bacterial type IV pilin sequences (Faguy *et al.,* 1994). Another remarkable feature of the archaeal flagellin amino acid sequences is the occurrence of an N-terminal signal sequence (Kalmokoff *et al.,* 1990). These data, taken together, have proposed that the export of flagellin molecules by archaea resembles more the type IV pilin export, involving signal peptide cleavage, than the export of bacterial flagellins through a central pore (Faguy *et al.,* 1994). Recently, an archaeal flagella-related putative gene product displaying similarity to type IV pilus accessory proteins was discovered (Bayley and Jarell, 1998). The occurrence of relatively narrow flagellar filaments also supports the type IV pili export hypothesis, since it questions the possibility of translocation of flagellins through a central pore.

2.2.2. REPORTED GLYCOSYLATED ARCHAEAL FLAGELLINS

In contrast to bacterial flagellins, glycosylation seems to be widespread in archaeal flagellins. In many of the cases, however, evidence is only given by sugar-staining reactions and no further characterizations were performed. The glycosylated flagellin of *Halobacterium halobium* is best characterized. (Wieland *et al.*, 1985). This flagellin carries sulfated oligosaccharides bound via an asparaginyl–glucose linkage. The oligosaccharides are of the type Glc4-1GlcU4-1GlcU4-1GlcU, where GlcU indicates glucuronic acid. A sulfate group is attached to each of the GlcU residues. The Asn is part of the (eukaryotic) N-glycosylation consensus acceptor sequence Asn-Xaa-Ser/Thr. *H. saccharovorum, H. salinarium,* and *H. volcanii* also were reported to contain glycosylated flagellins (Serganova *et al.*, 1995).

For the archaeal methanogens, *Methanococcus deltae* (Bayley *et al.*, 1993; Faguy *et al.*, 1992), *Methanospirillum hungatei* (Southam *et al.*, 1990), and *Methanothermus fervidus* (Faguy *et al.*, 1992) were reported to contain glycosylated flagellins. The glycosylation of the flagellins seems to be correlated with the sensitivity of the filament to low concentrations of Triton X-100 (Faguy *et al.*, 1992). The flagellar filaments from *Methanospirillum hungatei* are stable to temperatures up to 80°C and over a pH range from 4 to 10 (Faguy *et al.*, 1994). Incubation of *Methanococcus deltae* with bacitracin resulted in hypoglycosylated flagellins (Bayley *et al.*, 1993).

Natronobacterium magadii (Fedorov *et al.*, 1994) and possibly *N. pharaonis* (Serganova *et al.*, 1995) also are reported to have glycosylated flagellins. Fedorov *et al.* (1994) propose different subunit interactions in archaeal flagella as compared to bacterial flagella.

Glycosylation of the flagellin also was reported for the thermoacidophilic archaea *Sulfolobus shibatae* (Faguy *et al.*, 1996; Grogan, 1989) and *Thermoplasma volcanium* (Faguy *et al.*, 1996).

2.2.3. GLYCOSYLATION OF THE ARCHAEAL FLAGELLINS: OUTSIDE THE CYTOPLASMIC MEMBRANE

Different observations support the extracytoplasmic location of the glycosylation reaction in archaea. In *Halobacterium halobium* it was shown that EDTA, without entering the cell, is able to prevent the transfer of the sulfated oligosaccharides to the flagellins (Sumper, 1987; Sumper and Herrmann, 1978). This was due to the inhibition of a Mg^{2+}-dependent oligosaccharyltransferase. This influence of EDTA was not observed for the glycosylation of *M. deltae* flagellins (Bayley *et al.*, 1993). However, the level of glycosylation of the latter flagellins could be influenced by addition of bacitracin. This bacitracin also was shown to be able to prevent the addition of sulfated oligosaccharides to the S-layer glycoprotein of

H. halobium, without entering the cells (Mescher and Strominger, 1978). The influence of bacitracin on the glycosylation reaction indicates the involvement of a dolichol diphosphate lipid carrier. Furthermore, Lechner *et al.* (1985) reported the glycosylation of an artificial hexapeptide containing the consensus sequence for *N*-glycosylation Asn-Xaa-Ser/Thr, without the requirement for this peptide to enter the cell.

Results of Meyer and Schäfer (1992), who characterized a membrane-bound acid pyrophosphatase in *Sulfolobus,* and Zhu *et al.* (1995), who found plasma membranes of *Haloferax volcanii* to contain all enzyme activities for synthesis of *N*-linked glycoproteins, suggest the cytoplasmic membrane to be the archaeal counterpart of the eukaryotic endoplasmic reticulum and Golgi complex, where glycosylation reactions occur.

Jarrell *et al.* (1996) proposed a model for the assembly of archaeal flagella, taking into account the similarity with type IV pili and the extracytoplasmic site of glycosylation. In this model the flagellin is exported through the cytoplasmic membrane after cleavage of the leader peptide. Once outside the cell, the flagellin becomes glycosylated (by enzymes located in the cytoplasmic membrane) and is subsequently inserted into the base of the growing filament.

2.3. Glycosylated Flagellar Proteins: What Are the Sugar Chains For?

Several hypotheses concerning the function of the glycosylation of flagellar proteins have been proposed. Bacterial flagella are well known to induce immune responses. Glycosylation of the flagellar filament surface may contribute to antigenic variation, helping the bacterium to evade the immune response. Results of colonization experiments of the intestinal tract of rabbits by *Campylobacter* wild-type and a mutant in posttranslational modification of the flagellin (Guerry *et al.,* 1996) prompted the investigators to suggest that surface-exposed modifications of the flagellin are more critical in eliciting protection against subsequent challenge with the same bacterium than primary amino acid sequences. The sialic acid residues would block antibody production against the amino acid part of the flagellin, rendering the flagellum less detectable as foreign.

It has been proposed that glycosylation of archaeal flagellins plays a role in subunit interactions in the assembled flagellar filament or in the incorporation of the flagella into the cell envelope (Jarrell *et al.,* 1996). This was concluded from observations made with hypoglycosylated flagellins in *Methanococcus deltae* (Bayley *et al.,* 1993) and *Halobacterium halobium* (Wieland *et al.,* 1985), respectively. This raises the question, however, why glycosylation is not necessary in all archaeal flagellins.

Halobacteria have a different swimming behavior from the best-studied, peritrichously flagellated bacteria *Escherichia coli* and *Salmonella,* in which long runs are alternated with tumbles needed to change direction (Macnab, 1996). For the runs, the flagella turn counterclockwise, forming a bundle. For the tumbles, the flagella reverse to clockwise turning, resulting in a falling apart of the flagellar bundle. In *Halobacterium,* both clockwise and counterclockwise rotation results in a run, supported by bundled flagella. It has been proposed that this is due to the glycosylation of the flagellin, since it would make the flagella slide more smoothly to one another, without causing disassembling of the bundle when the direction of rotation changes (Alam and Oesterhelt, 1984). Similarly, the glycosylation on the periplasmic flagella of *Spirochaeta aurantia* could make them slide better over the periplasmic side of the membranes (Southam *et al.,* 1990). Glycosylation, however, was only reported for flagellar core polypeptides.

3. GLYCOPROTEINS IN BACTERIAL PILI, FIMBRIAE, AND CELL SURFACE FIBRILS

3.1. Bacterial Cell Surface Appendages

Bacteria can express several types of nonflagellar cell surface appendages (Ottow, 1975). They are variously called pili, fimbriae, or cell surface fibrils. They are mainly involved in adhesion processes (Jones and Isaacson, 1983).

Type IV pili occur in several bacterial species such as *Neisseria* and *Pseudomonas.* They are composed of several thousands of identical pilin subunits, arranged in a helical fashion, plus a few copies of pilus-associated proteins. Besides adhesion they are also implicated in other processes, such as twitching motility (Henrichsen, 1983) and bacteriophage adsorption (Bradley, 1972), mainly due to their ability to extend and retract. Pili are important immunogens and may undergo phase and antigenic variation (Seifert, 1996). *Neisseria* pili are known to be variable as a consequence of recombination between the pilin-encoding locus and a silent locus.

Type IV pilin amino acid sequences show high sequence conservation in the N-terminal part and display a C-terminal immunogenic region. The N-terminal part is important both for subunit to subunit interactions in the pilus fibre and for biogenesis of the pilus.

Several types of posttranslational modifications of pilin molecules have been reported including methylation (Strom and Lory, 1991), phosphorylation (Robertson *et al.,* 1977), and glycosylation (see Section 3.2). *Neisseria meningitidis* pilin displays three types of posttranslational modifications: an *O*-linked trisaccharide (see Section 3.2), an α-glycerophosphate, and a phosphoryl choline epitope (Virji, 1998) (Chapter 2, this volume).

3.2. Reported Glycoproteins in Pili, Fimbriae, and Fibrils

The glycosylated type IV pilins of *N. meningitidis* and *N. gonorrhoeae* are the best studied. In both cases, the structure of the glycan has been determined. Other reports mainly show the presence of glycosylated proteins in cell surface appendages by sugar-staining techniques only.

Pilins of *N. gonorrhoeae* and *N. meningitidis* are both *O*-glycosylated. The first report on glycosylation comes from Robertson *et al.* (1977). They found Gal and traces of Glc. Later (Virji *et al.*, 1993), *N*-glycosylation was proposed based on the occurrence of *N*-glycosylation consensus motifs. However, 2 years later, Stimson *et al.* (1995) provided evidence for an *O*-linked trisaccharide in *N. meningitidis,* and Parge *et al.* (1995) reported an *O*-linked disaccharide in *N. gonorrhoeae.* The *N. meningitidis* trisaccharide is of the type Gal(β1–4)Gal(α1–3)X, with X being a 2,4-diacetamido-2,4,6-trideoxyhexose that is *O*-glycosidically linked to Ser 63 (Marceau *et al.,* 1998). The *N. gonorrhoeae* glycan is of the form Gal(α1–3)Glc-NAc. This disaccharide also is covalently bound to Ser 63 (Parge *et al.,* 1995). This different glycosylation pattern, however, is not intrinsically a difference between all *N. meningitidis* and *N. gonorrhoeae* strains (Marceau *et al.,* 1998).

The type IV pilin of *P. aeruginosa* has been demonstrated to be glycosylated by an acidic carbohydrate moiety. Castric (1995) sequenced a region downstream of the pilin encoding gene. They found an open reading frame, *pilO,* with no significant sequence similarity to known genes. A *pilO* mutant produced a pilin with a lower apparent molecular weight and a more neutral isoelectric point (pI) value than the parental strain. This pilin failed to react with a sugar-specific reagent, which did react with the wild-type pilin.

Some *E. coli* pili also may contain covalently bound sugars. The presence of Glc (Brinton, 1971) and Glc, GlcN, and Gal (Tomoeda *et al.,* 1975) has been reported. Two types of *Myxococcus xanthus* fimbriae were purified by Dobson and McCurdy (1979) and both were reported to contain low but significant amounts of carbohydrate.

Cell surface fibrils of *Streptococcus sanguis* and *S. salivarius* were reported to consist of glycoproteins (Morris *et al.,* 1987; Weerkamp and Jacobs, 1982). The *S. salivarius* glycoprotein was first identified as a cell wall-associated protein antigen containing about 30% of neutral sugar and about 13% of amino sugar (Weerkamp and Jacobs, 1982). Later, Weerkamp *et al.* (1984) reported that this glycoprotein corresponds to cell surface fibrils. Morris *et al.* (1987) identified a *S. sanguis* glycoprotein isolated from cell surface fibrils.

Phormidium uncinatum is a filamentous gliding cyanobacterium. It contains an S-layer attached to the outer membrane and an array of parallel fibrils on top of the S-layer. These surface fibrils consist of a single protein, oscillin. Oscillin was shown to be a Ca^{2+}-binding, highly glycosylated protein (Hoiczyk and Baumeister, 1997), with a carbohydrate content of about 30% of the protein weight. The presence of Xyl, Glc, Rha, Fuc, Ara, and Gal was reported. Also, glycoproteins that are probably similar to oscillin have been reported in other cyanobacteria: in *Aphanothece halophytica* (Simon, 1981) and *Synechococcus* (Brahamsha, 1996).

3.3. Biosynthesis of Glycosylated Pilins

Type IV pili are known to be assembled by the general cellular secretion pathway involving signal peptide cleavage (Hultgren *et al.,* 1996; Pugsley, 1993). Pilins are synthesized as precursors that become processed at a highly conserved consensus cleavage site.

The high degree of hydrophobicity of PilO of *Pseudomonas aeruginosa* within the predicted transmembrane regions suggests that it resides in the cytoplasmic membrane (Castric, 1995). This location would be ideal if PilO functions catalytically on the periplasmic side of the cytoplasmic membrane to transfer carrier lipid-bound oligosaccharide subunits to emerging pilin monomers.

Other biosynthetic genes that have been identified are the *Neisseria galE* and *pglA* (Jennings *et al.,* 1998; Stimson *et al.,* 1995). The gene *galE* would encode a Gal epimerase needed for the production of UDP-Gal. It is also involved in lipopolysaccharide biosynthesis. The gene *pglA* probably encodes a galactosyl transferase. This enzyme is specific for the addition of Gal to pilin and is not involved in the production of lipopolysaccharides. PglA shows homology to glycosyltransferases involved in both lipopolysaccharide and capsular polysaccharide biosynthesis. Some of these enzymes have been shown to act on lipid intermediates. This suggests that pilin glycosylation may use a similar lipid intermediate pathway (Jennings *et al.,* 1998).

3.4. Significance of the Glycosylation of Nonflagellar Cell Surface Appendages

Numerous studies have shown that pili play an important role in the adhesion of *Neisseria* to both endothelial and epithelial cells. Besides pilin, two other proteins, PilC1 and PilC2, have been implicated in adhesion. They may be pilus tip-located adhesins. The influence of the glycosylation of pilin on adhesion has also been studied. Gubish *et al.* (1982) suggested that sugar moieties, and especially Gal, are required for optimal attachment of *N. gonorrheae* to their host cells, because galactosidase treatment of pili reduced attachment. For *N. meningitidis,* Virji *et al.* (1993) also found a correlation between the glycosylation status of pilin and adhesion. In later studies, however, a major role for the Gal residues was ruled out, and generally the significance of the whole sugar substitution in adhesion was questioned. Stimson *et al.* (1995) showed that a *galE* mutant of *N. menigitidis,* lacking the digalactosyl moiety, is similar to wild-type in adhesion and that wild-type and a hyperadherent variant have identical sugar substitutions. Jennings *et al.* (1998) also showed that a mutant in the gene for a glycosyltransferase, *pglA*, involved in the addition of Gal to the trisaccharide has no altered adhesion phenotype. Marceau *et al.* (1998) claimed that there is no major role for pilin glycosylation in piliation and subsequent pilus-mediated adhesion. Bacteria producing nonglycosylated pilin, by substituting Ser 63 by Ala, were slightly more piliated than wild-type strains, and this increase in piliation would be responsible for the moderate increase in adhesion that was observed. Their data demonstrated that glycosylation of the pilin facilitates solubilization of pilin monomers and/or individual pilus fibres.

Parge *et al.* (1995) determined the crystallographic structure of the *N. gonorrhoeae* pilin and proposed a pilus model in which carbohydrate and hypervariable regions protrude from a smooth cylinder. They suggest that these exposed structures are the bacterial "cloaking devices" against the host immune response. The major significance of glycosylation of pilin probably involves antigenic traits and not adhesion to host cells. This also has been suggested for other surface antigens such as glycosylated flagellar proteins (see Section 2).

The absence of a functional *pilO*, implicated in the glycosylation of the pilin of *P. aeruginosa* (Castric, 1995), did not influence twitching motility and phage sensitivity, both of which rely on extension and retraction of the pilus. Glycosylation of this pilin is thus not necessary for the formation of *Pseudomonas* pilus fibers and for the extension and retraction of these fibers.

For the gliding cyanobacterium *Phormidium,* the glycoprotein oscillin was shown to be necessary to move. It was suggested that the secretion of carbohydrates and the interaction of the resulting mucus and the glycoprotein surface of the organisms generates the thrust necessary for translocation (Hoiczyk and Baumeister, 1997). Also, other oscillinlike cell surface-associated glycoproteins have been proposed to be involved in cyanobacterial motility (Brahamsha, 1996; Simon, 1981).

4. GLYCOSYLATED ENZYMES AND COMPONENTS OF ENZYME COMPLEXES

4.1. Occurrence of Bacterial Glycosylated Enzymes and Components of Enzyme Complexes

Of all non–S-layer prokaryotic glycoproteins, the group of the bacterial glycoenzymes and glycosylated components of enzyme complexes is the one that is best characterized at the structural level. For quite a number of these proteins, the detailed structure of the glycan chain and the linkage unit has been identified. This group of glycoproteins consists mainly of secreted enzymes; however, cell wall-associated and intracellular glycoproteins also are reported.

The glycosylated enzymes and enzyme complexes are mainly involved in the degradation of cellulose and hemicellulose. This makes them important study objects for possible biotechnological applications. Most of the cellulases studied so far have been reported to be glycosylated. Also, some pectin-degrading, peptido-glycan-degrading, and proteolytic enzymes occur as glycoproteins.

4.2. Glycosylation Reported for Enzymes and Enzyme Complexes

Cellulose degradation by bacteria is mainly performed by extracellular multienzyme systems. In *Cellulomonas fimi*, different cellulases have been implicated in this process. Two cellulose-binding β-1,4-glucanases have been studied in more detail: CenA, an endoglucanase, and Cex, an exoglucanase. Both are glycoproteins (Gilkes *et al.*, 1984; Béguin and Eisen, 1978). The N-terminal part of CenA displays sequence similarity with the C-terminal part of Cex. These parts are proposed to comprise the cellulose-binding domains (Gilkes *et al.*, 1988). In both enzymes, the conserved region is separated from a nonconserved region by a sequence solely constituted by Pro and Thr residues.

Cex also is glycosylated when expressed in *Streptomyces lividans*. When the different domains of Cex were expressed separately, only the Pro–Thr linker was glycosylated, suggesting *O*-linked glycosylation of this particular domain of Cex. The glycans contained Man and Gal residues (Ong *et al.*, 1994).

Streptomyces lividans is reported to produce an intracellular glycosylated β-1,4-glucosidase (Mihoc and Kluepfel, 1990) and an extracellular glycosylated xylanase (Kluepfel *et al.*, 1990). Evidence is only based on positive sugar staining.

Glycoprotein components of the secreted cellulase and xylanase system of a mesophilic *Bacillus* species were reported by Paul and Varma (1992) based on lectin binding and sugar staining. The cellulase contained 11.5% of carbohydrate and the xylanase contained 20% of carbohydrate.

The cellulolytic bacterium *Fibrobacter succinogenes* produces a chloride-stimulated cellobiosidase that was shown to be a glycoprotein (Huang *et al.*, 1988). The reported carbohydrate content was estimated to be between 8 and 16%.

Calza *et al.* (1985) purified two β-1,4-endoglucanases from *Thermomono-spora fusca*. One was shown to be a glycoprotein containing 25% carbohydrate by weight.

Most of the cellulases of *Clostridium thermocellum* are organized into a multicomponent complex, called the cellulosome (Chapter 14, this volume). Cellulosomes occur in both extracellular and cell surface-associated forms. The cellulosome mediates strong adhesion of the bacterium to the substrate and consists of at least 14 polypeptide subunits. Most of these subunits are cellulases. Some of the subunits are glycoproteins. In particular, the largest subunit, S1, which exhibits no

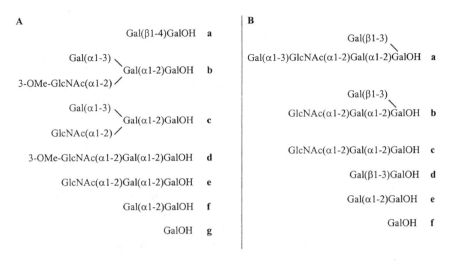

Figure 1. Structures of the oligosaccharide alditols generated by alkaline-borohydride treatment of the (A) cellulosome of *Clostridium thermocellum* and of the (B) cellulase complex of *Bacteroides cellulosolvens*. The alditols were fractionated via gel permeation chromatography and high-pressure liquid chromatography and subsequently analysed by monosaccharide analysis, methylation analysis, 500-MHz ^{1}H-nuclear magnetic resonance spectroscopy, and fast-atom-bombardment mass spectrometry (Gerwig *et al.*, 1989, 1991, 1992).

measurable cellulolytic activity and is probably involved in the structural organization of the cellulosome, has been estimated to contain 40% covalently bound sugars. Gerwig *et al.* (1989) characterized a tetrasaccharide and a disaccharide that are present in the cellulosome. The tetrasaccharide has been located on the S1 subunit. The structures of the oligosaccharide-alditols obtained after alkaline borohydride treatment are given in Fig. 1A (a and b). Later (Gerwig *et al.*, 1991), additional oligosaccharides were identified that were partial structures of the tetrasaccharide (Fig. 1A, c–g). The carbohydrate chains were split off using alkaline borohydride conditions, indicating *O*-linkages. This was confirmed by Gerwig *et al.* (1993), who analyzed glycopeptide fractions and determined that the sugar chains are *O*-linked by galactopyranose to a Thr residue.

The multiple cellulases-containing protein complex of *Bacteroides cellulosolvens*, which is probably similar to the cellulosome of *C. thermocellum*, also contains oligosaccharides that are *O*-linked, mainly to the largest subunit (Gerwig *et al.*, 1992). The structures of the oligosaccharide-alditols obtained after alkaline borohydride-treatment are given in Fig. 1B. They include a novel pentasaccharide and a series of closely related partial structures. The oligosaccharides are *O*-linked to Thr and partly to Ser by galactopyranose (Gerwig *et al.*, 1993).

Chryseobacterium meningosepticum secretes a variety of hydrolytic enzymes into the culture medium. Three of these enzymes, two endoglycosidases Endo F_2 and

Endo F_3, and a protease, P40 (also designated flavastacin), were shown to be glycosylated via an unusual O-linked oligosaccharide. The oligosaccharide is a heptasaccharide that is singly branched and contains uronic acids (glucuronic acid GlcU and 2-acetamido-2-deoxy-glucuronic acid GlcNAcU) and methylated sugars. Its structure has been shown to be (2-OMe)Man1–4GlcNAcU1–4GlcU1–4Glc1–4(2-OMe)GlcU-4[(2-OMe)Rha1–2]Man (Reinhold et al., 1995). Endo F_2 has three sites for the oligosaccharide; Endo F_3 and P40 have each one site. The oligosaccharide is attached via a Man residue to a Ser or Thr at consensus sites corresponding to Asp-Ser* or Asp-Thr*-Thr (Plummer et al., 1995). The oligosaccharide moiety of flavastacin is located distantly from the catalytic site (Tarentino et al., 1995).

    Thermoanaerobacterium thermosaccharolyticum produces an extracellular protein complex with pectin methylesterase and polygalacturonate hydrolase activity. The complex consists of two subunits. It was not possible to locate the activities on either subunit. The largest subunit contains 10% sugars. Monosaccharide analysis showed the presence of Gal and GalNAc (Van Rijssel et al., 1993).

    Bacillus circulans produces an enzyme that degrades guar gum, which is a galactomannosaccharide. This G-enzyme is a β-1,4-mannanase and is claimed to be a glycoprotein (Yoshida et al., 1998a). The G-enzyme was reported to contain GalNAc, Xyl, GlcNAc, Man, Fuc, and Gal, accounting for 1.8%, 0.9%, 11.0%, 6.1%, 1.1%, and 79.1%, respectively, of the total monosaccharides. Strangely, when this enzyme was expressed in *E. coli*, the recombinant enzyme seemed also glycosylated, with GalNAc, Xyl, GlcNAc, Man, Glc, and Gal accounting for 11.7%, 14.4%, 6.1%, 3.2%, 54.2%, and 10.4%, respectively, of the total monosaccharides. The enzymatic activity and optimal pH and temperature were similar for both enzymes.

    The autolytic N-acetylmuramoylhydrolase of *Enterococcus faecium* was purified by Kawamura and Shockman (1983). The autolysin has a high affinity for binding the cell wall. Lectin binding suggested that this enzyme is glycosylated. Sugar staining of gel electrophoresed cyanogen bromide cleaved protein stained three bands, indicating at least three sites of glycosylation. Carbohydrate analysis showed the presence of Glc.

    Bacillus megaterium produces a protein factor that stimulates peptidoglycan synthesis of *B. megaterium* cells treated with toluene and LiCl. This protein was extracted from toluene-treated cells using high concentrations of LiCl. The LiCl-treated cells had a decreased efficiency of peptidoglycan synthesis. Sugar staining showed that this protein is glycosylated. It is probably associated with the cell membrane (Taku and Fan, 1976).

4.3. Indications for the Function of Glycosylation of Enzymes or Components of Enzyme Complexes

    Several functions have been proposed for the glycosylation of enzymes or enzyme-associated proteins. They include influence on conformation of the enzyme, stability, activity, binding to the substrate, and protection from proteolysis. In several cases this has explicitly been studied by comparing glycosylated and nonglycosylated forms of the proteins. This is illustrated below by a few studies. Nonglycosylated forms may be produced as recombinant enzymes is in *E. coli*. However, it should be mentioned that the conclusions from the studies are not always unambiguous. The relevance of the glycan chains for some properties such as thermotolerance may differ from case to case. In general, the glycan chains seem to have no function in catalytic activity.

Thermomonospora fusca produces two endoglucanases, one of which is glycosylated (see Section 4.2). The nonglycosylated endoglucanase displayed the highest specific activity, was excreted normally, and was very stable. This led the authors to suggest that carbohydrates are not needed for these properties (Calza *et al.*, 1985).

Gilkes *et al.* (1984) observed that *C. fimi* cellulases had the same activity when they were produced in *E. coli*. Glycosylation thus seemed no absolute requirement for catalytic activity. Langsford *et al.* (1987) came to the same conclusion; in addition, they found that glycosylation does not significantly affect stability of the enzymes toward heat and pH effects.

The presence of glycans on the Pro–Thr linker of Cex, however, did increase its affinity for cellulose (Ong *et al.*, 1994), and the glycosylation protected both CenA and Cex from attack by a homologous protease when bound to cellulose (Gilkes *et al.*, 1988). In solution, they were cleaved slowly. Nonglycosylated cellulases were cleaved by the protease at specific sites (Gilkes *et al.*, 1988). The linker region in cellulases assumes in many cases an extended and stiff conformation, unlike the tightly folded globular domains, exposing the linker domain to protease digestion. Organisms may circumvent this problem by attaching sugars to this vulnerable region (Ong *et al.*, 1994).

A *Bacillus macerans* and a hybrid *B. amyloliquefaciens–B. macerans* β-(1,3–1,4)-glucanase gene were expressed in both *Saccharomyces cerevisiae* and *E. coli* (Olsen and Thomsen, 1991) in order to establish the relationship between enzyme stability and glycosylation. The enzymes secreted by *S. cerevisiae* were glycosylated, unlike their native forms in *Bacillus* or the form produced by *E. coli*. The glycosylated enzymes displayed a higher thermotolerance than their nonglycosylated counterparts from *E. coli* but had no different pH optimum or catalytic activity.

4.4. Biosynthetic Aspects

Most of the reported enzymes and enzyme-associated proteins occur extracellularly. In the case where the encoding gene is known, the presence of signal peptides has been reported (e.g., O'Neill *et al.*, 1986; Tarentino *et al.*, 1995). It is reasonable to suggest that these proteins become glycosylated when they pass the cytoplasmic membrane. How intracellular enzymes, such as a β-1,4-glucosidase of *S. lividans* (Mihoc and Kluepfel, 1990), become glycosylated remains obscure. For the addition of the individual sugar units to the glycan chains of the cellulosomes of *C. thermocellum* and *B. cellulosolvens,* a biosynthetic pathway has been proposed (Gerwig *et al.,* 1991, 1992). This is based on the family of sugar chains that have been found in both cases (see Fig. 1). The initiation could take place by transfer of a Gal to the protein core, followed by elongation through addition of the individual monosaccharides from their nucleotide-activated derivatives. This is a similar mechanism as known to occur in the synthesis of mucintype glycoproteins in higher organisms.

5. MISCELLANEOUS GLYCOPROTEINS IN BACTERIA AND ARCHAEA

5.1. Miscellaneous Non-S-Layer Glycoproteins

In this paragraph, studies of glycoproteins that cannot be classified in the previous groups are summarized. This includes excreted, cell envelope-associated, and other glycoproteins. "Cell envelope" includes all layers surrounding the cytoplasm, including the plasma membrane. "Other" means either glycoproteins from a specific location other than extracellularly or cell envelope-associated, or from an unknown location. In the latter case, the glycoproteins were detected in whole cell preparations. Most of the miscellaneous glycoproteins, however, reside extracellularly or in the cell envelope.

Both bacterial and archaeal miscellaneous glycoproteins are described. Only in a few cases, structural information on sugar composition and linkage unit is available.

5.2. Reports on Miscellaneous Excreted Glycoproteins

Mycobacterial antigens are intensively studied in view of their use as diagnostic reagents and/or vaccines for human and bovine tuberculosis. In the human pathogen, *Mycobacterium tuberculosis,* several prominent antigens are glycosylated. Glycosylation of mycobacterial antigens has already suggested, mainly based on lectin-binding experiments. A 45-kDa and a 19-kDa antigen, have been studied more in detail (Chapter 8, this volume).

Dobos *et al.* (1995, 1996) extensively purified the 45-kDa antigen before performing sugar analysis, since it is known that in mycobacterial extracts soluble lipoglycans and glycolipids predominate. Mass spectrometry of a peptide fragment generated by proteolytic digestion chemically proved covalent glycosylation (Dobos et al., 1995). A mannobiose unit was found to be linked to a Thr residue near the N-terminus of the protein. The presence of Man, Glc, Gal, and Ara in the glycoprotein was suggested (Dobos et al., 1995), but radioactive labeling showed that only Man is really covalently linked (Dobos et al., 1996). Monomeric mannose, mannobiose, and mannotriose occur at four sites, O-linked to Thr in Pro-rich domains (Dobos et al., 1996). All the Man-Man linkages are α(1–2).

The 19-kDa lipoprotein antigen of M. tuberculosis was shown to be glycosylated by lectin-binding techniques (Garbe et al., 1993). The 19-kDa antigen was expressed in E. coli and displayed a different apparent molecular weight than that of the native protein (Garbe et al, 1993). This posttranslational modification was said to be not essential for the expression of the 19-kDa antigenicity (Garbe et al., 1993). Herrmann et al. (1996) determined the peptide region that is required for glycosylation by constructing hybrids with PhoA. They found involvement of a Thr doublet and triplet, near the N-terminus. Based on experimental observations, they proposed a role for the glycosylation of these sites in regulation of cleavage of the proteolytically sensitive linker region close to the acylated N-terminus of the protein. When the 19-kDa antigen was expressed in the rapid-growing M. smegmatis, it also was glycosylated (Garbe et al., 1993).

The presence of glycosylated major antigens also was demonstrated by glycan detection and lectin-binding in M. bovis (Fifis et al., 1991). Monosaccharide analysis suggested the presence of mainly Man and Ara.

Erysipelothrix rhusiopathiae, an arthritogenic bacterium, also releases antigenic compounds into its culture supernatant. One of these is a protein of high molecular mass that is recognized by "inductive" monoclonal antibodies. Inductive antibodies can provide protection against arthritis when they are applied several days before challenge. Their working mechanism and especially their bacterial epitopes are not yet known and are under study. The high molecular mass protein was shown to be glycosylated by sugar staining on a protein gel (Meier et al., 1992). Long-term heat treatment of the protein did not diminish the reactivity by the monoclonal antibodies, which led the authors to suggest the presence of linear epitopes.

Clavibacter michiganense produces a phytotoxic glycopeptide that is able to produce wiltings in plant cuttings. This toxin bears a single carbohydrate chain that is a highly branched heteropolymer, consisting mainly of Man and Glc, but also small amounts of Rha and Gal are present (Strobel et al., 1972). β-Elimination suggests the presence of O-glycosylation to Thr. The linking sugar is Man (Strobel et al., 1972).

Webster et al. (1981) characterized an autolysin from Clostridium acetobutylicum. They purified it from an industrial-scale acetone–butanol fermentation as a proteinaceous substance with antibioticlike activity. Sugar staining showed that this autolysin is a glycoprotein.

CspA is the most abundant protein in the culture medium of *Clostridium acetobutylicum* NCIB 8052, but can also be detected in cell envelopes. It was previously thought to be an autolysin since a preparation containing 95% CspA displayed amidase activity on pneumococcal cell walls. Further analysis, however, revealed that this cell wall hydrolytic activity was not associated with CspA. CspA has a C-terminal choline-binding domain and specifically recognizes choline residues in the cell wall. CspA was positive in a glycan detection test and was shown to contain 12% of Rha (Sanchez-Beato *et al.*, 1995). Interestingly, when the *cspA* gene was expressed in *E. coli*, the molecular weight of the recombinant product was lower than that of the native *Clostridium* CspA, but higher than the molecular weight deduced from the amino acid sequence.

The copper-resistant methanogenic archaebacterium *Methanobacterium bryantii* secretes three proteins when exposed to copper. These copper response extracellular (CRX) proteins appear to be glycosylated since they displayed selective lectin binding, indicating the presence of terminal Man and/or Glc (Kim *et al.*, 1995). Sialidase treatment reduced the molecular weight, indicating the presence of *N*-acetyl neuraminic acid. Several observations, including identical N-terminal amino acid sequences, conserved epitopes, very similar electrophoretic and chromatographic properties, and Southern blot analyses, suggest the presence of only one *crx* gene. The three proteins seem to be different glycosylated forms of one protein encoded by a single gene (Kim *et al.*, 1995).

5.3. Glycoproteins That Occur Associated with the Cell Envelope

One major plasma membrane protein of the archaebacterium *Thermoplasma acidophilum* stains positive with sugar staining and was purified using lectin affinity chromatography. The carbohydrate portion of this glycoprotein was shown to consist mainly of Man. Also Glc, Gal, and traces of GlcN were detected. The carbohydrate content was 8 to 10%. Several techniques were combined in order to determine the glycan structure. This revealed a highly branched Man-rich glycan, probably *N*-glycosidically linked to Asn (Yang and Haug, 1979). *Thermoplasma acidophilum* is totally devoid of a cell wall and this membrane glycoprotein may play a role in the survival of this organism in its extreme environment (high temperature and low pH).

Streptococcus pyogenes produces a glycoprotein that displays antitumor properties (Yoshida *et al.*, 1998b). This protein is a surface-located antigen that was shown to contain sugars by sugar staining (Kanaoka *et al.*, 1987). The sugar components include allose (Yoshida *et al.*, 1998b).

The platelet aggregation-associated protein of *Streptococcus sanguis* is a cell wall protein. It contains 39% of carbohydrate, that is present in Rha-rich polysaccharides (Herzberg *et al.*, 1990). Also Gal, Glc, Man, Rib, GalNAc, and GlcNAc occur in the glycan chains (Erickson and Herzberg, 1993). Effects of treatment of this protein with glycosylation inhibitors and glycohydrolases suggested the presence of N-linked sugars. The occurrence of 16 different glycosylation sites bearing four to five different glycan chains (all bound either via GalNAc or GlcNAc to an Asn residue) was postulated from other experiments. Preliminary data coming from NMR spectroscopy on one glycopeptide fraction indicated the presence of an N-asparaginyl linkage via GlcNAc (Erickson and Herzberg, 1993). The same authors find it unlikely that the carbohydrates would participate in platelet interactions, since they have shown earlier that peptide fragments devoid of carbohydrate still retained biological activity.

The ability of *Pseudomonas syringae*, *P. fluorescens,* and *Pantoea agglomerans* to induce ice nucleation involves the products of the "ice genes." The genes of all three organisms are related and encode a protein with unique N-terminal and C-terminal ends and an internal repeating octapeptide. Posttranslational modification of this protein appears to be critical in ice nucleation. The ice nucleating component of *P. syringae* is best characterized, the *P. agglomerans* and *P. fluorescens* products are probably similar. The presence of three chemically distinct classes capable of inducing ice nucleation at different temperatures, as well as intermediates, has been shown on the surface of ice nucleation-active cells. For some reasons they cannot be readily isolated and separated and were identified using indirect approaches (Turner *et al.,* 1991). Class A structures, inducing ice nucleation at $-4.5°C$ and higher, were reported to contain the ice protein linked to phosphatidylinositol (PI) and Man, probably as a complex mannan, and possibly GlcN. The PI unit has been suggested to anchor the protein to the outer cell membrane. Class B structures, inducing ice nucleation between -6.0 and $-7.0°C$, were shown to contain only Man and GlcN. Class C structures, which have the poorest ice nucleation capacity and induce ice nucleation at $-8.0°C$ or lower, contained only a few Man residues (Turner *et al.,* 1991).

The coupling of the posttranslational modifications to the protein was studied and this revealed the presence of both N- and O-glycan linkages (Kozloff *et al.,* 1991). Man residues were found to be bound to the amide nitrogen of one or more Asn residues in the N-terminal part of the protein. These Man residues are involved in the attachment of PI to the protein. Additional sugar residues were shown to be O-linked to Thr and Ser in the repeating octapeptide. They include Gal, GlcN, and most likely additional Man. Evidence has been provided that the sugar modifications play a role in aggregating the ice gene lipoglycoprotein compound into larger aggregates, which are the most effective ice nucleation structures (Kozloff *et al.,* 1991). In the same report, a scheme illustrating the suggested sequential formation of the ice-nucleating structure from class C through class B to class A was drawn. Interestingly, introduction of ice genes in *E. coli* results in converting the *E. coli* phenotype from non–ice nucleating to ice nucleating, with the presence of all three classes of ice nucleation activity (Turner *et al.,* 1991). It has been suggested that the presence of the ice gene product stimulates the expression of otherwise cryptic genes, necessary to produce the ice nucleating compound.

Fibrobacter succinogenes produces several cellulose-binding proteins. One 180-kDa cellulose-binding protein displays antigenic cross-reactivity with numerous cell envelope proteins, including the previously described chloride-stimulated cellobiosidase (see Section 4.2). Periodate treatment of the 180-kDa protein and the cell envelope proteins resulted in loss of antibody binding, suggesting that the common epitope is carbohydrate in nature (Gong *et al.*, 1996). This 180-kDa cellulose-binding protein and/or related cross-reactive surface proteins were shown to have a role in the adhesion of *Fibrobacter* cells to cellulose (Gong *et al.*, 1996).

Lectin binding suggests the presence of many glycoproteins in whole cell lysates of *Borrelia burgdorferi* (Luft *et al.*, 1989; Coleman and Benach, 1988). Most of the reaction, however, was said to be nonspecific, with the possible exception of a few proteins, since co-incubation with the cognate sugars and treatment with specific glycosidases did not inhibit the lectin binding (Luft *et al.*, 1989; Coleman and Benach, 1988). Two major outer surface-exposed proteins (OspA and OspB) were shown to be glycosylated (Sambri *et al.*, 1992). They show a positive reaction with a glycan detection kit and N-glycosidase F treatment abolished this reaction. The latter result shows the presence of N-linked glycans, probably via the eukaryotic $N–N$ diacetylchitobiose unit. Periodate treatment of OspA and OspB abolished their reactivity with certain monoclonal antibodies, indicating carbohydrate epitopes. Furthermore, it was demonstrated that radioactively labeled GlcNAc was incorporated into the carbohydrate residues associated with OspA and OspB.

The green photosynthetic bacterium *Chloroflexus aurantiacus* produces three small blue copper proteins. They were designated auracyanins A, B-1, and B-2, and they are peripheral membrane proteins. Two of them were shown to be glycosylated by sugar staining. Neutral sugar analysis indicated that auracyanin B-1 contains $9.4 \pm 0.3\%$ sugar and auracyanin B-2 contains $3.3 \pm 0.2\%$ sugar (McManus *et al.*, 1992). Related plant blue copper proteins also are often reported to be glycosylated.

VGP is a prominent surface protein of vegetative *Myxococcus xanthus* cells. Sugar staining indicated that it is glycosylated. Furthermore, the presence of 13.5% sugars comprising primarily neutral sugars and smaller amounts of hexosamines and uronic acids was demonstrated. Lectin affinity chromatography indicated the presence of terminal GalNAc residues (Maeba, 1986).

Membranes of *Micrococcus luteus* are very rich in Man, most of which occurs in the form of a lipomannan. Some evidence was provided for the occurrence of other glycosylated molecules, probably glycoproteins. Doherty *et al.* (1982) showed that membranes of *M. luteus* were capable of catalyzing the glycosylation of a number of compounds in the presence of ^{14}C-labeled GDP-Man. The level of radioactivity associated with the labeled compounds decreased dramatically following protease digestion, and sodium dodecyl sulfate (SDS) treatment of the membranes prior to incubation with the radioactive precursors did not label endogenous membrane compounds.

5.4. Other Miscellaneous Glycoproteins

The archaebacterium *Methanosarcina mazei* undergoes major morphological changes during growth, involving unicellular and multicellular forms. Immunochemical properties were described that distinguish these two forms (Yao *et al.*, 1992). Lectin binding on extracts of whole cells indicated the presence of glycoproteins in both unicellular and multicellular forms. Both forms display a similar pattern of glycoproteins; however, the quantities may differ.

Bacillus thuringiensis produces an intracellular proteinaceous crystal toxin that is probably a glycoprotein. The reported carbohydrate content of purified *B. thuringiensis* crystals varies considerably, and it has even been suggested that the detected sugars are not covalently attached but are due to insufficient purification (Huber *et al.*, 1981). Pfannenstiel *et al.* (1987) provided evidence for the covalent attachment of the amino sugars GalNAc and GlcNAc to the toxin. They found that these sugars were still present after alkali solubilization, boiling in SDS, separation by polyacrylamide gel electrophoresis and transfer to nitrocellulose membranes, and therefore suggested a covalent attachment. These amino sugars were shown to be critical for the larvicidal action of the toxin (Muthukumar and Nickerson, 1987).

It was reported that *B. thuringiensis* sporangia contain glycoproteins different from the crystal proteins reported so far. These glycoproteins had a double localization in the sporangium occurring in both spores and membranes. Two glycoproteins were found, one of 72 kDa and one of 205 kDa. The 205-kDa glycoprotein was shown to be a multimer of the 72-kDa species. Deglycosylation resulted in a 54-kDa species. Three different oligosaccharides were found that occur *O*-linked to Ser. One of the oligosaccharides contained GalNAc at the reducing end, Rha, and a not yet identified compound (García-Patrone and Tandecarz, 1995).

Microcystis aeruginosa is a freshwater cyanobacterium that displays hemagglutinating activity. The lectin involved was purified from cell extracts and was shown to be a glycoprotein containing 7.8% neutral sugars (Yamaguchi *et al.*, 1998).

6. CONCLUDING REMARKS

The study of the glycosylation of proteins in prokaryotic organisms is a growing field. As can been seen when reading this chapter, information on structures, functions, and biosynthesis of non–S-layer glycoproteins is still fragmentary. It is therefore difficult at this moment to draw a general picture about the architecture of these glycosylated proteins. The linkage of the glycan chain with the protein via the amide nitrogen of Asn or via the hydroxyl group of Ser or Thr, however, is a structural property shared with eukaryotic glycoproteins. The linking sugar is more variable (see also Table 1 in Messner, 1997, and Chapter 4, this volume). Several proposals have been made concerning biological functions of the glycosylation. Extensive studies on this matter have been performed in the case of biotechnologically important enzymes. Similarities with functions of glycosylation in eukaryotic glycoproteins were found, but the results were not always unambiguous. Information on biosynthetic pathways is even more scarce and drawing parallels with the eukaryotic situation encounters difficulties, because of the lack of internal membranes in prokaryotes. It is plausible that the cell membrane can fulfill this function. But especially in the case of the biosynthesis of intracellular and bacterial flagellar glycoproteins, this hypothesis raises problems. Some biosynthetic information comes from studies of the cellulosomes of *Clostridium thermocellum* and *Bacteroides cellulosolvens.*

It is a general feeling that not all prokaryotes are able of glycosylating proteins. For instance, *E. coli* is used to express recombinant genes in order to obtain an unglycosylated protein that can be used to study the effect of glycosylation on that protein. However, the presence of glycosylated pili was suggested for *E. coli* (Tomoeda *et al.,* 1975), and some recombinant enzymes produced by *E. coli* appear to contain sugars (Yoshida *et al.,* 1998a). The most intruiging observation comes from the expression of the ice nucleation genes in *E. coli.* Ice nucleation is known to require posttranslational modifications of the gene products of the ice genes. Recombinant *E. coli* do show the three classes of ice nucleation, indicating the exact posttranslational modification of the ice gene products. The presence of cryptic genes involved in posttranslational modification was suggested (Kozloff *et al.,* 1991).

Some common bacterial enzymes can be shared by several biosynthetic pathways, including the enzymes required for glycosylation of proteins. This is the case for a phosphomannomutase, ExoC, of *A. brasilense* (Moens *et al.,* 1995), and a Gal epimerase, GalE, of *N. meningitidis* (Stimson *et al.,* 1995) where the corresponding mutants were shown to be affected in the synthesis of exopolysaccharides and lipopolysaccharides, respectively. Other enzymes, however, are specific for the incorporation of glycans in glycoproteins and not in other sugar-containing compounds. Examples of these enzymes are a glycosyltransferase, PglA, of *N. meningitidis* (Jennings *et al.,* 1998) and PtmB, a CMP-*N*-acetyl-neuraminic acid synthetase of *Campylobacter* (Guerry *et al.,* 1996).

ACKNOWLEDGMENTS

S. M. acknowledges J. Vanderleyden and M. Lambrecht for critical reading of the manuscript. S. M. is a recipient of a postdoctoral fellowship of the Fonds voor Wetenschappelijk Onderzoek-Vlaanderen.

REFERENCES

Alam, M., and Oesterhelt, D., 1984, Morphology, function and isolation of halobacterial flagella, *J. Mol. Biol.* **176**:459–475.

Ambler, R. P., and Rees, M. W., 1959, *N*-methyl-lysine in bacterial flagellar protein, *Nature* **184**:56–57.

Arnold, F., Bedouet, L., Batina, P., Robreau, G., Talobot, F., Lecher, P., and Malcoste, R., 1998, Biochemical and immunological analyses of the flagellin of *Clostridium tyrobutyricum* ATCC 25755, *Microbiol. Immunol.* **42**:23–31.

Baker, B. S., Smith, S. E., and McDonough, M. W., 1983, The presence of several residues of *N*-methyl lysine in *Proteus morganii* flagellin, *Microbios Lett.* **23**:7–12.

Bayley, D. P., and Jarrell, K. F., 1998, Further evidence to suggest that archaeal flagella are related to bacterial type IV pili, *J. Mol. Evol.* **46**:370–373.

Bayley, D. P., Kalmokoff, M. L., and Jarrell, K. F., 1993, Effect of bacitracin on flagellar assembly and presumed glycosylation of the flagellins of *Methanococcus deltae*, *Arch. Microbiol.* **160**:179–185.

Bedouet, L., Arnold, F., Robreau, G., Batina, P., Talbot, F., and Malcoste, R., 1998, Partial analysis of the flagellar antigenic determinant recognized by a monoclonal antibody to *Clostridium tyrobutyricum, Microbiol. Immunol.* **42**:87–95.

Béguin, P., and Eisen, H., 1978, Purification and partial characterization of three extracellular cellulases from *Cellulomonas* sp., *Eur. J. Biochem.* **87**:525–531.

Boller, T., and Felix, G., 1996, Olfaction in plants: Specific perception of common microbial molecules, in: *Biology of Plant–Microbe Interactions* (G. Stacey, B. Mullin, and P. M. Gresshoff, eds.), International Society for Plant-Microbe Interactions, St. Paul, Minnesota, pp. 1–8.

Bradley, D. E., 1972, Shortening of *Pseudomonas aeruginosa* pili after RNA-phage adsorption, *J. Gen. Microbiol.* **72**:303–319.

Brahamsha, B., 1996, An abundant cell-surface polypeptide is required for swimming by the nonflagellated marine cyanobacterium *Synechococcus, Proc. Natl. Acad. Sci. USA* **93**:6504–6509.

Brahamsha, B., and Greenberg, E. P., 1988, A biochemical and cytological analysis of the complex periplasmic flagella from *Spirochaeta aurantia, J. Bacteriol.* **170**:4023–4032.

Brimer, C. D., and Montie, T. C., 1998, Cloning and comparison of *fliC* genes and identification of glycosylation in the flagellin of *Pseudomonas aeruginosa* a-type strains, *J. Bacteriol.* **180**:3209–3217.

Brinton, C. C., 1971, The properties of sex pili, the viral nature of conjugal genetic transfer systems, and some possible approaches to the control of bacterial drug resistance, *Crit. Rev. Microbiol.* **1**:105–160.

Calza, R. E., Irwin, D. C., and Wilson, D. B., 1985, Purification and characterization of two β-1,4-endoglucanases from *Thermomonospora fusca, Biochemistry* **24**:7797–7804.

Castric, P., 1995, *pilO,* a gene required for glycosylation of *Pseudomonas aeruginosa* 1244 pilin, *Microbiology* **141**:1247–1254.

Coleman, J. L., and Benach, J. L., 1988, Lectin binding to *Borrelia burgdorferi, Ann. NY Acad. Sci.* **539**:372–375.

Craft, J. E., Fischer, D. K., and Shimamoto, G. T., 1986, Antigens of *Borrelia burgdorferi* recognized during Lyme disease, *J. Clin. Invest.* **78**:934–949.

Dobos, K. M., Swiderek, K., Khoo, K.-H., Brennan, P. J., and Belisle, J. T., 1995, Evidence for glycosylation sites on the 45-kilodalton glycoprotein of *Mycobacterium tuberculosis, Infect. Immun.* **63**:2846–2853.

Dobos, K. M., Khoo, K.-H., Swiderek, K. M., Brennan, P. J., and Belisle, J. T., 1996, Definition of the full extent of glycosylation of the 45-kilodalton glycoprotein of *Mycobacterium tuberculosis, J. Bacteriol.* **178**:2498–2506.

Dobson, W. J., and McCurdy, H. D., 1979, The function of fimbriae in *Myxococcus xanthus.* I. Purification and properties of *M. xanthus* fimbriae, *Can. J. Microbiol.* **25**:1152–1160.

Doherty, H., Condon, C., and Owen, P., 1982, Resolution and in vitro glycosylation of membrane glycoproteins in *Micrococcus luteus (lysodeikticus), FEMS Microbiol. Lett.* **15**:331–336.

Doig, P., Kinsella, N., Guerry, P., and Trust, T. J., 1996, Characterization of a post-translational modification of *Campylobacter* flagellin: Identification of a sero-specific glycosyl moiety, *Mol. Microbiol.* **19**:379–387.

Erickson, P. R., and Herzberg, M. C., 1993, Evidence for the covalent linkage of carbohydrate polymers to a glycoprotein from *Streptococcus sanguis, J. Biol. Chem.* **268**:23780–23783.

Faguy, D. M., Koval, S. F., and Jarrell, K. F., 1992, Correlation between glycosylation of flagellin proteins and sensitivity of flagellar filaments to Triton X-100 in methanogens, *FEMS Microbiol. Lett.* **90**:129–134.

Faguy, D. M., Koval, S. F., and Jarrell, K. F., 1994, Physical characterization of the flagella and flagellins from *Methanospirillum hungatei, J. Bacteriol.* **176**:7491–7498.

Faguy, D. M., Bayley, D. P., Kostyukova, A. S., Thomas, N. A., and Jarrell, K. F., 1996, Isolation and characterization of flagella and flagellin proteins from the thermoacidophilic archaea *Thermoplasma volcanium* and *Sulfolobus shibatae, J. Bacteriol.* **178**:902–905.

Fairchild, C. D., Jones, I. K., and Glazer, A. N., 1991, Absence of glycosylation on cyanobacterial phycobilisome linker polypeptides and rhodophytan phycoerythrins, *J. Bacteriol.* **173**:2985–2992.

Fedorov, O. V., Pyatibratov, M. G., Kostyukova, A. S., Osina, N. K., and Tarasov, V. Y., 1994, Protofilament as a structural element of flagella of haloalkalophilic archaebacteria, *Can. J. Microbiol.* **40**:45–53.

Fifis, T., Costopoulos, C., Radford, A. J., Bacic, A., and Wood, P. R., 1991, Purification and characterization of major antigens from a *Mycobacterium bovis* culture filtrate, *Infect. Immun.* **59**:800–807.

Garbe, T., Harris, D., Vordermeier, M., Lathigra, R., Ivanyi, J., and Young, D., 1993, Expression of the *Mycobacterium tuberculosis* 19-kilodalton antigen in *Mycobacterium smegmatis:* Immunological analysis and evidence of glycosylation, *Infect. Immun.* **61**:260–267.

García-Patrone, M., and Tandecarz, J. S., 1995, A glycoprotein multimer from *Bacillus thuringiensis* sporangia: Dissociation into subunits and sugar composition, *Mol. Cell. Biochem.* **145**:29–37.

Gerwig, G. J., de Waard, P., Kamerling, J. P., Vliegenthart, J. F. G., Morgenstern, E., Lamed,

R., and Bayer, E. A., 1989, Novel O-linked carbohydrate chains in the cellulase complex (cellulosome) of Clostridium thermocellum, J. Biol. Chem. 264:1027–1035.

Gerwig, G. J., Kamerling, J. P., Vliegenthart, J. F. G., Morag (Morgenstern), E., Lamed, R., and Bayer, E. A., 1991, Primary structure of O-linked carbohydrate chains in the cellulosome of different Clostridium thermocellum strains, Eur. J. Biochem. 196:115–122.

Gerwig, G. J., Kamerling, J. P., Vliegenthart, J. F. G., Morag, E., Lamed, R., and Bayer, E. A., 1992, Novel oligosaccharide constituents of the cellulase complex of Bacteroides cellulosolvens, Eur. J. Biochem. 205:799–808.

Gerwig, G. J., Kamerling, J. P., Vliegenthart, J. F. G., Morag, E., Lamed, R., and Bayer, E. A., 1993, The nature of the carbohydrate-peptide linkage region in glycoproteins from the cellulosomes of Clostridium thermocellum and Bacteroides cellulosolvens, J. Biol. Chem. 268:26956–26960.

Gilkes, N. R., Langsford, M. L., Kilburn, D. G., Miller, R. C. Jr., and Warren, R. A. J., 1984, Mode of action and substrate specificities of cellulases from cloned bacterial genes, J. Biol. Chem. 259:10455–10459.

Gilkes, N. R., Warren, R. A. J., Miller, R. C., Jr., and Kilburn, D. G., 1988, Precise excision of the cellulose binding domains from two Cellulomonas fimi cellulases by a homologous protease and the effect on catalysis, J. Biol. Chem. 263:10401–10407.

Glazer, A. N., DeLange, R. J., and Martinez, R. J., 1969, Identification of N-methyl-lysine in Spirillum serpens flagella and N-dimethyl-lysine in Salmonella typhimurium flagella, Biochim. Biophys. Acta 189:164–165.

Gong, J., Egbosimba, E., and Forsberg, C. W., 1996, Cellulose-binding proteins of Fibrobacter succinogenes and the possible role of a 180-kDa cellulose-binding glycoprotein in adhesion to cellulose, Can. J. Microbiol. 42:453–460.

Grogan, D. W., 1989, Phenotypic characterization of the archaebacterial genus Sulfolobus: Comparison of five wild-type strains, J. Bacteriol. 171:6710–6719.

Gubish, E. R. Jr., Chen, K. C. S., and Buchanan, T. M., 1982, Attachment of gonococcal pili to lectin-resistant clones of Chinese hamster ovary cells, Infect. Immun. 37:189–194.

Guerry, P., Doig, P., Alm, R. A., Burr, D. H., Kinsella, N., and Trust, T. J., 1996, Identification and characterization of genes required for post-translational modification of Campylobacter coli VC167 flagellin, Mol. Microbiol. 19:369–378.

Henrichsen, J., 1983, Twitching motility, Annu. Rev. Microbiol. 73:81–93.

Herrmann, J. L., O'Gaora, P., Gallagher, A., Thole, J. E. R., and Young, D. B., 1996, Bacterial glycoproteins: A link between glycosylation and proteolytic cleavage of a 19 kDa antigen from Mycobacterium tuberculosis, EMBO J. 15:3547–3554.

Herzberg, M. C., Erickson, P. R., Kane, P. K., Clawson, D. J., Clawson, C. C., and Hoff, F. A., 1990, Platelet-interactive products of Streptococcus sanguis protoplasts, Infect. Immun. 58:4117–4125.

Hoiczyk, E., and Baumeister, W., 1997, Oscillin, an extracellular, Ca^{2+}-binding glycoprotein essential for the gliding motility of cyanobacteria, Mol. Microbiol. 26:699–708.

Huang, L., Forsberg, C. W., and Thomas, D. Y., 1988, Purification and characterization of a chloride-stimulated cellobiosidase from Bacteroides succinogenes S85, J. Bacteriol. 170:2923–2932.

Huber, H. E., Luthy, P., Ebersold, H. R., and Cordier, J. L., 1981, The subunits of the parasporal crystal of *Bacillus thuringiensis:* Size, linkage and toxicity, *Arch. Microbiol.* **129:**14–18.

Hultgren, S. J., Jones, C. H., and Normark, S., 1996, Bacterial adhesins and their assembly, in: *Escherichia coli and Salmonella, Cellular and Molecular Biology,* 2nd ed. (F. C. Neidhardt, ed.), ASM Press, Washington DC, pp. 2730–2756.

Jarrell, K. F., Bayley, D. P., and Kostyukova, A. S., 1996, The archaeal flagellum: A unique motility structure, *J. Bacteriol.* **178:**5057–5064.

Jennings, M. P., Virji, M., Evans, D., Foster, V., Srikhanta, Y. N., Steeghs, L., van der Ley, P., and Moxon, E. R., 1998, Identification of a novel gene involved in pilin glycosylation in *Neisseria meningitidis, Mol. Microbiol.* **29:**975–984.

Jones, G. W., and Isaacson, R. E., 1983, Proteinaceous bacterial adhesins and their receptors, *Crit. Rev. Microbiol.* **10:**229–260.

Joys, T. M., and Kim, H., 1979, Identification of *N*-methyl-lysine residues in the phase-1 flagellar protein of *Salmonella typhimurium, Microbios Lett.* **7:**65–68.

Kalmokoff, M. L., Karnauchow, T. M., and Jarrell, K. F., 1990, Conserved N-terminal sequences in the flagellins of archaebacteria, *Biochem. Biophys. Res. Commun.* **167:** 154–160.

Kanaoka, M., Fukita, Y., Taya, K., Kawanaka, C., Negoro, T., and Agui, H., 1987, Antitumor activity of streptococcal acid glycoprotein produced by *Streptococcus pyogenes, Jpn. J. Cancer Res.* **78:**1409–1414.

Kawamura, T., and Shockman, G. D., 1983, Purification and some properties of the endogenous, autolytic *N*-acetylmuramoylhydrolase of *Streptococcus faecium,* a bacterial glycoenzyme, *J. Biol. Chem.* **258:**9514–9521.

Kelly-Wintenberg, K., Anderson, T., and Montie, T. C., 1990, Phosphorylated tyrosine in the flagellum filament protein of *Pseudomonas aeruginosa, J. Bacteriol.* **172:**5135–5139.

Kelly-Wintenberg, K. D., South, S. L., and Montie, T. C., 1993, Tyrosine phosphate in a- and b-type flagellins of *Pseudomonas aeruginosa, J. Bacteriol.* **175:**2458–2461.

Kim, B.-K., Pihl, T. D., Reeve, J. N., and Daniels, L., 1995, Purification of the copper response extracellular proteins secreted by the copper-resistant methanogen *Methanobacterium bryantii* BKYH and cloning, sequencing, and transcription of the genes encoding these proteins, *J. Bacteriol.* **177:**7178–7185.

Kimura, Y., and Stadtman, T. C., 1995, Glycine reductase selenoprotein A is not a glycoprotein: The positive periodic acid-Schiff reagent test is the result of peptide bond cleavage and carbonyl group generation, *Proc. Natl. Acad. Sci. USA* **92:**2189–2193.

Kluepfel, D., Vats-Metha, S., Aumont, F., Shareck, F., and Morosoli, R., 1990, Purification and characterization of a new xylanase (xylanase B) produced by *Streptomyces lividans* 66, *Biochem. J.* **267:**45–50.

Kozloff, L. M., Turner, M. A., and Arellano, F., 1991, Formation of bacterial membrane ice-nucleating lipoglycoprotein complexes, *J. Bacteriol.* **173:**6528–6536.

Langsford, M. L., Gilkes, N. R., Singh, B., Moser, B., Miller, R. C., Jr., Warren, R. A. J., and Kilburn, D. G., 1987, Glycosylation of bacterial cellulases prevents proteolytic cleavage between functional domains, *FEBS Lett.* **225:**163–167.

Lechner, J., Wieland, F., and Sumper, M., 1985, Transient methylation of dolichyl oligosaccharides is an obligatory step in halobacterial sulfated glycoprotein biosynthesis, *J. Biol. Chem.* **260:**8984–8989.

Li, Z., Dumas, F., Dubreuil, D., and Jacques, M., 1993, A species-specific periplasmic flagellar protein of *Serpulina (Treponema) hyodysenteriae, J. Bacteriol.* **175:**8000–8007.

Logan, S. M., Trust, T. J., and Guerry, P., 1989, Evidence for posttranslational modification and gene duplication of *Campylobacter* flagellin, *J. Bacteriol.* **171:**3031–3038.

Luft, B. J., Jiang, W., Munoz, P., Dattwyler, R. J., and Gorevic, P. D., 1989, Biochemical and immunological characterization of the surface proteins of *Borrelia burgdorferi, Infect. Immun.* **57:**3637–3645.

Macnab, R. M., 1996, Flagella and motility, in: *Escherichia coli and Salmonella, Cellular and Molecular Biology, 2nd* ed. (F. C. Neidhardt, ed.), ASM Press, Washington, DC, pp. 123–145.

Maeba, P. Y., 1986, Isolation of a surface glycoprotein from *Myxococcus xanthus, J. Bacteriol.* **166:**644–650.

Marceau, M., Forest, K., Béretti, J.-L., Tainer, J., and Nassif, X., 1998, Consequences of the loss of *O*-linked glycosylation of meningococcal type IV pilin on piliation and pilus-mediated adhesion, *Mol. Microbiol.* **27:**705–715.

McManus, J. D., Brune, D. C., Han, J., Sanders-Loehr, J., Meyer, T. E., Cusanovich, M. A., Tollin, G., and Blankenship, R. E., 1992, Isolation, characterization, and amino acid sequences of auracyanins, blue copper proteins from the green photosynthetic bacterium *Chloroflexus aurantiacus, J. Biol. Chem.* **267:**6531–6540.

Meier, B., Brunotte, C. M., Franz, B., Warlich, B., Petermann, M., Ziesenis, A., Schuberth, H.-J., Habermehl, G. G., Petzoldt, K., and Leibold, W., 1992, Isolation of a high-molecular mass glycoprotein from culture supernatant of an arthritogenic strain of the bacteria *Erysipelothrix rhusiopathiae* reacting with "inductive" monoclonal antibodies derived from rats with erysipelas polyarthritis, *Biol. Chem. Hoppe-Seyler* **373:**715–721.

Mescher, M. F., and Strominger, J. L., 1978, Glycosylation of the surface glycoprotein of *Halobacterium salinarium* via a cyclic pathway of lipid-linked intermediates, *FEBS Lett.* **89:**37–41.

Messner, P., 1997, Bacterial glycoproteins, *Glycoconj. J.* **14:**3–11.

Meyer, W., and Schäfer, G., 1992, Characterization and purification of a membrane-bound archaebacterial pyrophosphatase from *Sulfolobus acidocaldarius, Eur. J. Biochem.* **207:**741–746.

Mihoc, A., and Kluepfel, D., 1990, Purification and characterization of a β-glucosidase from *Streptomyces lividans* 66, *Can. J. Microbiol.* **36:**53–56.

Moens, S., and Vanderleyden, J., 1997, Glycoproteins in prokaryotes, *Arch. Microbiol.* **168:**169–175.

Moens, S., Michiels, K., and Vanderleyden, J., 1995, Glycosylation of the flagellin of the polar flagellum of *Azospirillum brasilense,* a gram-negative nitrogen-fixing bacterium, *Microbiology* **141:**2651–2657.

Morris, E. J., Ganeshkumar, N., Song, M., and McBride, B. C., 1987, Identification and preliminary characterization of a *Streptococcus sanguis* fibrillar glycoprotein, *J. Bacteriol.* **169:**164–171.

Muthukumar, G., and Nickerson, K. W., 1987, The glycoprotein toxin of *Bacillus thuringiensis* subsp. *israelensis* indicates a lectinlike receptor in the larval mosquito gut, *Appl. Env. Microbiol.* **53:**2650–2655.

Nachamkin, I., and Yang, X. H., 1989, Human antibody response to *Campylobacter jejuni* flagellin protein and a synthetic N-terminal flagellin peptide, *J. Clin. Microbiol.* **27:** 2195–2198.

Olsen, O., and Thomsen, K. K., 1991, Improvement of bacterial β-glucanase thermostability by glycosylation, *J. Gen. Microbiol.* **137:**579–585.

O'Neill, G., Goh, S. H., Warren, R. A. J., Kilburn, D. G., and Miller, R. C., Jr., 1986, Structure of the gene encoding the exoglucanase of *Cellulomonas fimi, Gene* **44:**325–330.

Ong, E., Kilburn, D. G., Miller, R. C., Jr., and Warren, R. A. J., 1994, *Streptomyces lividans* glycosylates the linker region of a β-1,4-glycanase from *Cellulomonas fimi, J. Bacteriol.* **176:**999–1008.

Ottow, J. C. G., 1975, Ecology, physiology, and genetics of fimbriae and pili, *Annu. Rev. Microbiol.* **29:**79–108.

Parge, H. E., Forest, K. T., Hickey, M. J., Christensen, D. A., Getzoff, E. D., and Tainer, J. A., 1995, Structure of the fibre-forming protein pilin at 2.6 Å resolution, *Nature* **378:**32–38.

Paul, J., and Varma, A. K., 1992, Glycoprotein components of cellulase and xylanase enzymes of a *Bacillus* sp., *Biotechnol. Lett.* **14:**207–212.

Penn, C. W., Bailey, M. J., and Cockayne, A., 1985, The axial filament antigen of *Treponema pallidum, Immunology* **54:**635–641.

Pfannenstiel, M. A., Muthukumar, G., Couche, G. A., and Nickerson, K. W., 1987, Amino sugars in the glycoprotein toxin from *Bacillus thuringiensis* subsp. *israelensis, J. Bacteriol.* **169:**796–801.

Plummer, T. H., Jr., Tarentino, A. L., and Hauer, C. R., 1995, Novel, specific O-glycosylation of secreted *Flavobacterium meningosepticum* proteins, *J. Biol. Chem.* **270:** 13192–13196.

Power, M. E., Guerry, P., McCubbin, W. D., Kay, C. M., and Trust, T. J., 1994, Structural and antigenic characteristics of *Campylobacter coli* Fla A flagellin, *J. Bacteriol.* **176:**3303–3313.

Pugsley, A. P., 1993, The complete general secretory pathway in gram-negative bacteria, *Microbiol. Rev.* **57:**50–108.

Reinhold, B. B., Hauer, C. R., Plummer, T. H., and Reinhold, V. N., 1995, Detailed structural analysis of a novel, specific O-linked glycan from the prokaryote *Flavobacterium meningosepticum, J. Biol. Chem.* **270:**13197–13203.

Robertson, J. N., Vincent, P., and Ward, M. E., 1977, The preparation and properties of gonococcal pili, *J. Gen. Microbiol.* **102:**169–177.

Sambri, V., Stefanelli, C., and Cevenini, R., 1992, Detection of glycoproteins in *Borrelia burgdorferi, Arch. Microbiol.* **157:**205–208.

Sanchez-Beato, A. R., Ronda, C., and Garcia, J. L., 1995, Tracking the evolution of the bacterial choline-binding domain: Molecular characterization of the *Clostridium acetobutylicum* NCIB 8052 *cspA* gene, *J. Bacteriol.* **177:**1098–1103.

Sandercock, L. E., MacLeod, A. M., Ong, E., and Warren, R. A. J., 1994, Non-S-layer glycoproteins in eubacteria, *FEMS Microbiol. Lett.* **118:**1–8.

Seifert, H. S., 1996, Questions about gonococcal pilus phase- and antigenic variation, *Mol. Microbiol.* **21:**433–440.

Serganova, I. S., Polosina, Y. Y., Kostyukova, A. S., Metlina, A. L., Pyatibratov, M. G., and

Fedorov, O. V., 1995, Flagella of halophilic archaea: Biochemical and genetic analysis, *Biochemistry* (Engl. Transl. *Biokhimiya*) **60**:953–957.

Simon, R. D., 1981, Gliding motility in *Aphanotheca halophytica:* Analysis of wall proteins in *mot* mutants, *J. Bacteriol.* **148**:315–321.

Southam, G., Kalmokoff, M. L., Jarrell, K. F., Koval, S. F., and Beveridge, T. J., 1990, Isolation, characterization, and cellular insertion of the flagella from two strains of the archaebacterium *Methanospirillum hungatei, J. Bacteriol.* **172**:3221–3228.

Stimson, E., Virji, M., Makepeace, K., Dell, A., Morris, H. R., Payne, G., Saunders, J. R., Jennings, M. P., Barker, S., Panico, M., Blench, I., and Moxon, E. R., 1995, Meningococcal pilin: A glycoprotein substituted with digalactosyl 2,4-diacetamido-2,4,6-trideoxyhexose, *Mol. Microbiol.* **17**:1201–1214.

Strobel, G. A., Talmadge, K. W., and Albersheim, P., 1972, Observations on the structure of the phytotoxic glycopeptide of *Corynebacterium sepedonicum, Biochim. Biophys. Acta* **261**:365–374.

Strom, M. S., and Lory, S., 1991, Amino acid substitutions in pilin of *Pseudomonas aeruginosa.* Effect on leader peptide cleavage, amino-terminal methylation, and pilus assembly, *J. Biol. Chem.* **266**:1656–1664.

Sumper, M., 1987, Halobacterial glycoprotein biosynthesis, *Biochim. Biophys. Acta* **906**:69–79.

Sumper, M., and Herrmann, G., 1978, Studies on the biosynthesis of bacterio-opsin. Demonstration of the existence of protein species structurally related to bacterio-opsin, *Eur. J. Biochem.* **89**:229–235.

Taku, A., and Fan, D. P., 1976, Purification and properties of a protein factor stimulating peptidoglycan synthesis in toluene- and LiCl-treated *Bacillus megaterium* cells, *J. Biol. Chem.* **251**:1889–1895.

Tarentino, A. L., Quinones, G., Grimwood, B. G., Hauer, C. R., and Plummer, T. H. Jr., 1995, Molecular cloning and sequence analysis of flavastacin: An *O*-glycosylated prokaryotic zinc metalloendopeptidase, *Arch. Biochem. Biophys.* **319**:281–285.

Taylor, D. E., 1992, Genetics of *Campylobacter* and *Helicobacter, Annu. Rev. Microbiol.* **46**:35–64.

Tomoeda, M., Inuzuka, M., and Date, T., 1975, Bacterial sex pili, *Prog. Biophys. Mol. Biol.* **30**:23–56.

Turner, M. A., Arellano, F., and Kozloff, L. M., 1991, Components of ice nucleation structures of bacteria, *J. Bacteriol.* **173**:6515–6527.

Van Rijssel, M., Gerwig, G. J., and Hansen, T. A., 1993, Isolation and characterization of an extracellular glycosylated protein complex from *Clostridium thermosaccharolyticum* with pectin methylesterase and polygalacturonate hydrolase activity, *Appl. Env. Microbiol.* **59**:828–836.

Virji, M., 1998, Glycosylation of the meningococcus pilus protein, *ASM News* **64**:398–405.

Virji, M., Saunders, J. R., Sims, G., Makepeace, K., Maskell, D., and Ferguson, D. J. P., 1993, Pilus-facilitated adherence of *Neisseria meningitidis* to human epithelial and endothelial cells: Modulation of adherence phenotype occurs concurrently with changes in primary amino acid sequence and the glycosylation status of pilin, *Mol. Microbiol.* **10**:1013–1028.

Wassenaar, T. M., Fry, B. N., and van der Zeijst, B. A. M., 1995, Variation of the flagellin gene locus of *Campylobacter jejuni* by recombination and horizontal gene transfer, *Microbiology* **141**:95–101.

Webster, J. R., Reid, S. J., Jones, D. T., and Woods, D. R., 1981, Purification and characterization of an autolysin from *Clostridium acetobutylicum*, *Appl. Env. Microbiol.* **41**:371–374.

Weerkamp, A. H., and Jacobs, T., 1982, Cell wall-associated protein antigens of *Streptococcus salivarius:* Purification, properties, and function in adherence, *Infect. Immun.* **38**:233–242.

Weerkamp, A. H., van der Mei, H. C., and Liem, R. S. B., 1984, Adhesive cell wall-associated glycoprotein of *Streptococcus salivarius* (K⁺) is a cell surface fibril, *FEMS Microbiol. Lett.* **23**:163–166.

Wieland, F., Paul, G., and Sumper, M., 1985, Halobacterial flagellins are sulfated glycoproteins, *J. Biol. Chem.* **260**:15180–15185.

Wilson, D. R., and Beveridge, T. J., 1993, Bacterial flagellar filaments and their component flagellins, *Can. J. Microbiol.* **39**:451–472.

Winstanley, C., and Morgan, J. A. W., 1997, The bacterial flagellin gene as a biomarker for detection, population genetics and epidemiological analysis, *Microbiology* **143**:3071–3084.

Yamaguchi, M., Jimbo, M., Sakai, R., Muramoto, K., and Kamiya, H., 1998, Purification and characterization of a *Microcystis aeruginosa* (freshwater cyanobacterium) lectin, *Comp. Biochem. Physiol.* [B] **119**:593–597.

Yang, L. L., and Haug, A., 1979, Purification and partial characterization of a procaryotic glycoprotein from the plasma membrane of *Thermoplasma acidophilum*, *Biochim. Biophys. Acta* **556**:265–277.

Yao, R., Macaria, A. J. L., and Conway de Macario, E., 1992, Immunochemical differences among *Methanosarcina mazei* S-6 morphologic forms, *J. Bacteriol.* **174**:4683–4688.

Yoshida, S., Sako, Y., and Uchida, A., 1998*a*, Cloning, sequence analysis, and expression in *Escherichia coli* of a gene coding for an enzyme from *Bacillus circulans* K-1 that degrades guar gum, *Biosci. Biotechnol. Biochem.* **62**:514–520.

Yoshida, J., Takamura, S., and Nishio, M., 1998*b*, Characterization of a streptococcal antitumor glycoprotein (SAGP), *Life Sci.* **62**:1043–1053.

Zhu, B. C. R., Drake, R. R., Schweingruber, H., and Laine, R. A., 1995, Inhibition of glycosylation by amphomycin and sugar nucleotide analogs PP36 and PP55 indicates that *Haloferax volcanii* β-glucosylates both glycoproteins and glycolipids through lipid-linked sugar intermediates: Evidence for three novel glycoproteins and a novel sulfated dihexosyl-archaeol glycolipid, *Arch. Biochem. Biophys.* **319**:355–364.

2

Glycans in Meningococcal Pathogenesis and the Enigma of the Molecular Decorations of Neisserial Pili

Mumtaz Virji

1. INTRODUCTION

Neisseria meningitidis (meningococcus) is a gram-negative bacterium that normally resides in the nasopharynx of healthy individuals but possesses the capacity to cause serious diseases (Fig. 1). Surface glycans such as capsule and the polysaccharide moieties of its outer membrane lipopolysaccharide (LPS) play important roles in the pathogenesis of the organism. Capsule exhibits antiphagocytic activity and only capsulate phenotypes survive in the blood. LPS is toxic via its lipid-A moiety. In addition, its surface-located polysaccharide side chains, which may be sialylated, impart a net negative charge to the bacterial surface and can act as a pseudo-capsule playing a similar role to the capsule. The only surface protein that is presented to the host in capsulate phenotypes is the filamentous structure, pilus [pili (plural) or fimbriae]. Pilus is a polymer of thousands of subunits arranged in a helical array to form a fiber. A bacterium may elaborate numerous fibers covering the entire surface. The fiber traverses the capsule and extends several microns beyond the surface of the bacterium. This proteinaceous structure was

Mumtaz Virji • Department of Pathology and Microbiology, School of Medical Sciences, University of Bristol, Bristol BS8 1TD, England.

Glycomicrobiology, edited by Doyle.
Kluwer Academic/Plenum Publishers, New York, 2000.

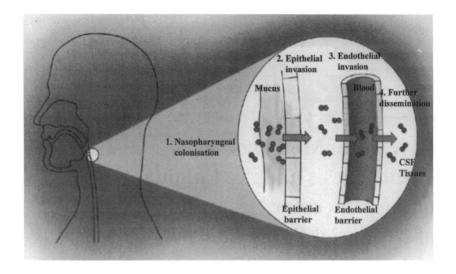

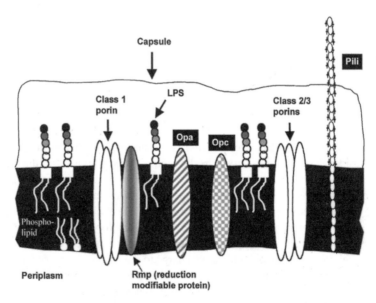

Figure 1. (Top) Stages in the pathogenesis of *Neisseria meningitidis*. Clinical observations suggest that blood dissemination occurs prior to further spread to the cerebrospinal fluid (CSF) and other tissues. Thus meningococcal translocation across the epithelial and endothelial barriers precedes bacteremia and meningitis and as such, interactions with these cells are central to pathogenesis. (Bottom) Schematic presentation of the outer membrane components of *N. meningitidis*. Opa, Opc, and pili are known to participate in interactions with host cells via specific receptors. Other components (porins, LPS, etc.) also interact with eukaryotic cells. Capsule and LPS (especially when sialylated) provide a protective coat masking antigens and ligands. Pili extend beyond these barriers, and thus are important for mediating adhesion. However, recent studies show surface-located glycans on pili, thus pili extend the sugar coating of the meningococcal surface.

relatively recently demonstrated to be glycosylated. The glycans on the fiber are surface located, and thus the pilus, coated with sugar molecules, extends the sugar coating of the pathogen (Figs. 1 and 2).

The pilus presents an enigma in being an apparently unique meningococcal protein with extensive modifications. Its constituents present unusual linkages and composition. It is an ever-changing molecule, since it undergoes primary sequence variations as well as phase variations (on–off–on switching of expression) of its glycan and other moieties independently of pilus phase variations. Thus, it is constantly antigenically variable, presumably to avoid host antibody recognition. Despite these structural modifications, it maintains its host receptor(s) adhesion function.

This chapter presents studies that characterize several distinct structural facets of this important and complex glycoprotein. It also discusses the problems of studying the protein and difficulties in interpreting the functional significance of its structural components. In addition, it deals with capsule and LPS, two other polysaccharide-containing surface structures important in meningococcal pathogenesis. Their functions, in particular their modulatory effects on the major adhesins and invasins of meningococci, have been discussed in light of the *in vitro* investigations and epidemiological observations. Although the chapter deals mainly with the meningococcus, its mechanisms of host colonization and pathogenesis are unlikely to be unique (Finlay and Falkow, 1997), and thus it represents a paradigm for pathogenic processes that lead to bacteremia and meningitis. Some novel structures on the pilus proteins are currently regarded as unique, but experience suggests that structurally or functionally similar decorations may well be present on other bacterial proteins and their discoveries imminent. Recent expansion of reports on prokaryotic glycoproteins is a testament to this (Messner, 1997; Moens and Vanderleyden, 1997; see also Chapters 1 and 4, this volume).

2. BACKGROUND

2.1. Meningococcal Colonization and Pathogenesis

N. meningitidis is the causative organism of one of the most rapidly progressive bacterial diseases and may result in death unless promptly treated with antibiotics. The organism, however, occurs in the nasopharynx of up to 30% of healthy individuals without causing adverse effects (Cartwright, 1995). Thus, in the main, it is a mucosal commensal. Indeed, being specific for its only host, man, its strategy of pathogenesis with high likelihood of resulting in death would be evolutionarily sterile. Therefore, its pathogenic potential may be regarded as an accidental or coincidental phenomenon. Although the organism has such an im-

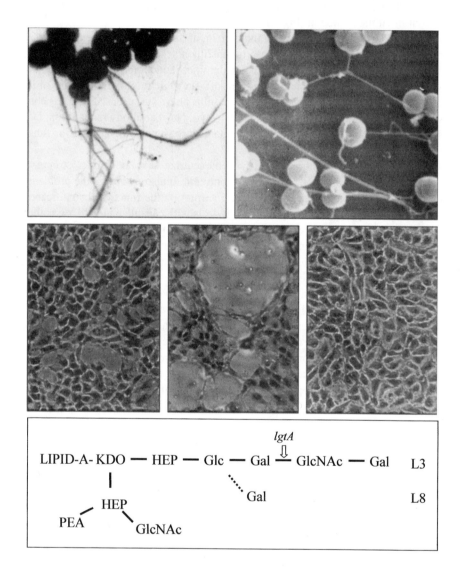

pressive ability to subdue its host, it lacks efficiency [attack rate of 5–25 cases per 10^5 individuals in developed countries (Ala'Aldeen and Griffiths, 1995; Cartwright, 1995)], suggesting that the host has developed efficient counterstrategies for defense. The immunological condition of the host therefore may be as important as the virulence potential of the organism in the initiation and progression of the disease. There is evidence to suggest that host immunity plays an important role in maintaining the organism in the nasopharynx. The most susceptible individuals are young children under 2 years of age during the time when maternal antibodies decline and before their own immunity is fully developed and adults with deficiency in the lytic components of complement. [It is believed that immunity to meningococci is acquired through exposure to the related commensal, *Neisseria lactamica,* that shares several immunogenic determinants with pathogenic neisseriae (Cartwright, 1995.)] However, other individuals also may become susceptible; these include young adults between 15 and 19 years. The precise reasons for their susceptibility is unclear, although carriage rates increase in this age group and may reflect greater transmission rates due to sociological behavior. It is important to note that colonization rarely leads to dissemination and disease. Thus, the carrier state means that the host is able to confine bacteria to the nasopharynx.

Epidemiological studies also suggest factors that damage mucosa, such as smoking, prior infection of the host (e.g., respiratory viral infections during winter months in Great Britain), or very dry atmospheric conditions (during dry seasons in Africa) may predispose the host to meningococcal infection (Achtman, 1995; Cartwright, 1995). Prior infection of the host with viruses or other agents could allow deeper ingress of meningococci as a result of damage to respiratory epithelial cells and exposure of extracellular matrix, components of which are targeted by meningococcal adhesins (see Section 2.2.3a). Alternately or in parallel,

---

Figure 2. Morphology of pili, adhesion, and toxicity. (Top) Transmission electron microscopy of negatively stained preparation of a piliated meningococcal strain (left). Hairlike filaments extend several microns from the bacterial cell surface. (Right) Scanning electron microscopy of piliated capsulate meningococci adherent to human umbilical vein endothelial cells. Ropelike structures consist of bundles of individual pilus fibers and can be seen making contact between different bacterial cells and between bacteria and the host cell surface. (Middle) Cytopathic effect of meningococci on human endothelial cells in culture. Confluent endothelial cell monolayers were inoculated with piliated or nonpiliated variants of a capsulate meningococcal strain or a capsulate clinical isolate of *Haemophilus influenzae.* No toxicity was observed in the latter case (right). Under the same condition, piliated meningococci caused significant damage (middle). Although some damage also was apparent when the endothelial cells were exposed to nonpiliated meningococci (left). The principal factor responsible for cytotoxicity is LPS (see Section 2.2.2). (Bottom) The structure of L3 and L8 immunotypes of meningococcal LPS. Neisserial LPS does not contain O chains typical of enteric bacteria, and as such also has been called lipo-oligosaccharide (LOS). Phase variation between the L3 and L8 immunotypes of LPS observed in some strains may be controlled via phase variation in the *lgtA* gene involved in the addition of *N*-acetylglucosamine (GlcNac) (Jennings *et al.,* 1995b). KDO: 2-Keto-3-deoxyoctulonic acid; PEA: phosphoethanolamine; HEP:heptose.

other processes involving host receptors may achieve a similar result. Many recent studies have shown that host components targeted by bacteria include hormone, cytokine, and adhesion receptors. Some host receptors are either not expressed constitutively or are expressed in low numbers and may be up-regulated by cells exposed to inflammatory cytokines and other factors. Alternately, receptors may be down-modulated during infection. For example, respiratory syncytial virus affects host cells such that they down-regulate adhesion receptors LFA-1 (lymphocyte function associated molecule-1) and ICAM-1 (intercellular adhesion molecule-1) on mononuclear cells, leading to compromised cellular defense. Other viruses such as parainfluenza virus type 2 up-regulate several receptors on human tracheal epithelial cells. Thus prior infections, during which inflammatory cytokines are produced and the host receptor repertoire is altered, may lead to a state that facilitates invasion by meningococci. Receptor density, multiple receptor occupancy, as well as the affinity of microbial ligands for target receptors may determine the status of a microorganism as a commensal or a pathogen (reviewed in Virji, 1996a,b).

2.2. Surface Glycans and Their Properties

2.2.1. CAPSULE

Capsule and LPS constitute two "shielding" molecules of meningococci that enhance bacterial survival during transmission, colonization, and dissemination. The polysaccharide capsule is believed to protect the organism against desiccation during airborne transmission between hosts. Although direct evidence of this is difficult to procure, some laboratory observations point to this role for the capsule. For example, a closely related organism, the gonococcus, with a lifestyle that does not involve airborne transmission, is acapsulate and more susceptible to desiccation under the same laboratory conditions compared with the capsulate meningococcus. During colonization of the nasopharynx, when the role of capsule is diminished, meningococci may become acapsulate, as has been demonstrated by the frequent isolation of acapsulate (nongroupable) meningococci from the nasopharynx of carriers (Cartwright, 1995). Capsule phase variation may be influenced by environmental conditions. However, two distinct genetic mechanisms of capsule expression have been demonstrated (see Section 2.2.3d).

Meningococci elaborate one of several capsular chemotypes (serogroups), designated A, B, C, 29E, H, I, K, L, W135, X, Y, and Z (Cartwright, 1995). Of these A, B, and C are most commonly encountered during disease. Serogroup A predominates in Africa and is responsible for epidemics, whereas serogroups B and C prevail in developed countries and are associated with sporadic disease outbreaks (Achtman, 1995). The precise reasons for this serogroup-dependent pathogenic potential and geographic distribution is not entirely clear. Host, socioeconomic, as well as climatic factors may determine these differences.

In contrast to colonizing isolates, meningococcal isolates from disseminated infections almost always express one of the polysaccharide capsules. Of the three common capsular types, capsules B and C are composed entirely of polysialic acids. Serogroup A is a polymer primarily of α1,6-linked 2-acetamido-2-deoxy mannose phosphate. Serogroup B capsule is a polymer of α2,8-linked polysialic acid (also present in *Escherichia coli* K1 capsule, another organism that causes bacterial meningitis in neonates). This structure is present on human neuronal adhesion molecules (N-CAM). As a result, the structure is poorly immunogenic in humans and poses a major problem in producing a polysaccharide-based vaccine against this serogroup. Serogroup C capsular polysaccharide is a polymer of α2,9-linked polysialic acid, and despite its small structural difference compared to the serogroup B capsule, it is an immunogen and is used in the current capsule-based vaccine against this serogroup. (A tetravalent vaccine, long in use, consists of polysaccharides from serogroups A, C, Y and W135.) The limitations of polysaccharides as vaccines arise from the fact that they tend to be T-independent antigens and as such evoke only short-term protection. The induction of T-cell response and memory is facilitated by conjugation of capsular polysaccharide with T-dependent antigens such as diphtheria or tetanus toxoid; this is the basis for a new generation of serogroup A and C conjugate vaccines (Ala'Aldeen and Griffiths 1995; Cartwright, 1995).

2.2.2. LIPOPOLYSACCHARIDES

Meningococcal LPS is toxic for human cells and its effect is augmented by pilus-mediated cellular interactions (see Section 3.3). The toxicity is serum (soluble CD14) dependent and can be abrogated by polymyxin B, an antibiotic known to bind specifically to lipid A (Dunn *et al.*, 1995). Thus lipid A of meningococcal LPS is primarily involved in this toxicity. Such toxic effect was not observed when using *Haemophilus influenzae,* another mucosal pathogen that can cause bacteremia and meningitis in humans (Fig. 2, in Virji *et al.*, 1991a). The potential roles of glycan side chains in influencing meningococcal toxicity was determined by the use of rough mutants lacking glycan residues of LPS. The studies indicate that the oligosaccharide portion does not play a significant role in LPS–CD14-human endothelial cell interactions (Jennings *et al.*, 1995a).

Meningococci elaborate many distinct lipopolysaccharides, each varying from the other in polysaccharide composition and structure. Several immunotypes (designated L1-L12) have been identified on serological basis (Ala'Aldeen and Griffiths, 1995; Cartwright, 1995). Chemical nature and enzymatic machinery of several structures have been identified (Kahler and Stephens, 1998; Jennings *et al.*, 1995b; Verheul *et al.*, 1993; DiFabio *et al.*, 1990). Several of the immunotypes are structurally closely related and arise due to phase variation of oligosaccharide epitopes as well as sialylation of terminal galactose moieties (especially that of lacto-*N*-neotetraose, Fig. 2). Thus, the surface of meningococci can be surrounded with two layers of negatively charged molecules, one afforded by LPS and another by capsule. We and others (McNeil and Virji, 1997; Hammershmidt *et al.*, 1996a; Virji *et al.*, 1995a; Moran *et al.*, 1994; Mackinnon *et al.*, 1993; Estabrook *et al.*, 1992) have investigated the influence of these glycan-determined charge barriers on bacterial survival and cellular interactions.

2.2.3. THE INTERPLAY BETWEEN SURFACE LIGANDS IN HOST INTERACTIONS, POSSIBLE MECHANISMS OF COLONIZATION, AND PATHOGENESIS

Capsular and LPS glycans inhibit opsonophagocytosis and complement detection by masking surface antigens, although the precise molecular mechanisms have not been satisfactorily unraveled (Moxon and Kroll, 1990). Their role in inhibiting bacteria–host cellular interactions and nonopsonic phagocytosis requires a brief introduction of the known meningococcal adhesins and their host cell receptors.

2.2.3a. Meningococcal Adhesins and Invasins

Three major adhesins and invasins have been identified in meningococci. In addition to pili, two transmembrane proteins, Opa and Opc (also known as class 5 or opacity proteins), are capable of mediating cellular interactions. Opa and Opc are both phase-variable basic proteins, but are distinct in structure. Opc is structurally largely invariable and encoded by a single gene, whereas Opa proteins are a family of related proteins and three to four distinct genes may be present in meningococcal isolates (Achtman, 1995; Sarkari *et al.*, 1994; Aho *et al.*, 1991). Opc appears to have the capacity to bind to multiple extracellular matrix (ECM) components and serum proteins. This property enables bacteria to target integrin receptors by using RGD (arginine-glycine-aspartic Acid)-containing ligands, such as vitronectin, as bridging molecules (Virji *et al.*, 1994). In addition, Opa proteins may target certain matrix proteins, and they have been shown to bind directly to the carcinoembryonic antigen (CEA, CD66) family of adhesion receptors (Virji *et al.*, 1996). In addition, Opa proteins may target alternate receptors, the heparan sulfate proteoglycans (van Putten and Paul, 1995). It is important to note that many of these receptors, particularly CD66, are expressed at low levels in resting cells and can be up-regulated (Dansky-Ullman, 1995).

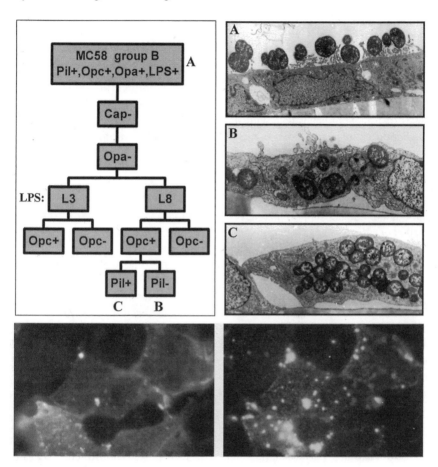

Figure 3. Studies on the interplay between surface components in meningococcal interactions with target cells. (Top: left) A family tree representing a library of variants of a serogroup B strain MC58 isolated by mutation (capsule and *opc* genes) or selection of naturally arising variants (LPS, Opa, pili). (Right) Transmission electron microscopy of human umbilical vein endothelial cells grown on polycarbonate filter supports and infected with distinct phenotypes of MC58. (A) A piliated, Opa-, and Opc-expressing sialylated bacteria can be seen adherent in large numbers to human endothelial cells. Despite the presence of OM invasins Opa and Opc, no cellular invasion occurs in this sialylated phenotype. The observed adhesion is pilus mediated since the nonpiliated counterpart is nonadherent (not shown). (B) Invasion of human endothelial cell by an asialylated phenotype expressing only Opc. (C) In such a phenotype, the additional expression of pili increases cellular invasion. (Bottom) Photomicrographs of COS (African Green monkey kidney) cells transiently transfected with cDNA encoding CD66a adhesion molecule and infected with capsulate, Opa-expressing bacteria. (Left) Monolayers were stained with FITC-labeled anti-CD66 monoclonal antibodies to detect the expression of these receptors on transfected cells. (Right) Adherent bacteria in the same field were detected with rhodamine-labeled anticapsular monoclonal antibodies. This demonstrates that meningococcal adhesion to cells expressing high levels of CD66 may occur via certain Opa proteins even in a sialylated phenotype.

2.2.3b. Interplay between Surface Glycans and Adhesive Ligands

In order to investigate the influence of the surface glycans on functions of integral outer membrane adhesins, a library of phenotypic variants and mutants starting from a capsulate serogroup B strain MC58 was created (Fig. 3) (Virji *et al.*, 1995a). The studies using these derivatives demonstrated that in sialylated phenotypes (capsulate or acapsulate with sialylated LPS), pili appear to be essential in mediating cellular adhesion (Fig. 3). In contrast, adhesion to and particularly invasion of human epithelial and endothelial cells are enhanced in acapsulate meningococci expressing an LPS immunotype that resists sialylation (L8 immunotype). The latter interaction requires the presence of one of the opacity proteins. Such results also were obtained using human monocytes and polymorphs, although pili do not support phagocytic interactions (McNeil and Virji, 1997; McNeil *et al.*, 1994). Thus surface glycans have a profound influence on bacterial interactions mediated via specific ligand-receptor interactions.

In a recent study, we observed for the first time that some capsulate meningococci expressing Opa proteins with high affinity for CD66 receptors were able to interact with transfected COS cells expressing CD66 (Fig. 3). The latter interaction, although lower than in acapsulate bacteria, nevertheless was significant particularly with cells expressing high levels of CD66 (Virji *et al.*, 1996). Therefore, the inhibitory effects of surface sialic acids may be overcome to some extent when appropriate ligand–receptor pairs are present at the required density.

2.2.3c. Phase Variation of Surface Glycans

Both capsular polysaccharide and the lacto-*N*-neotetraose structure on LPS, which is often modified by the addition of sialic acid, have been shown to be phase variable with a frequency of 10^{-3} to 10^{-4} per generation (Hammerschmidt *et al.*, 1996a; Jennings *et al.*, 1995b). This high-frequency reversible on–off phase variation results in changes in the number of nucleotides within polypyrimidine and polypurine stretches in the open reading frames of capsular (*siaD*) or LPS (*lgtA*) biosynthetic genes. The nucleotide repeats are thought to favor local DNA denaturation, and displacement of DNA duplex results in insertion or deletion of single nucleotides. This Rec A-independent slipped-strand mispairing event may occur during DNA replication or repair. Addition or deletion of a nucleotide(s) may result in a frameshift mutation leading to termination of translation. In the case of capsule phase variation, other mechanisms, such as reversible inactivation of *siaA*, required for the biosynthesis of sialic acid also may operate. It is proposed that insertion of a naturally occurring mobile genetic element IS1301 may be responsible for inactivation of the gene (Hammerschmidt *et al.*, 1996b).

2.2.3d. Phase Variation, Carriage, and Pathogenesis: A Hypothesis

The phase variable nature of surface glycans and the frequency of phase variation suggest that one mechanism of host invasion may involve phenotypic transitions: capsulate (transmission phenotype) to acapsulate and asialylated (coloniz-

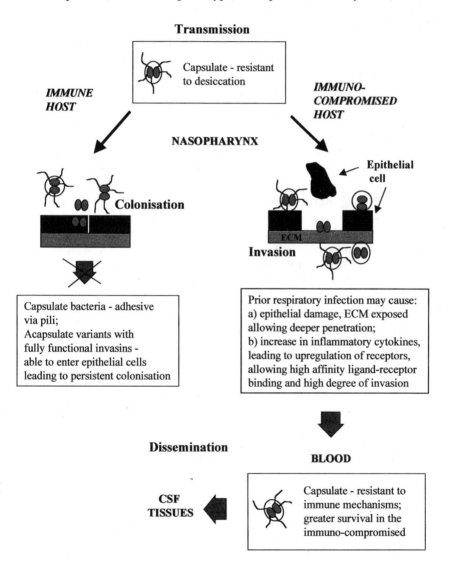

Figure 4. Possible mechanisms of meningococcal pathogenesis: a unifying hypothesis based on *in vitro* and epidemiological observations.

ing phenotype) back to capsulate and sialylated (disseminating phenotype). In such a model of meningococcal pathogenesis, opacity proteins may contribute significantly to host invasion. For example, during colonization, acapsulate, opacity protein expressing phenotype is often isolated. This invites the hypothesis that the loss of capsulation and sialic acid on LPS may help to establish long-term nasopharyngeal carriage, whereby becoming intracellular the bacterium is protected from host defenses. In addition to spontaneous phase variation, environmental factors may regulate capsule expression. Dissemination from the site of colonization would then require up-regulation of capsulation, since acapsulate bacteria are unlikely to survive in the blood. Alternatively, since blood provides an environment in which meningococci can grow rapidly, it is possible that a small number of capsulate organisms, arising as a result of natural phase variation, would be selected for in the blood.

Summarizing the above, recent advances in the molecular mechanisms of meningococcal interactions with human target cells are beginning to provide certain clues that may explain epidemiological observations. Several potential routes of invasion may exist. Acapsulate bacteria may target ECM exposed on epithelial damage following viral or other infections or they may invade epithelial cells. Further dissemination of such a phenotype requires selection of capsulate bacteria arising from natural phase variation. However, epidemiological evidence suggests that disease in susceptible individuals occurs soon after acquisition with no prolonged carriage. Recent observations that some receptor–ligand interactions may occur in capsulate bacteria provide a feasible rationale for an alternative, perhaps not exclusive, mechanism. In this scenario, targeting of cell adhesion molecules that are up-regulated by inflammatory cytokines may be the critical determinants of meningococcal invasion. Viral infections, or other conditions leading to inflammation, could result in increased expression of receptors that recruit meningococci via one or more ligands. In a host with inadequate immunological protection against meningococci, invasion would result in rapid growth and dissemination. This may constitute one of a number of mechanisms responsible in distinct circumstances for disease outbreaks (Fig. 4).

3. PILI, THE CAPSULE-TRAVERSING GLYCOPROTEINS

3.1. Structure

Neisseria meningitidis pili belong to type 4 structural class of pili present in numerous bacterial species including *Pseudomonas aeruginosa, Dichelobacter nodosus, Moraxella bovis, Eikenella corrodens, Kingella denitrificans,* and *Neisseria gonorrhoeae* (see *Gene*, 1997). *N. meningitidis* pili consist of identical subunits (pilin or PilE encoded by the *pilE* gene) of molecular weights of approximately 15–20 kDa. Pili can be easily observed in negatively stained preparations by transmission and scanning electron microscopy especially when they aggregate to form bundles (Fig. 2). As described above, pili are important in capsulate as well as acapsulate bacteria expressing sialylated LPS to mediate interactions with host cells. However, pili also participate in other important functions of *Neisseria* that make them successful human colonizers and pathogens. *Neisseria* are naturally transformable organisms and take up specific DNA (recognized by neisserial DNA uptake sequence) from the environment. Pili enhance transformation frequency, Pil E appears to play a central role in DNA uptake (Fussenegger *et al.*, 1997). In addition, neisserial pili are implicated in bacterial movement known as the twitching motility, as is the case with type 4 pili of *P. aeruginosa* (Darzins and Russell, 1997). Another consequence of piliation is a frequently observed phenomenon of bundling or autoagglutination. Interactions of pili between different organisms in a culture result in clumping of bacteria with several consequences. Bacteria in the center of a clump are protected from antibody and complement attack as well as exposure to antibiotics. Also, for a given number of receptors, *in vitro* studies show that more clumped bacteria may become localized on host cell surfaces compared with nonaggregated counterparts, and this could result in increased cellular toxicity or invasion. This is in contrast to aggregation of bacteria mediated by some antibodies that prevent colonization and increased mucosal clearance *in vivo*. Thus the phenomenon of bacterial agglutination, whether mediated by pili or other factors, is important in pathogenesis. Pili of the subclass type 4b exemplified by bundle-forming pili (BFP) and toxin-coregulated pili (TCP) of *E. coli* and *Vibrio cholerae* are well known for their lateral aggregation resulting in bundling morphology (Manning, 1997). Pili of some meningococcal strains aggregate in an array in a similar manner. However, nonaggregated pili also are frequently produced by meningococcal isolates and within a single meningococcal strain, variants may produce pili that are either individually elaborated or form bundles. The precise factors that determine pilus morphology have not been satisfactorily identified. Both primary sequence changes as well as glycosylation status of pilin have been implicated as factors responsible for pilus aggregation. A fuller discussion follows in Section 7.3.

Piliated phenotypes are prevalent *in vivo* and may be selected for in the course of disease, since piliation is rapidly lost during nonselective subculture *in vitro*. Antigenic variation during disease has been documented, and organisms belonging to a single strain isolated from distinct sites from individual patients were shown to express variant pilins (Heckels, 1989). Thus pili are subject to phase as well as antigenic/structural variations both *in vitro* and *in vivo*. Also, meningococcal strains express one of two major structural classes of pili known as class I and class II. These were initially identified as a result of differential reactivity of a monoclonal antibody SM1 raised against the related organism *N. gonorrhoeae.*

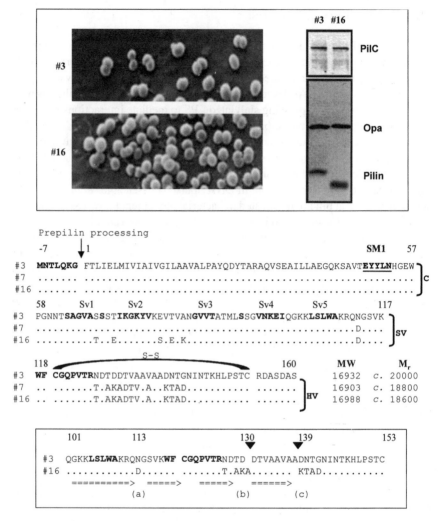

Figure 5. Adhesion, OM composition, and Pil E sequences of C311 clonal derivatives. (Top left) Clonal derivative #16 adheres in larger numbers to human endothelial cells compared to #3 (shown) and also to human epithelial cells (not shown). In analysis of surface components by SDS-PAGE and Western blotting, only pilins appear to vary between the derivatives. The Western blot was developed using antibodies specific for the OM components shown (right). (Middle) Deduced amino acid sequences of variant pilins of *N. meningitidis* strain C311. DNA sequences of the variants were obtained by PCR amplification and sequencing. The identical amino acids are marked by dots. Numbers 3 and 10 had identical sequences. C: Conserved N-terminal domains; SV: semivariable intermediate domains; HV: hypervariable C-terminal domains of pilins encompassing a disulfide loop (S-S). Molecular masses of the pilins predicted from amino acid sequences and their molecular weight obtained on SDS-PAGE gel electrophoresis are shown. EYYLN is the epitope recognized by class I-reactive monoclonal antibody SM1. (Bottom) By comparison of the sequences with the secondary structure of class I pilin (see Fig. 8) the sequence alignment of the hypervariable region has been adapted and the new alignment is shown (A. Hadfield and M. Virji, unpublished data). By allowing for insertion of Ala at position 131 and deletion at the position 138 (arrowheads), the secondary structure is maintained. The amino acid changes that appear only on the hyperadherent phenotype are restricted to three loops a, b, and c. This sequence has been used to build a model of hyperadherent pili (see Fig. 8)

SM1 reacts with an epitope EYYLN in the N-terminal region of pilin (Fig. 5). This epitope is present in all gonococcal pili and in class I pili of *N. meningitidis,* which are highly homologous to gonococcal pili (Virji *et al.,*1989). Recently class II pilins were cloned and the basis of structural variations identified (Aho *et al.,* 1997). Class II pili are similar to class I, but are truncated with a small deletion in the region spanning the SM1 epitope and a larger deletion in the disulfide loop. The intraclass and intrastrain structural variations of pili arise as a result of genomic Rec A-dependent recombination between the complete pilin expression gene locus (*pilE*) and incomplete silent pilin gene loci (*pilS*) that are present within neisserial genome (Seifert and So, 1988; Potts and Saunders, 1988). Besides intragenomic recombination, intergenomic recombinations are common, since *Neisseriae* are naturally competent and readily take up DNA from other lysed neisseriae.

3.2. Pilus-Associated Protein PilC

In 1991, a protein of 100 kDa, copurifying with pili, was shown to phase vary and its expression was required for pilus biogenesis (Jonsson *et al.,* 1991). Since then, other studies have reported that it also may be involved in cellular adhesion. Two *pilC* loci have been observed in the genomes of most meningococci and gonococci encoding related but not identical proteins, PilC1 and PilC2. The two proteins may be responsible, at least in meningococci, for different functions of pilus assembly, anchorage, or interactions with the host cell receptor; it can be found both on pili and in the outer membrane (Rahman *et al.,* 1997; Virji *et al.,* 1995b; Rudel *et al.,* 1995; Nassif *et al.,* 1994). Whether or not PilC is the ligand that determines the adhesiveness of pili, the primary structure of the major subunit, PilE, has considerable influence on pilus-mediated adhesion (Jonsson *et al.,* 1994; Nassif *et al.,* 1993; Rudel *et al.,* 1992; Virji *et al.,* 1991b, 1992, 1993) and also may carry a receptor-binding domain (Marceau *et al.,* 1995; Rothbard *et al.,* 1985; Gubish *et al.,* 1982).

3.3. Pilus Adhesion Function and Synergism with Other Ligands in Pathogenesis

In encapsulated–sialylated meningococci, pili not only mediate adhesion to target cells, but also impart both host and tissue tropism. Pili of either class are effective adhesins. Pili also contribute to the cytopathic effect of meningococci on human endothelial cells (Fig. 2). It has been shown that piliated bacteria cause greater damage to human umbilical vein endothelial cells (HUVECS) than nonpiliated bacteria and the damage is proportional to the level of bacterial adhesion mediated by structurally variant pili. Other adhesins cannot substitute for pili in this particular effect (Dunn *et al.,* 1995). Thus structural features that modulate pilus-mediated adhesion also may alter the severity of endothelial necrosis observed *in vivo.* The latter is a characteristic feature of disseminated meningococcal infection. *In vitro* studies also have shown that in acapsulate phenotypes, pili may synergize with Opc and increase cellular invasion of some human target cells (Fig. 3) (Virji *et al.,* 1995a). The identification of the structural components of pili that determine its functions is of importance in understanding meningococcal pathogenesis and ultimately for molecular targeting for intervention during the course of meningococcal infection. To date, the precise nature of the pilus epitopes or associated ligand(s) or ligand complex(es) that may be involved in interactions with distinct host cells is not clearly understood. The complexities of the structures that pili can elaborate make these studies particularly challenging.

During our investigations on the structure–function relationships of meningococcal pili, extensive posttranslational decorations of pili were discovered. The studies also emphasized that the structural makeup of pili is controlled at multiple levels. For example, variations in modifications can arise as a result of pilin sequence variations and/or genetic changes elsewhere in the chromosome not linked to *pilE,* such as *galE* mutations with resultant nonspecific effect on all galactose containing glycans or *pglA* mutations that affect glycosylation of pili specifically (Section 4.6). Structural analyses of the substituents have revealed unusual features suggesting their functional importance. The major part of the remainder of the chapter presents an overview of the approach in characterization of the posttranslational modifications of meningococcal pili and briefly explores some of the other known prokaryotic protein modifications and their proposed functions.

4. STRUCTURAL STUDIES OF MENINGOCOCCAL PILI

4.1. Discovery of Glycosylation

Multiple variants (clones) were derived from two serogroup B strains MC58 and C311, both expressing class I pili, by single-colony isolation from agar-grown bacteria with or without prior selection on host cells. Analysis of approximately 50 clonal cultures was sufficient to yield clones exhibiting both morphologically and functionally distinct pili. The studies demonstrate the frequency with which molecular structural variations are observed in meningococcal pili. On analysis of clonal variants in a functional assay for their adhesion to human cells, variants exhibiting strikingly different properties of adhesion to epithelial cells were readily isolated. These variants were shown to have similar expression of several known adhesins other than pili. Comparison of the variants of strain MC58 by sodium do-decyl sulfate–polyacrylamide gel electrophoresis (SDS-PAGE) showed that a pili-ated variant (#6), which was no longer adherent to epithelial cells, had pilins that migrated differently compared with the parental phenotype (#5). Also, pilins of three variants of strain C311 (#7, #10, and #16) migrated significantly farther than those of the rest of the clones and of clone #3, which represents the pre-dominant phenotype of the isolate C311 (e.g., with respect to migration of pilins in SDS-PAGE) (Fig. 5). Clonal variants #7 and #16 were hyperadherent for hu-man epithelial cells exhibiting more than tenfold increase in adhesion compared with the parental phenotype, whereas their increase in binding to endothelial cells was less pronounced (about threefold) and partly may be explained by increased agglutination (see Section 7.3). The differences in relative increase in adhesion to epithelial and endothelial cells suggest that there may be different mechanisms in-volved in pilus-mediated interactions with these cells. In order to assess the basis of the observed variations, the variant *pilE* genes were investigated.

On polymerase chain reaction (PCR) sequencing of the *pilE* genes of the vari-ants, no deletions were observed in any of the genes. Minor random amino acid substitutions were observed in pili of #7 and #16 (Fig. 5) but not of #10. Recent studies that employ molecular modeling are showing that these apparently random variations may be clustered in specific regions on the pilus (see Section 7.3).

Thus all variant pilins of strain C311 were of similar molecular weights as deduced from their DNA sequences, but migrated to different positions on SDS-PAGE (Fig. 5). Since migration on SDS-PAGE may alter as a result of posttrans-lational modifications of a protein, the observations suggested that meningococcal pilins may undergo modifications. Further, it was observed that the lipopolysac-charides of strain C311 clonal derivatives #3, 7, and 16 were identical in their mi-gration but that of #10 was truncated, raising the possibility that simultaneous variations in pili and LPS may be due to events elsewhere in the genome not linked to *pilE* in the derivative #10 (e.g., as found with *galE* mutations) (see Section 4.3). Therefore, both primary sequence changes and/or other unlinked events appear to determine the size of C311 pilins, possibly via posttranslational modifications.

4.2. Biochemical Demonstration of Pilus-Associated Glycans

Since glycosylation is a common posttranslational modification of surface proteins of eukaryotic cells, the possibility that pili may contain glycans was first investigated by biochemical methods. The monoclonal antibody SM1 was used to immunoprecipitate pilins from variants #3 and #16 of strain C311. These were electrophoretically separated and transferred to nitrocellulose. The blots were sub-jected to biotin hydrazide labeling under conditions that specifically incorporate biotin into carbohydrate moieties. After dissociation in standard SDS-PAGE dissociation buffer, the immunoprecipitated samples produced four major bands on gels corresponding to pilin, heavy and light chains of SM1, and an unidentified, contaminating protein. Only the heavy chains and pilins were labeled with biotin hydrazide (detected using avidin–alkaline phosphatase conjugate) but not the light chains or the contaminating protein. Since the heavy chains of immunoglobulins are frequently glycosylated unlike the light chains, the experiment provided internal controls and strongly suggested the association of carbohydrate moieties with pili. Glycosylation of pilins was further confirmed by chemical deglycosylation of whole cell lysates using anhydrous sulfonic acid. Pilins and LPSs were the only major constituents of whole cell lysates to be affected by this treatment (Virji *et al.*, 1993).

4.3. Genetic Evidence for Covalently Linked Glycans on Meningococcal Pili

Initial investigations were carried out to determine the sugar contents of pilin-derived tryptic fragments. Chromatographic analysis showed the presence of some sugar residues. Unlike glucose, galactose was present only in one of the pilin fragments, suggesting that it is a constituent of this pilin fragment and not merely a contaminant. Further proof that covalently linked galactose is present on pili was obtained by genetic manipulation of variants. Galactose epimerase (Gal E) is required in *N. meningitidis* for the production of UDP-galactose. Therefore, the absence of Gal E would be expected to result in the lack of incorporation of galactose into LPS (strain C311 LPS contains several galactose residues) as well as into pili, if these were decorated with galactose moieties. To test this hypothesis, the *galE* genes of the variants #3 and 16 were mutated. The resultant derivatives were analyzed for their possible concurrent alteration in the size and migration of pilin and LPS on SDS-PAGE. All Gal E mutants produced apparently truncated LPS and pili. In order to establish that the observed decrease in the size of pilins was not a result of deletion or any critical sequence changes in pilins, PCR sequencing of their *pilE* genes was carried out. Pilins from #3 and one of its *galE* mutants had identical predicted amino acid sequences but different migration on SDS-PAGE; the data indicated covalent linkage of galactose to pili (Stimson *et al.*, 1995).

4.4. Determination of the Structure of Pilin-Linked Glycans

To investigate the structure(s) of glycans on C311 pili, purified pili from the variants #3, #16, and their Gal E mutants were used. Pili were subjected to tryptic digestion and the resulting peptides purified by reverse phase high-pressure liquid chromatography (HPLC) and subjected to fast atom bombardment mass spectrometry (FAB-MS), electrospray mass spectrometry (ES-MS), and gas phase Edman sequencing. In most cases, FAB-MS produced signals that corresponded to the predicted masses of peptides. However, in the case of two peptides, no molecular ions were observed at their calculated masses. In the case of a major peptide [45]S-K[73] or [45]S-K[75], unassigned signals were observed at higher masses, indicating that this region of the protein might be posttranslationally modified. Further V8 protease digestion suggested that modification is within amino acids 50–73. On mild base treatment of the peptide, the modification was removed, suggesting that an O-linked glycan might be present. Reductive elimination and subsequent FAB-MS analysis of deuteroacetylated, acetylated, or permethylated products confirmed this. Parallel experiments of the parent and GalE mutant pilins demonstrated that the former contained a larger mass of 572 Da, while the corresponding moiety from the *galE* mutant was lower in mass equivalent to two hexose molecules (presumably two galactose residues). The GalE mutant contained a reducible residue of mass 228 Da. Sugar and linkage analyses showed that the disaccharide moiety was Gal1-4Gal, and the structure of the 228-Da residue was defined as 2,4-diacetamido-2,4,6-trideoxyhexose from the electron impact (EI) mass spectrum of its reduced trimethylsilyl derivative.

Enzymatic digestion with α- and β-galactosidases of different substrate specificities provided further structural information. *Choronia lampas* β-galactosidase digested a portion of the trisaccharide and the products of this digestion could be further digested with coffee bean α-galactosidase. These data and mass analysis of the saccharides taken together, are consistent with the structure Galβ1,

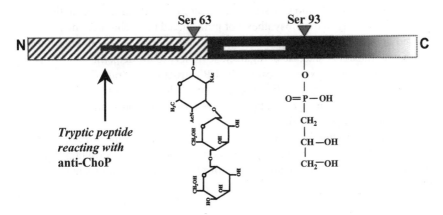

Figure 6. Posttranslational modifications of meningococcal strain C311 pili. Schematic presentation of the pilin subunit and positions of the three common modifications [Galβ1, 4Galα1,3 (2,4,diacetamido-2,4,6-trideoxyhexose), glycerol phosphate, and phosphorylcholine) identified on pilins of strain C311. Note: another modification is known to occur and is different in clones #3 and # 16, but its identity is unknown (Stimson *et al.,* 1995). Hatched area represents pilin-conserved domain. Solid bars represent regions of pili (amino acids 41–50; 69–84) (Rothbard *et al.,* 1985) that may be involved in adhesion in the homologous pili of gonococci.

4Galα1, 3 [2,4,diacetamido-2,4,6-trideoxyhexose] as a common constituent of variant pilins of strain C311 (Stimson *et al.*, 1995) (Fig. 6).

4.5. Expression of the Galβ1, 4Gal Structure in Meningococcal Isolates

GalE mutations were introduced in several isolates expressing class I or class II pili in order to investigate whether both classes of pili were modified with galactose-containing glycans. These studies have shown the presence of covalently linked galactose in a large proportion of the strains independently of the class of pili. To investigate the presence of a β1–4 digalactosyl moiety, we have raised monospecific antisera recognizing this structure. The antisera were raised by the use of synthetic disaccharides conjugated to keyhole-limpet hemocyanin (KLH) as immunogens. Antisera were then affinity purified on the disaccharide–sepharose columns. These antisera recognize the Galβ1, 4Gal disaccharide specifically (no binding to Galα1,3Gal structure was observed) and bind to pili of several meningococcal isolates (5/9 tested) including those expressing class II pili (unpublished studies). Thus the structure is widely distributed among meningococcal isolates.

4.6. Pgl A, a Galactosyl Transferase Specific for Neisserial Pili

Recently, using homology searching strategy, we identified a gene that is involved specifically in the biosynthesis of pilin-linked trisaccharide. The gene designated *pglA* (for *p*ilin *g*lycosylation) appears to encode a specific galactosyl transferase that transfers galactose or digalactose moiety to the deoxyhexose. Mutation in this gene has no apparent effect on the LPS structure of strain C311. Similarly, several enzymes involved in the biosynthesis of the lacto-*N*-neotetraose of LPS have no effect on pilin glycosylation, thus separating the biosynthetic machinery of LPS and pilin glycosylation.

Genetic investigations on *pglA* have revealed the presence of a homopolymeric tract of guanosine residues in the coding region of *pglA*, providing a mechanism for its phase variation and control of glycosylation of pili (Jennings *et al.*, 1998).

5. DETERMINATION OF A SECOND SUBSTITUTION, α-GLYCEROPHOSPHATE, COMMON TO C311 PILINS

Another tryptic peptide, [84]N-K[98], also did not produce molecular ions in FAB-MS or ES-MS at the calculated mass (mass:charge ratio: m:z 1478), instead, an intense, unassigned signal was observed at m:z 1632, suggesting the peptide modification with a substituent of mass 154 Da. Treatment of the peptide with 48% aqueous hydrogen fluoride under conditions known to cleave phosphodiester linkages removed the moiety, which was thus phosphodiester linked to pilin. The identity of the substituent was established as α-glycerophosphate by gas chromatography and EI-MS. Its site of attachment on peptide [84]N-K[98] was obtained by mass spectrometric analysis of elastase subdigestion products of the tryptic peptide. These data indicated that the α-glycerophosphate substituent is linked to Ser[93] (Stimson et al., 1996) (Fig. 6).

6. PHOSPHORYLCHOLINE: A FURTHER DECORATION OF NEISSERIAL PILI

Phosphorylcholine (ChoP) has been shown to be a common feature of the cell surface glycolipids of major pathogens of the human respiratory tract including *Streptococcus pneumoniae, H. influenzae,* and mycoplasma species (Mosser and Tomasz, 1970; Weiser et al., 1998a). We surveyed other gram-negative pathogens that frequently infect the human respiratory tract for the presence of the epitope using monoclonal antibodies specific for ChoP. The epitope was found on two further organisms, *P. aeruginosa* and *N. meningitidis.* Interestingly, in both these cases, the epitope was present on proteins that included a 43-kDa protein on all clinical isolates of *P. aeruginosa* and on several class I as well as class II pili of meningococcal isolates. ChoP also is present on pili of *N. gonorrhoeae* (Weiser et al., 1998b). As for many other ligands, the expression of the ChoP epitope on piliated neisseriae displayed phase variation, both linked to pilus expression and independently on fully piliated bacteria.

The specificity of the detection of ChoP epitope on pili by monoclonal antibodies was demonstrated by the use of analogues. Only phosphorylcholine inhibited the binding of anti-ChoP antibodies to pili. In recent studies using tryptic peptides of purified pili, the epitope was shown to be located on the N-terminal peptide that also carries the trisaccharide (Fig. 6). The epitope does not appear to be linked to glycans, since it is present on nonglycosylated mutants (unpublished observations).

In further studies we investigated whether ChoP epitope was also present on the pili of commensal *Neisseria,* such *N. lactamica* that elaborate structurally similar pili to meningococcal class II pili. These studies have revealed an interesting phenomenon. Despite the structural similarity, none of the strains of commensal *Neisseria* examined expressed the epitope on their pili; instead, the epitope was present on their LPS. A survey of 60 pathogenic *Neisseria* strains (30 gonococci and 30 meningococci) showed exclusive presence of ChoP on pili but not on LPS (L. Serino and M. Virji, 2000). The functional significance of the distinct sites of location of the phosphorylcholine epitope in commensalism and in pathogenesis remains to be investigated.

7. GLYCANS AS SUBSTITUENTS OF PROKARYOTIC PROTEINS

Until recently, glycosylation of proteins was regarded as uncommon in prokaryotes. It has been reported, however, in archebacterial and eubacterial S-layer proteins and a number of bacterial cellulases for some years (Messner, 1997). Their wider distribution among prokaryotes has been recognized relatively recently (Moens and Vanderleyden, 1997; Messner, 1997) (Chapters 1, 3, and 4, this volume).

7.1. Structural Diversity of Glycans

Pilins of the variants #3 and #16 of *N. meningitidis* strain C311 contain a covalently linked trisaccharide of the structure Galβ1, 4Galα1, 3 [2,4,diacetamido-2,4,6-trideoxyhexose], which can be released by reductive elimination, suggesting its attachment to Ser or Thr. Recently, Ser[63] was identified as the modified amino acid by site-directed mutagenesis of C311 pilins (Payne *et al.*, 1996). X-ray crystallographic studies on homologous gonococcal pilin (of strain MS11) also have shown glycosylation of Ser[63]. However, in this case, the *O*-linked glycan is a disaccharide of the structure Galα1, 3GlcNAc (Parge *et al.*, 1995). Our investigations using monospecific antibodies that recognize β1–4-linked diGal but not α1–3-linked diGal have demonstrated the presence of the former structure on pili of several meningococcal strains. But as may be predicted from the observations that glycans may be subject to phase variations via *pglA* (see Section 4.6), not all pili were recognized by these antisera. Another investigation using monoclonal antibodies that recognize digalactose structures on moraxella LPS also reported the presence of digalactose structure on an unrelated meningococcal strain to those employed by us (Rahman *et al.*, 1998). These studies and investigations in my laboratory have failed to detect a digalactose moiety on gonococcal MS11 pili, thus confirming the presence of several different glycan structures on distinct pili of neisserial strains. A recent study has identified a disaccharide structure on pili of a meningococcal strain that is similar to that of MS11 pili (Marceau *et al.*, 1998); thus the differences are strain dependent and not species specific. Whatever the nature of glycans on pili, *Neisseria* appear to decorate their principal adhesin with a variable glycan structure. The antigenic and phase-variable nature of the epitope and its location on pili in the vicinity of a putative cell-binding domain (Fig. 6) invites the hypothesis that glycans may provide a cloaking device against host immune response or protect the site from proteolytic cleavage.

Among structurally diverse sugar moieties found in prokaryotes are a range of hexoses, deoxy, and amino sugars. One of the first diamino sugar to be isolated was N-acetylbacillosamine (4-acetamido-2-amino-2,4,6-trideoxyglucose) (Zehavi and Sharon, 1973; Sharon and Jeanloz, 1960). Other diamino sugar residues, including diacetamidotrideoxyhexoses, have been found to be constituents of polysaccharides isolated from *P. aeruginosa, V. cholerae, E. coli,* and *Thiobacillus* spp. (Shashkov *et al.,* 1995; Whittaker *et al.,* 1994; Hermansson *et al.,* 1993; Tahara and Wilkinson, 1983). However, diacetamidotrideoxyhexoses as constituents of glycoproteins are rare. Because of its rarity in glycoconjugates and as a novel linkage sugar, the 2,4-diacetamido-2,4,6-trideoxyhexose found in *N. meningitidis* is especially interesting. Its biosynthetic machinery and functional importance remain to be determined.

7.2. Functional Implications of Glycans

The earliest indication of the possible presence of sugars and their likely role came from studies on gonococcal pili. The studies on purified pili of *N. gonorrhoeae* recorded approximately 1.3% (w/w) of galactose per pilin subunit (Robertson *et al.,* 1977). However, whether this was covalently linked was not described. Further studies from another laboratory indicated that treatment with β-galactosidase affected adherence properties of a cyanogen bromide fragment (CNBr1: amino acids 8–102) derived from *N. gonorrhoeae* F62 pilin (Gubish *et al.,* 1982). These regions of *N. gonorrhoeae* and *N. meningitidis* class I pilins are highly conserved and the studies described above have clearly demonstrated covalently linked galactose on *N. meningitidis* pili in this N-terminal region. However, in contrast to those of *N. gonorrhoeae* F62, pili of GalE mutant of meningococcal strain C311 are as effective as fully glycosylated pili in mediating host cell interactions (Stimson *et al.,* 1995).

The importance of the N-terminal conserved–semiconserved pilus region in neisserial interactions with target cells also was suggested in studies that used antibodies against pilin-derived peptides. Those against peptides 41–50 and 69–84 inhibited attachment of piliated *N. gonorrhoeae* to human endometrial cells (Rothbard *et al.,* 1985). Also, in *N. meningitidis* strain, MC58, a single amino acid substitution (Asn_{60} > Asp) has profound effects on bacterial interactions with epithelial cells (Virji *et al.,* 1992, 1993). However, variant gonococcal pili that do not show major changes in these regions (Nicolson *et al.,* 1987) apparently interact to variable extents with host epithelial cells (Virji *et al.,* 1982). Also, polyclonal antisera against variant pilins of gonococcal strain P9 inhibited adhesion of homologous variant but failed to inhibit heterologous variants (Virji *et al.,* 1982). In addition, only type-specific monoclonal antibodies against variable regions of pilin inhibit adhesion of homologous variants to epithelial cells (Virji and Heckels,

1984). Taken together, these data suggest that the spatial arrangement of pilin epitopes may be critical in determining the different roles of common and variable regions on pilin, and that both may contribute to domains interacting directly or indirectly with receptors on host cells (Section 7.3). The degree to which individual components of distinct pilins, with or without posttranslational modifications, participate in receptor interaction remains to be defined.

7.3. Glycosylation and Aggregation of Pili

During early investigations, we reported morphological differences in the pili elaborated by adhesion variants #3 and #16 of strain C311. Pili of clone #3 were individually elaborated whereas those of #16 were aggregated (Virji *et al.*, 1993) (Fig. 7). Since one of the functions assigned to glycans in eukaryotes is their effect on the solubility of proteins, it was reasonable to suppose that the observed differences were due to differences in the glycosylation of these variants. However, ensuing investigations have not supported this hypothesis unequivocally. For example, removal of galactose from either of the clones does not affect their morphology; *galE* mutant pili (containing only the deoxyhexose) of clone #3 are individually elaborated, whereas those of #16 form bundles (Fig. 7). If there are any differences in solubilities–aggregation, these are too subtle to be detected by electron microscopy. Similar investigations on nonglycosylated pili of another meningococcal strain were recently reported (Marceau *et al.*, 1998). In this study also, removal of glycans had no major effect on the morphology of pili. However, from aggregation of purified pili induced by pH change, the authors concluded that glycosylation may affect pilus solubility. Besides glycosylation, other factors may determine pilus agglutination. For example, variation in charge that may occur on sequence variation may increase lateral interactions between pilin subunits of distinct fibers, leading to clumping. This is supported by previous studies of Marceau *et al.* (1995): site-directed mutations in one surface-located amino acid appeared to alter the aggregation of pili without any apparent effect on glycosylation.

Our recent studies on molecular modeling have revealed, for example, that the apparently random changes on pilin sequences of the hyperadherent pili of the clones #7 and #16 (labeled a, b, and c in Fig. 5) cluster at one site (Fig. 8). These are apparently on the surface of assembled pili. The net alteration in charge (see Fig. 8) may change the lateral interactions between pili and may explain the distinct morphology of the parental phenotype and hyperadherent variants.

The observations outlined above on the multiple concurrent changes that occur on pili demonstrate the complexities and difficulties in interpretation of structure–function relationships of pili.

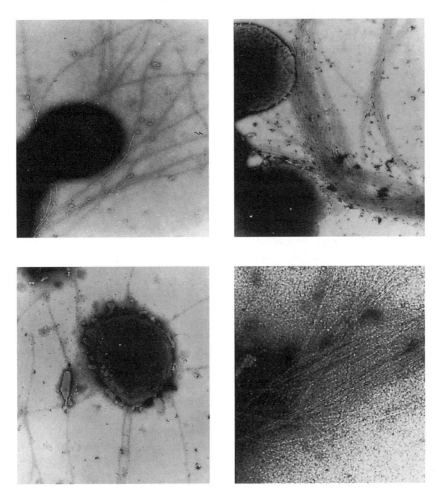

Figure 7.   TEM of pili of strain C311 adhesion variants. Electron microscopy of negatively stained preparations of clonal variants #3 (left) and #16 (right) and their Gal E mutants (bottom) showing the morphology of pili. Those of #3 and its galactose-deficient derivative are not agglutinated, whereas those of #16 and its derivative are produced in aggregates.

7.4. Other Possibilities

It has been shown that glycosylation may alter the function of both eukaryotic and prokaryotic proteins. Since the removal of galactose from pili of strain C311 does not abrogate their binding to human cells *in vitro*, the digalactosyl moi-

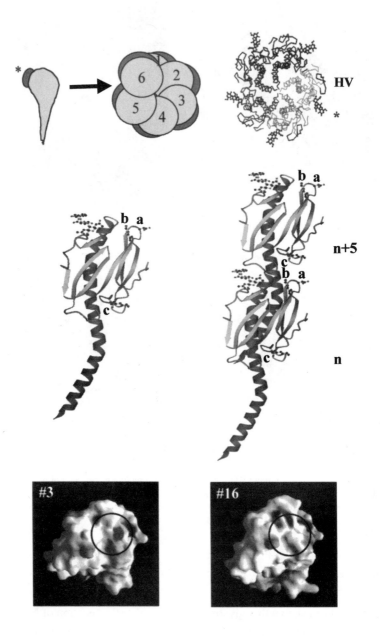

ety of the trisaccharide does not appear to be the ligand involved in these interactions. This, however, does not rule out the possibility that glycosyl residues may be important in modulating yet unknown or unexamined functions of pili.

Glycosylation of bacterial cellulases is known to increase resistance to proteolytic degradation of these proteins and to maintain conformational stability of others (see Moens and Vanderleyden, 1997; Virji, 1997). It has been recently suggested that glycans play similar roles of protection from proteolytic cleavage of the 19-kDa glycoprotein of mycobacteria (Herrmann *et al.,* 1996).

Since meningococcal pili are prominent components of disseminated isolates, it is tempting to speculate that glycans could influence pathogenesis. One potential effect of the presence of terminal galactose could be that pili could become sialylated, a process that, when it occurs on LPS, results in increased resistance to complement-mediated killing (Section 2). Moreover, normal human serum contains antibodies (anti-Gal) that react with terminal galactose structures on bacteria, including those on meningococcal pili (Hamadeh *et al.,* 1995). Epitopes recognized by anti-Gal antibodies are also present on brain glycoproteins in man (Jaison *et al.,* 1993). It is possible therefore that anti-Gal antibody may be carried on pili of bacteria entering the brain. The structural similarities between pili and

---

Figure 8. Molecular modeling of pili of the adhesion variants of strain C311. Three-dimensional structures of the variant pilins of strain C311 were based on that of *N. gonorrhoeae* MS11 pilin determined by X-ray crystallography (Parge *et al.,* 1995). The models were built with the help of structural databases and minimized using the program X-plor. Although the conformations of the side chains may not be exact, the backbone of the molecular model of meningococcal pilin protein is representative (A. Hadfield and M. Virji, unpublished data). (Top left) Schematic presentation of pilus made up of clublike pilin molecules arranged in a helical array with five monomers making a single helical turn (Parge *et al.,* 1995). (Top right) A cross-section through a model of a pilus shows five monomers arranged in a helical array such that the conserved N-terminal hydrophobic regions are centrally located and form the core of the fiber and the surface of the fiber is covered with hypervariable domains (HV) of pilin and glycans (*). (Middle left) A model of pilin of the variant #16. Our model indicates that the loops a and b around 113 and 130 (127–131) are closely positioned. The sequence comparisons suggest that the #16 pili will have a net change in charge of +1 on these two loops relative to parental, #3 pili. The loop c around 140 (139–142) is located at the other end of the β-sheet in the pilin monomer (middle left). There also is a change in charge of +1 in this loop in #16 pilin. Using the transformations suggested by Forest and Tainer (1997), a fiber model was constructed for #16 pili. In the fiber, pilin "n" of one helical turn and "n + 5" of the next are so juxtapositioned as to bring loops a and b of pilin "n" very close to loop c of pilin "n + 5" (middle right; only pilins n and n + 5 of a #16 pilus model are shown). Therefore, the three loops may present a single epitope on the surface of the fiber, which will be repeated many times along its length. Figures (top left and middle panel) produced using the drawing program of Kraulis (1991). (Bottom) The surface potentials of pilins of variants #3 and #16 were calculated using the program GRASP (Nicolls *et al.,* 1991). From this model, it appears that the change in charge distribution is displayed at the surface of the pilin protein, particularly for the substitution of aspartate (-1) in #3 for lysine (+1) in #16 at position 130. Here, negative charge seen as a dark spot in the middle of the circle (bottom left) is not present in the hyperadherent phenotype (bottom right).

brain tissue thus may enable meningococci to be localized to the central nervous tissue by bridging via anti-Gal antibodies.

8. α-GLYCEROPHOSPHATE: A UNIQUE MODIFICATION

α-Glycerophosphate (GolP) has been found in bacterial cell surface oligo- and polysaccharides, for example, the capsular polysaccharide of *Streptococcus pneumoniae* type 23F, membrane-derived oligosaccharides from *E. coli,* and the cyclic glucans of the *Rhizobiaceae* family. The occurrence of GolP alone previously as a substituent of prokaryotic or eukaryotic proteins, has not been reported previously, although it is found as a component of the amide-linked ethanolamine–phosphoglycerol moiety attached via Glu in elongation factor 1-α (EF-1α). The moiety is isolated from a wide range of species where it may play a role in modulating interactions of EF-1α with ribosomes. Phosphodiester-linked functional groups other than glycerol, important for biological function, are found in many bacteria. For example, phosphodiester linkages between the hydroxyl group of tyrosine residues and adenylyl and uridylyl groups have been reported in proteins such as glutamine synthetase. These modifications are reversible and are used to regulate the activity of glutamine synthetase, which plays a central role in the assimilation of ammonia in bacteria. Phosphodiester moiety can serve as a linker between carbohydrate and serine in a few glycoproteins.

The function of the GolP in meningococcal pilin remains to be defined. It is an intriguing possibility that glycerol could serve as a substrate for fatty acylation, thereby imparting membrane-anchoring properties to this posttranslational modification and of modulating pilus-mediated interactions with host cells (Stimson *et al.,* 1996).

9. PHOSPHORYLCHOLINE AND PATHOGENESIS

Phosphorylcholine, a relatively common component of glycolipids of mucosal commensals and pathogens, is recognized by naturally occurring anti-ChoP antibodies. Also, an acute-phase serum protein, C-reactive protein, binds to ChoP and serves as an opsonin for organism expressing this structure (Szalai *et al.,* 1996). ChoP appears to contribute to the adherence of the pneumococcus to human cells by acting as a ligand for the platelet-activating factor (PAF) receptor (Cundell *et al.,* 1995). The expression of ChoP has been shown to render *H. influenzae* sensitive to the bactericidal activity of serum (Weiser *et al.,* 1998a). In *H. influenzae,* the epitope is phase variable. This ability to turn off expression of phosphorylcholine may be important in colonization and infection to avoid detection by antibodies and C-reactive protein.

The expression of this epitope on pili of meningococci is likely to be of significance in pathogenesis, both from the point of view of serum sensitivity and because phosphorylcholine is a ligand for the PAF receptor and could be important in cellular interactions, perhaps as a second ligand. The significance of the exclusive presence of phosphorylcholine on pili of pathogenic *Neisseria* and LPS of commensal *Neisseria* remains to be investigated. It is possible that location on pili allows pathogenic strains to divert opsonins away from the immediate surface of bacteria, thus minimizing their effect. Parallel investigations of the pathogenic and commensal mucosal species, their mechanism of adhesion, and susceptibility to anti-ChoP antibodies should provide greater insight into the determinants of the pathogenic potential of meningococci.

10. CONCLUSIONS

Pili, which appear to be essential adhesins in capsulate *N. meningitidis,* are among the relatively few known prokaryotic proteins to be modified by glycans. More than one glycan structure may be present in different strains. Strain C311 contains an unusual trisaccharide structure. The Galβ1, 4Gal moiety is present on several of the strains investigated, suggesting that it is not unique to strain C311. In addition, pili of meningococci also contain other substitutions, one of these is α-glycerophosphate. Immunochemical studies also have shown the presence of an epitope reactive with antiphosphorylcholine antibodies. Such extensive modifications of this virulence determinant suggest that unique functions may be controlled or assisted by the decorations. The strategy behind such decorations and their functional significance and clinical importance are still mysteries waiting to be solved. Present investigations using variants with nonglycosylated pili have generated a further intrigue. We observed that Ser[63] mutations are unstable in strain C311 with a Rec A-replete background (Payne *et al.,* 1996). Rec A allows recombination between *pilE* and *pilS* loci, reinstating Ser[63], and these pili regain their glycosylation. These observations suggest that glycosylation is the preferred phase, at least in meningococcal strain C311. Recently, in an inducible Rec A mutant, Ser[63] has been successfully replaced by site-directed mutagenesis to produce a nonglycosylated phenotype whose functions remain to be investigated. These mutants and antibodies raised to synthetic analogues of the substituents should help clarify the functional significance of the extensive posttranslational modifications of meningococcal pili.

ACKNOWLEDGMENTS

M. V. is an MRC Senior Fellow and acknowledges the financial support from the MRC and the National Meningitis Trust.

REFERENCES

Achtman, M., 1995, Epidemic spread and antigenic variability of *Neisseria meningitidis,* *Trends Microbiol.* **3**(5):186–192.

Aho, E. L., Dempsey, J. A., Hobbs, M. M., Klapper, D. G., and Cannon, J. G., 1991, Characterisation of the *opa* (class 5) gene family of *Neisseria meningitidis. Mol. Microbiol.* **5**:1429–1437.

Aho, E. L., Botten, J. W., Hall, R. J., Larson, M. K., and Ness, J. K., 1997, Characterization of a class II pilin expression locus from *Neisseria meningitidis:* Evidence for increased diversity among pilin genes in pathogenic *Neisseria* species, *Infect. Immun.* **65**: 2613–2620.

Ala'Aldeen, D. A. A., and Griffiths, E., 1995, Vaccines against meningococcal diseases, in: *Molecular and clinical aspects of bacterial vaccine development* (D. A. A. Ala'Aldeen and C. E. Hormaeche, eds.), John Wiley and Sons, Chichester, pp. 1–39.

Cartwright, K., 1995 (ed.), *Meningococcal Disease,* John Wiley and Sons, New York.

Cundell, D. R., Gerard, N. P., Gerard, C., Idanpaan-Heikkila, I., and Tuomanen, E. I., 1995, *Streptococcus pneumoniae* anchor to activated human cells by the receptor for platelet-activating factor, *Nature* **377**:435–438.

Dansky-Ullman, C., Salgaller, M., Adams, S., Schlom, J., and Greiner, J. W., 1995, Synergistic effects of IL-6 and IFN-gamma on carcinoembryonic antigen (CEA) and HLA expression by human colorectal carcinoma cells:role for endogenous IFN-beta, *Cytokine* **7**:118–129.

Darzins, A. L., and Russell, M. A., 1997, Molecular genetic analysis of type-4 pilus biogenesis and twitching motility using *Pseudomonas aeruginosa* as a model system— A review, *Gene* **192**:109–115.

DiFabio, J. L., Michon, F., Brisson, J.-R., and Jennings, H. J., 1990, Structure of the L1 and L6 core oligosaccharide epitopes of *Neisseria meningitidis, Can. J. Chem.* **68**:1029–1034.

Dunn, K. L. R., Virji, M., and Moxon, E. R., 1995, Investigations into the molecular basis of meningococcal toxicity for human endothelial and epithelial cells: The synergistic effect of LPS and pili, *Microb. Pathog.* **18**:81–96.

Estabrook, M. M., Christopher, N. C., Griffiss, J. M., Baker, C. J., and Mandrell, R. E., 1992, Sialylation and human neutrophil killing of group C *Neisseria meningitidis, J. Infect. Dis.* **166**:1079–1088.

Finlay, B. B., and Falkow, S., 1997, Common themes in microbial pathogenicity revisited, *Microbiol. Mol. Biol. Rev.* **61**:136–169.

Forest, K. T., and Tainer, J. A., 1997, Type-4 pilus-structure: Outside to inside and top to bottom—A minireview, *Gene* **192**:165–169.

Fussenegger, M., Rudel, T., Barten, R., Ryll, R., and Meyer, T. F., 1997, Transformation competence and type-4 pilus biogenesis in *Neisseria gonorrhoeae*—A review, *Gene* **192**:125–134.

Gene, 1997, Proceedings of the workshop on Type IV pili, vol. 192.

Gubish, E. R., J. R., Chen, K. C. S., and Buchanan, T. M., 1982, Attachment of gonococcal pili to lectin-resistant clones of Chinese hamster ovary cells, *Infect. Immun.* **37**:189–194.

Hamadeh, R. M., Estabrook, M. M., Zhou, P., Jarvis, G. A., and Griffiss J. M., 1995, Anti-

Gal binds to pili of *Neisseria meningitidis:* The immunoglobulin A isotype blocks complement-mediated killing, *Infect. Immun.* **63**:4900–4906.

Hammershmidt, S., Muller, A., Sillmann, H., Muhlenhoff, M., Borrow, R., and Fox, A., 1996a, Capsule phase variation in *Neisseria meningitidis* serogroup B by slipped-strand mispairing in the polysialyltransferase gene (SiaD): Correlation with bacterial invasion and the outbreak of meningococcal disease, *Mol. Microbiol.* **20**:1211–1220.

Hammershmidt, S., Hilse, R., van Putten, J. P., Gerardy-Schahn, R., Unkmeir, A., and Frosch, M., 1996b, Modulation of cell surface sialic acid expression in *Neisseria meningitidis* via a transposable genetic element, *EMBO J.* **15**:192–198.

Heckels, J. E., 1989, Structure and function of pili of pathogenic *Neisseria* species, *Clin. Microbiol. Rev.* **2**:566–573.

Hermansson, K., Jansson, P., Holme, T., and Gustavsson, B., 1993, Structural studies of the *Vibrio cholerae* O:4 O-antigen polysaccharide, *Carbohydr. Res.* **248**:199–211.

Herrmann, J. L., 1996, Bacterial glycoproteins: A link between glycosylation and proteolytic cleavage of a 19-kDa antigen from *Mycobacterium tuberculosis, EMBO J.* **15**:3547–3554.

Jaison, P. L., Kannan, V. M., Geetha, M., and Appukuttan, P. S., 1993, Epitopes recognised by serum anti-α-galactoside antibody are present on brain glycoproteins in man, *J. Biosci.* **18**:187–193.

Jennings, M. P., Bisercic, M., Dunn, K. L. R., Virji, M., Martin, A., Wilks, K. E., Richards, J. C., and Moxon, E. R., 1995a, Cloning and molecular analysis of the *lsi1 (rfaF)* gene of *Neisseria meningitidis* which encodes a heptosyl -2-transferase involved in LPS biosynthesis: Evaluation of surface exposed carbohydrates in LPS mediated toxicity for human endothelial cells, *Microb. Pathog.* **19**:391–407.

Jennings, M. P., Hood, D. W., Peak, I. R. A., Virji, M., and Moxon, E. R., 1995b, Molecular analysis of a locus for the biosynthesis and phase variable expression of the lacto-N-neotetraose terminal LPS structure in *Neisseria meningitidis, Mol. Microbiol.* **18**:729–740.

Jennings, M. P., Virji, M., Evans, D., Foster, V., Srikhanta, Y. N., Steeghs, L., van der Ley, P., and Moxon, E. R., 1998, Identification of a novel gene involved in pilin glycosylation in *Neisseria meningitidis, Mol. Microbiol.* **29**:975–984.

Jonsson, A-B., Nyberg, G., and Normark, S., 1991, Phase variation of gonococcal pili by frame shift mutation in pilC, a novel gene for pilus assembly, *EMBO J.* **10**:35–43.

Jonsson, A-B., Ilver, D., Falk, P., Pepose, J., and Normark, S., 1994, Sequence changes in the pilus subunit lead to tropism variation of *Neisseria gonorrhoeae* to human tissue, *Mol. Microbiol.* **13**:403–416.

Kahler, C. M., and Stephens, D. S., 1998, Genetic basis for biosynthesis, structure, and function of meningococcal lipooligosaccharide, *Crit. Rev. Microbiol.* **24**:281–334.

Kraulis, P. J., 1991, MOLSCRIPT: A program to produce both detailed and schematic plots of protein structures, *J. Appl. Crystal.* **24**:946–950.

Mackinnon, F. G., Borrow, R., Gorringe, A. R., Fox, A. J., Jones, D. M., and Robinson, A., 1993, Demonstration of lipooligosaccharide immunotype and capsule as virulence factors factors for *Neisseria meningitidis* using an infant mouse intranasal infection model, *Microb. Pathog.* **15**:359–366.

Manning, P. A., 1997, The *tcp* gene cluster of *Vibrio cholerae, Gene* **192**:63–70.

Marceau, M., Beretti, J.-L., and Nassif, X., 1995, High adhesiveness of encapsulated *Neis-*

seria meningitidis to epithelial cells is associated with the formation of bundles of pili, *Mol. Microbiol.* **17**:855–863.

Marceau, M., Forest, K., Beretti, J. L., Tainer, J., and Nassif, X., 1998, Consequences of the loss of O-linked glycosylation of meningococcal type IV pilin on piliation and pilus-mediated adhesion, *Mol. Microbiol.* **27**:705–715.

McNeil, G., and Virji, M., 1997, Phenotypic variants of meningococci and their potential in phagocytic interactions: The influence of opacity proteins, pili, PilC and surface sialic acids, *Microb. Pathog.* **22**:295–304.

McNeil, G., Virji, M., and Moxon, E. R., 1994, Interactions of *Neisseria meningitidis* with human monocytes, *Microb. Pathog.* **16**:153–163.

Messner, P., 1997, Bacterial glycoproteins, *Glycoconj. J.* **14**:3–11.

Moens, S., and Vanderleyden, J.,1997, Glycoproteins in prokaryotes, *Arch. Microbiol.* **168**:169–175.

Moran, E. E., Brandt, B. L., and Zollinger, W. D., 1994, Expression of the L8 lipopolysaccharide determinant increases the sensitivity of *Neisseria meningitidis* to serum bactericidal activity, *Infect. Immun.* **62**:5290–5295.

Mosser, J. L., and Tomasz, A., 1970, Choline-containing teichoic acid as a structural component of pneumococcal cell wall and its role in sensitivity to lysis by an enzyme, *J. Biol. Chem.* **245**:287–298.

Moxon, E. R., and Kroll, J. S., 1990, The role of bacterial polysaccharide capsules as virulence factors, *Curr. Top. Microbiol. Immunol.* **150**:65–85.

Nassif, X., Lowy, J., Stenberg, P., O'Gaora, P., Ganji, A., and So, M., 1993, Antigenic variation of pilin regulates adhesion of *Neisseria meningitidis* to human epithelial cells, *Mol. Microbiol.* **8**:719–725.

Nassif, X., Beretti, J.-L., Lowy, J., Stenberg, P., O'Gaora, P., Pfeifer, J., Normark, S., and So, M., 1994, Roles of pilin and pilC in adhesion of *Neisseria meningitidis* to human epithelial and endothelial cells, *Proc. Natl. Acad. Sci. USA* **91**:3769–3773.

Nicholls, A., Sharp, K., and Honig, B., 1991, Protein folding and association–insights from the interfacial and thermodynamic properties of hydrocarbons, *Proteins, Structure, Function and Genetics,* **11**:281–296.

Nicolson, I. J., Perry, A. C. F., Virji, M., Heckels, J. E., and Saunders, J. R., 1987, Localization of antibody-binding sites by sequence analysis of cloned pilin genes from *Neisseria gonorrhoeae, J. Gen. Microbiol.* **133**:825–833.

Parge, H. E., Forest, K. T., Hickey, M. J., Christensen, D. A., Getzoff, E. D., and Tainer, J. A., 1995, Structure of the fibre-forming protein pilin at 2.6Å resolution, *Nature* **378**:32–38.

Payne, G., Virji, M., and Saunders, J. R., 1996, Genetic analysis of posttranslational modifications of meningococcal pilin, in: *Abstracts of the Tenth International Pathogenic Neisseria Conference* (W. D. Zollinger, C. E. Frasch, and C. D. Deal, eds.), National Institutes of Health, Bethesda, MD, pp. 393–394.

Potts, W. J., and Saunders, J. R., 1988, Nucleotide sequence of the structural gene for class I pilin from *Neisseria meningitidis:* Homologies with the *pilE* locus of *Neisseria gonorrhoeae, Mol. Microbiol.* **2**:647–653.

Rahman, M., Killstrom, H., Normark, S., and Jonsson, A-B., 1997, PilC of pathogenic *Neisseria* is associated with the bacterial cell surface, *Mol. Microbiol.* **25**:11–25.

Rahman, M., Jonsson, A-B., and Holm, T., 1998, Monoclonal antibodies to the epitope βGal-(1–4)α-Gal-(1- of *Moraxella catarrhalis* LPS react with a similar epitope in type IV pili of *Neisseria meningitidis, Microb. Pathog.* **24**:299–308.

Robertson, J. N., Vincent, P., and Ward, M. E., 1977, The preparation and properties of gonococcal pili, *J. Gen. Micobiol.* **102**:169–177.

Rothbard, J. B., Fernandez, R., Wang, L., Teng, N. N. H., and Schoolnik, G. K., 1985, Antibodies to peptides corresponding to a conserved sequence of gonococcal pilins block bacterial adhesion, *Proc. Natl. Acad. Sci. USA* **82**:915–919.

Rudel, T., van Putten, J. P. M., Gibbs, C. P., Haas, R., and Meyer, T. F., 1992, Interaction of two variable proteins (PilE and PilC) required for pilus-mediated adherence of *Neisseria gonorrhoeae* to human epithelial cells, *Mol. Microbiol.* **6**:3439–3450.

Rudel, T., Scheuerpflug, I., and Meyer, T. F., 1995, *Neisseria* PilC protein identified as type-4 pilus tip-located adhesin, *Nature* **373**:357–359.

Sarkari, J., Pandit, N., Moxon, E. R., and Achtman, M., 1994, Variable expression of the Opc outer membrane protein in *Neisseria meningitidis* is caused by size variation of a promoter containing poly-cytidine, *Mol. Microbiol.* **13**:207–217.

Seifert, H. S., and So, M., 1988, Genetic mechanisms of bacterial antigenic variation, *Microbiol. Rev.* **52**:327–336.

Serino, L., and Virji, M., 2000, Phosphorylcholine decoration of LPS differentiates commensal *Neisseriae* from pathogenic strains: Identification of *licA*-type genes in commensal *Neisseriae. Mol. Microbiol.*, in press.

Sharon, N., and Jeanloz, R. W., 1960, The diaminohexose component of a polysaccharide isolated from *Bacillus subtilis, J. Biol. Chem.* **235**:1–5.

Shashkov, A. S., Campos-Portuguez, S., Kochanowski, H., Yokota, A., and Mayer, H., 1995, The structure of the O-specific polysaccharide from *Thiobacillus sp.* IFO 14570, with three different diaminopyranoses forming the repeating unit, *Carbohydr. Res.* **269**: 157–166.

Stimson, E., Virji, M., Makepeace, K., Dell, A., Morris, H. R., Payne, G., Saunders, J., Jennings, M. P., Barker, S., Panico, M., Blench, I., and Moxon, E. R., 1995, Meningococcal pilin: A glycoprotein substituted with digalactosyl 2,4-diacetamido-2,4,6-trideoxyhexose, *Mol. Microbiol.* **17**:1201–1214.

Stimson, E., Virji, M., Panico, M., Blench, I., Barker, S., Moxon, E. R., Dell, A., and Morris, H. R., 1996, Discovery of a novel protein modification: (Glycerophosphate is a substituent of meningococcal pilin, *Biochem. J.* **316**:29–33.

Szalai, A. J.,, Briles, D. E. and Volanakis, J. E., 1996, Role of complement in C-reactive-protein-mediated protection of mice from *Streptococcus pneumoniae, Infect. Immun.* **64**:4850–4853.

Tahara, Y., and Wilkinson, S. G., 1983, The lipopolysaccharide from *Pseudomonas aeruginosa* NCTC 8505. Structure of the O-specific polysaccharide, *Eur. J. Biochem.* **134**:299–304.

van Putten, J. P., and Paul, S. M., 1995, Binding of syndecan-like cell surface proteoglycan receptors is required for *Neisseria gonorrhoeae* entry into human mucosal cells, *EMBO J.* **14**:2144–2154.

Verheul, A. F. M., Snippe, H., and Poolman, J. T.,1993, Meningococcal lipopolysaccharides: Virulence factor and potential vaccine component, *Microbiol. Rev.* **57**(1):34–49.

Virji, M., 1996a, Meningococcal disease: Epidemiology and pathogenesis, *Trends Microbiol.* **4**:466–469.

Virji, M., 1996b, Adhesion receptors in microbial pathogenesis, in: *Molecular Biology of Cell Adhesion Molecules* (M. A. Horton, eds.), John Wiley and Sons, Chichester, pp. 99–129.

Virji, M., 1997, Post-translational modifications of type 4 pili and functional implications, *Gene* **192**:141–147.

Virji, M., and Heckels, J. E., 1984, The role of common and type-specific pilus antigenic domains in adhesion and virulence of gonococci for human epithelial cells, *J. Gen. Microbiol.* **130**:1089–1095.

Virji, M., Everson, J. S., and Lambden, P. R., 1982, Effect of anti-pilus antisera on virulence of variants of *Neisseria gonorrhoeae* for cultured epithelial-cells, *J. Gen. Microbiol.* **128**:1095–1100.

Virji, M., Heckels, J. E., Potts, W. J., Hart, C. A., and Saunders, J. R., 1989, Identification of epitopes recognised by monoclonal antibodies SM1 and SM2 which react with all pili of *Neisseria gonorrhoeae* but which differentiate between two structural classes of pili expressed by *Neisseria meningitidis* and the distribution of their encoding sequences in the genomes of *Neisseria* spp., *J. Gen. Microbiol.* **155**:3239–3251.

Virji, M., Kahty, H., Ferguson, D. J. P., Alexandrescu, C., and Moxon, E. R., 1991a, Interactions of *Haemophilus influenzae* with cultured human endothelial cells, *Microb. Pathog.* **10**:231–245.

Virji, M., Kahty, H., Ferguson, D. J. P., Alexandrescu, C., Heckels, J. E., and Moxon, E. R., 1991b, The role of pili in the interactions of pathogenic *Neisseria* with cultured human endothelial cells, *Mol. Microbiol.* **5**:1831–1841.

Virji, M., Alexandrescu, C., Ferguson, D. J. P., Saunders, J. R., and Moxon, E. R., 1992, Variations in the expression of pili: the effect on adherence of *Neisseria meningitidis* to human epithelial and endothelial cells, *Mol. Microbiol.* **6**:1271–1279.

Virji, M., Saunders, J. R., Sims, G., Makepeace, K., Maskell, D., and Ferguson, D. J. P., 1993, Pilus-facilitated adherence of *Neisseria meningitidis* to human epithelial and endothelial cells: Modulation of adherence phenotype occurs concurrently with changes in primary amino acid sequence and the glycosylation status of pilin, *Mol. Microbiol.* **10**:1013–1028.

Virji, M., Makepeace, K., and Moxon, E. R., 1994, Distinct mechanisms of interaction of Opc-expressing meningococci at apical and basolateral surfaces of human endothelial cells; The role of integrins in apical interactions, *Mol. Microbiol.* **14**:173–184.

Virji, M., Makepeace, K., Peak, I. R. A., Ferguson, D. J. P., Jennings, M. P., and Moxon, E. R., 1995a, Opc- and pilus-dependent interactions of meningococci with human endothelial cells: Molecular mechanisms and modulation by surface polysaccharides, *Mol. Microbiol.* **18**:741–754.

Virji, M., Makepeace, K., Peak, I., Payne, G., Saunders, J. R., Ferguson, D. J. P., and Moxon, E. R., 1995b, Functional implications of the expression of PilC proteins in meningococci, *Mol. Microbiol.* **16**:1087–1097.

Virji, M., Watt, S. M., Barker, S., Makepeace, K., and Doyonnas, R., 1996, The N-domain of the human CD66a adhesion molecule is a target for Opa proteins of *Neisseria meningitidis* and *Neisseria gonorrhoeae, Mol. Microbiol.* **22**:929–939.

Weiser, J. N., Pan, N., McGowan, K. L., Musher, D., Martin, A., and Richards, J. C., 1998a, Phosphorylcholine on the lipopolysaccharide of *Haemophilus influenzae* contributes to persistence in the respiratory tract and sensitivity to serum killing mediated by C-reactive protein, *J. Exp. Med.* **187**:631–640.

Weiser, J. N., Goldberg, J. B., Pan, N., Wilson, L., and Virji, M., 1998b, The phosphoryl-choline epitope undergoes phase variation on a 43 kD protein in *Pseudomonas aeruginosa* and on pili of *Neisseria meningitidis* and *Neisseria gonorrhoeae, Infect. Immun.* **66**:4263–4267.

Whittaker, D. V., Parolis, L. A. S., and Parolis, H., 1994, *Escherichia coli* K48 capsular polysaccharide: A glycan containing a novel diacetamido sugar, *Carbohydr. Res.* **256:** 289–301.

Zehavi, U., and Sharon, N., 1973, Structural studies of 4-acetamido-2-amino-2,4,6-trideoxy-D-glucose (*N*-acetylbacillosamine), the *N*-acetyl-diamino sugar of *Bacillus licheniformis, J. Biol. Chem.* **248**:433–438.

3

Polypeptide Linkage to Bacterial Cell Envelope Glycopolymers

Howard F. Jenkinson

1. BACTERIAL CELL SURFACE STRUCTURE

In most bacteria the cytoplasmic membrane is protected by a coat of peptidoglycan, consisting of a repeating glycan backbone [N-acetylmuramic acid (MurNAc) in ß-1,4 linkage to N-acetylglucosamine (GlcNAc)] to which short peptide chains are linked. These are tetrapeptide units comprising L-Ala-D-Glu-L-R-D-Ala, where L-Ala may be substituted by L-Gly or L-Ser in some bacteria, and R can be *meso*- or LL-diaminopimelic acid (Dpm), L-Lys, L-Orn, L-diaminobutyric acid, or L-homoserine. The terminal D-Ala is involved in cross-linking of peptide chains, generally either by direct linkage from D-Ala to the D-carbon atom amino group of *meso*-Dpm in another chain (in gram-negative bacteria and gram-positive bacilli), or through a single amino acid or short peptide, such as the pentaglycine cross-link found in *Staphylococcus aureus*. The degree of cross-linking affects the size and number of fragments that result from hydrolysis of peptidoglycan by lysozyme, which cleaves the ß,1–4 link between MurNAc and GlcNAc. Analysis of wall fragments generated from lysozyme cleavage or following incubation with other enzymes such as N-acetylglucosaminidase, N-acetylmuramidase (e.g., mutanolysin), amidase, and lysostaphin have been instrumental in determination of structure and composition of bacterial peptidoglycans.

In gram-positive bacteria the peptidoglycan also carries covalently bound te-

Howard F. Jenkinson • Department of Oral and Dental Science, University of Bristol Dental Hospital and School, Bristol BS1 2LY, England.

Glycomicrobiology, edited by Doyle.
Kluwer Academic/Plenum Publishers, New York, 2000.

ichoic acid, a polymer of glycerol or ribitol phosphate usually containing sugars or amino sugars and D-Ala either as a substituents of glycerol or as components of the backbone chain. By contrast, lipoteichoic acids and lipoglycans are components of the cytoplasmic membrane of gram-positive bacteria (Fischer, 1994). Lipoteichoic acid (LTA) is classically a glycerol phosphate polymer with a hydrophobic membrane glycolipid anchor, and extends through the peptidoglycan. Glycosyl substituents are attached to the glycerol residues and interglycosidic linkages are found in the LTAs of enterococci and streptococci. Lipoglycans are linear or branched homo- or heteropolysaccharides that may carry monomeric glycerophosphate and do not occur together with LTA in the same organism. These macroamphiphiles, together with loosely associated polysaccharides and an array of polypeptides, comprise the exposed surface of the gram-positive bacterial cell (see Fig. 1B).

In gram-negative bacteria, short pieces of peptidoglycan are cross-linked into a gel that is formed between the outer membrane and cytoplasmic (inner) membrane (Fig. 1A). This periplasmic region contains proteins and anionic oligosaccharides (termed "membrane-derived oligosaccharides"). The periplasm functions to facilitate the traffic and processing of molecules entering or leaving the cell, whereas the peptidoglycan is a major determinant of cell shape, just as it is in gram-positive organisms. The outer membrane forms an asymmetric layer composed of mainly phospholipids on the inside face and a unique lipid species, lipopolysaccharide (LPS), making up the outer-facing leaf. The exposed surface of the gram-negative bacterial cell consists of LPS, cross-bridged by divalent cations, and associated proteins the major species of which act as pores to permit the passage of small hydrophilic molecules (see Fig. 1A). Outer membranes are also the anchoring points for certain external structures such as flagella, surface-layers (S-layers), capsules, and pili. One of the most significant functions of the outer membrane is to exclude a variety of environmental molecules, such as hydrophobic antibiotics and proteins, thus rendering the surface layers and peptidoglycan relatively inaccessible to extrinsic hydrolytic enzyme activities.

This chapter considers the structures of polypeptide–cell wall linkages in bacteria and the mechanisms by which polypeptides become linked to bacterial cell wall glycopolymers, principally to peptidoglycan. Polypeptides associated with the synthesis and function of bacterial LPS are discussed in Chapter 15, this volume. The various sites of the polypeptide–glycopolymer linkages considered in this chapter in relation to the topography of the bacterial cell surface layers are depicted in Fig. 1. Precursor polypeptides that are destined to become linked to bacterial cell wall generally carry an amino (N)-terminal leader peptide sequence (Izard and Kendall, 1994) that directs their secretion from the bacterial cell via the general export (Sec protein-dependent) pathway (Schatz and Beckwith, 1990). However, of major focus in this chapter is the recently discovered mechanism by which proteins are "sorted" and linked to peptidoglycan in gram-positive bacteria, compared with the sorting and linkage processes known for cell-wall-linked proteins in gram-negative bacteria.

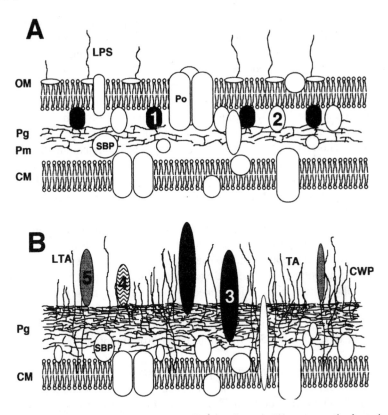

Figure 1. Diagrammatic representation of the cell surface layers in (A) gram-negative bacteria and (b) gram-positive bacteria. Proteins present within the surface layers are depicted as ellipsoids. Those polypeptides that have glycopolymer linkages and are discussed in the text are numbered: 1, murein (Braun's) lipoprotein (Lpp); 2, peptidoglycan-associated protein (PAL); 3, peptidoglycan-anchored polypeptide; 4, wall-polysaccharide-linked protein; 5, lipoteichoic acid-binding protein. Abbreviations: CM, cytoplasmic membrane; CWP, cell wall polysaccharide; LPS, lipopolysaccharide; LTA, lipoteichoic acid; OM, outer membrane; Pg, peptidoglycan; Pm, periplasmic region; Po, porin; SBP, solute-binding protein; TA, teichoic acid.

2. PEPTIDOGLYCAN-ASSOCIATED LIPOPROTEINS IN GRAM-NEGATIVE BACTERIA

2.1. Murein Lipoprotein

The first lipid-modified polypeptide to be purified from bacteria was murein lipoprotein, sometimes called Braun's lipoprotein (Hantke and Braun, 1973). This carries covalently linked lipid *N*-acyl-diacylglyceryl to N-terminal Cys, and identical linkages are found in a wide range of structurally and functionally diverse bacterial lipoproteins (Wu, 1996). The lipid portion of the lipoprotein determines its membrane location and contributes to the structural and immunogenic properties. It is generally accepted that the lipid at the N terminus is inserted into one leaflet of the lipid bilayer. Murein lipoprotein and other peptidoglycan-associated lipoproteins (PALs) in gram-negative bacteria are anchored to the inner leaflet of the outer membrane (Fig. 1A). Gram-positive bacterial lipoproteins, including the substrate-binding components of ATP-dependent solute transport (uptake) systems, are anchored to the outer leaflet of the cytoplasmic membrane (Sutcliffe and Russell, 1995) (Fig. 1B).

Lipoproteins are synthesized as precursors carrying a signal peptide with typical tripartite structure that is recognized by the Sec protein-exporting machinery. The C-terminal region of the prolipopolypeptide leader contains a lipobox sequence [-Leu(Ala,Val)-Leu-Ala(Ser)-Gly(Ala)-Cys-] defining the site of lipid modification and propolypeptide processing (Wu, 1996). Following generation of a diacylglyceryl prolipoprotein, the leader peptide is cleaved by a membrane-integral signal peptidase II between Gly(Ala) and thio-acylated Cys, and then fatty acid (usually palmitate) is amide-linked to the apolipoprotein (Fig. 2). The processing and modification requires the enzymic products of three unlinked genes, designated *lsp, lgt,* and *lnt* in *Escherichia coli,* functional Sec proteins (Schatz and Beckwith, 1990), but not Sec B, and an intact proton motive force (Kosic *et al.,* 1993).

Mature murein lipoprotein (Lpp) comprises 58 amino acid residues and exists in two forms, a free form in the outer membrane and a bound form covalently attached to the peptidoglycan. Twice the amount of free form is present, with the N-terminal lipid interacting with the outer membrane and the C-terminus of the protein interacting noncovalently with peptidoglycan (Braun and Rehn, 1969). The amino acid sequence of Lpp is highly repetitive, and an essentially α-helical coiled-coil structure is predicted for the polypeptide (Fig. 2), which forms trimers as suggested by chemical cross-linking experiments (Choi *et al.,* 1986). Attachment to the peptidoglycan occurs via the ϵ-amino group of the C-terminal Lys residue to the carboxyl group of the optical L-center of *meso*-Dpm within the pentapeptide side chain (Braun and Bosch, 1972) (Fig. 2). This resists chemical treatments that do not disrupt covalent bonds. Thus subjecting EDTA-treated *E. coli* cell envelopes to boiling in 4% sodium dodecyl sulfate (SDS) results in the generation of "rigid-layer" material comprising peptidoglycan to which Lpp remains covalently bound (Braun and Rehn, 1969). Distribution of Lpp over the peptidoglycan structure is essentially random (Hiemstra *et al.,* 1987) with protein molecules bound on average to every 10th to 12th Dpm residue (Braun and Bosch, 1972). The formation of the Lpp–peptidoglycan linkage does not appear to depend on either lipid modification or prolipoprotein cleavage (Zhang *et al.,* 1992). However, site-directed mutagenesis experiments have demonstrated that the three C-

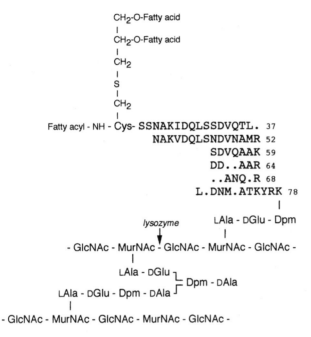

Figure 2. Structure of the Lpp–peptidoglycan complex in *E. coli* showing lipid and peptidoglycan attachment sites. Following linkage of diglyceride through a thioether bond to Cys-21 of the Lpp precursor, prolipoprotein signal peptidase II cleaves the 20 amino acid residue leader peptide, and a third fatty acid is attached to Cys via the N-terminal amino group. Peptidoglycan is bound to Lpp through the epsilon amino group of the C-terminal Lys-78 to the carboxyl group of the optical L-center of *meso*-diaminopimelate (Dpm). The Lpp sequence is presented with sequence gaps introduced to emphasize the repetitive nature of the amino acid sequence (possibly evoking a double or triple helix-stranded coiled coil structure). Numbers indicate amino acid residues from the N-terminal of the Lpp precursor.

terminal residues Tyr-76, Arg-77, and Lys-78 (Fig. 2), are important (Zhang and Wu, 1992). Although Lpp is one of the most abundant proteins in *E. coli* and much is known about the sequence and structural requirements for peptidoglycan linkage, neither the putative enzymic activity associated with forming this linkage (Lpp–peptidoglycan ligase) nor the gene encoding it has yet been identified.

2.2. Synthesis and Secretion of PALs

Lipoproteins of the same size and with identical lipid structure to Lpp are produced by all members of the *Enterobacteriaceae,* and homologous proteins with lower sequence similarities are produced by other gram-negative bacteria (Braun and Wu, 1994). They probably function, in concert with other PALs (see Section 2.3), to maintain the outer membrane structure and integrity (Suzuki *et al.*, 1978). Since PALs are found exclusively associated with the outer membrane, an aspect of much research has been to identify the intrinsic signals and macromolecular machinery that direct them there. Evidence strongly suggests that the Ser residue at position 2 of mature Lpp (Fig. 2) functions as the main outer membrane sorting signal. Substitution of this serine with negatively charged Asp causes inner membrane localization of Lpp (Yamaguchi *et al.*, 1988). This does not seriously affect the growth of *E. coli* in culture provided that in addition the C-terminal Lys residue involved in linkage to peptidoglycan is deleted. If C-terminal Lys is retained, then a covalent linkage between peptidoglycan and inner membrane localized Lpp is generated, and this is lethal for *E. coli* cells probably as a result of disrupting surface integrity and outer membrane function (Yakushi *et al.*, 1997). However, there are other structural determinants present within Lpp that also may influence membrane localization (Gennity *et al.*, 1992). In addition, recent work has demonstrated the presence of a 20-kDa periplasmic protein that forms a soluble complex with outer membrane-directed lipoproteins. This periplasmic protein probably acts as a carrier in translocation of lipoproteins from the inner membrane to the outer membrane (Matsuyama *et al.*, 1995).

2.3. Structure and Function of PAL Complexes

A second major lipoprotein that has been characterized extensively is PAL, originally identified in *Proteus mirabilis* (Mizuno, 1981) and found in a wide range of gram-negative bacteria. The precursor Pal polypeptide in *E. coli* contains 173 amino acid residues (Chen and Henning, 1987) and is encoded by the *excC* gene (Lazzaroni and Portalier, 1992). The protein carries the same lipid structure as Lpp and is found associated with the outer membrane. Although Pal proteins are generally very tightly associated with the peptidoglycan, they are not covalently linked to it, and therefore may be solubilized from cell envelope preparations by detergent treatments. Within the C-terminal 70 amino acid residues of Pal proteins from gram-negative bacteria is a consensus sequence associated with the formation of α-helix, which is predicted to associate with the peptidoglycan (Koebnik, 1995). Pal protein in *E. coli* forms part of a multiprotein complex involving periplasmic protein Tol B (Bouveret *et al.*, 1995), Lpp, and outer membrane proteins including Omp F, Omp C, and Omp A (depicted in Fig. 1A). The region of Pal that interacts with Tol B overlaps the binding region of Pal to peptidoglycan. The Pal–Tol complexes may facilitate some kind of association between the inner and outer membranes, possibly assisting the translocation of proteins destined for the outer membrane through the peptidoglycan (Clavel *et al.*, 1998).

It is likely that with new analytical techniques being applied to diverse microbial species, hitherto unrecognized mechanisms of polypeptide linkage to peptidoglycan will be identified. In this regard, a novel peptidoglycan-linked lipoprotein (Com L) has been identified in *Neisseria gonorrhoeae* that is essential for competence in DNA-mediated transformation (Fussenegger *et al.*, 1996). The protein is proposed to function by facilitating the translocation of incoming (transforming) DNA across the peptidoglycan. Evidence suggests that the 29-kDa Com L protein may be covalently linked to peptidoglycan, since it cannot be released from isolated murein by SDS, urea, formamide, or 2-mercaptoethanol treatments (Fussenegger *et al.*, 1996). Com L has no sequence similarity to Pal or to Lpp and does not contain C-terminus Lys, so either an alternative Lys residue is involved in peptidoglycan linkage or a novel linkage mechanism is present in *Neisseria.*

3. PEPTIDOGLYCAN-ASSOCIATED PROTEINS IN GRAM-POSITIVE BACTERIA

3.1. Cell Envelope Proteins

The proteins present on the gram-positive bacterial cells surface are, by analogy to the outer membrane proteins of gram-negative bacteria, the major environmental contact points for nutrient acquisition, signal transductions, and macromolecular translocations. Special interest in the wall-associated proteins in staphylococci, streptococci, and *Listeria* has arisen because of their involvement in the virulence properties of these bacteria. Virulence properties include adhesion to host cells, binding of human tissue proteins, antiphagocytic functions, and invasion of host tissues. Wall-associated proteins, many of which demonstrate antigenic variation, often carry extensive repeat blocks of amino acids and form extended surface structures (Jenkinson and Lamont, 1997). They have been characteristically difficult to purify intact from streptococcal and staphylococcal cells because they are relatively fragile yet firmly anchored to the bacterial cell wall. The observation that protein A, the major immunoglobulin-binding protein produced by *S. aureus,* could only be released from the surface of cells intact following enzyme (lysostaphin) treatment that hydrolyzed peptidoglycan (Sjoquist *et al.*, 1972), suggested that it was covalently bound to peptidoglycan. Evidence for covalent linkage of proteins to streptococcal cell walls was provided by Russell (1979) and by Nesbitt *et al.* (1980) who demonstrated that cell wall preparations contained tightly bound peptides resistant to exhaustive extraction with a range of protein solvents. This was later confirmed by Fischetti *et al.* (1985) who showed that the antiphagocytic M protein produced by *S. pyogenes* could be released intact from cell walls after treatment with muralytic enzymes. The cell wall-associated region of the M protein resides in the C-terminus. After proteolytic digestion, for example, with pepsin, of the exposed portion of the M protein molecule, the region embedded within the cell wall (and thus protected from degradation) was released by solubilizing the cell wall with a muralytic enzyme (Pancholi and Fischetti, 1988). The N-terminal amino acid residue of this segment released from the wall corresponded to amino acid residue 298 of the 441 residues of M protein predicted from the nucleotide sequence the *emm6* gene (Hollingshead *et al.,* 1986). Further analysis of this fragment showed that 25 amino acid residues at the C-terminus of the protein were missing (Pancholi and Fischetti, 1988), suggesting that this contained the sequence associated with cell surface protein anchorage.

Staphylococcal protein A (Spa) (Shuttleworth *et al.,* 1987) was the first and is the best characterized of what is now a large class of different gram-positive bacterial surface-anchored proteins. These proteins share a number of common structural features, deduced from the inferred amino acid sequences, that direct their secretion and cell surface anchorage. A cleavable N-terminal signal peptide directs cellular export, while a structurally conserved C-terminal domain consisting of C-terminus sequence containing four to six charged amino acids, followed by a sequence of up to 20 hydrophobic amino acids, is proposed to function as a secretion "pause" mechanism. Immediately adjacent N-terminal to the hydrophobic region of these surface proteins is a pentapeptide with the sequence LPxTG that is nearly 100% conserved, with all substitutions so far seen only in residues 4 and 5. It is this entire C-terminal region, designated the cell wall sorting signal (Schneewind *et al.,* 1993), that appears to be both necessary and sufficient for directing these proteins for anchorage to the cell surface of gram-positive bacteria.

3.2. Synthesis and Sorting of Cell Wall-Linked Polypeptides

Staphylococcal protein A is a model system for cell wall sorting and anchorage of polypeptides in gram-positive bacteria. Spa polypeptide is covalently linked to peptidoglycan and solubilized quantitatively from the staphylococcal cell surface only when the cell wall is digested with lysostaphin (Sjoquist *et al.,* 1972), an endopeptidase specific for the pentaglycine bridges of staphylococcal peptidoglycan (Fig. 3). The staphylococcal cell wall is resistant to treatment with 4% SDS, acid, and boiling, and protein A is not released by these treatments (Schneewind *et al.,* 1992). To begin to understand the C-terminal sequence requirements for and ultimately the mechanism of protein A–cell wall anchorage, a system was established whereby mutated *spa* genes were expressed from a plasmid in an isogenic derivative of *S. aureus* in which the chromosomal *spa* gene was inactivated. Mutants of protein A were constructed that were successively deleted for the C-terminal charged tail (RRREL), the hydrophobic region, and the LPxTG sequence, as well as a mutant in which the LPETGE sequence specifically was deleted and the effects of these deletions on protein A cellular compartment localization determined. Deletion of only the charged tail led to secretion of the protein into the medium, while deletion of the LPETG sequence led to protein being found cyto-

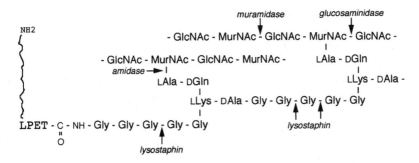

Figure 3. Structure of the staphylococcal peptidoglycan showing cell wall polypeptide-sorting signal attachment site and cleavage sites for some cell wall lytic enzymes. Linkage of the polypeptide occurs through the carboxyl group of C-terminal Thr (T) to the free amino group of N-terminal Gly (G) of the pentaglycine cross-bridge. Staphylococcal C-terminally anchored polypeptides are released from the wall following lysostaphin cleavage between the second and third glycyl residues of the pentaglycine cross-bridge (see text for details).

plasmically, in the extracellular medium, and in the cell wall fraction from where, unlike the wild-type protein, it could be solubilized without lysostaphin treatment (Schneewind *et al.,* 1992). It was noted that mutant proteins lacking the charged tail migrated more slowly on polyacrylamide gels than those that were released following lysostaphin treatment (Schneewind *et al.,* 1992). This was the first evidence indicating that cell wall anchorage of a polypeptide might involve a proteolytic cleavage occurring at the C-terminal, most likely through recognition of a conserved sequence LPxTG.

An obligate requirement for the C-terminal anchorage domain for polypeptide linkage to cell wall has subsequently been demonstrated for a number of other surface-anchored proteins (Schneewind *et al.,* 1993). For example, deletion of the C-terminal 250 amino acid residues of Csh A protein, a major cell surface protein in *Streptococcus gordonii* and *Streptococcus sanguis* and mediating adhesion of *S. gordonii* cells to a variety of oral cavity receptors (McNab *et al.,* 1994), results in secretion of the protein into the extracellular culture medium, with some remaining loosely associated with the cell surface. Loss of the bulk of the polypeptide from the *S. gordonii* surface results in reduced cell hydrophobicity and adhesion, properties normally conferred by Csh A, as well as loss of surface fibrils (McNab and Jenkinson, 1992) that are believed to be composed of Csh A polypeptide. The requirement for the C-terminal sorting signal for surface protein presentation was also shown for expression of internalin (Inl A) on *Listeria monocytogenes.* This 800 amino acid residue protein promotes listerial adhesion to host tissues and is required for entry of *Listeria* cells into intestinal epithelial cells. Removal of the sorting signal from the C-terminal region of Inl A results in nonretention of Inl A at the cell surface, secretion of the protein into the extracellular culture medium, and loss in ability of bacteria to invade host cells (Lebrun *et al.*, 1996). These examples demonstrate the importance of the C-terminal sorting signal in functional expression of proteins associated with adhesion and virulence.

The experiments first defining the sequences necessary for cell surface retention of protein A employed chimeric proteins of alkaline phosphatase (from *E. coli*) as enzyme reporter fused to the protein A leader peptide and C-terminal sequences. These initial experiments did not reveal whether or not the leader peptide of precursor protein A was required for any specific sorting function, bearing in mind that the sorting of lipoproteins to the outer membrane in *E. coli* depends on several signals within the N-terminal sequence (see Section 2.2). The current notion is that the leader peptide does not have any significant role to play in the cell wall anchorage mechanism, since it was found that staphylococcal enterotoxin B (SEB), a protein normally secreted into the medium, can be anchored to the cell wall by C-terminal fusion with the protein A C-terminal sorting signal (Schneewind *et al.*, 1993). Further experiments with SEB–protein A fusions demonstrated the importance of the charged tail for cell wall linkage. A serine scan experiment, in which serine replaced single as well as multiple residues present in the charged tail, revealed the major importance of the Arg (R) residues. Substituting the two C-terminal R residues (RSSEL) caused 80% protein release, while substitution of the first two R residues (SSREL) resulted in complete release of the fusion protein into the extracellular medium (Schneewind *et al.*, 1993).

The C-terminal sorting signals for more than 100 gram-positive bacterial surface proteins have now been determined, and although they are structurally conserved, the hydrophobic regions and charged tails vary somewhat in both sequence and length. Evidence suggests that, at least for protein A, the number of Arg residues required for cell wall sorting depends on the spacing between the LPxTG motif and the charged tail. The presence of two Arg residues following the hydrophobic domain at positions 31, 32, or 33 from the L (of LPxTG) are critical for surface retention of the protein A polypeptide (Schneewind *et al.*, 1993). Sorting signals in other polypeptides may contain up to five positively charged residues, with lysine or histidine residues present in addition to arginine, contributing to sorting function (Fig. 4). All the cell wall sorting signals contain, N-terminal to the charged tail, at least 15 hydrophobic amino acid residues (Fig. 4). This sequence is predicted to be sufficient to span the lipid bilayer of the cytoplasmic membrane; hence, the proposal is that this sequence acts in concert with the charged tail to brake the secretion process of the polypeptide.

3.3. Surface Protein–Peptidoglycan Anchorage

When the sequences necessary for cell wall sorting of polypeptides in gram-positive bacteria were delineated, neither the C-terminal end of an anchored sur-

```
Spa   413   GVHVVKPGDTVNDIAKANGTTADKIAADNKLADKNMIKPGQELVVDKKQPANHADANKAQALPET  GEENPFIGTTVFGGLSLALGAALLAGRRREL  508

Emm6  388   LKEQLAKQAEELAKLRAGKASDSQTPDAKPGNKVVPGKGQAPQAGTKPNQNKAPMKETKRQLPST  GETANPFFTAAALTVMATAGVAAVVKRKEEN  483

CshA  2410  TISASYTPRVTEIPVVPNRPSTPEQPKAPVIPVDPTVVVQTPKAEERVEPIYIDPKDEKGVLPRT  GSQTSDQTASGLLAAIASLTFFGLANRKKKSKED  2508

InlA  706   NPVAPPTTGGNTPPTTNNGGNTTPPSANIPGSDTSNTSTGNSASTTSTMNAYDPYNSKEASLPTT  GDSDNALYLLLGLLAVGTAMALTKKARASK  800

FimA  432   GANRDNQKDATARCYVLVETKAPAGYVLPAGDGAVTPVKIEVGAVTTDNVTIENTKQSVPGLPLT  GANGMLILTASGASLLMIAVGSVLVARYRERKQNANLAL  535
```

Figure 4.   Comparison of the inferred amino acid sequences of the C-terminal regions of five gram-positive bacterial cell surface proteins carrying the LPxTG motif (in bold type) and cell wall-sorting signal comprising membrane-associating sequence (underlined) and charged tail (positively charged residues in bold type). Proline residues located N-terminally to the LPxTG motif are underscored. Numbers represent amino acid residues from the N-terminus of the precursor polypeptide. Polypeptides are, from the top (GenBank accession numbers in parentheses): Spa, staphylococcal surface protein A (M18264); Emm 6, group A streptococcal type 6 M protein (M11338); Csh A, *S. gordonii* cell-surface hydrophobicity protein (X65164); Inl A, *L. monocytogenes* internalin (M67471); Fim A, *A. naeslundii* type II fimbrial subunit (AF019629).

face protein nor the nature of the chemical linkage to the bacterial cell wall had been determined. It is interesting to reflect that unlike the discovery of the Lpp–peptidoglycan linkage, which was made first through application of protein purification and biochemical analysis and then confirmed through recombinant DNA technology, the gram-positive protein–peptidoglycan linkage was inferred on the basis of gene cloning and sequencing and then confirmed more recently by biochemical analyses. Despite there now being available extensive genomic sequence information in the databases, no sequences similar to Lpp or to the peptidoglycan linkage region of Lpp have been revealed in gram-positive bacteria. Thus it seems likely that the protein–peptidoglycan linkage machinery genes evolved along with the genes encoding the wall synthesis and modification machinery in the gram-positive and gram-negative bacterial lineages.

A novel approach to identifying the C-terminal end of wall-linked protein A was devised whereby the N-terminus of the putatively cleaved distal (unanchored) fragment could be sequenced (Navarre and Schneewind, 1994). In order to achieve this, the cell wall sorting signal of Spa (comprising LPETG, hydrophobic region, and charged tail) was incorporated into a polypeptide chain, upstream of *E. coli* Mal E (maltose-binding protein) and downstream of SEB. Pulse-labeling of *S. aureus* cells expressing this hybrid protein revealed a large precursor protein that underwent cleavage into two products. These could be separated by immunoprecipitation with anti-SEB or anti-Mal E antibodies, the N-terminal SEB region being cell wall linked, while the C-terminal Mal E region was found to be present almost entirely within the cytoplasmic compartment (Navarre and Schneewind, 1994). N-terminal sequencing of the major 40-kDa Mal E fragment gave the sequence GEENPFI, which corresponded precisely to the sequence after the Thr residue of the LPxTG motif within Spa (see Fig. 4). On the basis of these data, it was proposed that a membrane-associated "sortase" machinery recognized the cell wall sorting signal of protein A and cleaved the precursor protein between T and G, resulting in anchorage of the N-terminal fragment to the bacterial cell wall (Navarre and Schneewind, 1994).

To identify the linkage chemical nature of the linkage between protein A and the staphylococcal cell wall, a sequence encoding maltose-binding protein Mal E was fused between an N-terminal SEB sequence and the protein A C-terminal LPETG-containing sorting signal (Schneewind *et al.*, 1995). A trypsin-susceptible sequence was also engineered between Mal E and the LPETG motif allowing cleavage of the hybrid protein from the putative cell wall anchorage. When expressed in *S. aureus* the hybrid protein (Mal E-Cws) became anchored to the cell wall and could only be solubilized by enzymic digestion of the peptidoglycan. Treatment of cell walls with muramidase solubilized Mal E-Cws, which appeared as a mixture of fragments of increasing mass corresponding to the attachment of variable numbers of cell wall peptidoglycan units. By contrast, lysostaphin treatment caused release of Mal E-Cws as a uniform species, indicating that the lysostaphin cleavage site was close to the anchorage point of the surface protein

(Schneewind *et al.*, 1995) (see Fig. 3). Mass spectrometric analyses of trypsin-derived Mal E-Cws–peptidoglycan fragments demonstrated that the Mal E-Cws was linked to the cell wall peptidoglycan through an amide bond between the C-terminal COO^- of threonine and the free amino group of the pentaglycine cross bridges. Although lysostaphin cleaves randomly synthetic glycylglycine pentapeptides, mass measurements of lysostaphin-solubilized Mal E-Cws were consistent with the presence of three glycine residues indicating that some steric hindrance effects of linked polypeptide may restrict lysostaphin cleavage to between the third and fourth glycine of the pentaglycine cross bridge (Ton-That *et al.*, 1997) (Fig. 3).

In summary, the current model for cell wall sorting and anchorage of protein A in *S. aureus* is as follows (see Fig. 5). During signal peptide-mediated translocation of the polypeptide precursor, the cell wall-sorting signal consisting of the LPxTG motif, hydrophobic domain, and charged tail is recognized by the "sortase" machinery. The hydrophobic region and charged tail acts as a secretion stop mechanism when the LPxTG motif is correctly positioned with respect to the pos-

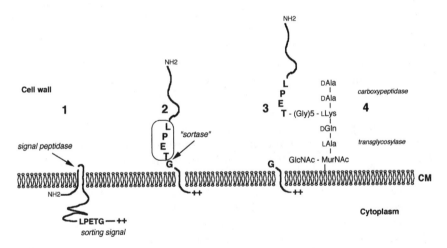

Figure 5.   Model for cell wall sorting of gram-positive bacterial polypeptides using staphylococcal protein A as a specific example. Precursor polypeptide with N-terminal leader and C-terminal sorting signal is secreted via the Sec machinery and the leader peptide cleaved by signal peptidase (1). The C-terminal hydrophobic region and positively charged tail function to brake polypeptide translocation, while the sorting signal is recognized by an extracytoplasmic (possible membrane-associated) machinery termed "sortase" (2). The polypeptide is cleaved between the T and G residues of the LPxTG motif and the N-terminal polypeptide becomes linked covalently via amide bond exchange to the free amino group of pentaglycine (see Fig. 3) within a peptidoglycan precursor unit. This reaction is analogous to the penicillin-sensitive transpeptidation reaction of cell wall cross-linking. The polypeptide–peptidoglycan precursor may then be incorporated into the cell wall by transglycosylation (4). At some stage, pre- or postlinkage of the polypeptide, carboxypeptidase cleaves the D-Ala–D-Ala bond of the pentapeptide to generate the final branched anchor peptide within the cell wall (see Fig. 3). The diagram was redrawn from Navarre and Schneewind (1994), with permission from Blackwell Science Ltd.

itively charged tail. Following recognition of this conformation by sortase, an enzymic reaction cleaves the precursor between T and G, and an amide bond is formed between the C-terminal Thr of the polypeptide and the amino of the pentaglycine cross bridge that is attached to the ε-amino of lysyl within a peptidoglycan tetrapeptide (Fig. 5). This branched anchor peptide is then linked to MurNAc, thus attaching surface polypeptides to the glycan chains of the staphylococcal wall. It is envisaged that proteins become linked to peptidoglycan precursors rather than to assembled wall, which is highly cross-linked and contains few free amino groups, and then are incorporated into the wall by a transglycosylation reaction (Ton-That *et al.*, 1997) (Fig. 5).

3.4. Cell Wall Distribution and Release of Surface Proteins

The sorting mechanism involving recognition of the LPxTG motif described for *S. aureus* protein A is believed to be a universal mechanism by which C-terminal sorting signal polypeptides are linked to the gram-positive bacterial cell wall. In those bacteria that do not form cell wall cross bridges, the amino group to which T becomes linked could be provided by the amino groups in the side chains of Lys or Dpm. A feature of the cell wall anchor structure of *S. aureus* protein A is that it cannot become cross-linked with neighboring wall peptides (Ton-That *et al.*, 1997). Thus the protein is not extensively cross-linked into the wall and can potentially be released from the cell surface by enzymic activity that cleaves the glycan strands. There has been some question as to whether the hydrophobic membrane-spanning region functions to retain a protein at the cell surface in the absence of LPxTG cleavage. It is interesting to note that Act A, which is involved in *L. monocytogenes*-induced actin assembly, is a 610 amino acid residue surface protein carrying a 22 amino acid residue region of hydrophobic amino acids at the C-terminal region without the LPxTG motif (Kocks *et al.*, 1992). This hydrophobic sequence appears sufficient to retain the protein at the cell surface, and since Act A can be released from cells following SDS treatment, no alternative covalent linkage is suggested. Thus it would seem that the hydrophobic region may be sufficient to retain a protein at the surface, and therefore this raises the possibility for temporal control of sortase-associated polypeptide cleavage and linkage to peptidoglycan.

The proposed mechanism by which proteins become inserted into the peptidoglycan assumes that delivery of the polypeptides occurs to the sites of new wall synthesis. In the accepted model for wall growth in gram-positive cocci (Higgins and Shockman, 1970) there is a single growth zone where peripheral and septal wall are synthesized by the controlled deposition of wall precursors. Material deposited initially in the cell septum becomes part of the peripheral wall as the cross wall separates at the base as a result of controlled autolytic action. The sortase machinery is probably functionally localized at the sites of new wall synthesis. Secreted proteins, most of which would be far too large to diffuse through the cell wall fabric (Demchick and Koch, 1996), therefore become wall linked or pass to the external medium through the inside-to-outside growth process in gram-positive bacteria (Kemper *et al.,* 1993). Owing to the mechanism of wall growth, the cell surface distribution of wall-anchored proteins in exponential phase cultures is not uniform. As demonstrated by Olmsted *et al.* (1993), who investigated by field emission scanning microscopy the expression of pheromone-induced cell-wall-linked aggregation factor protein Asc 10 on the surface of *E. faecalis,* some cells upon completion of division will appear devoid of surface protein. When the culture reaches late exponential phase and division slows, the proteins become more evenly distributed over the cell surface. Although the growth and division mechanisms for rod-shaped bacteria are fundamentally different from those operating in cocci, it is worth noting that wall-linked internalin (Inl A) displays a polarized distribution on the *Listeria* cell surface with fluorescent antibody staining weakly, being detectable at one pole with intensity of staining increasing toward the other pole (Lebrun *et al.,* 1996). A similar polarized localization is evident for membrane-linked Act A, implying that in *L. monocytogenes* the pole that has been formed during the previous cell division (the younger pole) is different from the older pole in that it does not express Act A (Kocks *et al.,* 1993).

There are still many facets of the sortase machinery that are not fully understood. In particular, the enzyme that cleaves the LPxTG motif has not yet been identified and the proposed polypeptide–peptidoglycan precursor intermediate has not been structurally confirmed. Another intriguing aspect of the C-terminal structure of sorting signal-containing polypeptides is that the sequence N-terminal to the LPxTG motif varies quite considerably. In some proteins, such as group A streptococcal M protein and Csh A from *S. gordonii* but not especially in protein A, the sequence immediately adjoining the sorting signal contains a high proportion of proline residues that are regularly spaced (Fig. 4). This would tend to favor formation of an extended secondary structure that might act as a "spacer" to allow folding of the polypeptide or presentation outside the confines of the cross-linked peptidoglycan. This region also could be involved in interaction with other cell wall components and contribute to retention of protein on the cell surface. Some evidence for this notion comes from analysis of the surface retention properties of truncated forms of the *Streptococcus mutans* surface protein P1. One truncated form, in which the complete sorting signal was deleted, was partially retained at the cell surface (Homonylo-McGavin and Lee, 1996), whereas when the N-terminally adjacent 44 amino acid residues containing regularly spaced prolines were also deleted, the protein was found exclusively in the extracellular medium.

Recent evidence indicates that the outcome of cleavage of the polypeptide between T and G may not necessarily result in its cell wall linkage. The *fimA* gene in *Actinomyces neaslundii* encodes a 535 amino acid residue precursor of the type 2 fimbrial subunit (Yeung *et al.*, 1998). The C-terminus of Fim A carries the LPxTG sorting signal and mature fimbriae comprise Fim A subunits that have been processed and lack the C-terminal signal. It is envisaged therefore that the Fim A precursor C-terminus is cleaved at the LPxTG motif, but that instead of becoming linked to cell wall peptidoglycan the Thr residue is linked to another Fim A molecule, either directly or indirectly through an oligopeptide or peptidoglycan fragment, for fimbrial assembly (Yeung *et al.*, 1998). Notably, the sorting signal from Fim P, the subunit of *A. naeslundii* type 1 fimbriae, does not direct efficient sorting of protein A in *S. aureus* when substituted for the protein A sorting signal. This may be due to suboptimal spacing between the LPxTG motif and the charged tail (Schneewind *et al.*, 1993). Thus it is possible there are additional sequence or structural signals within the C-terminal and adjacent regions of these various proteins that influence their retention, linkage, or assembly into macromolecular cell surface structures.

It has long been known that cell wall-linked gram-positive bacterial proteins differ somewhat in the extent to which they remain held at the cell surface or released into the extracellular medium. Protein A, for example, is rarely found in extracellular fluid under normal culture growth conditions (Schneewind *et al.*, 1992), whereas polypeptides such as M protein and the oral streptococcal antigen I/II protein adhesins have been often shown to be released from cells depending on culturing conditions. Enzymic activities are present in these streptococci that cause release of surface proteins. M protein is released from spheroplasts of *S. pyogenes* optimally at pH 7.4 via a Zn^{2+}-sensitive mechanism (Pancholi and Fischetti, 1989), while P1 (antigen I/II) protein is released from the surface of *S. mutans* optimally at pH 5.5 (Lee, 1995). It has not been determined whether release results from proteolytic cleavage at the C-terminal of the polypeptide or from wall glycan cleavage. It seems possible, however, that sequences immediately adjacent to the LPxTG motif may have a role in the protein release mechanisms. Several functions for protein release have been entertained: permitting escape from host immune surveillance through selective alteration of surface properties by shedding of bound antibodies (Lee, 1995); promoting cell detachment from a surface and dissemination to other sites (Lee *et al.*, 1996); and modulating host cell invasive potential (Lebrun *et al.*, 1996).

4. NONPEPTIDOGLYCAN-LINKED SURFACE PROTEINS IN GRAM-POSITIVE BACTERIA

4.1. Polysaccharide-Binding Proteins

Gram-positive bacterial cell walls contain a wide variety of polysaccharides attached to lipid or peptidoglycan. Structurally related linear phosphopolysaccha-

rides are produced by some species of oral streptococci and are composed of phosphodiester-linked hexa- or heptasaccharide repeating units containing GalNAc,
Glc, Gal, and Rha (Cisar *et al.,* 1997). Specific hostlike recognition motifs such as
GalNAcß1 → 3Gal and Galß1 → 3GalNAc are present within these polysaccharides,
which may act to reduce immune stimulation by bacteria, but which also act as
receptors for other bacteria. Although there is no direct evidence that polypeptides
are linked specifically to these polymers, Erickson and Herzberg (1993) described
a glycosylated form of cell surface platelet aggregation-associated protein (PAAP)
of *Streptococcus sanguis* that carried the sugars characteristically associated
with the cell wall phospopolysaccharides. Approximately 39% by weight of the
115-kDa PAAP isolated from spheroplasts was comprised of a Rha-rich *N*-
asparaginyl linked polysaccharide (Erickson and Herzberg, 1993). It was suggested that covalent linkage of polysaccharide units to proteins, such as PAAP, may represent a mechanism by which polysaccharide can be translocated through the bacterial surface layers prior to being cleaved from the protein and transferred to other
cell wall components such as peptidoglycan. Sequencing of the gene encoding
PAAP and genetic analysis of the glycosylation reaction will be necessary in order
to formally test this hypothesis and to determine whether PAAP is covalently linked
to cell wall polysaccharide in a novel surface protein anchorage mechanism.

4.2. Lipoteichoic Acid-Binding Proteins

The other major cell wall glycopolymers present in gram-positive bacteria are
teichoic acid and lipoteichoic acid. LTA binds fibronectin, serum proteins, as well
as mammalian cells (Hasty *et al.,* 1992) and interacts with bacterial cell surface
proteins in promoting bacterial adhesion to epithelia. Recent evidence for specific and functionally significant interactions of surface proteins with LTA comes
from work with *Streptococcus pneumoniae.* Pneumococci differ from other gram-
positive bacteria in that their LTA and wall teichoic acids have the same chain
structure. This is unusually complex, containing ribitol phosphate, a tetrasaccharide, and phosphorylcholine. Phosphorylcholine-substituted LTA serves to anchor
pneumococcal surface protein A (Psp A), the physiological function of which is
not yet fully understood, to the outer layer via an interaction with the C-terminal
region of Psp A (Yother and White, 1994). This noncovalent anchorage mechanism has now been found to account for the surface presentation in pneumococci
of a large class of proteins designated choline-binding proteins (CBPs) (Rosenow
et al., 1997). The cell wall hydrolase Lyt A is a CBP that functions in the separation of daughter cells during cell division, while Sps A is a surface CBP that also
binds human secretory immunoglobulin A (IgA) (Hammerschmidt *et al.,* 1997).
All CBPs possess at their C-terminus repeat blocks of uncharged amino acids that
are proposed to interact noncovalently with choline. These blocks of amino acid
each carry a YG motif (Fig. 6) that is present also within the C-terminal regions of

```
PspA 407   ENGMWYFYNTDGSMATGWLQ NNGSWYYLNSNGAMATGWLQ YNGSWYYLNANGAMATGWAK 466
SpsA 326   ENGMWYFYNTDGSMATGWLQ NNGSWYYLNANGAMATGWLQ NNGSWYYLNANGSMATGWLQ 385
CspA 501   TSAGWTYVKADGTKATGWLQ DGGAWYYLKADGTMATGWIQ DGATWYYLNGSGAMQTGWLN 550
SpaA 452   KDNKWFYIEKSGGMATGWKK VADKWYYLDNTGAIVTGWKK VANKWYYLEKSGAMATGWKK 511
LycA 247   DNEKWYYLKDNGAMATGWVL VGSEWYYMDDSGAMVTGWVK YKNNWYYMTNE......... 297
Gbp  165   KDGKWYYKKADGQLATGWQI IDGKQLYFNQDGSQVKGEIH V.................. 215
           ...wyy....G..•tGw•.   ....wyy....G..•tGw•.   ....wyy....G..•tGw•.
```

Figure 6.   Comparison of amino acid repeat block sequences present within some polypeptides that bind choline lipoteichoic acid or carbohydrate polymers. Only two or three repeat blocks are shown for the various polypeptides and each block conforms to the consensus for a YG repeat module (Giffard and Jacques, 1994), as indicated at the bottom of the comparison. Within each module, the central glycine (G) residue is fully conserved; small letters indicate residues that are at least 80% conserved; · designates the position of a usually hydrophobic residue. Other residues are generally poorly conserved. Polypeptides are, from the top (GenBank accession numbers in parentheses): Psp A, *S. pneumoniae* surface protein A, 619 amino acid residues containing 10 repeats (A41971); Sps A, *S. pneumoniae* secretory immunoglobulin A-binding protein, 539 amino acid residues with 4 repeats (AJ002054); Csp A, *Clostridium acetobutylicum* surface protein, 590 amino acid residues with 5 repeats (Z37723); Spa A, *Erysipelothrix rhusiopathiae* protective antigen protein, 606 amino acid residues with 7 repeats (AB012763); Lyc A, pneumococcal phage CP-1 lysin, 339 amino acid residues with 6 repeats (J03586); Gbp, *Streptococcus mutans* glucan-binding protein, 563 amino acid residues with 12 repeats (M30945). Numbers are amino acid residues from the N-terminus of the precursor polypeptide.

a range of nonpneumococcal surface proteins, including *Clostridium difficile* toxins (von Eichel-Streiber *et al.,* 1992), glucan-binding proteins from oral streptococci, and glucosyltransferase enzymes (Giffard and Jacques, 1994), all which bind to glycan polymers with repeating oligosaccharide units. Thus the YG repeat motif block, comprising between 18 and 24 amino acid residues, may represent a general structural theme for the binding of polypeptides to carbohydrate polymers (Fig. 6). The hydrophobic portions of the saccharide chains might interact with the aromatic residues within the carbohydrate-binding modules, while adjacent conserved residues may form hydrogen bonds to different H bond donors or acceptors. The carbohydrate-binding specificities of these proteins could be determined by the lengths of the repeat units and the spacing of conserved residues within these units, as well by as the identities of poorly conserved residues.

A related mechanism of cell wall polypeptide anchorage via noncovalent interactions with cell wall polymers is shown by the Inl B protein of *L. monocytogenes.* Inl B (630 amino acid residues) is a surface protein that, along with Inl A (internalin), is involved in bacterial entry into the host cell. Unlike Inl A, Inl B does not contain an LPxTG sorting signal and does not exhibit the YG motif. Instead, a 232 amino acid residue C-terminal domain containing three repeats of 80 amino acid residues starting with the dipeptide GW appears to sufficient to anchor the protein at the cell surface (Braun *et al.,* 1997). Ami, a newly identified *Listeria* surface protein with N-terminal homology to *S. aureus* bacteriolysin, also contains the GW module repeat at the C-terminal region, but carries eight copies arranged in tandem as opposed to the three module repeats in Inl B (Braun et al., 1997). Experimental data indicate that Ami is held more tightly at the cell surface than is Inl B, suggesting that the high number of repeats results in more efficient anchorage. Similarities to these sequences are found in the C-terminal regions of Lyt A (a staphylococcal phage amidase) and lysostaphin from S. simulans. The target cell specificity of lysostaphin is determined by the 92 amino acid residues at the C-terminus, which contains a single GW module (Baba and Schneewind, 1996). Precisely how these proteins recognize their cell wall target sequences is not known.

5. APPLICATIONS OF PROTEIN–CELL WALL LINKAGE

The presentation of biologically active proteins on the surface of microparticles has many applications in the fields of cell receptor biology, immunology, and biotechnology. The most familiar system is the use of filamentous bacteriophage for affinity selection (panning) of peptides or antibodies from libraries (Rader and Barbas, 1997). However, the ability to surface display biological molecules extends wider, from the use of recombinant bacteria to express antibody fragments for diagnostic purposes to the surface immobilization of ligands or enzymes to develop novel microbial biofilters or biocatalysis systems.

Single-chain Fv antibody fragments have been expressed and anchored to the outer membrane of E. coli by recombinant fusions with lipoproteins. Fluorescein-marked antigens were utilized to verify that the Fv fragments were functional and fluorescence-activated cell sorting (FACS) was used to enrich for Fv-expressing bacteria (Francisco et al., 1993). Despite the success of this strategy and the obvious possibility that this could provide an alternative to phage display technology for selection of peptides or antibody fragments, gram-positive bacteria have certain features that could make them more suitable for these kinds of applications. For example, there seems to be little restriction to the size or structure of proteins that may be expressed on the surface of gram-positive bacterial cells; gram-positive bacteria are more robust and a single membrane translocation step only is required for protein secretion. Both Staphylococcus xylosus and Staphylococcus carnosus have been developed for surface display of heterologous proteins, expressed from replicative plasmids and covalently linked to cell wall peptidoglycan (Gunneriusson et al., 1996). These systems have employed a variety of recombinant protein fusions of the Spa or staphylococcal lipase leader peptides to heterologous sequences, incorporating a segment of the streptococcal albumin-binding protein G or staphylococcal fibronectin-binding protein B, to the C-terminal cell wall sorting signal of S. aureus protein A. In addition to expressing active Fv fragments on the surface of these nonpathogenic staphylococcal species, enzymatically active lipase and semiactive ß-lactamase were expressed as surface-anchored molecules (Strauss and Gotz, 1996). Interestingly, the activity of lipase was enhanced if a spacer peptide of up to 90 amino acid residues was incorporated between the cell wall sorting signal and the C-terminal of lipase, suggesting that extension away from the cell wall layers assists the folding of the enzyme into a more active conformation. This kind of strategy is potentially applicable to a wide range of gram-positive bacteria because of the universal nature of the LPxTG cell wall sorting signal. Indeed, it recently has been demonstrated that heterologous expression of two streptococcal adhesins Ssp A and Ssp B on the surface of the food-grade organism *Lactococcus lactis* confer on lactococci the ability to adhere to human tissue proteins (Holmes *et al.*, 1998). The streptococcal proteins were sorted and covalently linked to the lactococcal cell surface through recognition by the lactococcal sortase machinery of streptococcal sorting signals. The staphylococcal protein A sorting signal is also recognized in *L. lactis* (Steidler *et al.*, 1998). Development of these surface display systems, especially those involving complex chimeric fusions, requires considerable fine-tuning to achieve maximal biological function of expressed protein, retention at the cell surface, and resistance to proteolytic modification. Notwithstanding current legislations and concerns surrounding the use of genetically modified material in human food production, one possible potential application of surface display might be in manipulating desirable surface traits of gram-positive bacteria that are utilized in production of fermented foods.

A common application for bacterial surface display of proteins is in the development of vaccine delivery systems. The use of gram-positive bacteria as live vector vaccines might go some way to overcoming concerns about utilizing disabled pathogenic organisms such as *Salmonella* (Chatfield *et al.*, 1993). Commensal streptococci and lactobacilli are considered to be suitable in this respect since they colonize naturally various mucosal surfaces (oral, intestinal, or vaginal) and potentially could be engineered to deliver antigens to elicit both enhanced local IgA responses in addition to immunoglobulin G and T-cell responses. By utilizing the M protein cell wall sorting signal as polypeptide anchor, fusions of various immunogenic sequences from human papilloma virus protein E7, white-faced hornet antigen Ag5.2, and HIV-1 envelope glycoprotein gp120 have been made to a chromosomally located cassette in *Streptococcus gordonii,* with resulting cell surface expression of these antigens (Medaglini *et al.*, 1995; Oggioni *et al.*, 1995; Pozzi *et al.*, 1994). Mice colonized vaginally or orally by these bacteria demonstrated both serum IgG and mucosal IgA responses to these various antigens (Medaglini *et al.*, 1997; Oggioni *et al.*, 1995). While this seems a promising strategy for vaccination, probably initially in animals, there are many issues to be resolved in developing live vector vaccines for human use. Some concerns are that individuals may be refractory to long-term colonization by exogenously applied bacteria; that bacteria such as *S. gordonii* have pathogenic potential; and that recombinant genes might become disseminated among the commensal population and be acquired by pathogens such as *S. pneumoniae.* Nevertheless, it seems that applications of the anchorage mechanisms linking polypeptides to the bacterial cell surface will feature widely in future developments in human and veterinary medicine and industrial and food technology.

REFERENCES

Baba, T., and Schneewind, O., 1996, Target cell specificity of a bacteriocin molecule: A C-terminal signal directs lysostaphin to the cell wall of *Staphylococcus aureus, EMBO J.* **15**:4789–4797.

Bouveret, E., Derouiche, R., Rigal, A., Lloubes, R., Lazdunski, C., and Benedetti, H., 1995, Peptidoglycan-associated lipoprotein–TolB interaction, *J. Biol. Chem.* **270**:11071–11077.

Braun, L., Dramsi, S., Dehoux, P., Bierne, H., Lindahl, G., and Cossart, P., 1997, InlB: An invasion protein of *Listeria monocytogenes* with a novel type of surface association, *Mol. Microbiol.* **25**:285–294.

Braun, V., and Bosch,V., 1972, Sequence of the murein–lipoprotein and the attachment site of the lipid, *Eur. J. Biochem.* **28**:51–69.

Braun, V., and Rehn, K., 1969, Chemical characterization, spatial distribution and function of a lipoprotein (murein-lipoprotein) of the *E. coli* cell wall. The specific effect of trypsin on the membrane structure, *Eur. J. Biochem.* **10**:426–438.

Braun, V., and Wu, H. C., 1994, Lipoproteins, structure, function, biosynthesis and model for protein export, in: *Bacterial Cell Wall* (J. -M. Ghuysen and R. Hakenbeck, eds.), Elsevier Science B. V., Amsterdam, pp. 319–341.

Chatfield, S., Roberts, M., Londono, P., Cropley, I., Douce, G., and Dougan, G., 1993, The development of oral vaccines based on live attenuated *Salmonella* strains, *FEMS Immunol. Med. Microbiol.* **7**:1–7

Chen, R., and Henning, U., 1987, Nucleotide sequence of the gene for the peptidoglycan-associated lipoprotein of *Escherichia coli* K12, *Eur. J. Biochem.* **163**:73–77.

Choi, D.-S., Yamada, H., Mizuno, T., and Mizushima, S., 1986, Trimeric structure and localization of the major lipoprotein in the cell surface of *Escherichia coli, J. Biol. Chem.* **261**:8953–8957.

Cisar, J. O., Sandberg, A. L., Reddy, G. P., Abeygunawardana, C., and Bush, C. A., 1997, Structural and antigenic types of cell wall polysaccharides from viridans group streptococci with receptors for oral actinomyces and streptococcal lectins, *Infect. Immun.* **65**:5035–5041.

Clavel, T., Germon, P., Vianney, A., Portalier, R., and Lazzaroni, J. C., 1998, TolB protein of *Escherichia coli* K-12 interacts with the outer membrane peptidoglycan-associated proteins Pal, Lpp and OmpA, *Mol. Microbiol.* **29**:359–367.

Demchick, P., and Koch, A. L., 1996, The permeability of the wall fabric of *Escherichia coli* and *Bacillus subtilis, J. Bacteriol.* **178**:768–773.

Erickson, P. R., and Herzberg, M. C., 1993, Evidence for the covalent linkage of carbohydrate polymers to a glycoprotein from *Streptococcus sanguis, J. Biol. Chem.* **268**:23780–23783.

Fischer, W., 1994, Lipoteichoic acids and lipoglycans, in: *Bacterial Cell Wall* (J. -M. Ghuysen and R. Hakenbeck, eds.), Elsevier Science B. V., Amsterdam, pp. 199–215.

Fischetti, V. A., Jones, K. R., and Scott, J. R., 1985, Size variation of the M protein in group A streptococci, *J. Exp. Med.* **161**:1384–1401.

Francisco, J. A., Campbell, R., Iverson, B. L., and Georgiou, G., 1993, Production and fluorescence-activated cell sorting of *Escherichia coli* expressing a functional antibody fragment on the external surface, *Proc. Natl. Acad. Sci. USA* **90**:10444–10448.

Fussenegger, M., Facius, D., Meier, J., and Meyer, T. F., 1996, A novel peptidoglycan-linked lipoprotein (ComL) that functions in natural transformation competence of *Neisseria gonorrhoeae, Mol. Microbiol.* **19**:1095–1105.

Gennity, J. M., Kim, H., and Inouye, M., 1992, Structural determinants in addition to the amino-terminal sorting sequence influence membrane localization of *Escherichia coli* lipoproteins, *J. Bacteriol.* **174**:2095–2101.

Giffard, P. M., and Jacques, N. A., 1994, Definition of a fundamental repeating unit in streptococcal glucosyltransferase glucan-binding regions and related sequences, *J. Dent. Res.* **73**:1133–1141.

Gunneriusson, E., Samuelson, P., Uhlen, M., Nygren, P-A., and Stahl, S., 1996, Surface display of a functional single-chain Fv antibody on staphylococci, *J. Bacteriol.* **178**: 1341–1346.

Hammerschmidt, S., Talay, S. R., Brandtzaeg, P., and Chhatwal, G. S., 1997, SpsA, a novel pneumococcal surface protein with specific binding to secretory immunoglobulin A and secretory component, *Mol. Microbiol.* **25**:1113–1124.

Hantke, K., and Braun, V., 1973, Covalent binding of lipid to protein. Diglyceride and amide-linked fatty acid at the N-terminal end of the murein-lipoprotein of the *Escherichia coli* outer membrane, *Eur. J. Biochem.* **34**:284–296.

Hasty, D. L., Ofek, I., Courtney, H. S., and Doyle, R. J., 1992, Multiple adhesins of streptococci, *Infect. Immun.* **60**:2147–2152.

Hiemstra, H., Nanninga, N., Woldringh, C. L., Inouye, M., and Witholt, B., 1987, Distribution of newly synthesised lipoprotein over the outer membrane and peptidoglycan sacculus of an *Escherichia coli lac–lpp* strain, *J. Bacteriol.* **169**:5434–5444.

Higgins, M. L., and Shockman, G. D., 1970, Model for cell wall growth of *Streptococcus faecalis, J. Bacteriol.* **101**:643–648.

Hollingshead, S. K., Fischetti, V. A., and Scott, J. R., 1986, Complete nucleotide sequence of type 6 M protein of the group A streptococcus: Repetitive structure and membrane anchor, *J. Biol. Chem.* **261**:1677–1686.

Holmes, A. R., Gilbert, C., Wells, J. M., and Jenkinson, H. F., 1998, Binding properties of *Streptococcus gordonii* SspA and SspB (antigen I/II family) polypeptides expressed on the cell surface of *Lactococcus lactis* MG1363, *Infect. Immun.* **66**:4633–4639.

Homonylo-McGavin, M. K., and Lee, S. F., 1996, Role of the C terminus in antigen P1 surface localization in *Streptococcus mutans* and two related cocci, *Infect. Immun.* **178**:801–807.

Izard, J. W., and Kendall, D. A., 1994, Signal peptides: Exquisitely designed transport promoters, *Mol. Microbiol.* **13**:765–773.

Jenkinson, H. F., and Lamont, R. J., 1997, Streptococcal adhesion and colonization, *Crit. Rev. Oral Biol. Med.* **8**:175–200.

Kemper, M. A., Urrutia, M. M., Beveridge, T. J., Koch, A. L., and Doyle, R. J., 1993, Proton motive force may regulate cell wall-associated enzymes of *Bacillus subtilis, J. Bacteriol.* **175:**5690–5696.

Kocks, C., Gouin, E., Tabouret, M., Berche, P., Ohayon, H., and Cossart, P., 1992, *L. monocytogenes*-induced actin assembly requires the *actA* gene product, a surface protein, *Cell* **68:**521–531.

Kocks, C., Hellio, R., Gounon, P., Ohayon, H., and Cossart, P., 1993, Polarized distribution of *Listeria monocytogenes* surface protein ActA at the site of directional actin assembly, *J. Cell. Sci.* **105:**699–710.

Koebnik, R., 1995, Proposal for a peptidoglycan-associating-alpha-helical motif in the C-terminal regions of some bacterial cell-surface proteins, *Mol. Microbiol.* **16:**1269–1270.

Kosic, N., Sugai, M., Fan, C.-K., and Wu, H. C., 1993, Processing of lipid-modified prolipoprotein requires energy and *sec* gene products *in vivo, J. Bacteriol.* **175:**6113–6117.

Lazzaroni, J.-C., and Portalier, R., 1992, The *excC* gene of *Escherichia coli* K-12 for cell envelope integrity encodes the peptidoglycan-associated lipoprotein (PAL), *Mol. Microbiol.* **6:**735–742.

Lebrun, M., Mengaud, J., Ohayon, H., Nato, F., and Cossart, P., 1996, Internalin must be on the bacterial surface to mediate entry of *Listeria monocytogenes* into epithelial cells, *Mol. Microbiol.* **21:**579–592.

Lee, S. F., 1995, Active release of bound antibody by *Streptococcus mutans, Infect. Immun.* **63:**1940–1946.

Lee, S. F., Li, Y. H., and Bowden, G. H., 1996, Detachment of *Streptococcus mutans* biofilm cells by an endogenous enzymatic activity, *Infect. Immun.* **64:**1035–1038.

Matsuyama, S., Tajima, T., and Tokuda, H., 1995, A novel periplasmic carrier protein involved in the sorting and transport of *Escherichia coli* lipoproteins destined for the outer membrane, *EMBO J.* **14:**3365–3372.

McNab, R., and Jenkinson, H. F., 1992, Gene disruption identifies a 290 kDa cell-surface polypeptide conferring hydrophobicity and coaggregation properties in *Streptococcus gordonii, Mol. Microbiol.* **6:**2939–2949.

McNab, R., Jenkinson, H. F., Loach, D. M., and Tannock, G. W., 1994, Cell-surface-associated polypeptides CshA and CshB of high molecular mass are colonization determinants in the oral bacterium *Streptococcus gordonii, Mol. Microbiol.* **14:**743–754.

Medaglini, D., Pozzi, G., King, T. P., and Fischetti, V. A., 1995, Mucosal and systemic immune responses to a recombinant protein expressed on the surface of the oral commensal bacterium *Streptococcus gordonii* after oral colonization, *Proc. Natl. Acad. Sci. USA* **92:**6868–6872.

Medaglini, D., Rush, C. M., Sestini, P., and Pozzi, G., 1997, Commensal bacteria as vectors for mucosal vaccines against sexually transmitted diseases: Vaginal colonization with recombinant streptococci induces local and systemic antibodies in mice, *Vaccine* **15:**1330–1337.

Mizuno, T., 1981, A novel peptidoglycan-associated lipoprotein (PAL) found in the outer membrane of *Proteus mirabilis* and other gram-negative bacteria, *J. Biochem. (Tokyo)* **89:**1039–1049.

Howard F. Jenkinson

Navarre, W. W., and Schneewind, O., 1994, Proteolytic cleavage and cell wall anchoring at the LPXTG motif of surface proteins in gram-positive bacteria, *Mol. Microbiol.* **14**:115–121.

Nesbitt, W. E., Staat, R. H., Rosan, B., Taylor, K. G., and Doyle, R. J., 1980, Association of protein with the cell wall of *Streptococcus mutans, Infect. Immun.* **28**:118–126.

Oggioni, M. R., Manganelli, R., Contorni, M., Tommasino, M., and Pozzi, G., 1995, Immunization of mice by oral colonization with live recombinant commensal streptococci, *Vaccine* **13**:775–779.

Olmsted, S. B., Erlandsen, S. L., Dunny, G. M., and Wells, C. L., 1993, High-resolution visualization by field emission scanning electron microscopy of *Enterococcus faecalis* surface proteins encoded by the pheromone-inducible conjugative plasmid pCF10, *J. Bacteriol.* **175**:6229–6237.

Pancholi, V., and Fischetti, V. A., 1988, Isolation and characterization of the cell-associated region of group A streptococcal M6 protein, *J. Bacteriol.* **170**:2618–2624.

Pancholi, V., and Fischetti, V. A., 1989, Identification of an endogenous membrane anchor-cleaving enzyme for group A streptococcal M protein, *J. Exp. Med.* **170**:2119–2133.

Pozzi, G., Oggioni, M. R., Manganelli, R., Medaglini, D., Fischetti, V. A., Fenoglio, D., Valle, M. T., Kunkl, A., and Manca, F., 1994, Human T-helper cell recognition of an immunodominant epitope of HIV-1 gp120 expressed on the surface *of Streptococcus gordonii, Vaccine* **12**:1071–1077.

Rader, C., and Barbas III, C. F., 1997, Phage display of combinatorial libraries, *Curr. Opin. Biotechnol.* **8**:503–508.

Rosenow, C., Ryan, P., Weiser, J. N., Johnson, S., Fontan, P., Ortqvist, A., and Masure, H. R., 1997, Contribution of novel choline-binding proteins to adherence, colonization and immunogenicity of *Streptococcus pneumoniae, Mol. Microbiol.* **25**:819–829.

Russell, R. R. B., 1979, Wall-associated protein antigens of *Streptococcus mutans, J. Gen. Microbiol.* **114**:109–115.

Schatz, P. J., and Beckwith, J., 1990, Genetic analysis of protein export in *Escherichia coli, Annu. Rev. Genet.* **24**:215–248.

Schneewind, O., Model, P., and Fischetti, V. A., 1992, Sorting of protein A to the staphylococcal cell wall, *Cell* **70**:267–281.

Schneewind, O., Mihaylova-Petkov, D., and Model, P., 1993, Cell wall sorting signals in surface proteins of gram-positive bacteria, *EMBO J.* **12**:4803–4811.

Schneewind, O., Fowler, A., and Faull, K. F., 1995, Structure of the cell wall anchor of surface proteins in *Staphylococcus aureus, Science* **268**:103–106.

Shuttleworth, H. L., Duggleby, C. J., Jones, S. A., Atkinson, T., and Minton, N. P., 1987, Nucleotide sequence analysis of the gene for protein A from *Staphylococcus aureus* Cowan 1 (NCTC8530) and its enhanced expression in *Escherichia* coli, *Gene* **58**:283–295.

Sjoquist, J., Meloun, B., and Hjelm, H., 1972, Protein A isolated from *Staphylococcus aureus* after digestion with lysostaphin, *Eur. J. Biochem.* **29**:572–578.

Steidler, L., Viaene, J., Fiers, W., and Remaut, E., 1998, Functional display of a heterologous protein on the surface of *Lactococcus lactis* by means of the cell wall anchor of *Staphylococcus aureus* protein A, *Appl. Env. Microbiol.* **64**:342–345.

Strauss, A., and Gotz, F., 1996, *In vivo* immobilization of enzymatically active polypeptides on the cell surface of *Staphylococcus carnosus, Mol. Microbiol.* **21**:491–500.

Sutcliffe, I. C., and Russell, R. R. B., 1995, Lipoproteins of gram-positive bacteria, *J. Bacteriol.* **177**:1123–1128.

Suzuki, H., Nishimura, Y., Yasuda, S., Nishimura, A., Yamada, M., and Hirota, Y., 1978, Murein-lipoprotein of *Escherichia coli:* A protein involved in the stabilization of bacterial cell envelope, *Mol. Gen. Genet.* **167**:1–9.

Ton-That, H., Faull, K. F., and Schneewind, O., 1997, Anchor structure of staphylococcal surface proteins, *J. Biol. Chem.* **272**:22285–22292.

von Eichel-Streiber, C., Sauerborn, M., and Kuramitsu, H. K., 1992, Evidence for a modular structure of the homologous repetitive C-terminal carbohydrate-binding sites of *Clostridium difficile* toxins and *Streptococcus mutans* glucosyltransferases, *J. Bacteriol.* **174**:6707–6710.

Wu, H. C., 1996, Biosynthesis of lipoproteins, in: *Escherichia coli and Salmonella* (F. C. Neidhardt, ed.), ASM Press, Washington, DC, pp. 1005–1014.

Yakushi, T., Tajima, T., Matsuyama, S., and Tokuda, H., 1997, Lethality of the covalent linkage between mislocalized major outer membrane lipoprotein and the peptidoglycan of *Escherichia coli*, *J. Bacteriol.* **179**:2857–2862.

Yamaguchi, K., Yu, F., and Inouye, M., 1988, A single amino acid determinant of the membrane localization of lipoproteins in *E. coli, Cell* **53**:423–432.

Yeung, M. K., Doonkersloot, J. A., Cisar, J. O., and Ragsdale, P. A., 1998, Identification of a gene involved in the assembly of *Actinomyces naeslundii* T14V type 2 fimbriae, *Infect. Immun.* **66**:1482–1491.

Yother, J., and White, J. M., 1994, Novel surface attachment mechanism of the *Streptococcus pneumoniae* protein PspA, *J. Bacteriol.* **176**:2976–2985.

Zhang, W. Y., and Wu, H. C., 1992, Alterations of the carboxy-terminal amino acid residues of *E. coli* lipoprotein affect the formation of murein-bound lipoprotein, *J. Biol. Chem.* **267**:19560–19564.

Zhang, W. Y., Inouye, M., and Wu, H. C., 1992, Neither lipid modification nor processing of prolipoprotein is essential for the formation of murein-bound lipoprotein in *Escherichia coli, J. Biol. Chem.* **267**:19631–19635.

4

Surface Layer Glycoproteins of Bacteria and Archaea

Paul Messner and Christina Schäffer

1. INTRODUCTION

Prior to the designation of the archaebacteria (archaea) as a second prokaryotic kingdom of life (Woese and Fox, 1977), glycoproteins were believed to be restricted to the eukaryotes (for reviews see, Montreuil, 1995; Johansen *et al.,* 1958). The observation of glycosylated cell envelope proteins of halobacteria has changed this perception (Mescher and Strominger, 1976a). Since then, an increasing number of reports have indicated the presence of glycoproteins in the domain archaea (Kandler, 1993, 1994). As a consequence, the occurrence of covalently linked glycan chains of bacterial proteins was regarded as a specific feature of archaebacteria (Mescher, 1981), constituting a significant difference to eubacteria. In the last decade, the presence of glycoproteins was also established in the domain bacteria. Thus, the ability of prokaryotic organisms to produce glycosylated proteins is not different in principle from that of higher organisms (Messner, 1996, 1997; Moens and Vanderleyden, 1997; Sumper and Wieland, 1995; Erickson and Herzberg, 1993; Messner and Sleytr, 1991; Lechner and Wieland, 1989). Glycosylation as an important secondary modification of proteins therefore exists in all domains of life (for a review, see Lis and Sharon, 1993).

While the surface layer (S-layer) (for reviews, see Sleytr *et al.,* 1996, 1999)

Paul Messner and Christina Schäffer • Zentrum für Ultrastrukturforschung und Ludwig Boltzmann-Institut für Molekulare Nanotechnologie, Universität für Bodenkultur Wien, A-1180 Wien, Austria.

Glycomicrobiology, edited by Doyle.
Kluwer Academic/Plenum Publishers, New York, 2000.

93

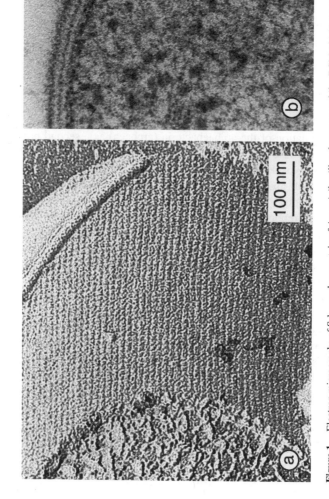

Figure 1. Electron micrographs of S-layer glycoprotein of *Aneurinibacillus thermoaerophilus* DSM 10155. (a) After freeze-etching of intact cells and metal-shadowing, the square S-layer lattice becomes visible. This type of preparation does not show the glycan chains, but does reveal the proteinaceous S-layer protomers. (b) Ultrathin section of high-pressure frozen and freeze-substituted cells intact cells where the glycan chains can be seen as filiform structures at the cell surface.

Table 1

Glycan Structures of Selected Archaeal and Bacterial S-layer Glycoproteins[a]

Bacteria

Bacillus stearothermophilus NRS 2004/3a (Messner and Sleytr, 1988; Christian *et al.*, 1986)

[→2)-α-L-Rhap-(1→2)-α-L-Rhap-(1→3)-β-L-Rhap-(1→]$_{n\sim50}$ -Rhap-(1→*N*)-Asn

Aneurinibacillus thermoaerophilus L420-91 (formerly *Bacillus thermoaerophilus*) (Kosma *et al.*, 1995a)

[→3)-α-D-Rhap-(1→3)-α-D-Rhap-(1→2)-α-D-Rhap-(1→2)-α-D-Rhap-(1→]$_{n\sim15}$ -GalNAc-(1→*O*)-Thr

α-D-Fucp3NAc α-D-Fucp3NAc
| |
1 1
↑ ↑
2 2

Aneurinibacillus thermoaerophilus GS4-97 (Schäffer *et al.*, 1999a)

[→3)-α-D-Rhap-(1→3)-α-D-Rhap-(1→2)-α-D-Rhap-(1→2)-α-D-Rhap-(1→]$_{n\sim15}$3)-[α-D-Rhap]$_{n=0-2}$-(1→3)-α-D-Rhap-(1→3)-β-D-GalNAc-(1→*O*)-Thr Ser

Gly

Asp

Thr

α-D-Fucp3NAc α-D-Fucp3NAc
| |
1 1
↑ ↑
2 2

Aneurinibacillus thermoaerophilus (DSM 10155) (formerly *Bacillus thermoaerophilus*) (Kosma *et al.*, 1995b; Wugeditsch *et al.*, 1999)

[→4)-α-L-Rhap-(1→3)-β-D-*glycero*-D-*manno*-Hepp-(1→]$_{n\sim18}$3)-[α-L-Rhap]$_{n=0-2}$-(1→3)-β-D-GalNAc-(1→*O*)-Thr/Ser

Paenibacillus alvei (CCM 2051) (formerly *Bacillus alvei*) (Messner *et al.*, 1995; Altman *et al.*, 1991)

α-D-Glcp
1
↓
6
β-D-Galp-(1→4)-β-D-ManpNAc-(1→[3)-β-D-Galp-(1→4)-β-D-ManpNAc-(1→]$_{n\sim20}$3)-α-L-Rhap-(1→3)-α-L-Rhap-(1→3)-α-L-Rhap-(1→3)-β-D-Galp-(1→*O*)-Tyr

α-D-Glcp
1
↓
6
β-D-Galp-(1→4)-β-D-ManpNAc-(1→[3)-β-D-Galp-(1→4)-β-D-ManpNAc-(1→]$_{n\sim20}$ 3)-α-L-Rhap-(1→3)-α-L-Rhap-(1→3)-α-L-Rhap-(1→3)-β-D-Galp-(1→*O*)-Tyr
4
↑
1
GroA-(2→*O*)-PO$_2$-(*O*→4)-β-D-ManpNAc

(continued)

Table I
(Continued)

Thermoanaerobacter thermohydrosulfuricus L111-69 and L110-69 (DSM 568) (formerly *Clostridium thermohydrosulfuricum*) (Bock et al., 1994; Christian et al., 1988)

3-OMe-α-L-Rhap-(1→4)-α-D-Manp-(1→[3]-α-L-Rhap-(1→4)-α-D-Manp-(1→]$_{n\sim27}$ 3)-α-L-Rhap-(1→3)-α-L-Rhap-(1→3)-β-D-Galp-(1→O)-Tyr

Thermoanaerobacter thermohydrosulfuricus S102-70 (formerly *Clostridium thermohydrosulfuricum*) (Christian et al., 1993; Messner et al., 1992a)

β-D-Galf-(1→3)-α-D-Galp-(1→2)-α-L-Rhap-(1→3)-α-D-Manp-(1→3)-α-L-Rhap-(1→3)-β-D-Glcp-(1→O)-Tyr

Thermoanaerobacter thermohydrosulfuricus L77-66 (DSM 569) and L92-71 (formerly *Clostridium thermohydrosulfuricum*) (Altman et al., 1992)

[1→3)-α-D-GalpNAc-(1→3)-α-D-GalpNAc-(1→]$_{n\sim25}$ O-glycosidic bond via Tyr ?
```
                      4
                      ↑
                      1
α-D-GlcpNAc-(1→2)-β-D-Manp
```

Thermoanaerobacterium thermosaccharolyticum D120-70 (formerly *Clostridium thermosaccharolyticum*) (Altman et al., 1990)

[1→3)-β-D-Manp-(1→4)-α-L-Rhap-(1→3)-α-D-Glcp-(1→4)-α-L-Rhap-(1→]$_{n}$ O-glycosidic bond via Tyr
```
                               2
                               ↑
                               1
                           α-D-Galp
```

Thermoanaerobacterium thermosaccharolyticum E207-71 (*Clostridium thermosaccharolyticum*) (Altman et al., 1995)

[1→4)-β-D-Galp-(1→4)-β-D-Glcp-(1→4)-α-D-Manp-(1→]$_{n\sim17}$ O-glycosidic bond via Tyr
```
                               3
                               ↑
                               1
                           α-L-Rhap
```

β-D-Quip3NAc-(1→6)-β-D-Galf-(1→4)-α-L-Rhap

Clostridium symbiosum HB25 (Messner et al., 1990)

[1→6)-α-D-ManpNAc-(1→4)-β-D-GalpNAc-(1→3)-α-D-BacpNAc-(1→4)-α-D-GalpNAc-(1→O)-PO$_2$-(O→]$_{n\sim15}$

Lactobacillus buchneri 41021/251 (Möschl et al., 1993)

α-D-Glcp-(1→6)-[α-D-Glcp-(1→6)-]$_{n=4\sim6}$ -α-D-Glcp-(1→O)-Ser

Archaea

Halobacterium halobium R₁M₁ (Sumper and Wieland, 1995)

$$\underset{\mid}{OSO_3^-} \qquad \underset{\mid}{OSO_3^-} \qquad Ala\text{-}NH_2$$

$$[\rightarrow 4)\text{-GlcNAc-}(1\rightarrow 4)\text{-GalA-}(1\rightarrow 3)\text{-GalNAc-}(1]_{n=10-15}\rightarrow N)\text{-Asn}$$

$$\underset{\mid}{\overset{6}{\underset{\uparrow}{}}} \qquad \underset{\mid}{\overset{3}{\underset{\uparrow}{}}} \qquad \underset{\mid}{Ala}$$

$$\underset{3\text{-}OMe\text{-GalA}}{1} \qquad \underset{Galf}{1} \qquad \underset{Ser}{\mid}$$

$$GlcA\text{-}(1\rightarrow 4)\text{-GlcA-}(1\rightarrow 4)\text{-GlcA-}(1\rightarrow 4)\text{-}\beta\text{-}D\text{-Glc-}(1\rightarrow N)\text{-Asn} \qquad \text{1/3 of GlcA residues can be replaced by IdA}$$

$$\underset{\mid}{\overset{3}{\underset{\uparrow}{}}} \qquad \underset{\mid}{\overset{3}{\underset{\uparrow}{}}} \qquad X$$

$$\underset{OSO_3^-}{} \qquad \underset{OSO_3^-}{} \qquad \underset{Thr/Ser}{\mid}$$

Haloferax volcanii DS2 (Sumper et al., 1990)

$$\beta\text{-}D\text{-Glc-}(1\rightarrow [4)\text{-}\beta\text{-}D\text{-Glc-}(1\rightarrow]_{n=8}\ 4)\text{-}\beta\text{-}D\text{-Glc-}(1\rightarrow N)\text{-Asn}$$

$$\alpha\text{-}D\text{-Glc-}(1\rightarrow 2)\text{-Gal-}(1\rightarrow O)\text{-Thr}$$

$$Glc\text{-}(1\rightarrow 2)\text{-Gal-}(1\rightarrow O)\text{-Thr}$$

Methanothermus fervidus V24S (Kärcher et al., 1993)

$$3\text{-}OMe\text{-}\alpha\text{-}D\text{-Manp-}(1\rightarrow 6)\text{-3-}OMe\text{-}\alpha\text{-}D\text{-Manp-}(1\rightarrow [2)\text{-}\alpha\text{-}D\text{-Manp-}(1\rightarrow]_{n=3}\ 4)\text{-}D\text{-GalNAc-}(1\rightarrow N)\text{-Asn}$$

Methanosaeta soehngenii FE (formerly *Methanothrix soehngenii*) (Pellerin et al., 1990)

Oligosaccharide-Rha-(1→N)-Asn

[a]*Abbreviations*: Glc*p*, glucose (pyranose form); Gal*f*, galactose (furanose form); Man, mannose; Rha, rhamnose; GlcNAc, *N*-acetylglucosamine; GalNAc, *N*-acetylgalactosamine; ManNAc, *N*-acetylmannosamine; Fuc3NAc, 3-*N*-acetylfucosamine; Qui3NAc, 3-*N*-acetylquinovosamine (3-acetamido-3,6-dideoxyglucose); BacNAc, *N*-acetylbacillosamine (2-acetamido-4-amino-2,4,6-trideoxyglucose); GlcA, glucuronic acid; GalA, galacturonic acid; IdA, iduronic acid; ManA, mannuronic acids; 3-*O*Me-GalA, 3-*O*-methylgalacturonic acid; *O*Me, *O*-methyl; SO₄²⁻, sulfate; PO₄²⁻, phosphate; Asn, asparagine; Thr, threonine; Ser, serine; Tyr, tyrosine; Ala, alanine; Gly, glycine; Asp, aspartic acid; X, interchangeable amino acid.

glycoprotein of *Halobacterium halobium* was the first prokaryotic glycoprotein to be analyzed in detail (Mescher and Strominger, 1976a,b), the occurrence of glycosylated prokaryotic proteins is not limited to S-layers. Glycoproteins have been reported to occur at different locations in the bacterial cell. They include, for example, cytoplasmic membrane glycoproteins of *Thermoplasma acidophilum* (Yang and Haug, 1979), exoenzymes such as cellulases and xylanases (Sandercock *et al.*, 1994), glycosylated flagella (Virji, 1998; Jarrell *et al.*, 1996; Stimson *et al.*, 1995), or secreted antigens of *Mycobacteria* (Dobos *et al.*, 1996; Herrmann *et al.*, 1996). A number of reviews have been published on occurrence, structure, function, biosynthesis, and genetics of prokaryotic glycoproteins, providing a general picture of the architecture of prokaryotic glycoproteins, particularly of S-layer glycoproteins of archaea and bacteria (Kandler and König, 1998; Messner, 1997; Moens and Vanderleyden, 1997; Sumper and Wieland, 1995; Sandercock *et al.*, 1994; Erickson and Herzberg, 1993; Messner and Sleytr, 1991; Lechner and Wieland, 1989; König, 1988; Sumper, 1987). As far as they have been examined, the structures of prokaryotic glycoproteins are very different from those of eukaryotic glycoproteins (for reviews, see Kobata, 1984; Kornfeld and Kornfeld, 1980). A list of known S-layer glycan structures is given in Table I and includes the various linkages of S-layer glycans to the S-layer polypeptide chains. While the N-glycosidic linkage between N-acetylglucosamine (GlcNAc) and asparagine is highly conserved in all eukaryotic organisms (Kobata, 1984; Kornfeld and Kornfeld, 1980), a greater variety of N-glycosidic linkages has been found in S-layer glycoproteins (Fig. 1, Table I). Similarly, a wider variety exists in the O-linked glycoprotein glycans of prokaryotic origin (for review, see Messner, 1997). Thus, glycosylation of the S-layer proteins may partly reflect the evolutionary adaptation of prokaryotic organisms to different environments, hostile habitats, and ecological niches.

In this chapter, we summarize the present knowledge of the structure, chemistry, function, genetics, and biotechnological applications of bacterial S-layer glycoproteins ranging from a historical perspective to new applications in the future. See Chapter 1, this volume for a review of current knowledge concerning non-S-layer glycoproteins.

2. BACTERIAL S-LAYER GLYCOPROTEINS

2.1. Chemical Composition and Structure

At the time of the first report of an archaeal S-layer glycoprotein (Mescher and Strominger, 1976a) initial evidence was provided for the existence of glycosylated S-layers in bacteria (Sleytr and Thorne, 1976). The authors had examined

a   S-Layer Glycoproteins   **b**   LPS-O-Antigens

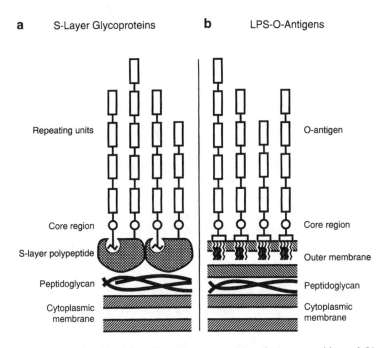

Figure 2. Schematic drawing of the cell envelope composition of (a) gram-positive and (b) gram-negative bacteria underlining structural similarities between S-layer glycoproteins and lipopolysaccharides. Modified from Messner (1996).

the hexagonal and square S-layer lattices of *Thermoanaerobacter* (formerly *Clostridium*) *thermohydrosulfuricus* strain L111–69 and *Thermoanaerobacterium* (formerly *Clostridium*) *thermosaccharolyticum* strain D120–70. By sodium dodecyl sulfate–polyacrylamide gel electrophoresis (SDS-PAGE) analysis of both strains, only the S-layer protein band was found to be glycosylated. Chemical analyses have revealed rhamnose and hexoses as sugar constituents (Sleytr and Thorne, 1976). Since then, glycosylated bacterial S-layer proteins have been reported almost exclusively from organisms belonging to the *Bacillaceae,* including *Sulfobacillus thermosulfidooxidans* (Severina *et al.,* 1993). However, among the gram-positive organisms, exceptions are known, such as *Lactobacillus buchneri* 41021/251 (Möschl *et al.,* 1993) and possibly *Corynebacterium glutamicum* (Peyret *et al.,* 1993). Whether S-layer glycoproteins are also common in gram-negative organisms [e. g., *Aquaspirillum sinuosum* (Smith and Murray, 1990)] remains to be established.

So far, complete structural analyses of S-layer glycoprotein glycans from gram-positive bacilli have been described only by our group (Messner, 1996; Messner and Sleytr, 1991, 1992). All characterizations of glycan chains by nuclear magnetic resonance (NMR) techniques were performed on glycopeptides derived by exhaustive pronase digestion of the respective S-layer glycoproteins. For purification of these materials, gel permeation chromatography, ion exchange chromatography, chromatofocusing, and reversed phase high-pressure liquid chromatography (HPLC) were used (for an example, see Bock *et al.*, 1994). The glycan structures of all bacterial and archaeal S-layer glycoproteins investigated so far are listed in Table I. It can be concluded from these data that the glycans of most bacterial S-layer glycoproteins consist of up to 50 repeating units. Their structures, however, differ considerably (see Messner, 1996; Messner and Sleytr, 1991, 1992) (Table I). It is interesting to note that among the unusual monosaccharide constituents of S-layer glycans, sugars have been found that typically occur in *O*-antigens of lipopolysaccharides of gram-negative bacteria (Fig. 2) (for reviews see Knirel and Kochetkov, 1994; Rocchetta *et al.*, 1999). Examples are quinovosamine (Altman *et al.*, 1995), D-rhamnose and *N*-acetyl-D-fucosamine (Schäffer *et al.*, 1999a; Kosma *et al.*, 1995a). Further support for the notion that S-layer glycoproteins of gram-positive bacteria and lipopolysaccharides of gram-negative bacteria are at least structurally related came from the recent observation of heptose residues as components of S-layer glycans. The repeating unit of *Aneurinibacillus* (formerly *Bacillus*) *thermoaerophilus* strain DSM 10155 (Fig. 1) corresponds to the disaccharide →4)-α-L-Rha*p*-(1 → 3)-β-D-*glycero*-D-*manno*-Hep*p*-(1→ (Kosma *et al.*, 1995b). In contrast to the common L-*glycero*-D-*manno*-heptose occurring in lipopolysaccharide (LPS) cores, this heptose residue is in the D-*glycero*-D-*manno*-configuration, which has been observed frequently in *O*-antigens of LPS (Knirel and Kochetkov, 1994). In LPS biosynthesis, D-*glycero*-D-*manno*-heptose is the precursor of L-*glycero*-D-*manno*-heptose. The former is converted to the final product by epimerization through the action of ADP-L-*glycero*-D-*manno*-heptose-6-epimerase (Rfa D) (for reviews, see Raetz, 1996; Schnaitman and Klena, 1993).

        Characterization of the carbohydrate–protein linkage regions of different thermophilic and mesophilic *Bacillus* and *Thermoanaerobacter* (formerly *Clostridium*) strains also led to the characterization of novel linkage types. The first observation of a tyrosine-linked glycan chain had been reported in *Thermoanaerobacter thermohydrosulfuricus* S102–70 (Messner *et al.*, 1992a). Its glycan structure was determined by a combination of ^{1}H and ^{13}C NMR experiments on proteolytically derived glycopeptides. In contrast to the extended polysaccharide chains of most bacterial strains (Table I), the S-layer glycans of this organism consist of short hexasaccharide chains that are attached to the S-layer polypeptide by alkali-stable *O*-glycosidic linkages between β-D-glucose and tyrosine residues. Whether the short carbohydrate chains represent only a core structure without repeating units, which originates either from a mutation or an enzyme defect, is not known. Calculation of the number of glycosylation sites indicated that four to five heterosaccharide chains are present on one S-layer protein protomer (Christian *et al.*, 1993). The number of glycosylation sites was substantiated by the results of analyses on glycans from *T. thermohydrosulfuricus* L111–69 and L110–69 (Bock *et al.*, 1994). After proteolytic digestion of the intact S-layer glycoprotein, four glycopeptides have been isolated that differ in the amino acid compositions of their peptide portions. However, tyrosine was always present as the linkage amino acid. Based on their different hydrophobicity, the glycopeptides were separated by reversed phase HPLC (Bock *et al.*, 1994). The glycan chains have identical constituents, but the number of disaccharide repeats of the structure →(3)-α-L-Rha*p*-(1 → 4)-α-D-Man*p*-(1→ varies between 23 and 33, with a maximum at 27 repeats. At the nonreducing end, the chain is capped by a disaccharide unit with a modified rhamnose residue (3-*O*-methylrhamnose). The same type of capping with a terminal 3-*O*-methylrhamnose residue was observed in the S-layer glycan of *Aneurinibacillus* (formerly *Bacillus*) *thermoaerophilus* GS4–97 (Schäffer *et al.*, 1999a). The presence of 3-*O*-methylated sugar residues seems to play a role as termination signal for chain elongation, as was also discussed with other carbohydrate chains (for example, Gerwig *et al.*, 1991). In *T. thermohydrosulfuricus* L111–69 the polysaccharide chains are linked to the S-layer protein via a core consisting of three α1,3-linked rhamnoses and a reducing terminal β-D-galactose residue. The latter is *O*-linked to tyrosine residues of the S-layer protein (Bock *et al.*, 1994). The β-D-galactose-tyrosine linkage has not been observed before in glycoproteins. Subsequently, tyrosine-linked glycan chains also were found in the S-layer glycoprotein of *Thermoanaerobacter* (formerly *Acetogenium*) *kivui;* the glycan structure and linkage sugars of this glycoprotein have not been reported (Lupas *et al.*, 1994; Peters *et al.*, 1992). For this organism, four glycosylation sites per S-layer protomer were deduced from sequencing experiments. The structures of the linkage regions known so far are indicated in Table I.

Of particular interest was the observation that, in *Paenibacillus* (formerly *Bacillus*) *alvei* CCM 2051, the S-layer glycan is linked to the polypeptide via a similar core structure as in *T. thermohydrosulfuricus* L111–69, although the structures of the repeating units of both organisms are completely different (Messner *et al.*, 1995). The glycan examination was performed on S-layer glycopeptides by a combination of proton, carbon, and phosphorus NMR techniques. While the signals attributable to the repeating units are generally large, the very small signals from the core structures can cause serious problems with the interpretation of the data.

As a general principle the structural organization of bacterial S-layer glycoproteins can be seen to resemble the architecture of the polysaccharide portions of LPS (Fig. 2) (for reviews, see Whitfield, 1995; Schnaitman and Klena, 1993; Raetz, 1990). Strain-specific oligosaccharide chains are attached to the S-layer protein via identical cores. However, not all bacterial S-layer glycans are linked to the respective S-layer proteins via tyrosine residues. In *Aneurinibacillus thermoaerophilus* strains L420–91 (Kosma *et al.*, 1995a), GS4–97 (Schäffer *et al.*, 1999a) (Fig. 3), and strain DSM 10155 (Wugeditsch *et al.*, 1999), for example, the glycans are linked to the S-layer polypeptides by novel *O*-glycosidic linkages between β-D-GalNAc and threonine/serine residues. This is in contrast to eukaryotic glycoproteins where only α-D-GalNAc-residues have been found to exist as the linkage sugars of *O*-linked glycans (for review, see Vliegenthart and Montreuil, 1995). Additionally, for the first time, an unexpected heterogeneity was observed in the sugar composition of the core structures of different *Aneurinibacillus* strains (Schäffer *et al.,* 1999a; Wugeditsch *et al.,* 1999). Up to now, among S-layer glycoproteins, heterogeneity was observed only regarding the number of repeating units of the glycan chains (Bock *et al.,* 1994). In *A. thermoaerophilus* strain DSM 10155, for example, not only 1,3-linked complete core structures consisting of two α-L-rhamnose and the β-D-GalNAc residues are present, but also, truncated forms of the core were found in almost equal amounts. They consisted either of one Rha and the GalNAc residue or GalNAc alone, linked either to a threonine or a serine residue of the S-layer polypeptide (Wugeditsch *et al.,* 1999). It is speculated that, despite the observed heterogeneity, the transfer of the individual sugar units could be effected by a relatively simple enzyme system such as a single transferase because, for α-L-Rha*p* and β-D-Gal*p*NAc, the geometry for attachment to carbon 3 of either sugar is very similar (Shashkov *et al.,* 1988). So far, variability of core structures was only observed on S-layer glycoproteins of *A. thermoaerophilus* strains and may reflect a genus-specific feature (Wugeditsch *et al.,* 1999; Schäffer *et al.,* 1999a).

The S-layer glycan chains protruding some 30 to 40 nm from the surface of *A. thermoaerophilus* DSM 10155 were directly demonstrated by electron microscopy of thin sections of Lowicryl-embedded whole cells of this organism (Wugeditsch, 1998). Comparable results had been obtained earlier by labeling of the glycan chains of *T. thermohydrosulfuricus* L111–69 with polycationic ferritin after derivatizing the sugars with carboxylate groups by succinylation. Subsequent to this modification, two to three ferritin molecules were bound to the carbohydrate chains. This is in good agreement with an extension of 30 to 40 nm above the cell surface (Sára *et al.,* 1989).

In some bacterial strains such as *Bacillus stearothermophilus* NRS 2004/3a (Messner *et al.,* 1987) and *Thermoanaerobacterium* (formerly *Clostridium*) *thermosaccharolyticum* strains D120–70 (Altman *et al.,* 1990) and E207–71 (Altman *et al.,* 1996), more than one carbohydrate structure was found in the respective S-layer glycoprotein preparations. In previous studies, Wieland and co-workers (for review, see Sumper and Wieland, 1995) had demonstrated that in the archaea *Halobacterium halobium* and *Haloferax volcanii* several structurally different glycan chains are covalently linked to the S-layer proteins. Therefore, a similar assembly principle was supposed to be present in bacterial S-layers. Careful reexaminations of the linkage regions of the bacterial S-layer glycoproteins demonstrated that in these strains only one of the glycan chains represents a true glycoprotein, that is, is covalently linked to the S-layer polypeptide. The other

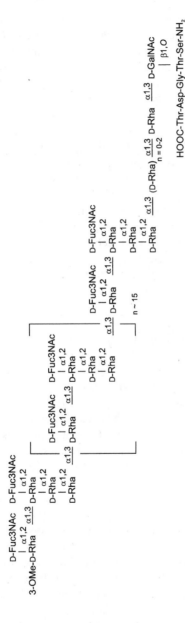

Figure 3. Structure of a typical bacterial S-layer glycoprotein. As an example the S-layer glycoprotein of *Aneurinibacillus thermoaerophilus*, strain GS4–97 is presented (Schäffer *et al.*, 1999a). Reprinted with permission of Oxford University Press.

saccharide chain is that of an accessory secondary cell wall polymer associated with the peptidoglycan sacculus. For example, the diacetamidodideoxyuronic acid-containing glycan of *Bacillus stearothermophilus* NRS 2004/3a (Messner *et al.*, 1987), which on average is composed of six repeating units, is linked via pyrophosphate bridges to about 20–25% of the muramic acid residues of the peptidoglycan sacculus (Schäffer *et al.*, 1999b). The identified linkage region of the Man$p$A2,3(NAc)$_2$-containing glycan, together with the observation that this glycoconjugate can be separated from the S-layer glycoprotein by gel permeation chromatography, confirms that this glycan is indeed an accessory cell wall polymer. Since similar Man$p$A2,3(NAc)$_2$-containing glycans also have been observed in other *B. stearothermophilus* strains such as strains PV72/p6 and ATCC 12980 (Egelseer *et al.*, 1998), it can be assumed that this type of cell wall polymer represents a genus-specific secondary cell wall polymer of all *B. stearothermophilus* wild-type strains. This observation implies that, in bacteria, only one type of glycan is covalently linked to the S-layer protein. Possible biological functions of the secondary cell wall polymers in bacilli recently have been discussed in the context of mediating binding of the S-layer to the peptidoglycan sacculus of *B. stearothermophilus* PV72/p2 (Sára *et al.*, 1998; Ries *et al.*, 1997). Whether this is also the case with strain *B. stearothermophilus* NRS 2004/3a remains to be investigated.

For several *Bacillus* strains, secondary cell wall polymers containing cores with the structure →3)-β-D-Man$p$NAc-(1 → 4)-β-D-Glc$p$NAc-(1→ between the oligosaccharide chain and the pyrophosphate group have been described (Araki and Ito, 1989; Kojima *et al.*, 1985; Kaya *et al.*, 1984). Similar cores were also found in typical ribitol and glycerol teichoic acids of gram-positive organisms (for review, see Munson and Glaser, 1981). In two *T. thermosaccharolyticum* strains (Altman *et al.*, 1990, 1996), glycans with identical backbone structures but strain-specific side chain sugars have been found. Currently, their linkage regions are investigated in detail, but from the data accumulated, it is highly probable that they also represent secondary cell wall polymers and not S-layer glycoproteins with different glycan structures.

The S-layer of *Sulfobacillus thermosulfidooxidans* is a glycoprotein containing approximately 10% carbohydrate (Severina *et al.*, 1993). Mannose is the main component, but additionally, glucosamine, glucose, xylose and galactose have been found in amounts decreasing in that order. The amino acid composition was found to be typical for an S-layer protein with a large proportion of hydrophobic residues, 21.5% acidic and 8.2% basic amino acids (Sleytr and Messner, 1983).

In other gram-positive bacteria, characterizations of glycosylated S-layer proteins have been performed in somewhat lesser detail. For example, chemical characterization of a peritrichously flagellated organism with a hexagonal S-layer lattice from St. Lucia hot springs by SDS-PAGE and subsequent staining by Alcian blue or thymol–sulfuric acid methods has indicated the presence of an S-layer glycoprotein with an apparent molecular weight of 200,000 (Karnauchow *et al.*,

1992). Physiological tests revealed that this organism belongs to the genus *Clostridium*. The characterization of the *cspB* gene encoding PS2, the S-layer protein of *Corynebacterium glutamicum,* indicated a molecular weight of approximately 63,000 for the mature protein (Peyret *et al.,* 1993). By contrast, the calculated molecular mass for the 510 amino acid-polypeptide is 55,426 Da. Seven potential glycosylation sites are present in the S-layer sequence. However, the actual presence of glycan chains, possibly explaining the differences between calculated and observed molecular masses, remains to be established.

Recently, two major outer membrane proteins were extracted from the gram-negative vent prosthecate bacterium *Hyphomonas jannaschiana*. The results of the examination suggest that p116 and p29 are S-layer glycoproteins, with p116 being a tetramer of p29 (Shen and Weiner, 1998).

2.2. Biosynthesis

S-Layer glycans and LPS *O*-antigens may be expected to generate comparable hydrophilic environments on the cell surfaces of the respective bacterial cells (Fig. 2). Investigations of the glycosylation of S-layer proteins also should include aspects of functional similarities between these structures. Presently, studies have been initiated to identify similarities between the biosynthetic pathways of S-layer glycoprotein glycans from gram-positive bacteria and *O*-antigens of LPS from gram-negative bacteria.

The characterization of the biosynthesis of S-layer glycan chains in bacteria had been initiated on *Paenibacillus* (formerly *Bacillus*) *alvei* CCM 2051. Characterization of the isolated intermediates showed that, in addition to nucleotide-bound monosaccharides, nucleotide-activated oligosaccharides were also present in the cytoplasm (Hartmann *et al.,* 1993). This observation is in agreement with data from archaeal S-layer glycoproteins (Hartmann and König, 1989). However, it is in contrast to other glycoconjugates where oligosaccharide intermediates are found only in the lipid-bound state [e.g., LPS biosynthesis (Raetz, 1990)]. C_{55}-Dolichol was identified as the lipid carrier in place of the common prokaryotic lipid carrier, undecaprenol. In the activated precursors additional sugars such as three GlcNAc residues have been found that are not present in the mature glycan of the S-layer glycoprotein of *P. alvei* (Altman *et al.,* 1991). The location of the presumed trimming reaction is presently not known (Hartmann *et al.,* 1993).

Based on these experiments, growing cells of *P. alvei* CCM 2051 were metabolically labeled with [^{14}C]glucose and analyzed for radioactive compounds. After pronase digestion of S-layer glycoproteins, S-layer glycopeptides were obtained that contained all radioactivity in the glycan chains. The labeled sugars were identified after hydrolysis of this material and anion exchange chromatography on a Dionex system. Only Glc, Gal, and ManNAc were labeled, which are present in the repeating units of the glycan chain (Altman *et al.,* 1991). When crude cell extracts of *P. alvei* were used as a source of enzymes, radiolabeled sugar nucleotides (UDP–[^{14}C]glucose and UDP–[^{14}C]galactose) were incorporated into the same lipophilic substance. This was demonstrated by thin layer chromatography using organic solvent systems. Analysis of the nucleotide-activated monosaccharides had shown that the activated form of ManNAc is GDP-ManNAc. To have the nucleotide sugars from the sugars of the mature S-layer glycan available for biochemical assays, GDP-ManNAc was chemically synthesized for the first time (Klaps *et al.*, 1996). During the studies we experienced considerable interference with the activated sugar compounds required for the synthesis of the secondary cell wall polymer (C. Neuninger and P. Messner, unpublished observations). Therefore, no conclusive results on S-layer glycan synthesis in *P. alvei* CCM 2051 are presently available.

Similar biosynthetic routes for the assembly of S-layer glycan chains also have been suggested for other bacteria such as *Thermoanaerobacterium thermosaccharolyticum* E207–71 (Schäffer *et al.*, 1996; Altman *et al.*, 1995), but definitive data are not yet available. So far, only the nucleotide-bound repeating unit structure has been isolated and characterized. The hexasaccharide [β-D-Qui*p*3NAc-(1 → 6)-β-D-Gal*f*-(1 → 4)-α-L-Rha*p*-(1 → 3)-]α-D-Man*p*-(1 → 4)-β-D-Gal*p*-(1 → 4)-β-D-Glc*p* is linked via the Man residue to guanosine diphosphate (GDP) (Schäffer *et al.*, 1996). Additional transient sugars, as observed in the intermediates of *P. alvei,* were not detected. The only general conclusion that presently can be drawn is that for the biosynthesis of bacterial S-layer glycans nucleotide-activated oligosaccharides play an important role. This, however, is in contrast to the well-investigated biosynthetic routes of eukaryotic glycoproteins (Kornfeld and Kornfeld, 1985).

In a parallel set of experiments, the isolation of lipid-bound precursors of the S-layer glycan biosynthesis of *T. thermosaccharolyticum* E207–71 was initiated. Isolation of lipid-linked compounds from intact or broken cells was attempted by extraction with organic solvents. Due to the complex nature of the repeating unit of this organism (see Table I) and difficulties with the mass-spectroscopic analysis, a final characterization of the lipid-linked substances has not been achieved (C. Schäffer and P. Messner, unpublished observations). However, one conclusion that already can be drawn is that short dolichols (11 to 12 prenyl units) rather than undecaprenol are the common lipid carriers in eubacterial S-layer glycoprotein biosynthesis (for reviews, see Sumper and Wieland, 1995; König *et al.*, 1994; Hartmann *et al.*, 1993).

2.3. Molecular Biology

So far, no genetic analyses have been performed of enzymes involved in the glycosylation of either archaeal or bacterial S-layer proteins. When we initiated our comparison of glycan formation in S-layer glycoproteins of gram-positive bacteria to that of O-antigens in LPS of gram-negative bacteria (see Fig 2) (Schäffer *et al.*, 1996), we were able to benefit from the considerable progress of molecular characterization of lipopolysaccharide biosynthesis in the past decade (for reviews, see Heinrichs *et al.*, 1998; Raetz, 1996; Whitfield 1995; Reeves, 1994; Schnaitman and Klena, 1993). Therefore, we chose rhamnose for our investigations, a sugar that is common to S-layer glycoproteins and LPS O-antigens (see Fig. 2, Table I) (Messner, 1996; Knirel and Kochetkov, 1994).

The nucleotide-activated form of L-rhamnose is dTDP-L-rhamnose. Its biosynthesis in enterobacteria was investigated and found to proceed in a four-step reaction sequence (Glaser and Kornfeld, 1961; Kornfeld and Glaser, 1961). The genetic studies in S-layer glycoprotein-carrying gram-positive bacteria are performed on *Aneurinibacillus thermoaerophilus* DSM 10155, which possesses a rhamnose- and heptose-containing S-layer glycoprotein with the repeating unit structure [→4)-α-L-Rhap-(1 → 3)-β-D-*glycero*-D-*manno*-Hep*p*-(1→] (Kosma *et al.*, 1995b). For comparative reasons, the four enzymes involved in dTPP-L-rhamnose formation, glucose-1-phosphate thymidyltransferase (RmlA), dTDP-D-glucose 4,6-dehydratase (RmlB), dTDP-4-dehydrorhamnose 3,5-epimerase (RmlC), and dTDP-4-dehydrorhamnose reductase (RmlD) were first overexpressed in *Escherichia coli* (Graninger *et al.*, 1999). Initial results indicate that the reaction sequence is analogous in gram-positive and gram-negative bacteria. However, Southern hybridization using polymerase chain reaction (PCR)-amplified probes for *rmlA* from *Salmonella enterica* serovar Typhimurium LT2 (Jiang *et al.*, 1991) and *A. thermoaerophilus* DSM 10155 did not result in a positive hybridization reaction (Graninger and Messner, in preparation). Recently, *rmlA* from the latter organism has been cloned by use of a degenerate probe. The analysis of the remaining three enzymes of the *rml* operon and of other enzymes of S-layer glycan biosynthesis in *A. thermoaerophilus* is under way to determine the organization of this gene cluster.

2.4. Application Potentials

The wealth of information existing on the general principle of S-layers has revealed a broad application potential. Particularly their repetitive physicochemical properties down to the subnanometer scale make S-layers unique structures for functionalization of surfaces and interfaces (for review, see Sleytr *et al.*, 1999; Pum and Sleytr, 1998; Sleytr and Sára, 1997). One specific example is the use of S-layers as combined carrier/adjuvants for vaccination and immunotherapy. For the immobilization of haptens and antigens, not only S-layer proteins but also S-layer glycoproteins have been used (for review, see Sleytr *et al.*, 1999; Unger *et al.*, 1997; Jahn-Schmid *et al.*, 1996; Malcolm *et al.*, 1993a,b). The immunogens have been attached either to the protein portion of the S-layer or to the glycan chains of S-layer glycoproteins by specific immobilization reactions such as periodate oxidation and reductive amination, divinylsulfone activation, and so on (Messner *et al.,* 1992b). This method would allow the preparation of multivalent vaccines.

From studies of the biosynthesis of S-layer glycoproteins, recombinant enzymes are now available, for example, for the production of nucleotide-activated sugars such as dTDP-L-rhamnose (Graninger *et al.,* 1999). An additional advantage should come from the use of thermophilic bacteria such as *Aneurinibacillus thermoaerophilus.* In comparison to glycan biosynthetic enzymes from mesophilic organisms such as *Salmonella enterica* we expect increased thermal stability for the enzymes from the thermophilic strains (for review, see Adams and Kelly, 1998; Danson and Hough, 1998). With the increasing knowledge of glycosylation pathways of S-layer glycoproteins and by the use of the tools of molecular biology, economically feasible approaches to intermediate products will be available in the future. Frequently, their production with recombinant enzymes will be more cost-effective than organic chemical syntheses (for review, see Ichikawa *et al.,* 1992).

3. ARCHAEAL S-LAYER GLYCOPROTEINS

3.1. Chemical Composition and Structure

The archaeal domain, positioned between prokaryotes and eukaryotes, features a host of unusual organisms, with the major groups being extreme halophiles, methanogens, and a variety of sulfur-dependent, thermophilic, and hyperthermophilic organisms. This class of organisms has unusual membrane lipids. Many of the individual strains lack a cell-shape-determining/maintaining structure equivalent to the murein sacculus of bacteria. Archaeal cells are devoid of organelles, yet they synthesize and export N-linked and O-linked glycoproteins utilizing only the cytoplasmic membrane. Among archaea, the S-layer often is the only wall component outside the cytoplasmic membrane. Although glycosylation is not obligatory in S-layer biosynthesis, most archaeal S-layer proteins are true glycoproteins (Messner, 1996). For an overview of structural information about archaeal S-layer glycoproteins, see Table I. The architecture of archaeal S-layer glycans generally comprises short linear chains of up to ten sugar residues, N-glycosidic linkages such as Glc-Asn, GalNAc-Asn, and Rha-Asn, and O-glycosidic linkages via threonine. The occurrence of conserved core-regions has not been reported (for review, see Messner, 1997).

The most detailed information is available from the S-layer glycoprotein of *Halobacterium halobium,* which is the first prokaryotic glycoprotein to be described (Mescher *et al.,* 1974). *Hb. halobium* is an extreme halophile with a salt requirement of 4 to 5 M NaCl. The cell wall consists only of an S-layer with hexagonally arranged glycoprotein subunits and this S-layer is very tightly joined to the cytoplasmic membrane (Sumper, 1993). Its extremely acidic character results from a heavily sulfated saccharide portion that is attached to the asparagine residue in position 2 of the mature S-layer glycoprotein. This saccharide, present in one copy per glycoprotein molecule, consists of pentasaccharide repeating units composed of a linear backbone of GalNAc-GalA-GalNAc repeats to which a methylated galacturonic acid and a galactofuranose are bound peripherally (Paul and Wieland, 1987). Each pentasaccharide bears two sulfate residues. The overall chain length of this glycosaminoglycanlike polysaccharide ranges between 10 and 20 repeats. The GalNAc residue at the reducing end is linked directly to an asparagine within the typical *N*-glycosylation acceptor sequence Asn-X-Thr(Ser). This glycosaminoglycanlike structure represents the only high-molecular-mass S-layer-attached saccharide chain described in archaea, so far. Ten copies of another type of sulfated saccharide with a low molecular mass are *N*-glycosidically linked to the S-layer polypeptide (Lechner *et al.,* 1985a). The linkage unit Asn-Glc is extended by a linear chain of two or three glucuronic acids, each substituted with sulfate. About one third of these glucuronic residues are found to be replaced by iduronic acid. So the glycoprotein of *Hb. halobium* offers the unique situation of being equipped with two different types of *N*-glycosidic linkages. A total of 12 potential *N*-glycosylation sites are found throughout the polypeptide chain. The third S-layer glycan of this organism is a neutral disaccharide of Glc-Gal repeats, present in about 20 copies. It occurs in highly clustered arrangements, *O*-glycosidically linked via Thr (Wieland *et al.,* 1982).

Identical *O*-glycosidically linked neutral disaccharide chains are present in *Haloferax volcanii.* The outer surface of this moderate halophile, requiring up to 2.4 M NaCl and 0.25 M Mg^{2+}, is covered with a hexagonally packed S-layer glycoprotein. The *N*-glycosidically bound saccharides consist of only nine to ten glucose residues attached to seven potential *N*-glycosylation sites via the linkage unit asparaginylglucose (Mengele and Sumper, 1992). The replacement of the highly charged sulfated oligosaccharides found in the extremely halophilic glycoprotein by completely uncharged saccharides, resulting in a drastic change of the net surface charge, most probably reflects the adaptation from a moderately to an extremely halophilic environment (Sumper, 1993). Highly negatively charged loops are required for the stabilization of the S-layer protein in high salt concentrations. Upon removal of the repeating unit saccharide in position 2, which results in a reduction of the surface charge, the bacteria are no longer rod-shaped but grow as spheres. This observation indicates a functional role of the glycosaminoglycan chain in maintaining the structure of the S-layer glycoprotein (Sumper and Wieland, 1995).

A recent microscopic study revealed hexagonal arrays on the cell surface of

Haloarcula japonica strain TR-1. The 180-kDa S-layer glycoprotein, with a total carbohydrate content of 5%, seems to be important in maintaining the characteristic triangular disk shape of this halophilic archaeon (Nakamura *et al.,* 1995; Nishiyama *et al.,* 1992). The amino acid sequence of the *Ha. japonica* cell surface glycoprotein showed 52% and 43% homologies with those from *Hb. halobium* and *Hf. vocanii,* respectively. Five potential *N*-glycosylation sites were identified in the mature cell surface polypeptide, different from those found in the two other organisms mentioned (Wakai *et al.,* 1997). So far, no compositional or structural data on the carbohydrate portion of the cell surface protein from *Ha. japonica* have been published.

Glycosylated S-layer proteins have been reported for a number of different methanogen species. The cell walls of gram-positive methanogens consist of pseudomurein or methanochondroitin. In some species an S-layer is present in addition. The gram-negative methanogenic bacteria have cell walls composed of single-layered crystalline protein or glycoprotein subunits, forming an S-layer (König *et al.,* 1993). However, in most cases, no detailed information is available on the S-layer glycan chains, as the characterizations are limited to carbohydrate-staining reactions in SDS-PAGE. Examples are the 138-kDa S-layer glycoprotein of *Methanoculleus marisnigri* (Bayley and Koval, 1994) and the 135-kDa hexagonal S-layer glycoprotein of *Methanoplanus limicola* (Cheong *et al.,* 1991). Carbohydrates were also detected in *Methanoculleus liminatans* (Zellner *et al.,* 1990), *Methanocorpusculum* spp. (Zellner *et al.,* 1989c), and *Methanolacinia paynerti* (Zellner *et al.,* 1989a). For *Methanosaeta soehngenii* (formerly *Methanothrix soehngenii*), an oligosaccharide was demonstrated to be bound via asparaginyl-rhamnose to the S-layer polypeptide (Pellerin *et al.,* 1990).

The extremely halophilic archeon *Methanothermus fervidus* possesses a double-layered cell envelope. The inner pseudomurein sacculus is covered by an S-layer with hexagonal symmetry, which is assembled from glycoprotein subunits (König *et al.,* 1993; Nußer *et al.,* 1988). The components of the branched repeating unit saccharide were determined to be 3-*O*-methylmannose and mannose in the molar ratio of 2:3. This heterosaccharide is linked via *N*-acetylglucosamine to an asparagine residue of the peptide moiety (Kärcher *et al.,* 1993). The mature S-layer polypeptide contains a total of 20 sequons, that is, potential *N*-glycosylation sites (Bröckl *et al.,* 1991).

Some of the extremely thermophilic, sulphur-metabolizing archaea contain glycoproteins. Thus, a 126-kDa S-layer glycoprotein has been described in *Pyrodictium abyssi* (Pley *et al.,* 1991). The S-layer glycoprotein of *Sulfococcus mirabilis* contains 10% of carbohydrates (Bashkatova *et al.,* 1991). The double S-layer of *Thermococcus stetteri* is assembled from two major glycoproteins with apparent molecular weights of 80,000 and 210,000 (Gongadze *et al.,* 1993; Miroshnichenko *et al.,* 1989). The cells of *Archaeoglobus fulgidus* are also covered by a glycoprotein S-layer (Zellner *et al.,* 1989b).

In a recent study, the cell envelope of *Sulfolobus* spp. was characterized as a complex of interacting proteins and membrane components (Grogan, 1996a). The S-layer sacculi of various isolates consist of at least two dissimilar glycoprotein subunits, SP1 and SP2. They differ significantly in regard to their electrophoretic mobilities, amino acid compositions, and apparent molecular masses (65 kDa and 135 kDa, respectively) (Grogan, 1989, 1996b). The primary structural role of SP1 is to form the highly ordered S-layer covering the cell; it is associated with the cell envelope by polar (protein–protein) interactions only. SP2, which is attached to the *S. acidocaldarius* sacculus by strong polar and hydrophobic interactions, anchors the S-layer to the cell membrane (Grogan, 1996a).

The scaffold of the surface layer covering *Staphylothermus marinus* is formed by an extended filiform glycoprotein complex, termed tetrabrachion. This complex is anchored to the cell membrane through one end of a 70-nm stalk. At the other end, it branches into four arms of 24-nm length (Peters *et al.,* 1996). The arms, forming a canopylike meshwork, enclose a quasi-periplasmic space, which is also observed in other hyperthermophiles. The tetrabrachion was shown to be composed of two highly glycosylated polypeptides with estimated molecular masses of 85 kDa and 92 kDa, respectively. Two molecules of a 150-kDa protease are associated with the tetrabrachion. The tetrabrachion–protease complex with an overall carbohydrate content of 38% forms a right-handed coiled coil (Peters *et al.,* 1995). To date, no structural information is available on the glycan portion of this thermostable complex.

3.2. Biosynthesis

Most of the studies of S-layer glycoprotein biosynthesis have been performed on archaea. As this class of organisms is devoid of organelles, it is apparent that the biosynthetic machinery for glycoproteins must have evolved much earlier than cellular compartments such as the endoplasmic reticulum or the Golgi apparatus. Thus, plasma membranes of archaea are likely to contain a rudimentary machinery for glycosylation of proteins. This would include proteins for transport and elongation of sugar chains. It can be assumed that the archaeal glycosylation pathway, including N-glycosylation, O-glycosylation, and glycolipid synthesis, takes place at the outside of the plasma membrane, analogous to the lumen of the endoplasmic reticulum in eukaryotes (Zhu and Laine, 1996; Sumper, 1987). Activities for the formation of lipid-linked sugar compounds have been unequivocally demonstrated in the archaeal pathway of S-layer glycoprotein biosynthesis. The report of the conserved Asn-X-Thr (Ser) code for *N*-glycosylation in archaea indicates that this code appeared before the evolutionary divergence of archaea and bacteria (Woese and Fox, 1977). Apparently, in some species of archaea, such as *Haloferax volcanii,* typical prokaryotic mechanisms of glycosylation are operating, whereas others seem to be much closer to eukaryotes in this regard.

Functionally, the cell surface of halobacteria resembles the lumen of the en-

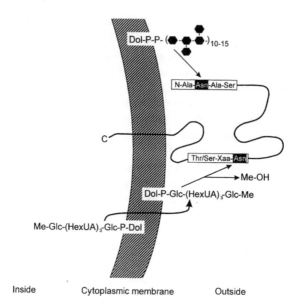

Figure 4. Proposed biosynthesis of the N-linked S-layer glycoproteins of *Halobacterium halobium* (Sumper and Wieland, 1995). Reprinted with permission of the author and Elsevier Science Publishers.

doplasmic reticulum of eukaryotes: Proteins are first translocated through a membrane and thereafter *N*-glycosylated. The concluding scheme of the biosynthesis of the sulfated S-layer glycoprotein of halobacteria proposed by Sumper and Wieland (1995) is depicted in Fig. 4.

As already mentioned, two different *N*-glycosidic linkages are synthesized within the complex glycoprotein of *Hb. halobium*, and the two biosynthetic pathways differ in the type of saccharide precursors involved. The repeating unit saccharide chain is completed in a lipid-linked state including sulfatation, and then transferred *en bloc* to the nascent protein chain (Wieland *et al.*, 1981). This implies that the glycosaminoglycan is assembled on the lipid carrier by polymerization of preformed pentasaccharides, a mechanism similar to the one described for the biosynthesis of the lipopolysaccharide *O*-antigen. The lipid used most probably is a dolichol diphosphate. The transfer of the carbohydrate chain to the Asn-Ala-Ser acceptor peptide results in the linkage unit GalNAc-Asn (Paul *et al.*, 1986). In the case of the second *N*-glycosidic linkage unit in halobacteria, Glc-Asn, completely sulfated lipid-linked precursors are produced on the cytosolic side of the cell membrane. The reducing sugar is linked to a C_{60}-dolichol carrier via a monophosphate (Sumper and Wieland, 1995). Thereafter, the oligosaccharides are translocated to the cell surface, possibly by a mechanism that involves a transient methylation of their peripheral glucose residues (Lechner *et al.*, 1985b). Finally, for both *N*-glycans, transfer to the protein with the generation of the *N*-glycosyl linkages would occur at the cell surface (Lechner and Wieland, 1989). With the mechanism described (Fig. 4), a nonglycosylated protein core can be translocated through the cell membrane, and the corresponding glycoconjugates may pass this membrane in their lipid-linked stage. Nevertheless, questions concerning the translocation of the glycosaminoglycan to the cell surface and its blockwise assembly still remain. The mechanism of the assembly of the O-glycosyl units in halobacteria also is unknown. In eukaryotes, glycoconjugate glycans are covalently modified at the protein-linked level in the Golgi apparatus, where trimming, further glycosylation, sulfatation, or epimerization occurs. Interestingly, despite the absence of compartments and of a secretory pathway for glycoproteins, halobacterial biosynthesis is essentially similar to glycoprotein biosynthesis in eukaryotes (Sumper and Wieland, 1995).

In a recent study, glycosyl transfer enzymes from *Haloferax volcanii* were shown to possess unique properties: they require high salt, 2.5 M NaCl and 0.25 M $MgCl_2$ gave optimal activity, low salt conditions irreversibly denature the enzymes, and unlike other transferases these enzymes are strongly inhibited by detergents. They are presumed to be located on the inner face of the plasma membrane where sugar nucleotides would be available (Zhu *et al.*, 1995). *Hf. volcanii* seems to utilize UDP-Glc for the synthesis of Glc-phosphoryl-polyisoprenol, which then functions as a donor of glucose to the S-layer glycoprotein acceptor. The occurrence of C_{55}-phosphodolichol-linked oligosaccharides was also confirmed in *Hf. volcanii*. It is very likely that the lipid-activated oligosaccharides are transferred to a suitable S-layer protein acceptor sequence. The transfer of glucose from polyprenyl intermediates to glycoprotein and glycolipid products is inhibited by amphomycin and by two recently described sugar nucleotide analogues, PP36 and PP55 (Zhu *et al.*, 1995). In eukaryotes, all three inhibitors are reported to block the transfer of sugar from UDP sugars to phosphopolyisoprenols. In *Hf. volcanii*, small amounts of dolichol phosphate-linked monosaccharides were detected. This leads to the assumption that the availability of these compounds rather than of free dolichol phosphate limits the process of glycosylation in halobacteria, as has been suggested by Potter *et al.* (1981). Thus, the use of dolichol as saccharide carrier is a common feature of halobacteria and eukaryotes. However, there are some differences (Kuntz *et al.*, 1997). First, in the dolichol from *Hf. volcanii*, the ω-terminal isoprene unit is saturated. Furthermore, the oligosaccharides are found to be linked via a monophosphate bridge, whereas in eukaryotes oligosaccharides are exclusively linked via pyrophosphate. Finally, the halobacterial cells possess a limited spectrum of dolichol species containing only 11 or 12 isoprene units. This is in contrast to the broad spectrum of long-chain dolichols (14 to 23 isoprene residues) present in eukaryotes.

Based on isolated precursors, a biosynthetic pathway for the S-layer glycoprotein of *Methanothermus fervidus* was proposed by Hartmann and König (1989). The biosynthesis of the glycan chain starts with the formation of the C1 phosphate derivatives of the constituting sugars, which are converted into the corresponding nucleoside diphosphate-activated derivatives. Besides these precursors, oligosaccharides with rather complex compositions have been found to occur as UDP-linked derivatives. Processing (methylation) of the oligosaccharides seems to take place at the UDP-linked level. But unlike in halobacteria, this methylation is not transient; it is found as a stoichiometric component in the mature S-layer glycoprotein. At a subsequent stage, the oligosaccharides are probably transferred from UDP to short-chain C_{55}-dolichol diphosphate. Finally, a transient glycosylation with six or eight glucose residues seems to occur of the lipid-linked oligosaccharide. These sugar residues are removed during further processing, as they are absent in the mature S-layer glycoprotein. Thus, the transient addition of glycosyl units during glycoprotein biosynthesis is a feature common to archaea and eukaryotes.

3.3. Molecular Biology

Until now, the genes encoding the S-layer glycoproteins of *Hb. halobium, Hf. volcanii, M. fervidus,* and *Ha. japonica* have been cloned and sequenced. By the identification and sequence analysis of the corresponding genes, the primary structures of the S-layer glycoproteins from those organisms were deduced.

Lechner and Sumper (1987) cloned and sequenced the halobacterial S-layer gene. The polypeptide chain of the mature glycoprotein from *Hb. halobium* consists of 818 amino acids. The open reading frame encodes an N-terminal leader peptide of 34 amino acid residues, reminiscent of a typical signal peptide. The mature S-layer polypeptide from *Hf. volcanii* contains 793 amino acids and its amino acid sequence was deduced from the cloned cDNA (Sumper *et al.,* 1990). The 90-kDa proteins from both these species consist mainly of polar and negatively charged amino acids; they show distinct regions of homology, particularly a 21 amino acid hydrophobic domain at the C-terminus, which is presumed to serve as a membrane anchor. Close to this membrane binding domain, a cluster of 14 threonine residues was identified, all of which are *O*-glycosylated. As known from eukaryotic glycoproteins, the amino acid sequence Asx-X-Ser(Thr) always constitutes the *N*-glycosylation site. The homology between the two proteins becomes less towards their N-termini, that is, toward the extracellular part most distant from the cell surface. Possibly, the regions of highly conserved sequences indicate sites of essential protein–protein interactions (Sumper and Wieland, 1995). The overall homology is estimated to be approximately 40%, but there are marked differences in the pattern of glycosylation between the two species. Based on the molecular information, a model for the S-layer of *Hb. halobium* and *Hf. volcanii* has been proposed, in which the threonine-rich string of amino acids would form a spacer region just above the cell membrane with the bulk of the protein protruding above the cell surface and away from the cell (Kessel *et al.,* 1988).

The *slgA* gene of the S-layer glycoprotein from *M. fervidus* was cloned and sequenced by Bröckl *et al.* (1991). It encodes for a precursor of the mature S-layer protein containing 593 amino acid residues, resulting in a molecular mass of 65 kDa with a putative N-terminal sequence of 22 amino acids. The deduced protein sequence contains 20 sequon structures for *N*-glycosylation. Compared to mesophilic S-layer glycoproteins, this hyperthermophilic glycoprotein contains significantly higher amounts of isoleucine, asparagine, and cysteine residues, resulting in higher average hydrophobicity and isoelectric points. Predicted secondary structures indicate a high content of β-sheet structure (44%) and only 7% α-helix structures. As β-structures interact intermolecularly as well as intramolecularly, these high amounts of β-sheets may stabilize the proteins and contribute to the formation of regular crystalline arrays.

Cloning and sequencing of the complete gene encoding the cell surface glycoprotein from *Ha. japonica* was performed by Wakai *et al.* (1997). The gene has an open reading frame of 2586 base pairs and a potential archaeal promotor sequence of approximately 150 base pairs upstream of the ATG initiation codon. The mature S-layer glycoprotein is composed of 828 amino acids, corresponding to a molecular mass of 87 kDa, and is preceded by a signal sequence of 34 amino acids. A hydrophobic stretch at the C-terminus probably serves as a transmembrane domain, a feature shared with *Hb. halobium* and *Hf. volcanii.* In *Ha. japonica,* five glycosylation sites are recognized, fewer than in *Hb. halobium* and *Hf. volcanii.* Furthermore, the localization of the potential glycosylation sites in *Ha. japonica* is quite different from that in the other halobacterial S-layer glycoproteins.

4. CONCLUSIONS AND PERSPECTIVES

Over the last two decades a significant change of perception has taken place regarding the existence of prokaryotic glycoproteins. This is the case in particular with surface layer (S-layer) glycoproteins. With the designation of archaea as a second prokaryotic domain of life, the occurrence of glycosylated S-layer proteins had been considered a taxonomic criterion for differentiation between bacteria and archaea. However, extensive structural investigations by our group have demonstrated that S-layer glycoproteins not only are present in archaea but also in bacteria. Among gram-positive bacteria, glycosylated S-layer proteins have unambiguously been identified only in members of the *Bacillaceae*. In gram-negative organisms their presence is still not fully investigated; presently, there is no indication for their existence in this class of bacteria.

Extensive biochemical studies on the S-layer glycoprotein (cell surface glycoprotein) of *Halobacterium halobium* by Wieland, Sumper, and co-workers at least in part have unraveled the glycosylation pathway in archaea. While differences to the established eukaryotic biosynthesis pathways exist, the major reactions follow known routes. Genetic analyses of the glycosylation pathways in archaea have not been performed so far. Other significant observations concern the existence of unusual linkage regions both in archaeal and bacterial S-layer glycoproteins. Regarding bacteria, much of the work of the last few years was focused on the structural characterization of S-layer glycans. Only recently, genetic analyses of the glycosylation pathways of S-layer glycoproteins of thermophilic bacilli have been initiated.

In addition to basic research work on S-layer glycoproteins, their biotechnological application potential has been explored. One example is their use as carrier–adjuvants for vaccine development. With the development of relatively staightforward molecular biological methods, new and fascinating possibilities become available for expression of prokaryotic glycoproteins. S-layer glycoprotein research has opened up opportunities for the production by recombinant enzymes of large quantities of activated carbohydrate intermediates that are commercially not yet available. In the future, these bacterial systems may provide economical technologies for the production of medically important glycan structures.

ACKNOWLEDGMENTS

We thank Prof. Frank M. Unger for useful discussions and critical reading of the manuscript. The work was supported by the Austrian Science Fund, projects S7201-MOB and P12966-MOB (to P.M.), the Austrian Federal Ministry of Science and Transportation, and the Hochschuljubiläumstiftung der Stadt Wien, project H-109/97 (to C.S.).

REFERENCES

Adams, M. W. W., and Kelly, R. M., 1998, Finding and using hyperthermophilic enzymes, *Trends Biotechnol.* **16:**329–332.

Altman, E., Brisson, J.-R., Messner, P., and Sleytr, U. B., 1990, Chemical characterization of the regularly arranged surface layer glycoprotein of *Clostridium thermosaccharolyticum* D120–70, *Eur. J. Biochem.* **188:**73–82.

Altman, E., Brisson, J.-R., Messner, P., and Sleytr, U. B., 1991, Chemical characterization of the regularly arranged surface layer glycoprotein of *Bacillus alvei* CCM 2051, *Biochem. Cell Biol.* **69:**72–78.

Altman, E., Brisson, J.-R., Gagné, S. M., Kolbe, J., Messner, P., and Sleytr, U. B., 1992, Structure of the glycan chain from the surface layer glycoprotein of *Clostridium thermohydrosulfuricum* L77–66, *Biochim. Biophys. Acta* **1117:**71–77.

Altman, E., Schäffer, C., Brisson, J.-R., and Messner, P., 1995, Characterization of the glycan structure of a major glycopeptide from the surface layer glycoprotein of *Clostridium thermosaccharolyticum* E207–71, *Eur. J. Biochem.* **229**:308–315.

Altman, E., Schäffer, C., Brisson, J.-R., and Messner, P., 1996, Isolation and characterization of an amino sugar-rich glycopeptide from the surface layer glycoprotein of *Thermoanaerobacterium thermosaccharolyticum* E207–71, *Carbohydr. Res.* **295**:245–253.

Araki, Y., and Ito, E., 1989, Linkage units in cell walls of gram-positive bacteria, *Crit. Rev. Microbiol.* **17**:121–135.

Bashkatova, N. A., Severina, L. O., Golovacheva, R. S., and Mityushina, L. L., 1991, Surface layers of extremely thermoacidophilic archaebacteria of the genus *Sulfolobus, Mikrobiologiya* **60**:90–94.

Bayley, D. P., and Koval, S. F., 1994, Membrane association and isolation of the S-layer protein of *Methanoculleus marisnigri, Can. J. Microbiol.* **40**:237–241.

Bock, K., Schuster-Kolbe, J., Altman, E., Allmaier, G., Stahl, B., Christian, R., Sleytr, U. B., and Messner, P., 1994, Primary structure of the *O*-glycosidically linked glycan chain of the crystalline surface layer glycoprotein of *Thermoanaerobacter thermohydrosulfuricus* L111–69. Galactosyl tyrosine as a novel linkage unit, *J. Biol. Chem.* **269**:7137–7144.

Bröckl, G., Behr, M., Fabry, S., Hensel, R., Kaudewitz, H., Biendl, E., and König, H., 1991, Analysis and nucleotide sequence of the genes encoding the surface-layer glycoproteins of the hyperthermophilic methanogens *Methanothermus fervidus* and *Methanothermus sociabilis, Eur. J. Biochem.* **199**:147–152.

Cheong, G.-W., Cejka, Z., Peters, J., Stetter K. O., and Baumeister, W. 1991, The surface protein layer *of Methanoplanus limicola:* Three-dimensional structure and chemical characterization, *System. Appl. Microbiol.* **14**:209–217.

Christian, R., Schulz, G., Unger, F. M., Messner, P., Küpcü, Z., and Sleytr, U. B., 1986, Structure of a rhamnan from the surface layer glycoprotein of *Bacillus stearothermophilus* strain NRS 2004/3a, *Carbohydr. Res.* **150**:265–272.

Christian, R., Messner, P., Weiner, C., Sleytr, U. B., and Schulz, G., 1988, Structure of a glycan from the surface-layer glycoprotein of *Clostridium thermohydrosulfuricum* strain L111–69, *Carbohydr. Res.* **176**:160–163.

Christian, R., Schulz, G., Schuster-Kolbe, J., Allmaier, G., Schmid, E. R., Sleytr, U. B. and Messner, P., 1993, Complete structure of the tyrosine-linked saccharide moiety from the surface layer glycoprotein of *Clostridium thermohydrosulfuricum* S102–70. *J. Bacteriol.* **175**:1250–1256.

Danson, M. J., and Hough, D. W., 1998, Structure, function and stability of enzymes from the Archaea, *Trends Microbiol.* **6**:307–314.

Dobos, K. M., Khoo, K.-H., Swiderek, K., Brennan, P. J., and Belisle, J. T., 1996, Definition of the full extent of glycosylation of the 45-kilodalton glycoprotein of *Mycobacterium tuberculosis, J. Bacteriol.* **178**:2498–2506.

Egelseer, E., Leitner, K., Jarosch, M., Hotzy, C., Zayni, S., Sleytr, U. B., and Sára, M., 1998, The S-layer proteins of two *Bacillus stearothermophilus* wild-type strains are bound via their N-terminal region to a secondary cell wall polymer of identical chemical composition, *J. Bacteriol.* **180**:1488–1495.

Erickson, P. R., and Herzberg, M. C., 1993, Evidence for the covalent linkage of carbohydrate polymers to a glycoprotein from *Streptococcus sanguis, J. Biol. Chem.* **268:** 23780–23783.

Gerwig, G. J., Kamerling, J. P., Vliegenthart, J. F. G., Morag (Morgenstern), E., Lamed, R., and Bayer, E. A., 1991, Primary structure of *O*-linked carbohydrate chains in the cellulosome of different *Clostridium thermocellum* strains, *Eur. J. Biochem.* **196:**115– 122.

Glaser, L., and Kornfeld, S., 1961, The enzymatic synthesis of thymidine-linked sugars. II. Thymidine diphosphate L-rhamnose, *J. Biol. Chem.* **236:**1795–1799.

Gongadze, G. M., Kostyukova, A. S., Miroshnichenko, M. L., and Bonch-Osmolovskaya, E. A., 1993, Regular proteinaceous layers of *Thermococcus stetteri* cell envelope, *Curr. Microbiol.* **27:**5–9.

Graninger, M., and Messner, P., Detection and analysis of enzymes involved in the dTDP-L-rhamnose biosynthesis pathways in *Aneurinibacillus thermoaerophilus* DSM 10155, *J. Bacteriol.,* in preparation.

Graninger, M., Nidetzky, B., Heinrichs, D. E., Whitfield, C., and Messner, P., 1999, Characterization of dTDP-4-dehydrorhamnose 3,5-epimerase and dTDP-4-dehydrorhamnose reductase, required for dTDP-L-rhamnose biosynthesis in *Salmonella enterica* serovar Typhimurium LT2, *J. Biol. Chem.* **274:**25069–25077.

Grogan, D. W., 1989, Phenotypic characterization of the archaebacterial genus *Sulfolobus:* Comparison of five wild-type strains, *J. Bacteriol.* **171:**6710–6719.

Grogan, D. W., 1996a, Organization and interactions of cell envelope proteins of the extreme thermoacidophile *Sulfolobus acidocaldarius, Can. J. Microbiol.* **42:**1163–1171.

Grogan, D. W., 1996b, Isolation and fractionation of cell envelope from the extreme thermoacidophile *Sulfolobus acidocaldarius, J. Microbiol. Methods* **26:**35–43.

Hartmann, E., and König, H., 1989, Uridine and dolichyl diphosphate activated oligosaccharides are intermediates in the biosynthesis of the S-layer glycoprotein of *Methanothermus fervidus, Arch. Microbiol.* **151:**274–281.

Hartmann, E., Messner, P., Allmaier, G., and König, H., 1993, Proposed pathway for biosynthesis of the S-layer glycoprotein of *Bacillus alvei, J. Bacteriol.* **175:**4515–4519.

Heinrichs, D. E., Yethon, J. A., and Whitfield, C., 1998, Molecular basis for structural diversity in the core regions of the lipopolysaccharides of *Escherichia coli* and *Salmonella enterica, Mol. Microbiol.* **30:**221–232.

Herrmann, J. L., O'Gaora, P., Gallagher, A., Thole, J. E. R., and Young, D. B., 1996, Bacterial glycoproteins: A link between glycosylation and proteolytic cleavage of a 19 kDa antigen from *Mycobacterium tuberculosis, EMBO J.* **15:**3547–3554.

Ichikawa, Y., Look, G. C., and Wong, C.-H., 1992, Enzyme-catalyzed oligosaccharide synthesis, *Anal. Biochem.* **202:**215–238.

Jahn-Schmid, B., Messner, P., Unger, F. M., Sleytr, U. B., Scheiner, O., and Kraft, D., 1996, Towards selective elicitation of T_H1 controlled vaccination responses: Vaccine applications of bacterial surface layer proteins, *J. Biotechnol.* **44:**225–231.

Jarrell, K. F., Bayley, D. P., and Kostyukova, A. S., 1996, The archaeal flagellum: A unique motility structure, *J. Bacteriol.* **178:**5057–5064.

Jiang, X.-M., Neal, B., Santiago, F., Lee, S. J., Romana, L. K., and Reeves, P. R., 1991, Structure and sequence of the *rfb* (O antigen) gene cluster of *Salmonella* serovar typhimurium (strain LT2), *Mol. Microbiol.* **5:**695–713.

Johansen, P., Marshall, R. D., and Neuberger, A., 1958, Carbohydrate peptide complex from egg albumin, *Nature* **181:**1345–1346.

Kandler, O., 1993, Archaea (Archaebacteria), *Progr. Botany* **54**:1–24.

Kandler, O., 1994, Cell wall biochemistry and three-domain concept of life, *System. Appl. Microbiol.* **16**:501–509.

Kandler, O., and König, H., 1998, Cell wall polymers in archaea (archaebacteria), *Cell. Mol. Life Sci.* **54**:305–308.

Kärcher, U., Schröder, H., Haslinger, E., Allmaier, G., Schreiner, R., Wieland, F., Haselbeck, A., and König H., 1993, Primary structure of the heterosaccharide of the surface glycoprotein of *Methanothermus fervidus, J. Biol. Chem.* **268**:26821–26826.

Karnauchow, T. M., Koval, S. F., and Jarrell, K. F., 1992, Isolation and characterization of three thermophilic anaerobes from a St. Lucia hot spring, *System. Appl. Microbiol.* **15**:296–310.

Kaya, S., Yokoyama, K., Araki, Y., and Ito, E., 1984, N-Acetylmannosaminyl(1–4) N-acetylglucosamine, a linkage unit between glycerol teichoic acid and peptidoglycan in cell walls of several *Bacillus* strains, *J. Bacteriol.* **158**:990–996.

Kessel, M., Wildhaber, I., Cohen, S., and Baumeister, W., 1988, Structure of the surface glycoprotein from *Halobacterium volcanii* as revealed by electron microscopy, in: *Crystalline Bacterial Cell Surface Layers* (U. B. Sleytr, P. Messner, D. Pum, and M. Sára, eds.), Springer, Berlin, pp. 79–82.

Klaps, E., Neuninger, C., Messner, P., and Schmid, W., 1996, Synthesis of GDP-2-acetamido-2-deoxy-α-D-mannose (GDP-ManNAc), *Carbohydr. Lett.* **2**:97–100.

Knirel, Y. A., and Kochetkov, N. K., 1994, The structure of lipopolysaccharides of gramnegative bacteria. III. The structure of O-antigens, *Biochemistry (Moscow)* **59**:1325–1383.

Kobata, A., 1984, The carbohydrates of glycoproteins, in: *Biology of Carbohydrates*, Vol. 2 (V. Ginsburg and P. W. Robbins, eds.), John Wiley & Sons, New York, pp. 87–161.

Kojima, N., Araki, Y., and Ito, E, 1985, Structure of the linkage units between ribitol teichoic acids and peptidoglycan, *J. Bacteriol.* **161**:299–306.

König, H., 1988, Archaebacterial cell envelopes, *Can. J. Microbiol.* **34**:395–406.

König, H., Hartmann, E., Bröckl, G., and Kärcher, U., 1993, The unique chemical formats and biosynthetic pathways of methanogenic surfaces, in: *Advances in Paracrystalline Surface Layers* (T. J. Beveridge and S. F. Koval, eds.), Plenum Press, New York, pp. 99–108.

König, H., Hartmann, E., and Kärcher, U., 1994, Pathways and principles of the biosynthesis of methanobacterial cell wall polymers, *System. Appl. Microbiol.* **16**:510–517.

Kornfeld, R., and Kornfeld, S., 1980, Structure of glycoproteins and their oligosaccharide units, in: *The Biochemistry of Glycoproteins and Proteoglycans* (W. J. Lennarz, ed.), Plenum Press, New York, pp. 1–34.

Kornfeld, R., and Kornfeld, S., 1985, Assembly of asparagine-linked oligosaccharides, *Annu. Rev. Biochem.* **54**:631–664.

Kornfeld, S., and Glaser, L., 1961, The enzymatic synthesis of thymidine-linked sugars. I. Thymidine diphosphate glucose, *J. Biol. Chem.* **236**:1791–1794.

Kosma, P., Neuninger, C., Christian, R., Schulz, G., and Messner, P., 1995a, Glycan structure of the S-layer glycoprotein of *Bacillus* sp. L420–91, *Glycoconjugate J.* **12**:99–107.

Kosma, P., Wugeditsch, T., Christian, R., Zayni, S., and Messner, P., 1995b, Glycan structure of a heptose-containing S-layer glycoprotein of *Bacillus thermoaerophilus, Glycobiology* **5**:791–796.

Kuntz, C., Sonnenbichler, J., Sonnenbichler, I., Sumper, M., and Zeitler, R., 1997, Isolation and characterization of dolichol-linked oligosaccharides from *Haloferax volcanii*, *Glycobiology* **7**:897–904.

Lechner, J., and Sumper, M., 1987, The primary structure of a procaryotic glycoprotein, *J. Biol. Chem.* **262**:9724–9729.

Lechner, J., and Wieland, F., 1989, Structure and biosynthesis of prokaryotic glycoproteins, *Annu. Rev. Biochem.* **58**:173–194.

Lechner, J., Wieland, F., and Sumper, M., 1985a, Biosynthesis of sulfated oligosaccharides N-glycosidically linked to the protein via glucose, *J. Biol. Chem.* **260**:860–866.

Lechner, J., Wieland, F., and Sumper, M., 1985b, Transient methylation of dolichyl oligosaccharides is an obligatory step in halobacterial sulphated glycoprotein synthesis, *J. Biol. Chem.* **260**:8984–8989.

Lis, H., and Sharon, N., 1993, Protein glycosylation. Structural and functional aspects, *Eur. J. Biochem.* **218**:1–27.

Lupas, A., Engelhardt, H., Peters, J., Santarius, U., Volker, S., and Baumeister, W., 1994, Domain structure of the *Acetogenium kivui* surface layer revealed by electron crystallography and sequence analysis, *J. Bacteriol.* **176**:1224–1233.

Malcolm, A. J., Best, M. W., Szarka, R. J., Mosleh, Z., Unger, F. M., Messner, P., and Sleytr, U. B., 1993a, Surface layers from *Bacillus alvei* as a carrier for a *Streptococcus pneumoniae* conjugate vaccine, in: *Advances in Bacterial Paracrystalline Surface Layers* (T. J. Beveridge and S. F. Koval, eds.), Plenum Press, New York, pp. 219–233.

Malcolm, A. J., Messner, P., Sleytr, U. B., Smith, R. H., and Unger, F. M., 1993b, Crystalline bacterial cell surface layers (S-layers) as combined carrier/adjuvants for conjugate vaccines, in: *Immobilised Macromolecules: Application Potentials* (U. B. Sleytr, P. Messner, D. Pum, and M. Sára, eds.), Springer-Verlag, London, pp. 195–207.

Mengele, R., and Sumper, M., 1992, Drastic difference in glycosylation of related S-layer glycoproteins from moderate and extreme halophiles, *J. Biol. Chem.* **267**:8182–8185.

Mescher, M. F., 1981, Glycoproteins as cell-surface structural components, *Trends Biochem. Sci.* **6**:97–99.

Mescher, M. F., and Strominger, J. L., 1976a, Purification and characterization of a prokaryotic glycoprotein from the cell envelope of *Halobacterium salinarium*, *J. Biol. Chem.* **251**:2005–2014.

Mescher, M. F., and Strominger, J. L., 1976b, Structural (shape-maintaining) role of the cell surface glycoprotein of *Halobacterium salinarium*, *Proc. Natl. Acad. Sci. USA* **73**: 2687–2691.

Mescher, M. F., Strominger, J. L., and Watson, S. W., 1974, Protein and carbohydrate composition of the cell envelope of *Halobacterium salinarium*, *J. Bacteriol.* **120**:945–954.

Messner, P., 1996, Chemical composition and biosynthesis of S-layers, in: *Crystalline Bacterial Cell Surface Proteins* (U. B. Sleytr, P. Messner, D. Pum, and M. Sára, eds.), R. G. Landes/Academic Press, Austin, TX, pp. 35–76.

Messner, P., 1997, Bacterial glycoproteins, *Glycoconjugate J.* **14**:3–11.

Messner, P., and Sleytr, U. B., 1988, Asparaginyl-rhamnose: A novel type of protein–carbohydrate linkage in a eubacterial surface-layer glycoprotein, *FEBS Lett.* **228**:317–320.

Messner, P., and Sleytr, U. B., 1991, Bacterial surface layer glycoproteins, *Glycobiology* **1**:545–551.

Messner, P., and Sleytr, U. B., 1992, Crystalline bacterial cell-surface layers, *Adv. Microb. Physiol.* **33:**213–275.

Messner, P., Sleytr, U. B., Christian, R., Schulz, G., and Unger, F. M., 1987, Isolation and structure determination of a diacetamidodideoxyuronic acid-containing glycan chain from the S-layer glycoprotein of *Bacillus stearothermophilus* NRS 2004/3a, *Carbohydr. Res.* **168:**211–218.

Messner, P., Bock, K., Christian, R., Schulz, G., and Sleytr, U. B., 1990, Characterization of the surface layer glycoprotein of *Clostridium symbiosum* HB25, *J. Bacteriol.* **172:**2576–2583.

Messner, P., Christian, R., Kolbe, J., Schulz, G., and Sleytr, U. B., 1992a, Analysis of a novel linkage unit of *O*-linked carbohydrates from the crystalline surface layer glycoprotein of *Clostridium thermohydrosulfuricum* S102–70. *J. Bacteriol.* **174:**2236–2240.

Messner, P., Mazid, M. A., Unger, F. M., and Sleytr, U. B., 1992b, Artificial antigens. Synthetic carbohydrate haptens immobilized on crystalline bacterial surface layer glycoproteins, *Carbohydr. Res.* **233:**175–184.

Messner, P., Christian, R., Neuninger, C., and Schulz, G., 1995, Similarity of "core" structures in two different glycans of tyrosine-linked eubacterial S-layer glycoproteins, *J. Bacteriol.* **177:**2188–2193.

Miroshnichenko, M. L., Bonch-Osmolovskaya, E. A., Neuner, A., Kostrikina, N. A., Chernych, N. A., and Alekseev, V. A., 1989, *Thermococcus stetteri* sp. nov., a new extremely thermophilic marine sulfur-metabolizing archaebacterium, *System. Appl. Microbiol.* **12:**257–262.

Moens, S., and Vanderleyden, J., 1997, Glycoproteins in prokaryotes, *Arch. Microbiol.* **168:**169–175.

Montreuil, J., 1995, The history of glycoprotein research: A personal view, in: *Glycoproteins* (J. Montreuil, J. F. G. Vliegenthart, and H. Schachter, eds.), Elsevier, Amsterdam, pp. 1–12.

Möschl, A., Schäffer, C., Sleytr, U. B., Messner, P., Christian, R., and Schulz, G., 1993, Characterization of the S-layer glycoproteins of two lactobacilli, in: *Advances in Bacterial Paracrystalline Surface Layers* (T. J. Beveridge and S. F. Koval, eds.), Plenum Press, New York, pp. 281–284.

Munson, R. S., and Glaser, L., 1981, Teichoic acid and peptidoglycan assembly in gram-positive organisms, in: *Biology of Carbohydrates,* Vol. 1 (V. Ginsburg and P. Robbins, eds.), Wiley, New York, pp. 91–122.

Nakamura, S., Mizutani, S., Wakai, H., Kawasaki, H., Aono, R., and Horikoshi, K., 1995, Purification and partial characterization of cell surface glycoprotein from extremely halophilic archaeon *Haloarcula japonica* strain TR-1, *Biotechnol. Lett.* **17:**705–706.

Nishiyama, Y., Takashina, T., Grant, W. D., and Horikoshi, K., 1992, Ultrastructure of the cell wall of the triangular halophilic archaebacterium *Haloarcula japonica* strain TR-1, *FEMS Microbiol. Lett.* **99:**43–48.

Nußer, E., Hartmann, E., Allmeier, H., König, H., Paul, G., and Stetter, K. O., 1988, A glycoprotein surface layer covers the pseudomurein sacculus of the extreme thermophile *Methanothermus fervidus,* in: *Crystalline Bacterial Cell Surface Layers* (U. B. Sleytr, P. Messner, D. Pum, and M. Sára, eds.), Springer Verlag, Berlin, pp. 21–25.

Paul, G., and Wieland, F., 1987, Sequence of the halobacterial glycosaminoglycan, *J. Biol. Chem.* **262:**9587–9593.

Paul, G., Lottspeich, F., and Wieland, F., 1986, Asparaginyl-*N*-acetylgalactosamine: Linkage unit of halobacterial glycosaminoglycan, *J. Biol. Chem.* **261**:1020–1024.

Pellerin, P., Fournet, B., and Debeire, P. 1990, Evidence for the glycoprotein nature of the cell sheath of *Methanosaeta*-like cells in the culture of *Methanothrix soehngenii* strain FE, *Can. J. Microbiol.* **36**:631–636.

Peters, J., Rudolf, S., Oschkinat, H., Mengele, R., Sumper, M., Kellermann, J., Lottspeich, F., and Baumeister, W., 1992, Evidence for tyrosine-linked glycosaminoglycan in a bacterial surface protein, *Biol. Chem. Hoppe-Seyler* **373**:171–176.

Peters, J., Nitsch, M., Kühlmorgen, B., Golbik, R., Lupas, A., Kellermann, J., Engelhardt, H., Pfander, J. P., Müller, S., Goldie, K., Engel A., Stetter, K. O., and Baumeister, W., 1995, Tetrabrachion—A filamentous archaebacterial surface protein assembly of unusual structure and extreme stability, *J. Mol. Biol.* **245**:385–401.

Peters, J., Baumeister, W., and Lupas, A., 1996, Hyperthermostable surface layer protein tetrabrachion from the archaebacterium *Staphylothermus marinus:* Evidence for the presence of a right-handed coiled coil derived from the primary structure, *J. Mol. Biol.* **257**:1031–1041.

Peyret, J. L., Bayan, N., Joliff, G., Gulik-Krzywicki, T., Mathieu, L., Shechter, E., and Leblon, G., 1993, Characterization of the *cspB* gene encoding PS2, an ordered surface-layer protein in *Corynebacterium glutamicum, Mol. Microbiol.* **9**:97–109.

Pley, U., Schipka, J., Gambacorta, A., Jannasch, H. W., Fricke, H., Rachel, R., and Stetter, K. O., 1991, *Pyrodictium abyssi* sp. nov. represents a novel heterotrophic marine archaeal hyperthermophile growing at 110°C, *System. Appl. Microbiol.* **14**:245–253.

Potter, J. E. R., James, M. J., and Kandutsch, A. A., 1981, Sequential cycles of cholesterol and dolichol synthesis in mouse spleens during phenylhydrazine-induced erythropoiesis, *J. Biol. Chem.* **256**:2371–2376.

Pum, D., and Sleytr, U. B., 1999, The application of bacterial S-layers in molecular nanotechnology, *Trends Biotechnol.* **17**:8–12.

Raetz, C. R. H., 1990, Biochemistry of endotoxins, *Annu. Rev. Biochem.* **59**:129–170.

Raetz, C. R. H., 1996, Bacterial lipopolysaccharides: a remarkable family of bioactive macroamphiphiles, in: *Escherichia coli* and *Salmonella,* in: *Cellular and Molecular Biology,* 2nd. Ed., Vol. 1 (F. C. Neidhardt, R. Curtis, III, J. L. Ingraham, E. C. C. Lin, K. B. Low, B. Magasanik, W. S. Reznikoff, M. Riley, M. Schaechter, and H. E. Umbarger, eds.), ASM Press, Washington, DC, pp. 1035–1063.

Reeves, P. R., 1994, Biosynthesis and assembly of lipopolysaccharide, *New Comp. Biochem.* **27**:281–314.

Ries, W., Hotzy, C., Schocher, I., Sleytr, U. B., and Sára, M., 1997, Evidence that the N-terminal part of the S-layer protein from *Bacillus stearothermophilus* PV72/p2 recognizes a secondary cell wall polymer, *J. Bacteriol.* **179**:3892–3898.

Rocchetta, H. L., Burrows, L. L., and Lam, J. J., 1999, Genetics of O-antigen biosynthesis in *Pseudomonas aeruginosa, Microbiol. Mol. Biol. Rev.* **63**:523–553.

Sandercock, L. E., MacLeod, A. M., Ong, E., and Warren, R. A. J., 1994, Non-S-layer glycoproteins in eubacteria, *FEMS Microbiol. Lett.* **118**:1–8.

Sára, M., Küpcü, S., and Sleytr, U. B., 1989, Localization of the carbohydrate residue of the S-layer glycoprotein from *Clostridium thermohydrosulfuricum* L111–69, *Arch. Microbiol.* **151**:416–420.

Sára, M., Dekitsch, C., Mayer, H. F., Egelseer, E. M., and Sleytr, U. B., 1998, Influence of the secondary cell wall polymer on the reassembly, recrystallization, and stability properties of the S-layer protein from *Bacillus stearothermophilus* PV72/p2, *J. Bacteriol.* **180**:4146–4153.

Schäffer, C., Wugeditsch, T., Neuninger, C., and Messner, P., 1996, Are S-layer glycoproteins and lipopolysaccharides related? *Microbial Drug Resistance* **2**:17–23.

Schäffer, C., Müller, N., Christian, R., Graninger, M., Wugeditsch, T., Scheberl, A., and Messner, P., 1999a, Complete glycan structure of the S-layer glycoprotein of *Aneurinibacillus thermoaerophilus* GS4–97, *Glycobiology* **9**:407–414.

Schäffer, C., Kählig, H., Christian, R., Schulz, G., Zayni, S., and Messner, P., 1999b, The diacetamidodideoxyuronic-acid-containing glycan chain of *Bacillus stearothermophilus* NRS 2004/3a represents the secondary cell wall polymer of wild-type *B. stearothermophilus* strains, *Microbiology* **145**:1575–1583.

Schnaitman, C. A., and Klena, J. D., 1993, Genetics of lipopolysaccharide biosynthesis in enteric bacteria, *Microbiol. Rev.* **57**:655–682.

Severina, L. O., Senyushkin, A. A., and Karavaiko, G. I., 1993, Ultrastructure and chemical composition of the S-layer of *Sulfobacillus thermosulfidooxidans, Dokl. Akad. Nauk.* **328**:633–636.

Shashkov, A. S., Lipkind, G. M., Knirel, Y. A., and Kochetkov, N. K., 1988, Stereochemical factors determining the effects of glycosylation on the ^{13}C chemical shifts in carbohydrates, *Magn. Res. Chem.* **26**:735–747.

Shen, N., and Weiner, R. M., 1998, Isolation and characterization of S-layer proteins from a vent prosthecate bacterium, *Microbios* **93**:7–16.

Sleytr, U. B., and Messner, P., 1983, Crystalline surface layers on bacteria, *Annu. Rev. Microbiol.* **37**:311–339.

Sleytr, U. B., and Sára, M., 1997, Bacterial and archaeal S-layer proteins: Structure–function relationships and their biotechnological applications, *Trends Biotechnol.* **15**:20–26.

Sleytr, U. B., and Thorne, K. J. I., 1976, Chemical characterization of the regularly arrayed surface layers of *Clostridium thermosaccharolyticum* and *Clostridium thermohydrosulfuricum, J. Bacteriol.* **126**:377–383.

Sleytr, U. B., Messner, P., Pum, D., and Sára, M. (eds.), 1996, *Crystalline Bacterial Cell Surface Proteins,* R. G. Landes/Academic Press, Austin, TX.

Sleytr, U. B., Messner, P., Pum, D., and Sára, M., 1999, Crystalline bacterial cell surface layers (S-layers): From supramolecular cell structure to biomimetics and nanotechnology, *Angew. Chem. Int. Ed.* **38**:1034–1054.

Smith, S. H., and Murray, R. G. E., 1990, The structure and associations of the double S-layer on the cell wall of *Aquaspirillum sinuosum, Can. J. Microbiol.* **36**:327–335.

Stimson, E., Virji, M., Makepeace, K., Dell, A., Morris, H. R., Payne, G., Saunders, J. R., Jennings, M. P., Barker, S., Panico, M., Blench, I., and Moxon, E. R., 1995, Meningococcal pilin: A glycoprotein substituted with digalactosyl 2,4-diacetamido-2,4,6-trideoxyhexose, *Mol. Microbiol.* **17**:1201–1214.

Sumper, M., 1987, Halobacterial glycoprotein biosynthesis, *Biochim. Biophys. Acta* **906**:69–79.

Sumper, M., 1993, S-layer glycoproteins from moderately and extremely halophilic archaeobacteria, in: *Advances in Paracrystalline Surface Layers* (T. J. Beveridge and S. F. Koval, eds.), Plenum Press, New York, pp. 109–117.

Paul Messner and Christina Schäffer

Sumper, M., and Wieland, F. T., 1995, Bacterial glycoproteins, in: *Glycoproteins* (J. Montreuil, J. F. G. Vliegenthart, and H. Schachter, eds.), Elsevier, Amsterdam, pp. 455–473.

Sumper, M., Berg, E., Mengele, R., and Strobl, I., 1990, Primary structure and glycosylation of the S-layer protein of *Haloferax volcanii, J. Bacteriol.* **172**:7111–7118.

Unger, F. M., Messner, P., Jahn-Schmid, B., and Sleytr, U. B., 1997, Vaccine applications of crystalline bacterial surface layer proteins (S-layer), *FEMS Microbiol. Rev.* **20**:157–158.

Virji, M., 1998, Glycosylation of the meningococcus pilus protein, *ASM News* **64**:398–405.

Vliegenthart, J. F. G., and Montreuil, J., 1995, Primary structure of glycoprotein glycans, in: *Glycoproteins* (J. Montreuil, J. F. G. Vliegenthart, and H. Schachter, eds.), Elsevier, Amsterdam, pp. 13–28.

Wakai, H., Nakamura, S., Kawasaki, H., Takada, K., Mizutani, S., Aono, R., and Horikoshi, K., 1997, Cloning and sequencing of the gene encoding the cell surface glycoprotein of *Haloarcula japonica* strain TR-1, *Extremophiles* **1**:29–35.

Wieland, F., Lechner, J., Bernhardt, G., and Sumper, M., 1981, Sulphatation of a repetitive saccharide in halobacterial cell wall glycoprotein, *FEBS Lett.* **132**:319–323.

Wieland, F. T., Lechner, J., and Sumper, M., 1982, The cell wall glycoprotein of halobacteria: structural, functional and biosynthetic aspects, *Zbl. Bakt. Hyg., I. Abt. Orig. C* **3**:161–170.

Whitfield, C., 1995, Biosynthesis of lipopolysaccharide O antigens, *Trends Microbiol.* **3**:178–185.

Woese, C. R., and Fox, G. E., 1977, Phylogenetic structure of the prokaryotic domain: The primary kingdoms, *Proc. Natl. Acad. Sci. USA* **77**:5088–5090.

Wugeditsch, T., 1998, Strukturanalyse des S-Schichtglykoprotein Glykans und Zellwand-Aminozuckerpolymers von *Aneurinibacillus thermoaerophilus* DSM 10155. Doctoral Thesis, University of Agricultural Sciences Vienna.

Wugeditsch T., Zachara, N. E., Puchberger, M., Kosma, P., Gooley, A. A., and Messner, P., 1999, Structural heterogeneity in the core oligosaccharide of the S-layer glycoprotein from *Aneurinibacillus thermoaerophilus* DSM 10155, *Glycobiology* **9**:787–795.

Yang, L. L., and Haug, A., 1979, Purification and partial characterization of a procaryotic glycoprotein from the plasma membrane of *Thermoplasma acidophilum, Biochim. Biophys. Acta* **556**:265–277.

Zellner, G., Messner, P., Kneifl, H., Tindall, B. J., Winter J., and Stackebrandt, E., 1989a, *Methanolacinia* gen. nov., incorporating *Methanomicrobium paynteri* as *Methanolacinia paynteri* comb. nov., *J. Gen. Appl. Microbiol.* **35**:185–202.

Zellner, G., Stackebrandt, E., Kneifel, H., Messner, P., Sleytr, U. B., Conway de Macario, E., Zabel, H.-P., Stetter, K. O., and Winter, J., 1989b, Isolation and characterization of a thermophilic, sulfate reducing archaebacterium, *Archaeoglobus fulgidus* strain Z, *System. Appl. Microbiol.* **11**:151–160.

Zellner, G., Stackebrandt, E., Messner, P., Tindall, B. J., Conway de Macario, E., Kneifel, H., Sleytr, U. B., and Winter, J., 1989c, *Methanocorpusculaceae* fam. nov., represented by *Methanocorpusculum parvum, Methanocorpusculum sinense* spec. nov. and *Methanocorpusculum bavaricum* spec. nov, *Arch. Microbiol.* **151**:381–390.

Zellner, G., Sleytr, U. B., Messner, P., Kneifl., H., and Winter, J., 1990, *Methanogenium lim-*

inatans spec. nov., a new coccoid, mesophilic methanogen able to oxidize secondary alcohols, *Arch. Microbiol.* **153:**287–293.

Zhu, B. C. R., and Laine, R. A., 1996, Dolichyl-phosphomannose synthase from the archae *Thermoplasma acidophilum, Glycobiology* **6:**811–816.

Zhu, B. C. R., Drake, R. R., Schweingruber, H., and Laine, R., 1995, Inhibition of glycosylation by amphomycin and sugar nucleotide analogs PP36 and PP55 indicates that *Haloferax volcanii* β-glycosylates both glycoproteins and glycolipids through lipid-linked sugar intermediates: Evidence for three novel glycoproteins and a novel dihexosyl-archaeol glycolipid, *Arch. Biochem. Biophys.* **319:**355–364.

5

Assembly Pathways for Biosynthesis of A-Band and B-Band Lipopolysaccharide in *Pseudomonas aeruginosa*

Lori L. Burrows, Heather L. Rocchetta, and Joseph S. Lam

1. INTRODUCTION

Pseudomonas aeruginosa is among the most frequently isolated nosocomial pathogens (Jarvis and Martone, 1992; Spencer, 1996), particularly in burn wound units (de Vos *et al.,* 1997). This organism is also the leading cause of morbidity and mortality in persons afflicted with cystic fibrosis (CF). The success of this opportunistic pathogen can be attributed to many elements, including its intrinsic resistance to antibiotics and its ability to elaborate numerous virulence factors. Among its virulence determinants are the cell surface polysaccharides alginate (a mucoid substance or slime) (Govan and Deretic, 1996), rhamnolipid (a biosurfactant) (Ochsner *et al.,* 1996), and two distinct, co-produced forms of lipopolysaccharide (LPS), called A-band LPS and B-band LPS (Lightfoot and Lam, 1991). Of these surface polysaccharides, only LPS is expressed constitutively, and thus is involved in the majority of interactions between the bacterium and its environment.

LPS is an integral part of the outer membrane of gram-negative organisms. It

Lori L. Burrows, Heather L. Rocchetta, and Joseph S. Lam • Canadian Bacterial Diseases Network, Department of Microbiology, University of Guelph, Guelph, Ontario, N1G 2W1, Canada. *Glycomicrobiology,* edited by Doyle. Kluwer Academic/Plenum Publishers, New York, 2000.

is composed of a phosphorylated diglucosamine moiety, covalently substituted with acyl chains (lipid A) that anchors LPS in the outer leaflet of the outer membrane. Lipid A is capped by a hetero-oligosaccharide moiety, called the core, containing hexoses, heptoses, and octoses (Schnaitman and Klena, 1993). The octose residues are the unique sugar, 3-deoxy-D-manno-octulosonic acid (Kdo), which is found in all gram-negative bacteria and, rarely, in bioactive pectins from some plants (Hirano *et al.*, 1994). Lipid A and the inner portion of the core oligosaccharide, containing heptose and octose moieties, are highly conserved among gram-negatives. Attached to the core and extending some distance from the cell surface is a polysaccharide of heterogeneous lengths whose composition is remarkably varied, not only among species but among strains of a single species (Schnaitman and Klena, 1993). The various monosaccharide constituents, their linkages to one another, and the lengths of the polysaccharide chains contribute to the induction of a specific serum response in the host. This terminal portion of LPS, called the O-polysaccharide or O-antigen, dictates the serotype of a particular strain.

We have been exploring the genetics underlying the biosynthesis of A-band and B-band LPS in *P. aeruginosa*. The relatively short polysaccharide region of A-band LPS is composed of a repeating, trisaccharide unit of D-rhamnose (D-Rha), which is expressed by most serotypes of *P. aeruginosa* (Lam *et al.*, 1989; Arsenault *et al.*, 1991). Various pseudomonads other than *P. aeruginosa* also produce a D-Rha homopolymer with linkages similar to that of the A-band polymer (Smith *et al.*, 1985). *Pseudomonas syringae* pathovars *morsprunorum* C28 and *cerasi* 435 express a D-Rha homopolymer composed of trisaccharide repeating units that possess the same linkages as A-band LPS (Smith *et al.*, 1985; Vinogradov *et al.*, 1991). One serotype of a related pseudomonad, *Stenotrophomonas (Xanthomonas) maltophilia* O7, has an O antigen identical to A band (Winn and Wilkinson, 1998).In addition, a CF isolate of *Burkholderia cepacia* has been shown to produce two types of LPS, designated "major" and "minor" LPSs. The polysaccharide region of the minor LPS contains trisaccharide repeats of D-Rha with the same linkages as A-band LPS (Cerantola and Montrozier, 1997). *S. maltophilia* and *B. cepacia* are multidrug-resistant, emerging human pathogens that colonize CF patients and can cause fulminant infections characterized by a rapid and fatal clinical deterioration (Govan and Deretic, 1996; Quinn, 1998). The conservation of A-band LPS in isolates of *P. aeruginosa, B. cepacia,* and *S. maltophilia* warrants further investigation of its role in host–pathogen interactions.

B-band LPS is a longer heteropolymer that masks underlying A-band LPS, and therefore serves as the predominant serotype-specific surface antigen, or O-antigen. There are various serotyping schemes for the classification of *P. aeruginosa* strains, but the most extensive and internationally accepted scheme [the International Antigenic Typing Scheme (IATS)] is composed of 20 or more distinct serotypes (B-band O-antigens) of *P. aeruginosa* (Liu and Wang, 1990). A recent review listed 31 separate "chemotypes" of *P. aeruginosa* B-band LPSs (Stanislavsky and Lam, 1997). B-band O-antigens consist of repeating units of di- to pentasaccharides, containing uronic acids, amino sugars, and some peculiar monosaccharides, such as pseudaminic acid, not found elsewhere (Knirel and Kochetkov, 1994).

LPS is a complex molecule whose assembly requires a number of specific proteins. It has become clear in the last decade that the assembly of homopolymeric and heteropolymeric O-antigens are fundamentally different (reviewed in Whitfield, 1995). We now have substantial evidence showing that this paradigm is true also for *P. aeruginosa,* and that A-band and B-band LPS are assembled via separate pathways. In this chapter we review those pathways, discuss their differences, and indicate points of convergence that are unique to *P. aeruginosa*. Those areas that require further study will be highlighted.

2. ASSEMBLY OF THE HOMOPOLYMER, A-BAND LPS

Lipopolysaccharide molecules that possess a single type of sugar in their O-polysaccharide repeating unit are referred to as homopolymers. Examples of such homopolymeric O-polysaccharides include *P. aeruginosa* A band, *Escherichia coli* O8 and O9, *Yersinia enterocolitica* O:3, *Klebsiella pneumoniae* O1 and O8, *Serratia marscescens* O16, and *Salmonella enterica* serovar Borreze O:54 (reviewed in Knirel and Kochetkov, 1994; Popoff and LeMinor, 1985). The sugar backbone of these O-polysaccharides is more simplistic than that of heteropolymers, and their mechanisms of synthesis are therefore quite different.

2.1. Initiation of A-Band Polymer Synthesis

In the case of both homopolysaccharides and heteropolysaccharides, sugar nucleotide precursors are synthesized within the cell cytoplasm (Shibaev, 1986) and used as donor molecules for assembly of the O-polysaccharide region. An initiating glycosyltransferase serves to transfer the first sugar residue onto a carrier lipid molecule, identified as the C_{55} polyisoprenoid alcohol derivative undecaprenol phosphate (Und-P) (Wright *et al.,* 1967). Und-P also serves as a scaffold for peptidoglycan biosynthesis (Fuchs-Cleveland and Gilvarg, 1976). Synthesis of homopolysaccharides requires the activity of an initiating glycosyltransferase that adds only the initial, nonhomopolymeric sugar onto Und-P. This sugar acts as a primer and does not form part of the O-repeating unit (Whitfield, 1995). In contrast, heteropolysaccharides require the initiating glycosyltransferase for the formation of each O-repeating unit on Und-P. Thus, the initiating sugar becomes the first sugar of every O-unit.

Glycosyltransferases that are known to initiate the biosynthesis of homopolymers in the *Enterobacteriaceae* include Wec A (Rfe) (Alexander and Valvano, 1994; Rick *et al.*, 1994) and Wba P (Wang *et al.*, 1996). During initiation of *E. coli* O8 and O9 homopolymer synthesis, the glycosyltransferase Wec A catalyzes the transfer of *N*-acetylglucosamine-1-phosphate (GlcNAc-1-P) to Und-P (Rick *et al.*, 1994; Kido *et al.*, 1995). Wec A-dependency also has been reported for the homopolymeric O-polysaccharides produced by *K. pneumoniae* O1 and O8 (Clarke and Whitfield, 1992; Clarke *et al.*, 1995), *S. marcescens* O16 (Szabo *et al.*, 1995), and *S. enterica* serovar Borreze O:54 (Keenleyside and Whitfield, 1996). In heteropolysaccharide biosynthesis, Wec A transfers either a GlcNAc-1-P (Alexander and Valvano, 1994) or an *N*-acetylgalactosamine-1-phosphate (GalNAc-1-P) moiety to Und-P (Zhang *et al.*, 1997; Amor and Whitfield, 1997). These studies indicate that Wec A is flexible with respect to both the substrates it recognizes and the polymers it initiates.

A homologue of Wec A, designated Wbp L (Burrows *et al.*, 1996), is encoded within the B-band O-antigen gene cluster in *P. aeruginosa* serotype O5 (mapped to 37 minutes on the 75-minute map of strain PAO1) (Lightfoot and Lam, 1993). Wbp L, like Wec A, is predicted to be a hydrophobic, integral membrane protein with multiple membrane-spanning domains. These structural properties correlate with the requirement of Wec A and Wbp L to interact with the hydrophobic acceptor molecule Und-P. Analysis of *wbpL* chromosomal mutants in serotype O5 has demonstrated that Wbp L is required for synthesis of both A-band and B-band LPS (Rocchetta *et al.*, 1998), showing that it is functionally analogous to the initiating transferase, Wec A. Complementation of a *wbpL* mutant of *P. aeruginosa* (which is deficient in the synthesis of both A-band and B-band LPS) with *wecA* from *E. coli* restores only A-band but not B-band LPS synthesis (Rocchetta *et al.*, 1998). These results show that Wbp L is similar to Wec A. However, Wbp L has a broader substrate specificity, since it is capable of transferring Fuc2NAc-1-P to initiate B-band LPS synthesis, and GlcNAc-1-P to initiate A-band synthesis (Fig. 1A).

2.2. A-Band O-Polysaccharide Assembly

Postinitiation reactions involve the activities of particular glycosyltransferases that act sequentially to form the O-polysaccharide repeating unit. These transferases catalyze specific glycosidic linkages and recognize certain donor and

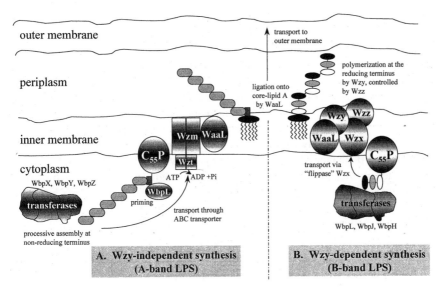

Figure 1. Two distinct biosynthetic pathways and the components associated with O-antigen biosynthesis in *Pseudomonas aeruginosa*. (A) The Wzy-independent pathway for the homopolymeric A-band LPS synthesis. Features characteristic of this pathway include the transfer of a priming sugar by Wbp L, the processive assembly of the D-rhamnan polymer by rhamnosyltransferases, and the transport of the A-band O-antigen polymer across the cytoplasmic membrane by the ABC-transporter, Wzm and Wzt. (B) The Wz-dependent pathway for the heteropolymeric B-band LPS synthesis. Wbp L acts as the first glycosyltransferase. Wbp J and Wbp H will then act as nonprocessive glycosyltransferases for the assembly of the next two sugar residues in the O-unit. Other important components in this pathway include Wzx (flippase/translocase), Wzy (O-polymerase) and Wzz (modulator of O-antigen chain length). Waa L has not yet been identified in the *P. aeruginosa* genome. Adapted from Rocchetta *et al.* (1998), Rocchetta and Lam (1997), Burrows *et al.* (1996, 1997), de Kievit *et al.* (1995), and Whitfield (1995).

acceptor molecules. Such specificity preserves the structure of the O-polysaccharide repeat. Three additional glycosyltransferases have been identified for assembly of the A-band D-rhamnan polymer in *P. aeruginosa* (Rocchetta *et al.,* 1998). These rhamnosyltransferases, Wbp X, Wbp Y, and Wbp Z, are encoded within the A-band O-polysaccharide gene cluster (Rocchetta *et al.,* 1998), which maps between 10.5 and 13.3 minutes on the PAO1 chromosome (Lightfoot and Lam, 1993). Wbp X, Wbp Y, and Wbp Z each contain a motif (EX$_7$E) identified among retaining glycosyltransferases that catalyze α-glycosidic linkages (Geremia *et al.,* 1996). This motif also has recently been identified within glycosyltransferases that catalyze β-glycosidic linkages (Heinrichs *et al.,* 1998), and appears to be important for glycosyltransferase activity. Chromosomal mutations in each of *wbpX*, *wbpY*, and *wbpZ* result in a loss of A-band LPS biosynthesis, while B-band LPS is unaffected (Rocchetta *et al.*, 1998). An assembly scheme has been proposed for these D-rhamnosyltransferases (Rocchetta *et al.*, 1998), based on similarities with *E. coli* O9a mannosyltransferases (Kido et al., 1997) and on the chemical structure of the A-band O-polysaccharide region (containing $\alpha1,2, \alpha1,3, \alpha1,3$ linkages). Figure 1A illustrates the proposed mechanism for A-band O-polysaccharide assembly. Wbp Z is predicted to add the first D-Rha residue onto the GlcNAc-PP-Und acceptor, previously formed through the action of Wbp L. WbpY recognizes this terminal D-Rha moiety and is thought to add two D-Rha residues via $\alpha1,3$ linkages. The terminal $\alpha1,3$ linked D-Rha is in turn recognized by Wbp X, which subsequently adds an $\alpha1,2$ linked D-Rha. The latter two transferases, Wbp Y and Wbp X, are similar to processive transferases and are predicted to alternate in their activities for sequential assembly of the A-band polymer (Fig. 1A). Future experiments utilizing chemically synthesized substrates are necessary to definitively assign transferase specificity to these enzymes.

2.3. Transport of A-Band O-Polysaccharide

An ATP-binding cassette (ABC) transport system serves to export most homopolymeric O-polysaccharides to the periplasm for ligation to core lipid A. Systems for homopolymer export have been identified in *E. coli* O9a (previously designated serotype O9) (Kido *et al.*, 1997), *K. pneumoniae* O1, *S. marcescens* O16, *Y. enterocolitica* O:3, and *V. cholerae* O1 (Kido *et al.*, 1995; Bronner *et al.*, 1994; Szabo *et al.*, 1995; Zhang *et al.*, 1993; Manning *et al.*, 1995). These transporters are typically composed of two components, a hydrophilic ATP-binding protein and an integral membrane protein, arranged as paired homodimers (Fath and Kolter, 1993). The ATP-binding component contains a highly conserved ATP-binding motif, while the hydrophobic component is composed of six membrane-spanning domains, with the N- and C-terminus of the protein localized to the cytoplasm (Higgins, 1992).

This type of transport system, known as ATP transporter dependent (Keenleyside and Whitfield, 1996), has been demonstrated to be necessary for export of the A-band O-polysaccharide (Rocchetta and Lam, 1997). In *P. aeruginosa*, the two transport components, Wzm (integral membrane protein) and Wzt (ATP-binding protein), are encoded within the A-band gene cluster (Rocchetta and Lam, 1997). Interestingly, chromosomal *P. aeruginosa wzm* and *wzt* mutants produce A-band LPS with a faster rate of mobility in sodium dodecyl sulfate–polyacrylamide gel electrophoresis (SDS-PAGE) gels than that of the parent strain (Rocchetta and Lam, 1997). Lack of an active transport system in these mutants prevents translocation of the polymer to the periplasm for ligation to core lipid A. As a result, A-band O-polysaccharides linked to Und-P accumulate within the cell cytoplasm. The absence of core residues in these accumulating polymers results in a decrease in molecular weight and is reflected as an increase in LPS migration on SDS-PAGE gels. Immunoelectron microscopy studies using an A-band-specific monoclonal antibodies (MAb) confirmed the lack of surface-associated A-band LPS in the *wzm* and *wzt P. aeruginosa* mutants and revealed small amounts of A-band LPS within the cell cytoplasm (Rocchetta and Lam, 1997). Synthesis and assembly of B-band LPS in the *wzm* and *wzt* mutants was unaffected, implying that A-band and B-band LPS synthesis proceeds via separate pathways (Fig. 1A,B).

Recent studies on the export system of the *E. coli* homopolymeric K1 capsule have led to a proposed mechanistic model for ABC transporters by Bliss and Silver (1996, 1997). This model involves association of the ATP-binding protein, Kps T, with the K1 polymer and suggests that Kps T undergoes a conformational change upon ATP binding. The Kps T–polymer complex is then thought to insert into the membrane at sites defined by the integral membrane component Kps M. Following ATP hydrolysis, Kps T is thought to return to its original conformation allowing deinsertion from the membrane and release of the polymer to the periplasm. In the case of this K1 export system, two periplasmic proteins, Kps D and Kps E, as well as an outer membrane protein, participate in capsule export to the cell surface. No such proteins have been identified for the export of LPS molecules, and it is likely that a more general mechanism exists for export of completed LPS molecules to the cell surface.

3. ASSEMBLY OF THE HETEROPOLYMER, B-BAND LPS

The mechanism of heteropolymer assembly differs in many respects from that of homopolymers. In the case of heteropolymers, each O-repeat unit is assembled at the cytoplasmic face of the inner membrane by nonprocessive glycosyltransferases, and is translocated to the periplasmic face via the action of the integral membrane protein Wzx (formerly Rfb X). At present, the mechanism by which this translocation, or "flipping," of O-units occurs is poorly defined. No ATP-dependent transporter is required for export of individual O-units. On the periplasmic face of the cytoplasmic membrane, individual O-units are polymerized into chains by the O-antigen polymerase, Wzy (formerly Rfc), to a strain-specific range of lengths determined by the O-antigen chain-length regulator, Wzz (formerly Rol or Cld). The completed chains are covalently attached to core lipid A by the O-antigen ligase, Waa L (Fig. 1B). This type of pathway, which relies on the activity of Wzy to produce long O-chains, is called Wzy dependent (Whitfield, 1995).

3.1. Synthesis and Transport of B-Band O-Units

Work in our laboratory has shown that B-band LPS of serotype O5, which contains a repeating trisaccharide of di-*N*-acetylmannuronic acid and *N*-acetyl-6-deoxygalactose (Fuc2NAc) (Knirel and Kochetkov, 1994), is synthesized through the action of specific, nonprocessive glycosyltransferases. Three glycosyltransferases are predicted to be necessary for assembly of the trisaccharide, and three putative glycosyltransferase genes have been identified in the B-band LPS biosynthetic cluster (Burrows *et al.*, 1996). Wbp L, as described above, is the initiating glycosyltransferase, attaching Fuc2NAc-1-P to Und-P. Wbp L has a flexible substrate specificity, allowing it to recognize both Fuc2NAc-1-P and GlcNAc-1-P (or GalNAc-1-P) to initiate B-band and A-band LPS biosynthesis, respectively. Wbp J and Wbp H are predicted to be mannosaminuronyl transferases and have 23% identity to one another. This observation is consistent with the fact that both proteins are thought to transfer di-*N*-acetylmannuronic acid residues to complete the O-unit. Interestingly, initial analysis of the LPS from *wbpJ*::Gm and *wbpH*::Gm mutants showed that a small amount of O-antigen, recognized by anti-serotype O5 MAbs, continues to be produced by Wbp J-minus mutants (Burrows *et al.*, unpublished data). The phenotype of *wbpJ*::Gm mutants suggests that the homology between Wbp J and Wbp H may allow Wbp H to partially compensate for the loss of Wbp J.

After assembly of the O-antigen unit by the action of the glycosyltransferases, the completed O-unit must be translocated from the cytoplasmic to the periplasmic face of the inner membrane. The most likely candidate for a protein with translocase or "flippase" activity is Wzx (Liu *et al.*, 1996). Previous studies in *Salmonella* demonstrated that strains carrying a mutated copy of Wzx accumulated Und-P-linked O-units in the cytoplasm (Liu *et al.*, 1996). Mutation of *wzx* in *P. aeruginosa* abrogated B-band LPS biosynthesis, and interestingly also caused a significant lag in A-band LPS biosynthesis (Burrows and Lam, 1999). It seemed that the inability of the cells to successfully translocate completed O-units led to sequestering of WbpL, which is necessary for initiation of synthesis of both types of LPS. The delay in A-band synthesis could be alleviated by supplying multiple copies of *wbpL in trans* (Burrows and Lam, 1998). The results of these studies raise the possibility that the glycosyltransferases responsible for synthesis of the O-unit are only recycled or released upon translocation of the completed O-unit to the periplasm.

3.2. Assembly of the B-Band O-Antigen

Following transfer of completed O-units to the periplasmic face of the cytoplasmic membrane, they are polymerized by Wzy, the O-polymerase (Kanegasaki and Wright, 1970; Whitfield, 1995). In contrast to those linkages catalyzed by typical glycosyltransferases, Wzy transfers the growing polymer to the nascent subunit, creating the glycosidic linkage at the nonreducing end of the polymer. This mechanism of synthesis is reminiscent of protein and lipid biosynthesis (Robbins *et al.*, 1967). To date, *in vitro* studies on the mechanism of Wzy activity have been hampered by the inability to express Wzy, due to its poor ribosome-binding site, hydrophobicity, and the presence of multiple rare codons within the *wzy* open reading frame (Daniels *et al.*, 1998). Primary amino acid sequences can be poorly conserved among Wzy proteins both inter- and intraspecies, since they recognize only their cognate O-unit, or close approximations thereof (Schnaitman and Klena, 1993; de Kievit et al., 1997). This lack of homology has made it difficult to identify potential active sites in these proteins. It is possible (and heretical) that Wzy does not directly catalyze the formation of the glycosidic linkage between O-units, but acts instead as a scaffold on which the linkage is formed by another, yet undescribed, protein. In *P. aeruginosa,* specific knockout mutation of *wzy* resulted in cells elaborating only a single B-band O-antigen unit attached to lipid A core (de Kievit *et al.*, 1995, 1997), while A-band LPS biosynthesis was unaffected. This semirough phenotype is consistent with that seen in *wzy* mutants of other bacteria (Schnaitman and Klena, 1993).

Two other genes, *wzx* and *wzz,* found in the B-band LPS gene cluster, are commonly associated with Wzy-dependent O-antigen biosynthesis (Burrows *et al.*, 1996). The activities of both gene products are intimately associated with that of Wzy (Whitfield, 1995). By constructing null mutants, we confirmed that Wzx is involved in B-band O-unit synthesis (Burrows and Lam, 1998) and that Wzz modulates O-antigen chain length (Burrows *et al.*, 1997). These studies indicate that B-band O-antigen is assembled via the classical Wzy-dependent pathway (Fig. 1B). An unusual feature of *P. aeruginosa* serotype O5 is the presence of two separate *wzz* genes on the chromosome; one located adjacent to the B-band LPS biosynthetic genes (Burrows *et al.*, 1996), now designated *wzz1,* and a second, unlinked version, designated *wzz2* (Burrows *et al.*, unpublished data). There is only limited identity (20.8%) (Burrows *et al.*, unpublished data) between the two proteins. Wzz 1 was shown to influence the modulation of LPS in the related serotypes O5 and O16 (Burrows *et al.*, 1997). In that study, the presence of a second copy of *wzz* was predicted based on the continued modulation of LPS length in *wzz1*::Gm^R mutants of serotype O5 (Burrows *et al.*, 1997). Similarly, serotype O16 likely contains two or more copies of *wzz,* since it also continues to synthesize chain-length-modulated LPS following inactivation of *wzz1* (Burrows *et al.*, 1997). Further analysis is underway to determine the contribution of each Wzz to the length of B-band O antigen in serotype O5, a feature that relates to the biological properties of the LPS (Hong and Payne, 1997). The presence of multiple *wzz* genes may represent useful redundancy in terms of evasion of host immune system mechanisms, since loss of chain length modulation has been shown to cause loss of serum resistance in the strains examined (Burns and Hull, 1998).

4. FUTURE DIRECTIONS

4.1. Attachment of O-Antigens to the Core

There are several areas of *P. aeruginosa* LPS assembly that remain to be explored. In particular, the latter stages of assembly, such as attachment of O-polysaccharide to the core oligosaccharide, are not well characterized. In *Enterobacteriaceae*, an enzyme called O-antigen ligase (Waa L) is responsible for attachment of a variety of polysaccharides to core lipid A (Whitfield *et al.*, 1997). This protein, which is usually encoded within the core oligosaccharide gene cluster, has not yet been identified in *P. aeruginosa*. Although the complete chemical structure of the *P. aeruginosa* B-band core oligosaccharide (Sadovskoya *et al.*, 1998), and almost the entire genome sequence of strain PAO1 (www.pseudomonas.com) are now available, less than half of the genes predicted to be involved in synthesis and assembly of the core oligosaccharide have been identified. In addition, while the point of attachment of B-band O antigen to the core oligosaccharide has been identified through structural studies as a side-branch glucose (Sadovskaya *et al.*, 1998), the attachment point of the A-band O-polysaccharide remains to be ascertained. Interestingly, based on analysis of column fractionated LPS (Rivera and Mc-Groarty, 1989), the structure of the core region of A-band LPS may differ from that of B-band LPS, containing sulfate, rather than phosphate substitutions. Monoclonal antibody analysis (Rivera *et al.*, 1992) revealed additional differences between the outer core regions of A-band and B-band LPS, although the chemical basis for these differences is not known.

4.2. Translocation of O-Units and Completed LPS Molecules

It has proved to be technically challenging to characterize the activity of Wzx, the putative O-unit translocase, likely due to some of the same reasons hampering studies of Wzy. Wzx is an integral membrane protein that is not amenable to overexpression using current systems (Liu *et al.*, 1996; Burrows and Lam, 1999). In addition, mutation of Wzx in enteric organisms is deleterious, which has hampered the generation of specific mutants (Schnaitman and Klena, 1993; Macpherson *et al.*, 1995; Liu *et al.*, 1996). The mechanism(s) by which A band or B band is translocated to the cell surface after ligation to core lipid A is unknown. The appearance of newly synthesized LPS molecules on the cell surface has been shown in *E. coli* to occur at multiple points, the majority of which appeared to represent zones of adhesion between the inner and outer membranes (Mühlradt *et al.*, 1973). The manner in which completed LPS molecules cross the periplasm is not understood, and the existence of adhesion zones (also called Bayer's junctions, or Bayer's bridges) (Bayer, 1991) is still a matter of debate. However, it is tempting to speculate that such regions of intimate contact between the inner and outer membrane would be a likely site for translocation of LPS between membranes.

4.3. Mechanisms of Regulation of O-Chain Length

An aspect of heteropolymer synthesis that remains poorly understood is the manner in which Wzz is able to influence O-antigen chain length. Recent analysis of heterologous Wzz expression has shown that chain length is determined by Wzz irrespective of the O-antigen structure being modulated (Franco *et al.*, 1998). Comparison of Wzz proteins specifying short, medium or long chains led to the tentative identification of residues characteristic of each chain length type. Mutant Wzz proteins have been generated through site-directed mutagenesis of critical amino acid residues and construction of chimeric Wzz proteins (Franco *et al.*, 1998). Modification of amino acid residues thought to be crucial in determining chain length resulted in O-chain length alterations, suggesting specificity is at least partially dependent on the primary sequence of Wzz (Franco *et al.*, 1998). However, the exact mechanism by which a particular Wzz protein controls polymerization in order to generate its preferred chain length is still unclear.

Another unsolved mystery involves the way in which the chain length of homopolymers is controlled. Although homopolyaccharides possess polymers with a clearly defined modal distribution, synthesis of homopolymers, including A-band LPS, is known to proceed independently of Wzz (Whitfield, 1995; Franco *et al.*, 1996; Dodgson *et al.*, 1996; Burrows *et al.*, 1997). The length of A-band LPS has been found to be similar among the various serotypes of *P. aeruginosa* (Lam *et al.*, 1989). Therefore, an alternative mechanism must exist that regulates the chain length of such O-polysaccharides. Data from chemical analysis have identified 3-*O*-methyl sugar residues at the nonreducing terminus of some homopolysaccharides. For example, the mannan O-polysaccharides of *E. coli* O8 (Jansson *et al.*, 1985) and *K. pneumoniae* O5 (Lindberg *et al.*, 1972) have been found to terminate in 3-*O*-methyl-D-mannose residues, while the *Campylobacter fetus* serotype B rhamnan O-polysaccharide terminates in a 3-*O*-methyl-D-rhamnose moiety (Senchenkova *et al.*, 1996). The presence of such 3-*O*-methyl sugars at the nonreducing terminus likely prevents subsequent chain elongation, and thus may regulate homopolymer chain length. Our group has detected the presence of 3-*O*-methyl-rhamnose in A-band polysaccharide (Arsenault *et al.*, 1991). Although it has not yet been determined whether this 3-*O*-methyl-rhamnose occupies the terminal position of the A-band polymer, it seems likely, in light of the above examples. In future it will be of interest to determine how this 3-*O*-methyl sugar is synthesized, as well as when and where it is added to the A-band polysaccharide.

5. CONCLUSION

Studies of these and other areas of uncertainty in LPS biosynthesis are fundamental to our knowledge of bacterial physiology. Ultimately, we intend to investigate the biological significance and mechanisms by which *P. aeruginosa* is able to express multiple and appropriate cell surface molecules in response to specific environmental conditions/signals. Insight in this area will be crucial for the rational design of antimicrobial agents directed against *P. aeruginosa* and other medically significant gram-negative bacteria.

ACKNOWLEDGMENTS

The results reviewed here have been supported by research grants to J. S. L. from the Canadian Cystic Fibrosis Foundation (CCFF), Canadian Bacterial Diseases Network, a consortium of the Networks of Centers of Excellence program, and the Medical Research Council (grant #MT14687). L. L. B. is the recipient of a CCFF Kinsmen Fellowship, and H. L. R. is a recipient of a CCFF studentship.

REFERENCES

Alexander, D. C., and Valvano, M. A., 1994, Role of the *rfe* gene in the biosynthesis of the *Escherichia coli* O7-specific lipopolysaccharide and other O-specific polysaccharides containing *N*-acetylglucosamine, *J. Bacteriol.* **176:**7079–7084.

Amor, P. A., and Whitfield, C., 1997, Molecular and functional analysis of genes required for expression of group IB K antigens in *Escherichia coli:* Characterization of the *his*-region containing gene clusters for multiple cell-surface polysaccharides, *Mol. Microbiol.* **26:**145–161.

Arsenault, T. L., Huges, D. W., MacLean, D. B., Szarek, W. A., Kropinski, A. M., and Lam, J. S., 1991, Structural studies on the polysaccharide portion of "A-band" lipopolysaccharide from a mutant (AK1401) of *Pseudomonas aeruginosa* strain PAO1, *Can. J. Chem.* **69:**1273–1280.

Bayer, M. E., 1991, Zones of membrane adhesion in the cryofixed envelope of *Escherichia coli, J. Struct. Biol.* **107:**268–280.

Bliss, J. M., and Silver, R. P., 1996, Coating the surface: a model for expression of capsular polysialic acid in *Escherichia coli* K1, *Mol. Microbiol.* **21:**221–231.

Bliss, J. M., and Silver, R. P., 1997, Evidence that KpsT, the ATP-binding component of an ATP-binding cassette transporter, is exposed to the periplasm and associates with polymer during translocation of the polysialic acid capsule of *Escherichia coli* K1, *J. Bacteriol.* **179:**1400–1403.

Bronner, D., Clarke, B. R., and Whitfield, C., 1994, Identification of an ATP-binding cassette transport system required for translocation of lipopolysaccharide O-antigen sidechains across the cytoplasmic membrane of *Klebsiella pneumoniae* serotype O1, *Mol. Microbiol.* **14:**505–519.

Burns, S. M., and Hull, S. I., 1998, Comparison of loss of serum resistance by defined lipopolysaccharide mutants and an acapsular mutant of uropathogenic *Escherichia coli* O75:K5, *Infect. Immun.* **66:**4244–4253.

Burrows, L. L., and Lam, J. S., 1999, Effects of a *wzx* (*rfbX*) mutation on A-band and B-band lipopolysaccharide biosynthesis in *Pseudomonas aeruginosa* serotype O5. *J. Bacteriol.* **181:**973–980.

Burrows, L. L., Charter, D. F., and Lam, J. S., 1996, Molecular characterization of the *Pseudomonas aeruginosa* serotype O5 (PAO1) B-band lipopolysaccharide gene cluster, *Mol. Microbiol.* **22:**481–495.

Burrows, L. L., Chow, D., and Lam, J. S., 1997, *Pseudomonas aeruginosa* B-band O-antigen chain length is modulated by Wzz (Rol), *J. Bacteriol.* **179:**1482–1489.

Cerantola, S., and Montrozier, H., 1997, Structural elucidation of two polysaccharides present in the lipopolysaccharide of a clinical isolate of *Burkholderia cepacia, Eur. J. Biochem.* **246:**360–366.

Clarke, B. R., and Whitfield, C., 1992, Molecular cloning of the *rfb* region of *Klebsiella pneumoniae* serotype O1:K20: The *rfb* gene cluster is responsible for synthesis of the D-galactan I O polysaccharide, *J. Bacteriol.* **174:**4614–4621.

Clarke, B. R., Bronner, D., Keenleyside, W. J., Severn, W. B., Richards, J. C., and Whitfield, C., 1995, Role of Rfe and RfbF in the initiation of biosynthesis of D-galactan I, the lipopolysaccharide O antigen from *Klebsiella pneumoniae* O1, *J. Bacteriol.* **177:**5411–5418.

Daniels, C., Vindurampulle, C., and Morona, R., 1998, Overexpression and topology of the *Shigella flexneri* O-antigen polymerase, *Mol. Microbiol.* **28:**1211–1222.

de Kievit, T. R., Dasgupta, T., Schweizer, H., and Lam, J. S., 1995, Molecular cloning and characterization of the *rfc* gene of *Pseudomonas aeruginosa* (serotype O5), *Mol. Microbiol.* **16:**565–574.

de Kievit, T. R., Staples, T., and Lam, J. S., 1997, *Pseudomonas aeruginosa rfc* genes of serotypes O2 and O5 could complement O-polymerase-deficient semi-rough mutants of either serotype. *FEMS Microbiol. Lett.* **147:**251–257.

De Vos, D., Lim Jr, A., Pirnay, J. P., Duinslaeger, L., Revets, H., Vanderkelen, A., Hamers, R., and Cornelis, P., 1997, Analysis of epidemic *Pseudomonas aeruginosa* isolates by isoelectric focusing of pyoverdine and RAPD-PCR: modern tools for an integrated antinosocomial infection strategy in burn wound centres, *Burns* **23:**379–386.

Dodgson, C., Amor, P., and Whitfield, C., 1996, Distribution of the *rol* gene encoding the regulator of lipopolysaccharide O-chain length in *Escherichia coli* and its influence on the expression of group I capsular K antigens, *J. Bacteriol.* **178:**1895–1902.

Fath, M. J., and Kolter, R., 1993, ABC transporters: Bacterial exporters, *Microbiol. Rev.* **57:**995–1017.

Franco, A. V., Liu, D., and Reeves, P. R., 1996, A Wzz (Cld) protein determines the chain length of K lipopolysaccharide in *Escherichia coli* O8 and O9 strains, *J. Bacteriol.* **178:**1903–1907.

Franco, A. V., Liu, D., and Reeves, P. R., 1998, The *wzz* (*cld*) protein in *Escherichia coli:* amino acid sequence variation determines O-antigen chain length specificity. *J. Bacteriol.* **180:**2670–2675.

Fuchs-Cleveland, E., and Gilvarg, C., 1976, Oligomeric intermediate in peptidoglycan biosynthesis in *Bacillus megaterium, Proc. Natl. Acad. Sci. USA* **73:**4200–4204.

Geremia, R. A., Petroni, E. A., Ielpi, L., and Henrissat, B., 1996, Towards a classification of glycosyltransferases based on amino acid sequence similarities: Prokaryotic α-mannosyltransferases, *Biochem. J.* **318:**133–138.

Govan, J. R. W., and Deretic, V., 1996, Microbial pathogenesis in cystic fibrosis: Mucoid *Pseudomonas aeruginosa* and *Burkholderia cepacia, Microbiol. Rev.* **60:**539–574.

Heinrichs, D. E., Valvano, M. A., and Whitfield, C., 1999, Biosynthesis and genetics of lipopolysaccharide core, in: *Endotoxin in health and disease* (H. Brade, D. C. Morrison, S. M. Opal, and S. N. Vogel, eds.), Marcel Dekker, New York, pp. 305–330.

Higgins, C. F., 1992, ABC transporters: From microorganisms to man, *Annu. Rev. Cell. Biol.* **8:**67–113.

Hirano, M., Kiyohara, H., Yamada, H., 1994, Existence of a rhamnogalacturonan II-like region in bioactive pectins from medicinal herbs, *Planta. Med.* **60:**450–454.

Hong, M., and Payne, S. M., 1997, Effect of mutations in *Shigella flexneri* chromosomal and plasmid-encoded lipopolysaccharide genes on invasion and serum resistance, *Mol. Microbiol.* **24:**779–791.

Jansson, P.-E., L^nngren, J., and Widmalm, G., 1985, Structural studies of the O-antigen polysaccharides of *Klebsiella pneumoniae* O5 and *Escherichia coli* O8, *Carbohydr. Res.* **145:**59–66.

Jarvis, W. R., and Martone, W. J., 1992, Predominant pathogens in hospital infections, *J. Antimicrob. Chemother.* (Suppl. A) **29:**19–24.

Kanegasaki, S., and Wright, A., 1970, Mechanism of polymerization of the *Salmonella* O-antigen: Utilization of lipid-linked intermediates, *Proc. Natl. Acad. Sci. USA* **67:**951–958.

Keenleyside, W. J., and Whitfield, C., 1996, A novel pathway for O-polysaccharide biosynthesis in *Salmonella enterica* serovar Borreze, *J. Biol. Chem.* **271:**28581–28592.

Kido, N., Torgov, V. I., Sugiyama, T., Uchiya, K., Sugihara, H., Komatsu, T., Kato, N., and Jann, K., 1995, Expression of the O9 polysaccharide of *Escherichia coli:* Sequencing of the *E. coli* O9 *rfb* gene cluster, characterization of mannosyl transferases, and evidence for an ATP-binding cassette transport system, *J. Bacteriol.* 177:2178–2187.

Kido, N., Morooka, N., Paeng, N., Ohtani, T., Kobayashi, H., Shibata, N., Okawa, Y., Suzuki, S., Sugiyama, T., and Yokochi, T., 1997, Production of monoclonal antibody discriminating serological differences in *Escherichia coli* O9 and O9a polysaccharides. *Microbiol. Immunol.* **41:**519–525.

Knirel, Y. A., and Kochetkov, N. K., 1994, The structure of lipopolysaccharides of gram-negative bacteria. III. The structure of O-antigens: A review, *Biochemistry* **59:**1325–1382.

Lam, M. Y. C., McGroarty, E. J., Kropinski, A. M., MacDonald, L. A., Pedersen, S. S., Høiby, N., and Lam, J. S., 1989, Occurrence of a common lipopolysaccharide antigen in standard and clinical strains of *Pseudomonas aeruginosa, J. Clin. Microbiol.* **27**:962–967.

Lightfoot, J., and Lam, J. S., 1991, Molecular cloning of genes involved with expression of A-band lipopolysaccharide, an antigenically conserved form of *P. aeruginosa, J. Bacteriol.* **173**:5624–5630.

Lightfoot, J., and Lam, J. S., 1993, Chromosomal mapping, expression and synthesis of lipopolysaccharide in *Pseudomonas aeruginosa:* A role for guanosine diphospho (GDP)-D-mannose, *Mol. Microbiol.* **8**:771–782.

Lindberg, B., Lönngren, J., and Nimmich, W., 1972, Structural studies on *Klebsiella* O group 5 lipopolysaccharides, *Acta Chem. Scand.* **26**:2231–2236.

Liu, D., Cole, R. A., and Reeves, P. R., 1996, An O-antigen processing function for Wzx (RfbX): A promising candidate for O-unit flippase, *J. Bacteriol.* **178**:2102–2107.

Lui, P. V., and Wang, S., 1990, Three new major somatic antigens of *Pseudomonas aeruginosa, J. Clin. Microbiol.* **28**:922–925.

Macpherson, D. F., Manning, P. A., and Morona, R., 1995, Genetic analysis of the rfbX gene of *Shigella flexneri. Gene* **155**:9–17.

Manning, P. A., Stroeher, U. H., Karageorgos, L. E., and Morona, R., 1995, Putative O-antigen transport genes within the *rfb* region of *Vibrio cholerae* O1 are homologous to those for capsule transport, *Gene* **158**:1–7.

Mühlradt, P. F., Menzel, J., Golecki, J. R., and Speth, V., 1973, Outer membrane of *Salmonella.* Sites of export of newly synthesized lipopolysaccharide on the bacterial surface, *Eur. J. Biochem.* **35**:471–481.

Ochsner, U. A., Hembach, T., and Fiechter, A., 1996, Production of rhamnolipid biosurfactants, *Adv. Biochem. Eng. Biotechnol.* **53**:89–118

Popoff, M. Y., and LeMinor, L., 1985, Expression of antigenic factor O:54 is associated with the presence of a plasmid in *Salmonella, Ann. Inst. Pasteur Microbiol.* **136B**:169–179.

Quinn, J. P., 1998, Clinical problems posed by multiresistant nonfermenting gram-negative pathogens, *Clin. Infect. Dis.* 27 (Suppl 1):S117–S124.

Rick, P. D., Hubbard, G. L., and Barr, K., 1994, Role of the *rfe* gene in the synthesis of the O8 antigen in *Escherichia coli* K-12, *J. Bacteriol.* **176**:2877–2884.

Rivera, M., and McGroatry, E. J., 1989, Analysis of a common-antigen lipopolysaccharide from *Pseudomonas aeruginosa, J. Bacteriol.* **171**:2244–2248.

Rivera, M., Chivers, T. R., Lam, J. S., and McGroarty, E. J., 1992, Common antigen lipopolysaccharide from *Pseudomonas aeruginosa* AK1401 as a receptor for bacteriophage A7, *J. Bacteriol.* **174**:2407–2411.

Robbins, P. W., Bray, D., Dankert, M., and Wright, A., 1967, Direction of chain growth in polysaccharide synthesis. *Science* **158**:1536–1542.

Rocchetta, H. L., and Lam, J. S., 1997, Identification and functional characterization of an ABC transport system involved in polysaccharide export of A-band lipopolysaccharide in *Pseudomonas aeruginosa, J. Bacteriol.* **179**:4713–4724.

Rocchetta, H. L., Burrows, L. L, Pacan, J. C., and Lam, J. S., 1998, Three rhamnosyltransferases responsible for assembly of the A-band D-rhamnan polysaccharide in

Pseudomonas aeruginosa: A fourth transferase, WbpL, is required for initiation of both A-band and B-band LPS synthesis, *Mol. Microbiol.* **28:**1103–1119.

Sadovskaya, I., Brisson, J. R., Lam, J. S., Richards, J. C., and Altman, E., 1998, Structural elucidation of the lipopolysaccharide core regions of the wild-type strain PAO1 and O-chain-deficient mutant strains AK1401 and AK1012 from *Pseudomonas aeruginosa* serotype O5, *Eur. J. Biochem.* **255:**673–684.

Schnaitman, C. A., and Klena, J. D., 1993, Genetics of lipopolysaccharide biosynthesis in enteric bacteria, *Microbiol. Rev.* **57:**5670–5679.

Senchenkova, S. N., Shashkov, A. S., Knirel, Y. A., McGovern, J. J., and Moran, A. P., 1996, The O-specific polysaccharide chain of *Campylobacter fetus* serotype B lipopolysaccharide is a D-rhamnan terminated with 3-*O*-methyl-D-rhamnose (D-acofriose), *Eur. J. Biochem.* **239:**434–438.

Shibaev, V. L., 1986, Biosynthesis of bacterial polysaccharide chains composed of repeating units, *Adv. Carbohydr. Chem. Biochem.* **44:**277–339.

Smith, R. W., Zamze, S. E., Munro, S. M., Carter, K. J., and Hignett, R. C., 1985, Structure of the side chain of lipopolysaccharide from *Pseudomonas syringae* pv. *morsprunorum* C28, *Eur. J. Biochem.* **149:**73–78.

Spencer, R. C., 1996, Predominant pathogens found in the European prevalence of infection in intensive care study, *Eur. J. Clin. Microbiol. Infect. Dis.* **15:**281–285.

Stanislavsky, E. S., and Lam, J. S., 1997, *Pseudomonas aeruginosa* antigens as potential vaccines, *FEMS Microbiol. Rev.* **21:**243–277.

Szabo, M., Bronner, D., and Whitfield, C., 1995, Relationships between *rfb* gene clusters required for biosynthesis of identical D-galactose-containing O antigens in *Klebsiella pneumoniae* serotype O1 and *Serratia marcescens* serotype O16, *J. Bacteriol.* **177:**1544–1553.

Vinogradov, E. V., Shashkov, A. S., Knirel, Y. A., Zdorovenko, G. M., Solyanik, L. P., and Gvozdyak, R. I., 1991, Somatic antigens of pseudomonads: Structure of the O-specific polysaccharide chain of *Pseudomonas syringae* pv. *syringae* (*cerasi*) 435 lipopolysaccharide, *Carbohydr. Res.* **212:**295–299.

Wang, L., Liu, D., and Reeves, P. R., 1996, C-terminal half of *Salmonella enterica* WbaP (RfbP) is the galactosyl-1-phosphate transferase domain catalyzing the first step of O-antigen synthesis, *J. Bacteriol.* **178:**2598–2604.

Whitfield, C., 1995, Biosynthesis of lipopolysaccharide O antigens, *Trends Microbiol.* **3:**178–185.

Whitfield, C., Amor, P. A., and Köplin, R., 1997, Modulation of the surface architecture of gram-negative bacteria by the action of surface polymer:lipid A-core ligase and by determinants of polymer chain length, *Mol. Microbiol.* **23:**629–638.

Winn, A. M., and Wilkinson, S. G., 1998, The O7 antigen of *Stenotrophomonas maltophilia* is a linear D-rhamnan with a trisaccharide repeating unit that is also present in polymers for some *Pseudomonas* and *Burkholderia* species, *FEMS Microbiol. Lett.* **166:**57–61.

Wright, A., Dankert, M., Fennessey, P., and Robbins, P. W., 1967, Characterization of a polyisoprenoid compound functional in O-antigen biosynthesis, *Proc. Natl. Acad. Sci. USA* **57:**1798–1803.

Zhang, L., Al-Hendy, A., Toivanen, P., and Skurnik, M., 1993, Genetic organization and sequence of the *rfb* gene cluster of *Yersinia enterocolitica* serotype O:3: Similarities to the dTDP-L-rhamnose biosynthesis pathway of *Salmonella* and to the bacterial polysaccharide transport systems, *Mol. Microbiol.* **9:**309–321.

Zhang, L., Radziejewska-Lebrecht, J., Krajewska-Pietrasik, D., Toivanen, P., and Skurnik, M., 1997, Molecular and chemical characterization of the lipopolysaccharide O-antigen and its role in the virulence of *Yersinia enterocolitica* serotype O:8, *Mol. Microbiol.* **23:**63–76.

6

Interactions of Bacterial Lipopolysaccharide and Peptidoglycan with Mammalian CD14

Roman Dziarski, Artur J. Ulmer, and Dipika Gupta

1. STRUCTURE OF LPS AND PEPTIDOGLYCAN

Lipopolysaccharide (LPS) is the main amphiphilic component of the outer membrane present in gram-negative bacteria. LPS is typically composed of three regions: lipid A, core polysaccharide, and O-specific polysaccharide (Fig. 1) (Rietschel *et al.*, 1994). The structure of lipid A is highly conserved among eubacteria. Lipid A is composed of a β-(1 → 6)-linked D-glucosamine disaccharide substituted at positions 4′ and 1 by phosphomonoester groups, with fatty acids linked to the remaining hydroxyl and amino groups. In enterobacteria, the amide- and ester-linked D-3-hydroxy fatty acids consist of 14 carbon β-hydroxymyristic acids, with their C3-OH positions often further esterified with saturated fatty acids. The inner core polysaccharide is attached to the hydroxyl group at C6′, and typically consists of 2-keto-3-deoxy-octulosonic acid (KDO), heptose, and phosphate. The outer core of LPS in enteric bacteria is composed of hexoses (such as glucose, galactose, and N-acetylglucosamine). The O-specific polysaccharide is highly

Roman Dziarski and Dipika Gupta • Northwest Center for Medical Education, Indiana University School of Medicine, Gary, Indiana 46408. *Artur J. Ulmer* • Division of Cellular Immunology, Research Center Borstel, 23845 Borstel, Germany.

Glycomicrobiology, edited by Doyle.
Kluwer Academic/Plenum Publishers, New York, 2000.

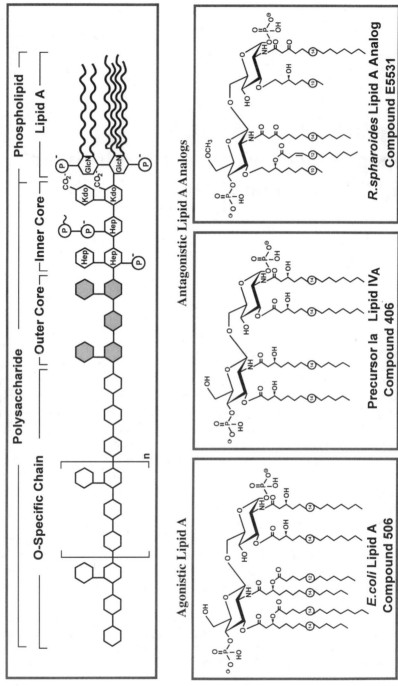

Figure 1. Chemical structures of *Enterobacteriaceae* LPS, lipid A, and lipid A antagonists.

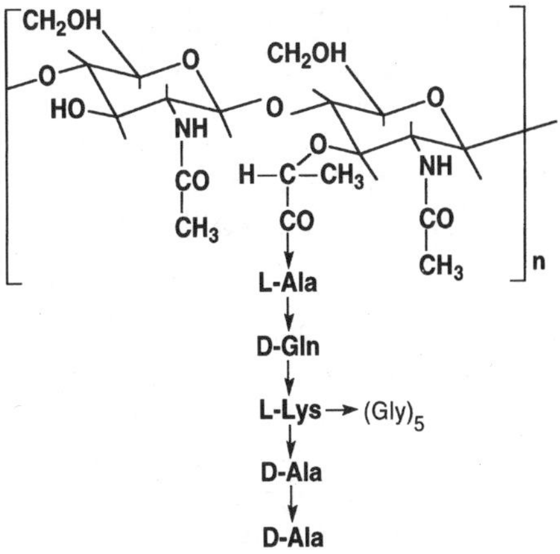

Soluble peptidoglycan
from *S. aureus*

Figure 2. Chemical structure of un-crosslinked *S. aureus* PG.

- [→4)-ß-D-GlcNAc*p*-(1→4)-ß-D-MurNAc*p*- 3→
→L-Ala→D-Gln→L-Lys-[(Gly)$_5$]→D-Ala→D-Ala-(1-]$_n$-

variable among species and even among strains and is typically composed of 10 to more than 100 repeating tri- or tetraoligosaccharide units.

Peptidoglycan (PG) is found in the cell walls of virtually all bacteria and is especially abundant in the cell walls of gram-positive bacteria. PG is composed of a glycan backbone of up to 100 alternating units of β-(1 → 4)-linked *N*-acetylglucosamine and *N*-acetylmuramic acid (MurNAc), with short peptides linked to the lactyl group of the MurNAc residues (Fig. 2). The general structure of the peptide is L-alanine-D-glutamic acid-a diamino acid-D-alanine-D-alanine. The diamino acid in position 3 is typically lysine in gram-positive cocci and diaminopimelic acid in gram-positive bacilli and Gram-negative bacteria. In the cell wall, D-alanine in position 4 in the peptide of one chain is often cross-linked to the diamino acid in position 3 of a different glycan chain either directly (in gram-positive bacilli and gram-negative bacteria) or through a peptide bridge (in gram-positive cocci, e.g., pentaglycine bridge in *Staphylococcus aureus*) (Schleifer and Kandler, 1975). This peptide cross-linking results in the formation of an enormous basketlike macromolecule surrounding the cytoplasmic membrane. In the cell wall, numerous macromolecules, such as cell wall teichoic acid, polysaccharides, and proteins are often covalently bound to PG (Rosenthal and Dziarski, 1994).

When gram-positive bacteria grow in the presence of β-lactam antibiotics (*in vitro* or *in vivo*), they secrete soluble polymeric un–cross-linked PG fragments of 50 to 100 disaccharide units, owing to continued synthesis of PG chains and inhibition of transpeptidation and lack of incorporation of this newly synthesized PG into the existing cell wall (Rosenthal and Dziarski, 1994).

2. OVERVIEW OF BIOLOGICAL ACTIVITIES OF LPS AND PEPTIDOGLYCAN

LPS has an extraordinary array of biological activities that influence virtually every tissue and organ in the body. Despite quite different chemical structure, PG has many similar activities (Dziarski, 1986; Heymer *et al.,* 1985), although usually it is not as active as LPS. This large number of biological effects exerted by a single molecule and the large number of similar biological effects exerted by two chemically different molecules are due to the indirect induction of these effects by LPS and PG, through the release of various mediators from host cells (Table I). Thus, the similarity of the biological effects of LPS and PG results from induction of the same mediators from the same target cells.

The release of these mediators accounts for the ability of LPS and PG to reproduce all major signs and symptoms of bacterial infections, including fever, inflammation, hypotension, leukocytosis, sleepiness, decreased appetite, malaise, and arthritis (Table II). It also should be remembered that, in addition to the direct targets of LPS and PG listed in Table I, there are numerous secondary targets that are affected by the mediators induced by LPS and PG. Moreover, primary mediators listed in Tables I and II also induce secondary mediators or more of the primary mediators. For example, tumor necrosis factor-α (TNF-α) and interleukin-1 (IL-1) induce secretion of IL-6 and IL-8 or expression of adhesion molecules (E- and P-selectins, intercellular adhesion molecule (ICAM-1), and vascular cellular adhesion molecule (VCAM-1)) on endothelial cells.

Despite the general similarity of the major biological effects of LPS and PG (Tables 1 and 2), there are two major differences:

1. Most of the major biological effects of LPS are induced at concentrations that are several orders of magnitude lower than the concentrations of PG needed to induce similar effects. For example, there is a 4 log difference between the effective macrophage-activating concentrations (on per weight basis) between LPS and PG (Dziarski *et al.,* 1998; Weidemann *et al.,* 1994). This difference, as will be discussed below in more detail, mainly applies to the macrophage-mediated effects that occur through the CD14 receptor and are potentiated (for LPS, but not for PG) by LPS- binding protein. Only in few cases is PG more potent than LPS, mainly in the induction of arthritis and chronic inflammation, because of the ability of PG or PG-containing cell walls to persist in tissues for an extended period of time.

Table I

Cellular and Humoral Targets of LPS and PG and Mediators Produced

Target	Mediators produced[a]	Stimulants[b]	
		LPS	PG
Macrophage	TNF-α	Yes[1-4]	Yes[5-7]
	IL-1	Yes[4,8]	Yes[9-11]
	IL-6	Yes[4,12]	Yes[11]
	IL-8	Yes[13-15]	?[16]
	IL-10	Yes[17-19]	?
	IL-12	Yes[20]	Yes[21]
	CSF	Yes[22]	Yes[10,23]
	NO	Yes[24]	?[25,26]
	PAF	Yes[27,28]	?
	Tissue factor	Yes[29]	?
	GRO and other C-X-C chemokines	Yes[15,30]	?
	MIP-1 and other C-C chemokines	Yes[31]	?
Neutrophil	PAF	Yes[27,28]	?
	MIP-1	Yes[32]	?
Platelets	Vasoactive amines	Yes[33]	?[34,35]
	PAF	Yes[27,28]	?
Endothelial and some epithelial cells (+ scD14)	IL-6	Yes[36,37]	No[38]
	IL-8	Yes[13,37]	No[38]
	PAF	Yes[27,28]	?
	NO	Yes[24]	?
	E-selectin, P-selectin	Yes[36]	No[40]
	ICAM-1	Yes[37,39]	No[40]
	VCAM-1	Yes[37]	No[38]
Complement	C3a	Yes[41]	Yes[42]
	C5a	Yes[41]	Yes[42]
Hageman factor	Kinins (bradykinin)	Yes[43]	Yes[44-46]

[a]Abbreviations: CSF, colony-stimulating factor; GRO, growth-related peptide; ICAM, intercellular adhesion molecule; IL, interleukin; MIP, macrophage-inflammatory protein; NO, nitric oxide; PAF, platelet-activating factor; TNF, tumor necrosis factor; VCAM, vascular cellular adhesion molecule; ?, the effect was not studied with PG.
[b]References: 1. Beutler and Cerami (1988); 2. Vassalli (1992); 3. Beutler and Grau (1993); 4. Rietschel et al. (1994); 5. Mathison et al. (1992); 6. Timmerman et al. (1993); 7. Gupta et al. (1995); 8. Durum et al. (1985); 9. Vacheron et al. (1983); 10. Gold et al. (1985); 11. Weidemann et al. (1994); 12. Van Snick (1990); 13. Baggiolini (1989); 14. Schroder et al. (1990); 15. LaRosa et al. (1992); 16. Yes for PG-polysaccharide complex; Vowels et al. (1995); 17. Luster and Leder (1993); 18. de Vries (1995); 19. Ziegler-Heitbrock (1995); 20. Trinchieri (1995); 21. Lawrence and Nauciel (1998); 22. Metcalf (1991); 23. Dokter et al. (1994); 24. MacMicking et al. (1997); 25. Yes for PG-polysaccharide complex; Kissin et al. (1997); 26. PG enhances lipoteichoic acid (LTA)- and IFN-γ-induced NO; De Kimpe et al. (1995), and Kengatharan et al. (1998); 27. Camussi et al. (1995); 28. Kruse-Elliott et al. (1996); 29. Mackman et al. (1991); 30. Haskill et al. (1990); 31. Wolpe and Cerami (1989); 32. Kasma et al. (1993); 33. Mannel and Grau (1997); 34. induces platelet aggregation and lysis; Ryc and Rotta (1975); 35. Kessler et al. (1991); 36. Frey et al. (1992); 37. Pugin et al. (1993a); 38. Jin et al. (1998); 39. Haziot et al. (1993b); 40. R. Dziarski, unpublished results; 41. Cooper (1991); 42. Heymer et al. (1985); 43. Proud and Kaplan (1988); 44. Kalter et al. (1983); 45. DeLaCadena et al. (1991); 46. Blais et al. (1997).

Roman Dziarski *et al.*

Table II
Main Biological Effects of Mediators Induced by LPS and PG

Biological effect	Main mediators[a]	Stimulants[b]	
		LPS	PG
Fever	IL-1, TNF-α, IL-6	Yes[1-6]	Yes[7]
Inflammation	TNF-α, IL-1, IL-6, IL-8, NO, PAF, C3a, C5a, eicosanoids, adhesion molecules, and other	Yes[1-6]	Yes[7]
Acute-phase response	IL-6, TNF-α, IL-1	Yes[1-6,8]	?
Hypotension	PAF, NO, TNF-α, IL-1, bradykinin, eicosanoids	Yes[1-6,8]	Yes[9,10]
Decreased peripheral circulation and perfusion	TNF-α, IL-1, IL-6, NO, PAF, C3a, C5a, eicosanoids, kinins	Yes[1-6,8]	?
Circulatory shock and death	TNF-α, IL-1, IL-6, NO, PAF, C3a, C5a, eicosanoids, kinins	Yes[1-6,8]	No[9]
Leukopenia followed by leukocytosis	IL-1, TNF-α, CSF	Yes[1-4,6,8]	Yes[7]
Sleepiness	IL-1, TNF-α	Yes[1-4,6,8]	Yes[11]
Decreased appetite	IL-1, TNF-α	Yes[1-4,8]	Yes[12]
DIC, thrombosis	Tissue factor, Hageman factor, PAF, platelet aggregation	Yes[6,8]	Yes[13]
Thrombocytopenia	PAF, clotting factors	Yes[6,8]	Yes[10,14]
Arthritis	IL-1, TNF-α, kinins	Yes[15,16]	Yes[7,17]
Immune adjuvant	IL-1	Yes[18]	Yes[7]

[a]Abbreviations, see Table I.
[b]References: 1. Beutler and Cerami (1988); 2. Vassalli (1992); 3. Beutler and Grau (1993); 4. Durum *et al.* (1985); 5. Van Snick (1990); 6. Cortran *et al.* (1994); 7. Heymer *et al.* (1985); 8. Young (1995); 9. De Kimpe *et al.* (1995); 10. Verhoef and Kalter (1985); 11. Johannsen (1993); 12. Biberstine and Rosenthal (1994); 13. Kessler *et al.* (1991); 14. Spika *et al.* (1982); 15. Matsukawa *et al.* (1993); 16. Noyori *et al.* (1994); 17. Blais *et al.* (1997); 18. Alving (1993).

Table III
Biological Effects Exhibited by LPS but Not by PG[a]

Circulatory shock and death[1,2]
Toxicity in galactosamine-treated mice[3]
Toxicity in adrenalectomized mice[4]
Enhancement of cell activation by LBP[5-7]
Generalized Shwartzman reaction[8]
Gelation of *Limulus* lysate[9,10]

[a]References: 1. Redl *et al.* (1989); 2. De Kimpe *et al.* (1995); 3. J.T. Ulrich (unpublished data); 4. Dziarski and Dziarski (1979); 5. Mathison *et al.* (1992); 6. Weidemann *et al.* (1994); 7. Dziarski *et al.* (1998); 8. Heymer *et al.* (1985); 9. Wildfeuer *et al.* (1975); 10. Rosenthal and Dziarski (1994).

Table IV
Biological Effects Exhibited by PG but Not by LPS[a]

B cell and T cell mitogenicity for human PBL[1-3]
Induction of polyclonal antibodies in human PBL[2,3]
Activation of B cells and macrophages from C3H/HeJ and C57BL/10ScCR mice[4-8]
Activation of insect hematocytes and production of antibacterial proteins[9]
Activation of prophenol oxidase cascade in insects[10]

[a]References: 1. Dziarski and Dziarski (1979); 2. Rasanen and Arvilommi (1981); 3. Levinson *et al.* (1983); 4. Saito-Taki *et al.* (1980a); 5. Saito-Taki *et al.* (1980b); 6. Guenounou *et al.* (1982); 7. Vacheron *et al.* (1983); 8. Vacheron *et al.* (1986); 9. Dunn *et al.* (1985); 10. Yoshida *et al.* (1996).

2. There are some effects that are exhibited by LPS but not by PG and vice versa (Tables III and IV). Interestingly, most of the effects exhibited by LPS, but not by PG, (except shock and death) are unique for LPS (Table III), whereas the effects exhibited by PG, but not by LPS (Table IV), are not unique for PG and are exhibited by other bacterial or nonbacterial products.

One of the most significant differences between the biological effects of LPS and PG is that PG by itself, in contrast to LPS, is not lethal and does not induce circulatory shock. However, PG may act synergistically with other bacterial stimulants (e.g., lipoteichoic acid or superantigenic toxins) in induction of shock and death (De Kimpe *et al.*, 1995).

The molecular basis for the similarities and the differences in the effects of LPS and PG, for example, activation of macrophages through the same receptor (CD14), will be discussed in the subsequent sections of this chapter. It should be noted, however, that both LPS and PG are heterogeneous molecules and that although the effects listed in Tables 1–4 are typical for these compounds, these effects are not always uniformly induced by all LPS or all PG.

3. CD14 AS THE RECEPTOR FOR LPS

CD14 is a cell surface glycosylphosphatidylinositol (GPI)-linked 55-kDa glycoprotein expressed predominantly on myelomonocytic cells (including monocytes, macrophages, and Langerhans cells) and also at lower levels on neutrophils (Barclay *et al.*, 1997). Its structure contains 10 repeats with some similarity to the leucine-rich glycoprotein repeats. Soluble CD14 (sCD14) is also present in normal serum and in urine of nephrotic patients (Barclay *et al.*, 1997).

3.1. Evidence for the Function of CD14 as the LPS Receptor

The discovery that CD14 represents the prominent cellular binding site for LPS was based on experiments showing that binding of LPS-coated erythrocytes or [^{125}I]-LPS to monocytes can be blocked by anti-CD14 monoclonal antibody (MAb) (Ulmer *et al.*, 1992; Wright *et al.*, 1990). The interaction of LPS with CD14 is facilitated by a serum protein, LPS-binding protein (LBP) (Wright *et al.*, 1990; Schumann *et al.*, 1990). The direct interaction of LPS with CD14 expressed on monocytes was demonstrated after incubation of THP-1 cells with [^{125}I]-ASD-LPS, followed by cross-linking of LPS:CD14 complexes (Tobias *et al.*, 1993) or by incubation of monocyte cell membranes with LPS and subsequent co-immunoprecipitation of LPS:CD14 complexes with anti-LPS MAb (El-Samalouti *et al.*, 1997). Physical interaction of LPS with purified sCD14 was demonstrated by a gel shift assay (Hailman *et al.*, 1994).

Additional evidence for binding of LPS to CD14 was provided by CD14-transfected cell lines: Chinese hamster ovary (CHO) fibroblasts or murine pre-B cell line 70Z/3, both bind LPS–LBP complexes after transfection with cDNA encoding human CD14 and expression of CD14 on the surface (Stelter *et al.*, 1997; Golenbock *et al.*, 1993; Lee *et al.*, 1992). These results suggested that CD14 is the primary LPS-binding protein on monocytes and macrophages.

The function of CD14 as the LPS cell-activating receptor was first established by the finding that anti-CD14 MAbs block LPS-induced cytokine production (Haziot *et al.*, 1993a; Wright *et al.*, 1990). Furthermore, anti-CD14 MAbs cause activation of monocytes, indicating that CD14 is indeed capable of generating intracellular signals (Lauener *et al.*, 1990; Schütt *et al.*, 1988). Unequivocal evidence was provided by CD14 transfected cells (e.g., CHO or 70Z/3 cells): LPS-unresponsive cells became highly responsive following transfection and expression of mCD14 (Han *et al.*, 1993; Golenbock *et al.*, 1993; Lee *et al.*, 1992). Moreover, monocytes and macrophages from CD14-knockout mice were 1,000 to 10,000 less responsive to LPS and were also resistant to LPS-induced lethal shock *in vivo* (Haziot *et al.*, 1996).

3.2. Binding of LPS to Membrane and Soluble CD14

Binding of LPS to CD14, either membrane (mCD14) or soluble (sCD14), is a complex phenomenon that is not resolved in all aspects. Investigation of this binding is complicated by the amphiphilic nature of LPS and formation of supramolecular structures, like liposomes, in aqueous environment. In contrast, CD14 is a hydrophilic monomolecular structure. For this reason a rather high concentration of LPS is necessary for the binding to CD14 in the absence of further

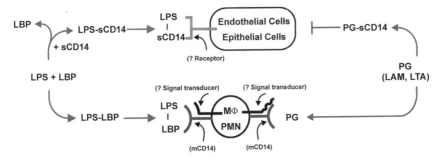

Figure 3. Role of mCD14, sCD14, and LBP in activation of mCD14-positive and -negative cells by LPS and PG.

catalytic helper molecules. This high concentration exceeds the pathophysiological concentrations of LPS that are reached during infection with gram-negative bacteria. At low concentrations LPS binding to CD14 requires a catalytic helper molecule, LBP (Wright *et al.*, 1990) (Fig. 3). LBP is a 60-kDa acute-phase protein present at about 100 ng/ml in normal serum, and its concentration increases more than 100-fold during an acute phase reaction (Tobias *et al.*, 1986).

LBP recognizes the lipid A moiety of LPS, dissociates LPS aggregates, and catalytically transfers LPS monomers from these aggregates to sCD14 or mCD14 (Yu and Wright, 1996; Tobias *et al.*, 1989, 1993, 1995; Hailman *et al.*, 1994; Wright *et al.*, 1990) (Fig. 3). In the first step, LBP forms a complex with LPS with a 1:1 stoichiometry and a dissociation constant (K_D) of about 1 nM (Tobias *et al.*, 1989). However, at high ratios of LPS to LBP large complexes with multiple numbers of aggregated LPS molecules predominate (Tobias *et al.*, 1995). The fate of the LBP–LPS complexes depends on whether they will interact with sCD14 or mCD14. Ternary complexes of LBP–LPS and mCD14 were observed under physiological concentrations of these molecules in CD14-transfected CHO cells (Gegner *et al.*, 1995), resulting in a molar ratio of bound LPS to mCD14 of about 8:1 (Kirkland *et al.*, 1993). However, sCD14 only forms complexes with LPS at a ratio of one to two molecules of LPS per single sCD14, even at high multiples of LPS to sCD14. Stable ternary LBP–LPS–sCD14 complexes were not observed (Tobias *et al.*, 1995; Hailman *et al.*, 1994), but may occur transiently (Yu and Wright, 1996).

Therefore, it is believed that LBP functions catalytically as a lipid transfer protein (Hailman *et al.*, 1994). Indeed, LBP is able to catalytically transfer LPS not only to sCD14 but also to high-density lipoprotein and phospholipid membranes (Wurfel *et al.*, 1994; Wurfel and Wright, 1997; Schromm *et al.*, 1996). LPS binds to LBP with about 10 times higher affinity (K_D = 3.5 nM) (Tobias *et al.*, 1995), than to CD14 (K_D = 27 − 32 nM) (Dziarski *et al.*, 1998; Stelter *et al.*,

1997; Tobias *et al.*, 1995; Kirkland *et al.*, 1993). As a consequence of these differ- ent dissociation constants, in normal serum, where LBP and sCD14 are in equal concentrations, or in acute phase serum, where LBP concentrations are more ele- vated than sCD14, LPS should be predominantly associated with LBP and not with sCD14 (Tobias *et al.*, 1995).

It has been shown that not only LBP but also sCD14 is able to transfer LPS to mCD14, a reaction that does not depend on the presence of LBP (Kitchens and Munford, 1998; Hailman *et al.*, 1996). Furthermore, the active transfer of LPS to mCD14 by sCD14 resulted in a 30- to 100-fold increase in the response of neu- trophils or macrophages (Hailman *et al.*, 1996).

3.3. Structural Requirements of LPS for Binding to CD14 and Cell Activation

The successful chemical synthesis of lipid A and corresponding lipid A par- tial structures, such as *Escherichia coli*-type lipid A (named compound 506 or LA- 15-PP) or precursor Ia (named compound 406, LA-14-PP, or lipid IVa) has pro- vided the experimental basis to determine the structure requirements for the bioactivity of LPS and lipid A (Rietschel *et al.*, 1994). The chemical structures of LPS, compound 506, and two prominent antagonistic lipid A analogues are shown in Fig. 1.

Full biological activity is already expressed by a lipid A molecule with two gluco-configurated hexosamine residues, two phosphoryl groups, and six fatty acids, as present in *E. coli*-type lipid A or the synthetic compound 506 (Rietschel *et al.*, 1994). Lipid A partial structures deficient in one of these elements are less active or even nonactive in inducing of monokines in human monocytes. For in- stance, the 1-dephospho (compound 504) and the 4'-dephospho (compound 505) synthetic lipid A partial structures were less active than compound 506, indicating the importance of the phosphoryl groups for the biological activity of lipid A. The lipid A precursor Ia (compound 406), which is only tetra-acylated, is inactive in inducing IL-1, IL-6, and TNF-α release in human monocytes and does not acti- vate human T lymphocytes (Rietschel *et al.*, 1994). Additionally, the highly acy- lated hepta-acyl lipid A (*S. minnesota* lipid A) shows less bioactivity than com- pound 506. The location of the secondary acyl residues is also important, as shown by the low bioactivity of compound LA-22-PP, which, in contrast to compound 506, has a symmetrical distribution of the fatty acids.

Therefore, the biological activity of lipid A depends on the phosphorylation and acylation pattern of the hexosamine disaccharide. Maximal monokine-induc- ing activity is displayed by the bisphosphorylated lipid A with six acyl residues, which structurally corresponds to *E. coli*-type lipid A (compound 506).

Some of the nonbioactive disaccharide lipid A analogues are efficient antagonists of the endotoxic activity of LPS, exemplified by synthetic lipid A precursor Ia (also known as compound 406, lipid IVa, or LA-14-PP). Precursor Ia is able to inhibit LPS-induced monokine production in human monocytes, as well as in endothelial and smooth muscle cells (Rietschel *et al.*, 1994). Precursor Ia exerts its inhibitory effect on monokine release, as well as on LPS-induced protein phosphorylation (Heine *et al.*, 1995) and induction of IL-1 and TNF-α mRNA (Rietschel *et al.*, 1994). Binding of LPS (Rietschel *et al.*, 1994) is also blocked by precursor Ia.

Similar antagonistic effects were also described for a penta-acyl diphosphoryl lipid A isolated from *Rhodobacter sphaeroides* (Golenbock *et al.*, 1991; Takayama *et al.*, 1989), enzymatically deacylated LPS (dLPS) (Kitchens and Munford 1995; Kitchens *et al.*, 1992), nonactive LPS of *Rhodobacter capsulatus* (with five acyl residues bound to the lipid A backbone) (Loppnow *et al.*, 1990), and synthetic compound E5531 (antagonistic penta-acyl lipid A analogue based on *R. capsulatus* lipid A). Compound E5531 protected mice from endotoxin-induced lethality and when administered together with an antibiotic from the lethal outcome of an *E. coli*-induced peritonitis (Christ *et al.*, 1995).

These findings raised the question of the mechanism of this inhibitory action. It is evident that not only smooth and rough LPS but also lipid A and antagonistic lipid A analogues bind to both CD14 and LBP (Tobias *et al.*, 1989). Lineweaver–Burk plot analyses provided evidence for a competitive inhibition of LPS binding to its receptor, presumably CD14, on human monocytes by lipid A analogues (Heine *et al.*, 1994). A similar competition between dLPS and LPS for binding to mCD14 on THP-1 cells in the presence of LBP excess was also observed (Kitchens and Munford 1995). At suboptimal concentrations of LBP, a competition between dLPS and LPS for engaging LBP was suggested. Comparable results were obtained when inhibition of formation of LPS–LBP and LPS–sCD14 complexes by diphosphoryl lipid A from *R. sphaeroides* was measured (Jarvis *et al.*, 1997).

These findings seemed to indicate that inhibition of the bioactivity of LPS by antagonistic LPS or lipid A analogues was simply mediated by a competitive inhibition of binding to CD14 or LBP. However, this assumption has to be questioned, since it was shown that precursor Ia or dLPS is able to block cytokine release in the human monocytic cell line THP under conditions where binding of LPS was not affected (Kitchens and Munford, 1995; Kitchens *et al.*, 1992). Moreover, it is not resolved why all these lipid A and LPS analogues are antagonistic and not agonistic, although they bind strongly to the relevant soluble LPS-binding proteins and to the LPS-receptors on the responding cells. All antagonistic lipid A and LPS analogues have a lamellar three-dimensional supramolecular structure in an aqueous environment (Seydel *et al.*, 1993).

Subsequent studies proved that CD14 cannot discriminate between agonistic and antagonistic LPS structures. All lipid A analogues, that is, compound 406, E5531 as well as lipid A from *R. sphaeroides,* exhibit LPS antagonistic properties in human cells, but only E5531 and lipid A from *R. sphaeroides* inhibit LPS-induced responses in murine cells, whereas compound 406 is stimulatory for mouse cells. However, HT-1080 human fibrosarcoma cells do not respond to any of these analogues (compound 406, E5531, and lipid A from *R. sphaeroides*), regardless whether they are transfected with human or murine CD14. Moreover, murine 70Z/3 cells transfected with either human or murine CD14 respond to compound 406 but not to E5531 or *R. sphaeroides* lipid A, whereas similarly transfected hamster CHO cells respond to all three compounds (Delude *et al.*, 1995). Therefore, these results clearly demonstrate that the target which discriminates between agonistic and antagonistic structures is not mCD14 but another so far unidentified downstream molecule (Delude *et al.*, 1995).

3.4. The Regions of CD14 Involved in Binding to LPS and Cell Activation

The regions of CD14 involved in binding to LPS and cell activation are characterized using three approaches: (1) MAbs against CD14 epitopes; (2) short synthetic peptides corresponding to the amino acid sequences of CD14; and (3) the CD14 deletion or amino acid substitution mutants.

An extensive immunological characterization of epitopes of human CD14 was published in the CD14 cluster workshop report of the Fifth International Workshop and Conference on Human Leukocyte Differentiation Antigens, Boston, 1993 (Schütt *et al.*, 1995; Goyert *et al.*, 1995). Two most prominent MAbs, MEM-18 and MY-4, inhibit binding of LPS to the responding cells (mCD14) as well as to sCD14 and block the activation of monocytes and endothelial cells. This indicates that these MAbs recognize CD14 epitopes that are necessary for LPS binding (Fig. 4).

Other MAbs, for example, 63D3 and biG6, bind to CD14 but do not block the binding of LPS or the activation of LPS responsive cells, and therefore indicate the presence of epitopes that are irrelevant for the biological activity of CD14.

Some other MAbs do not influence the binding of LPS to CD14 but affect the response of cells to LPS (clones 18E12, RPA-M1, GRS1, X8, biG-2, and biG-4) (Gegner *et al.*, 1995; Viriyakosol and Kirkland, 1995). The existence of anti-CD14 MAbs that inhibit LPS-induced cell activation but not LPS binding indicates that different CD14 epitopes are involved in LPS binding and transmission of the cell-activating signal.

A series of nine peptides, 15 amino acids in length with overlapping amino acid sequences of the N-terminal end of human CD14, were used for a biochemical analysis of the LPS-binding domain (Shapiro *et al.*, 1997). A peptide corresponding to amino acid residues 47–61 competed with the binding of LPS to immobilized sCD14, suggesting that this region is important for the binding of LPS to CD14.

```
Epitope
for
                                                                 50
        TTPEPCELDD EDFRCVCNFS EPQPDWSEAF QCVSAVEVEI HAGGLNLEPF
3C10            _____
60BCA                                           _____
RPA                                             _____
biG 4                                                _____
biG 14                                                  _____
MY-4                                                 _____
deletions       _____           _____          _____

LPS             _____           _____          _____
PG                                              _____

                                                                100
        LKRVDADADP RQYADTVKAL RVRRLTVGAA QVPAQLLVGA LRVLAYSRLK
MEM-18          _____
CRIS-6          _____
deletions       _____

LPS             _____
PG              _____

                                                                150
        ELTLEDLKIT GTMPPLPLEA TGLALSSLRL RNVSWATGRS WLAELQQWLK
Leu M3                                          _____

PG                                              _____
```

Figure 4. N-terminal amino acid sequence of CD14 and regions of mCD14 involved in LPS and PG binding and cell activation.

Further advancements were made with the use of CD14 mutants. The N-terminal fragment corresponding to less than half of sCD14 or mCD14 (amino acids 1–152) had full LPS-binding and cell-activating capacity (Viriyakosol and Kirkland, 1996; Juan *et al.*, 1995c). The use of deletion or alanine substitution CD14 mutants further revealed that amino acids 57–64 of sCD14 were required for binding of both LPS and MEM-18 MAb and for activation of U373 cells and polymorphonuclear leukocytes (PMN) by LPS (Juan *et al.*, 1995a). By contrast, alanine substitution of amino acids 7–10 did not interfere with LPS binding but impaired the capacity of sCD14 to activate cells (Juan *et al.*, 1995b) and to bind 3C10 MAbs (Juan *et al.*, 1995c).

However, different epitopes are involved in the activation of cells through mCD14. Deletion of amino acids 35–39 or 22–25 (but not 59–63) abolished LPS responsiveness through mCD14 (Gupta *et al.*, 1996; Viriyakosol and Kirkland, 1995). Similarly, alanine substitutions revealed that the LPS-binding region is located between amino acid 39 and 44, whereas [Ala9-Ala13]mCD14, as well as [Ala 57, Ala59, Ala61–63]mCD14, still were able to bind LPS and to activate transfected CHO cells (Stelter *et al.*, 1997).

Altogether, these results indicate that conformational (rather than linear) epitopes are involved in LPS binding and activation (Fig. 4), and that somewhat different epitopes are needed for cell activation induced through sCD14 and mCD14.

4. CD14 AS THE RECEPTOR FOR PEPTIDOGLYCAN

4.1. Evidence for the Function of CD14 as the Peptidoglycan Receptor

The first indication that CD14 may serve not only as an LPS receptor but also as a PG receptor came from the experiments showing that anti-CD14 MAbs inhibit not only LPS-induced but also PG-induced production of cytokines (IL-1 and IL-6) in human monocytes (Weidemann et al., 1994). Moreover, synthetic LPS partial structures that act as LPS antagonists (compounds 406 and 606) also inhibited PG-induced activation of cytokine secretion by PG in human monocytes (Weidemann et al., 1994).

More evidence for the function of CD14 as a PG receptor was provided by the experiments with CD14 transfectants. 70Z/3 cells, which are CD14-negative immature mouse B cells, do not respond to PG. However, 70Z/3-hCD14 transfectants (70Z/3 cells stably transfected with human CD14) respond to PG, as evidenced by activation of a ubiquitous transcription factor nuclear faction κB (NF-κB) accompanied by degradation of its inhibitor IκB-α, followed by differentiation into surface immunoglobulin M (IgM)-expressing B cells (Gupta et al., 1996). Moreover, activation of 70Z/3-hCD14 transfectants by PG (similarly to LPS) is inhibited by anti-CD14 MAbs (Gupta et al., 1996).

4.2. Binding of Peptidoglycan to Membrane and Soluble CD14

To prove that CD14 is a genuine PG receptor, it was necessary to show that PG directly binds to CD14. The first indication that PG may bind to CD14 came from experiments showing that specific binding of PG to human monocytes was inhibited by anti-CD14 MAbs (Weidemann et al., 1997). PG binding was also inhibited by LPS and an LPS antagonist, compound 406 (Weidemann et al., 1997). Specific binding of PG to membrane CD14 was then confirmed by photoaffinity cross-linking and immunoprecipitation of PG–CD14 complexes with anti-CD14 MAbs (Dziarski et al., 1998).

PG also binds to sCD14 at a ratio of PG to sCD14 of approximately 1–1, as demonstrated by a difference in electrophoretic mobility between sCD14 and PG–sCD14 complexes in native electrophoresis (Weidemann et al., 1997). Binding of sCD14 to PG was also confirmed by three other assays: binding of ^{32}P-labeled sCD14 to agarose-immobilized PG, photoaffinity cross-linking, and binding of biotin-labeled PG to immobilized sCD14 in an enzyme-linked immunosorbent assay (ELISA) (Dziarski et al., 1998).

Binding of sCD14 to PG (immobilized to agarose) was slower than to immobilized ReLPS but of higher affinity (K_D = 25 nM for PG vs. 41 nM for ReLPS) (Dziarski et al., 1998). LBP increased the binding of sCD14 to PG by adding another lower-affinity K_D and another higher B_{max}, in contrast to ReLPS, for which LBP increased the affinity of binding by yielding two K_D with significantly higher affinity (7 and 27 nM). LBP also enhanced inhibition of sCD14 binding to immobilized PG and LPS by soluble LPS, ReLPS, and lipid A, but not by soluble PG. Therefore, LBP enhances the binding of PG and LPS to CD14 by different mechanisms. For LPS, it increases the affinity of binding and lowers the cell-activating concentration of LPS. By contrast, for PG, LBP only increases its low-affinity binding (Dziarski et al., 1998), which does not enhance CD14-dependent cell activation by PG (Weidemann et al., 1994; Mathison et al., 1992).

4.3. Structural Requirements of Peptidoglycan for Binding to CD14 and Cell Activation

The exact structural requirements of PG for binding to CD14 have not yet been determined. It is known, however, that polymeric PG is needed for the binding, since CD14 binds to insoluble polymeric PG and to soluble polymeric PG but does not bind to soluble synthetic or natural PG fragments, such as monomeric muramyl dipeptide (MDP), or PG pentapeptide or dimeric GlcNAc-MDP, and since digestion of PG with PG-lytic enzymes reduces the binding of PG to CD14 proportionately to the extent of digestion (Dziarski et al., 1998). CD14, however, binds to agarose-immobilized MDP or GlcNAc-MDP (Dziarski et al., 1998) but not to agarose-immobilized PG pentapeptide. These results suggest that solid-phase-bound MDP or GlcNAc-MDP mimic the CD14-binding polymeric PG structure and also indicate that the glycan part of PG (but not the entire peptide) is essential for the binding to CD14.

4.4. The Regions of CD14 Involved in Binding to Peptidoglycan and Cell Activation

Similar to LPS, less than half of the CD14 molecule, that is, the N-terminal 152 amino acids, are sufficient both for PG binding (with similar affinity as the full-length CD14) and for CD14-mediated cell activation by PG (Dziarski et al.,

1998; Gupta *et al.*, 1996). These results indicate that both the binding and cell activation domains for both LPS and PG are located within the N-terminal 152 amino acid fragment of CD14.

The exact binding site for both LPS (Stelter *et al.*, 1997; Juan *et al.*, 1995a) and PG (Dziarski *et al.*, 1998) seems to be conformational. The sequences that are most critical for CD14 binding to both LPS and PG (Fig. 4) are located between amino acids 51–64 (the binding site of anti-CD14 Mab MEM-18), because anti-CD14 Mab MEM-18 is most efficient in inhibiting CD14 binding to both ReLPS (by over 95%) and to PG (by over 80%). MEM-18 is also most efficient (out of 14 anti-CD14 Mabs) in inhibiting cell activation by both LPS and PG, indicating that the same epitope on CD14 is of primary importance for both binding and cell activation by both LPS and PG.

This region, however, may not be sufficient for binding of LPS and PG, because Mabs specific to other regions of CD14 also partially inhibit binding of LPS and PG and cell activation (Dziarski *et al.*, 1998; Stelter *et al.*, 1997). However, these other sequences that contribute to LPS and PG binding and cell activation are at least partially different, because there are several anti-CD14 Mabs (directed to more N-terminal regions of CD14) that inhibit LPS binding but not PG binding, and one Mab (directed to a more C-terminal region of CD14) that inhibits PG binding but not LPS binding (Dziarski *et al.*, 1998). Therefore, it appears that LPS and PG bind to conformational rather than linear CD14 epitopes that are partially similar (amino acids 51–64) and partially different.

In general, a similar conclusion can be reached from cell activation studies with CD14 deletion mutants, that is, the domains of CD14 most critical for CD14-mediated cell activation by both LPS and PG are located within the N-terminal 65 amino acids (Gupta *et al.*, 1996). However, the specific amino acid sequences responsible for cell activation by LPS and PG are not identical, since some CD14 deletion mutants were still responsive to PG, but unresponsive to LPS (Gupta *et al.*, 1996).

5. CD14 AS THE RECEPTOR FOR OTHER BACTERIAL AND NONBACTERIAL POLYMERS

In addition to LPS and PG, other bacterial cell wall compounds, such as lipoarabinomannan, lipoteichoic acid, lipopeptides, sphingolipids, various other bacterial cell wall preparations, and also nonbacterial and synthetic polymers induce the production of cytokines in monocytes and macrophages; several of these compounds can stimulate these cells in a CD14-dependent manner (Tables V and VI). It should be noted, however, that CD14 is not involved in the stimulation of macrophages by every cell wall component. For example, glycosphingolipid from

Table V
CD14-Dependent and -Independent Inflammatory Bacterial Cell Wall Components

Cell wall compound	Source	CD14-dependence[a]
LPS	Gram-negative bacteria	Yes[1]
β1,4-D-mannuronic acid (poly M)	Gram-negative bacteria	Yes[2]
Glycosphingolipids	Gram-negative bacteria	No[3]
Peptidoglycan, insoluble	Gram-positive bacteria	Yes[4]
Peptidoglycan, soluble	Gram-positive bacteria	Yes[4–7]
Peptidoglycan, monomer	Gram-positive bacteria	No[7]
Monomeric muramyl dipeptide	Gram-positive bacteria	No[7]
Lipoteichoic acid	Gram-positive bacteria	Yes[8, 9]
Cell walls, insoluble	Gram-positive bacteria	Yes[10,11]
Rhamnose–glucose polymer	Gram-positive bacteria	Yes[12]
Lipoarabinomannan	Mycobacteria	Yes[4,10,13,14]
Lipoprotein, lipopeptide	Spirochetes	Yes[15,16]

[a]References: 1. Wright *et al.* (1990); 2. Espevik *et al.* (1993); 3. Krziwon *et al.* (1995); 4. Gupta *et al.* (1996); 5. Weidemann *et al.* (1994); 6. Weidemann *et al.* (1997); 7. Dziarski *et al.* (1998); 8. Cleveland *et al.* (1996); 9. Kusunoki *et al.* (1995); 10. Pugin *et al.* (1994); 11. Medvedev *et al.* (1998); 12. Soell *et al.* (1995); 13. Zhang *et al.* (1993); 14. Savedra *et al.* (1996); 15. Wooten *et al.* (1998); 16. Sellati *et al.* (1998).

Sphingiomonas paucibilis induces CD14-independent cytokine production in human monocytes (Krziwon *et al.*, 1995).

The first report on the involvement of CD14 in the bioactivity of an inflammatory compound other than LPS was demonstration of CD14-dependent binding and activation of human monocytes by uronic acid polymers (Espevik *et al.*, 1993). However, the epithelial-like astrocytoma cell line U373 was unable to respond to the polyuronic acid even in the presence of serum containing sCD14, indicating

Table VI
CD14-Dependent Nonbacterial Compounds

Compound	Source	CD14-dependence[a]
High M alginate	*Ascophyllum nodosum*	Yes[1]
Chitosans	Arthropods	Yes[2]
WI-1 (cell wall Ag)	*Blastomyces dermatitidis*	Yes[3]
Fucoidan	*Fucus vesiculosus*	Yes[4]
β1,4-glucuronic acid	Synthetic	Yes[1]
Phospholipids	Synthetic	Yes[5]
Taxol	*Taxus brevifolia*	Yes[6]
IL-2	T lymphocytes	Yes[7]
Apoptotic cell membranes	Apoptotic cells	Yes[8]

[a]References: 1. Espevik *et al.* (1993); 2. Otterlei *et al.* (1994); 3. Newman *et al.* (1995); 4. Cavaillon *et al.* (1996); 5. Yu *et al.* (1997); 6. Perera *et al.* (1997); 7. Bosco *et al.* (1997); 8. Devitt *et al.* (1998).

that these uronic acid polymers interact with mCD14 but are unable to activate CD14 negative cells through sCD14.

In the following section we have focused on two most studied compounds, that is, mycobacterial lipoarabinomannan (LAM) and LTA.

5.1. CD14 Interaction with Lipoarabinomannan

The lipoglycan LAM, a major antigen of the mycobacterial cell wall, stimulates cytokine production in human and murine monocytes and macrophages; in 1993, it was found that both anti-CD14 MAb and lipid IV_A inhibit LAM-induced cytokine release in human THP-1 cell line (Zhang *et al.*, 1993). Later it also was shown that transfection of 70Z/3 cells with CD14 makes these cells responsive to LAM (Gupta *et al.*, 1996; Pugin *et al.*, 1994). LAM that lacks terminal mannosyl units is reactive with CD14, whereas terminally mannosylated LAM is a poor activator of cytokines and preferentially binds to the macrophage mannose receptor (Bernardo *et al.*, 1998).

LAM directly interacts with CD14, as shown by inhibition of changes in fluorescence intensity of fluorescein isothiocyanate (FITC)-LPS induced by the binding to sCD14 (Pugin *et al.*, 1994). However, CD14-negative cells (such as U373 epithelial cells) are not activated by sCD14–LAM complexes (Savedra *et al.*, 1996), although the same cells can be activated by LAM through mCD14 (Orr and Tobias, 1998).

5.2. CD14 Interaction with Lipoteichoic Acids and Related Compounds

Lipoteichoic acids (LTA) are amphiphilic glycolipids present in the cell wall of gram-positive but not gram-negative bacteria. LTA can activate macrophages to secrete cytokines or nitric oxide; *in vivo*, they also act synergistically with PG to induce multiple organ failure and shock (Kengatharan *et al.*, 1998; De Kimpe *et al.*, 1995). LTA induce CD14-dependent secretion of IL-12, which is inhibited by anti-CD14 MAb MY-4, as well as by an LPS antagonist, *Rhodobacter sphaeroides* LPS (Cleveland *et al.*, 1996). LTA also inhibits binding of both LPS and PG to CD14 (Dziarski *et al.*, 1998).

It should be noted however, that the concept of biologically active LTA has been questioned (Takada *et al.*, 1995), because chemically synthesized structures resembling LTA of *Enterococcus hirae* were not active in inducing cytokines and antitumor activity. Therefore, it appears that the biological activity of purified natural LTA is not due to the main constituent of the LTA preparations, but to other, so far unknown, compounds (Suda *et al.*, 1995).

This conclusion is supported by the finding that the CD14-dependent cytokine-inducing factor in *S. aureus* LTA preparation, fractionated on a reverse-phase column, was distinct from the main LTA fraction, and the main purified LTA fraction failed to stimulate IL-6 release in human monocytes and U373 cells (Kusunoki *et al.*, 1995). However, this main purified LTA fraction could block the sCD14-dependent activation of U373 cell by LPS and also could bind directly to CD14, as shown by a shift in the electrophoretic mobility of sCD14 in a native electrophoresis (Kusunoki *et al.*, 1995). Therefore, LTA preparations are heterogeneous and contain both agonistic and antagonistic LTA-like molecules.

6. ACTIVATION OF CD14-NEGATIVE CELLS BY SOLUBLE CD14–LPS COMPLEXES

sCD14 lacks the GPI anchor but has the same amino acid sequence as mCD14 and is present in normal human serum at 4–6 μg/ml. Cells that do not express mCD14, such as vascular endothelial cells, epithelial cells, vascular smooth muscle cells, fibroblasts, and astrocytes, can be activated by complexes of LPS with sCD14 (Fig. 3). Formation of sCD14–LPS complexes is greatly enhanced by LBP and, thus activation of endothelial and other mCD14-negative cells by LPS and sCD14 is greatly enhanced by LBP (Loppnow *et al.*, 1995; Arditi *et al.*, 1993; Read *et al.*, 1993; Haziot *et al.*, 1993b; Pugin *et al.*, 1993a; Frey *et al.*, 1992).

Neither the exact mechanism of activation of these mCD14-negative cells by sCD14–LPS complexes nor the binding sites for these complexes on CD14-negative cells are clearly understood. Although identification of binding sites for sCD14–LPS complexes on endothelial cells has been reported (Vita *et al.*, 1997), these results could not be confirmed by other investigators (Tapping and Tobias, 1997). The latest results show LBP-dependent binding of sCD14–LPS–LBP complexes to nonmyeloid mCD14-negative cells; however, this binding resulted in LPS internalization and was not directly involved in cellular activation (Tapping and Tobias, 1997).

Vascular endothelial cells participate in inflammation, organ failure, and shock by being both a producer of and a target for proinflammatory mediators. Stimulation of endothelial cells by sCD14–LPS complexes induces production of proinflammatory cytokines (e.g., IL-1 and IL-6) and chemokines (e.g., IL-8) and expression of adhesion molecules (Table 1). These adhesion molecules promote inflammation by enhancing attachment of leukocytes to vascular endothelium and extravasation of leukocytes and their migration into the inflamed tissues (Bevilacqua, 1993; Wahl *et al.*, 1996).

Vascular endothelial and other mCD14-negative cells are also strongly activated by proinflammatory cytokines, such as TNF-α and IL-1. In fact, these cells are activated much more efficiently by LPS in the presence of small amounts of whole blood (2–4%) than in the presence of serum or purified scD14 and LBP. This indirect activation is mediated by TNF-α and IL-1 that originate from LPS-activated monocytes and macrophages (Pugin *et al.,* 1993b, 1995). It is therefore likely that *in vivo,* similarly to *ex vivo,* indirect activation of endothelial cells by cytokines induced by LPS from monocytes and macrophages is of primary importance, although it is also possible that cytokines *in vivo* have numerous other targets and can be quickly "used up" before being able to fully activate endothelial and epithelial cells.

7. LACK OF DIRECT ACTIVATION OF CD14-NEGATIVE CELLS BY SOLUBLE CD14–PG COMPLEXES

Although PG binds to scD14 with high affinity (Dziarski *et al.,* 1998) and forms stable complexes with scD14 (Dziarski *et al.,* 1998; Weidemann *et al.,* 1997), the scD14–PG complexes do not activate endothelial and epithelial cells to secrete cytokines, to express adhesion molecules, or to activate NF-κB (Jin *et al.,* 1998) (Fig. 3). Similar results also were obtained with some other bacterial non-LPS CD14 ligands, such as mycobacterial lipoarabinomannan (Savedra *et al.,* 1996). However, this finding does not apply to all non-LPS CD14 ligands, because spirochetal lipoprotein–scD14 or lipopeptide–scD14 complexes do activate endothelial cells (Wooten *et al.,* 1998). These findings underscore the differences in the function of scD14 as the facilitator of cell activation by LPS and various other bacterial CD14 ligands, even though the molecular and biochemical reasons responsible for these differences are currently not understood.

However, as was the case with LPS, PG can induce very strong activation of endothelial and epithelial cells indirectly in the presence of even small amounts (2–4%) of whole blood (Jin *et al.,* 1998). Again, similarly to LPS, the secretion of both TNF-α and IL-1 from blood monocytes is responsible for this activation (Jin *et al.,* 1998).

8. FUNCTION OF CD14 AS A "PATTERN RECOGNITION RECEPTOR"

Given a large number of structurally different ligands that can bind to CD14, three models for the function of CD14 as a cell-activating receptor have been proposed. According to the first model (Pugin *et al.,* 1994), CD14 serves as a "pattern recognition receptor" that can recognize shared features of microbial cell surface components and can enable host cells to respond to pathogenic bacteria but not to a great variety of other nonpathogenic or nonmicrobial polysaccharides. This model implied that CD14 can discriminate between different ligands and can control the specificity of macrophage responses. How this discrimination was achieved, however, was not clear. The second alternative model (Wright, 1995) proposed that CD14 does not have the recognition specificity and merely serves as an albuminlike carrier molecule that transfers ligands to an as yet unidentified recognition–cell-activating molecule(s). The latter model was supported by the inability of CD14 to discriminate between agonistic and antagonistic derivatives of LPS (Delude *et al.,* 1995). Finally, in the third "combinatorial" model, both CD14 and the putative recognition–cell-activating molecule would contribute to the specificity of cell activation. Other models, which proposed that CD14 could be a component of a heteromeric receptor complex and not directly bind other ligands, were less likely because of direct high-affinity binding of PG and other non-LPS ligands to CD14 (Dziarski *et al.,* 1998).

It is still not possible to definitively discriminate among the first three models. However, it seems that CD14 has at least some specificity, because it binds some molecules with high affinity (e.g., LPS, PG, LTA) and does not bind other similar molecules (e.g., ribitol teichoic acid, dextran, dextran sulfate, or heparin) (Dziarski *et al.,* 1998). These findings would at least partially support CD14 function as a pattern recognition receptor. It is still not clear, however, what are the chemical features of the "pattern" that CD14 recognizes. It seemed that most ligands recognized by CD14 have polymeric carbohydrates often with closely located carbonyl residues (such as LPS or PG). However, not all CD14 ligands have these structural features, because CD14 also binds ligands that do not have carbohydrates, such as lipoproteins (Sellati *et al.,* 1998; Wooten *et al.,* 1998) or phospholipids (Yu *et al.,* 1997). Therefore, it can be postulated that CD14 recognizes glycoconjugates or phospholipids via distinct patterns of ionic charges (Ulmer *et al.,* 1999). The specificity for these ligands is still tightly controlled, because several other charged polymers, for example, dextran sulfate or heparin, do not bind to CD14 with affinity similar to LPS or PG (Dziarski *et al.,* 1998).

Recent results also suggest that recognition of different patterns is encoded in somewhat different regions of CD14, and that the binding sites are conformational and composed of several regions, partially identical and partially different for different ligands (Dziarski *et al.,* 1998). Such a multifunctional binding site could then much more easily accommodate specific binding to a variety of structurally different ligands.

9. MECHANISM OF CELL ACTIVATION BY CD14

It is still not exactly known how CD14 transmits the signal into the cytoplasm and activates cells. Because mCD14 is a GPI-linked molecule, it does not have a

A

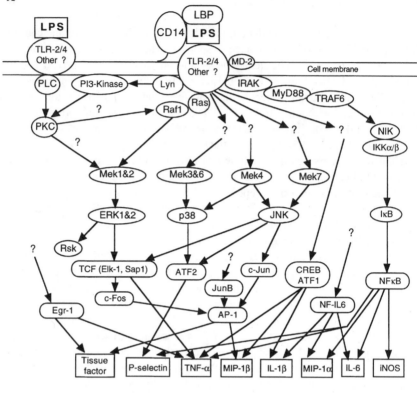

B

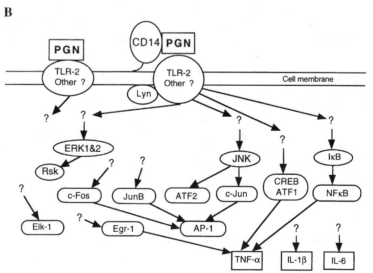

Figure 5. Signal transduction pathways, transcription factors, and main genes activated by (A) LPS and (B) PG.

transmembrane domain and by itself cannot transmit the activating signal into the cell. For this reason (and for other reasons explained in the preceding section), it has been proposed that either CD14–ligand complexes associate with other molecules or CD14 ligands are transferred from CD14 to other cell-activating or coreceptor molecule(s) (Figs. 3 and 5) (Ulevitch and Tobias, 1995; Haziot *et al.*, 1993b; Pugin *et al.*, 1993a; Frey *et al.*, 1992).

Two members of the Toll-like receptor (TLR) family, TLR-2 and TLR-4, have been recently identified as the most likely candidates for this cell-activating coreceptor for gram-negative LPS (Chow *et al.*, 1999; Hoshino *et al.*, 1999; Qureshi *et al.*, 1999; Poltorak *et al.*, 1998; Kirschning *et al.*, 1998; Yang *et al.*, 1998). TLRs are type I transmembrane proteins with leucine-rich repeats in their extracellular domains, and cytoplasmic domains with sequence homology to the IL-1 receptor. TLR-2 protein binds LPS, expression of TLR-2 converts LPS-unresponsive cells into LPS-responsive cells, and the responsiveness is enhanced by co-expression of CD14 (Kirschning *et al.*, 1998; Yang *et al.*, 1998). However cells that express CD14 alone and do not express TLRS are unresponsive to LPS. TLR-2 associates with CD14 on the cell surface, and LPS enhances the oligomerization of TLR-2 (Yang *et al.*, 1999). LPS-induced responsiveness through TLR-4 requires a cell-surface helper molecule, MD-2 (Shimazu *et al.*, 1999), and a mutation in the *tlr4* gene that substitutes histidine for proline at position 712 is responsible for LPS unresponsiveness of C3H/HeJ mice (Hoshino *et al.*, 1999; Qureshi *et al.*, 1999; Poltorak *et al.*, 1998).

TLR-2 also has been recently identified as the cell-activating receptor or coreceptor for gram-positive bacteria and their PG and LTA components (Yoshimura *et al.*, 1999; Schwandner *et al.*, 1999), as well as for other CD14 ligands, such as mycobacteria, spirochetes, and their lipoprotein components (Means *et al.*, 1999; Hirschfeld *et al.*, 1999; Aliprantis *et al.*, 1999; Brightbill *et al.*, 1999).

GPI-linked cell surface molecules, including CD14, associate with protein tyrosine kinases (Stefanova *et al.*, 1991), and CD14 has been shown in macrophage lysates to co-immunoprecipitate with Lyn, a member of the Src family of receptor-associated tyrosine kinases (Stefanova *et al.*, 1993). Since it is not known whether mCD14 associates with Lyn through TLR-2, it is not clear how CD14, linked to the cell surface by a GPI anchor, can associate with a tyrosine kinase that is present inside the cell or on the inner surface of the cell membrane.

GPI-linked molecules are highly mobile on the cell surface, and polyvalent polymeric ligands can easily induce their clustering or association with other receptors. However, it is still not clear whether CD14 clustering is required for cell activation.

In addition to triggering cell activation, CD14 also promotes internalization of its ligands. Indeed, mCD14-bound ligands (including LPS and PG), as well as sCD14–LPS complexes that bind to cells, are rapidly (within minutes) internalized (Kitchens and Munford, 1998; Tapping and Tobias, 1997; A. J. Ulmer, unpublished). However, such internalization (at least for LPS) is independent of and not needed for cell activation, because there are some MAbs that inhibit cell activation but not LPS internalization, and some other MAbs that inhibit LPS internalization but not activation (Tapping and Tobias, 1997; Gegner *et al.*, 1995). Moreover, highly aggregated LPS is preferentially internalized as compared to less aggregated or monomeric LPS, whereas aggregation does not enhance LPS-induced cell activation (Kitchens and Munford, 1998). In fact, monomeric LPS is the preferred cell-activating species of LPS (Takayama *et al.*, 1994).

10. CD14-INDEPENDENT CELL ACTIVATION BY LPS AND PEPTIDOGLYCAN

There is no doubt that CD14 is the main macrophage and neutrophil receptor for cell activation and induction of cytokine production by low concentrations of LPS, since CD14-knockout mice are more than 1000 times less sensitive to LPS than their wild-type littermates (Haziot *et al.*, 1996). However, there also are CD14-independent mechanisms of cell activation by LPS, because cells from CD14-knockout mice do respond to high concentrations of LPS with production of cytokines (Perera *et al.*, 1997). Moreover, CD14-deficient mice show no alteration in the induction of acute phase proteins *in vivo* by LPS or lipid A (Haziot *et al.*, 1998). Therefore, some effects of LPS are CD14-dependent, but other effects can be partially or totally CD14 independent.

The mechanism of these CD14-independent responses to LPS, however, is still poorly understood. As mentioned in the preceding section, TLR-2 can function as an LPS receptor without the presence of CD14 (Kirschning *et al.*, 1998; Yang *et al.*, 1998). β2-Integrins, the complement receptors CR3 (CD11b/CD18) and CR4 (CD11c/CD18), can act as cell-activating LPS receptors, because transfection of CR3 or CR4 cDNA into CR3- and CR4-negative CHO cells confers on them the ability to respond to LPS (Ingalls *et al.*, 1997; Ingalls and Golenbock, 1995). Similarly, transfection of CHO cells with CD55 [also known as decay-accelerating factor (DAF) a GPI-linked cell surface molecule that protects cells from complement-mediated damage] makes these cells responsive to LPS, which indicates that CD55 can also serve as a CD14-independent LPS receptor (Hamann *et al.*, 1998). It is not known, however, whether β2-integrins or CD55 serve as the major CD14-independent LPS receptors. There is little evidence for the receptor function of other LPS-binding proteins, such as, for example, the scavenger receptor, which most likely serves to remove and detoxify LPS (Haworth *et al.*, 1997).

Even less is known about the mechanism of CD14-independent cell activation by other CD14 ligands. These ligands, for example, PG, can activate cells through CD14-independent pathways, because cells from CD14-knockout mice are only five to ten times less sensitive to PG than the cells from CD14-expressing wild-type littermates (S. M. Goyert, unpublished). TLR-2 has been recently identified as the CD-14-independent receptor for PG and other bacterial components (Brightbill *et al.,* 1999; Yoshimura *et al.,* 1999; Schwander *et al.,* 1999). It is not known whether CR3 and CR4 serve as receptors for PG, but it is known that by themselves they do not function as receptors for streptococcal cell walls (another CD14-dependent cell activator), even though they may enhance CD14-mediated responses to these cell walls (Medvedev *et al.,* 1998).

11. SIGNAL TRANSDUCTION PATHWAYS ACTIVATED BY LPS AND PEPTIDOGLYCAN

Since a macrophage is the main target cell activated by LPS and PG, and since CD14 is its main LPS and PG receptor, most of the signal transduction pathways described in this section will deal with CD14-mediated activation of macrophages, as summarized in Fig. 5. In reviewing these data, however, three points need to be remembered. First, that the CD14-dependent signal transduction pathways are likely to be actually initiated by an as yet unidentified coreceptor (possibly TLR-2; see Section 9), that may or may not be the same for LPS and PG. Second, that some effects may be initiated through CD14-independent mechanisms, and the receptors involved here may or may not be the same as the putative CD14 coreceptor. Third, that in different cells (such as CD14-transfected nonmacrophage cells, or endothelial or epithelial cells that are responsive to sCD14–LPS complexes, or even macrophages from different species and/or different tissues) different signal transduction pathways may be activated.

The earliest intracellular signal transduction event that can be detected within 1 min of macrophage activation by LPS is increased tyrosine phosphorylation and activation of Lyn, and possibly two other members of the Src family of tyrosine kinases, Hck and Fgr (Gupta *et al.,* 1995; Beaty *et al.,* 1994; Stefanova *et al.,* 1993) (Fig. 5A). This is reminiscent of the activation of the tyrosine kinases associated with other receptors, such as T- and B-cell antigen receptors or some cytokine receptors. PG also induces increased tyrosine phosphorylation of Lyn, but increased phosphorylation of other Src kinases could not be detected (Gupta *et al.,* 1995) (Fig. 5B).

Although tyrosine phosphorylation or activation of Lyn by LPS has been now confirmed by several groups (Herrera-Velit and Reiner, 1996; Gupta *et al.,* 1996; Henricson *et al.,* 1995; Beaty *et al.,* 1994; Stefanova *et al.,* 1993), it is still not clear what signal transduction pathways are activated by Lyn in LPS-stimulated cells. Typically in other systems, Src kinases activate Syk or Vav, which in turn can lead to the activation of a small GTP-binding protein, Ras, which can then activate Raf, which then triggers activation of mitogen-activated protein kinases (MAP kinases). Such a connection of Lyn to the Ras → Raf → MAP kinase pathway so far has not been convincingly shown for LPS, even though one group did show LPS-induced increased tyrosine phosphorylation of Vav (English *et al.,* 1997), and another group showed LPS-induced tyrosine phosphorylation of a protein (p145) that associates with Syk and a second adaptor protein, Shc (Crowley *et al.,* 1996). However, activation of Syk by LPS has not been shown, and Syk-deficient cells show normal responses to LPS (Crowley *et al.,* 1997). Activation of Lyn by LPS has been shown to activate phosphatidylinositol 3-kinase (PI3-kinase) in a CD14-dependent manner (Herrera-Velit and Reiner, 1996), which in turn activates protein kinase C (PKC)-ζ (Herrera-Velit *et al.,* 1997). PKC is known to activate the Raf → Mek → ERK (extracellular signal-related kinase) pathway, but this link so far has not been shown for LPS. None of these pathways have been so far shown for PG.

The second issue is the significance of Lyn activation for the biological effects of LPS. Even though Lyn and other Src kinases may participate in the cell activation by LPS and even though the Hck⁻/Fgr⁻ double knockout mice are resistant to endotoxic shock (Lowell and Berton, 1998), it is now clear that these Src kinases are *not* required for several major effects of LPS, because macrophages from Lyn⁻/Hck⁻/Fgr⁻ triple knockout mice had no major defects in LPS-induced stimulation of nitrate production, IL-1, IL-6, and TNF-α secretion, as well as activation of ERK1/2 and c-Jun NH_2-terminal kinase (JNK) MAP kinases and transcription factor NF-κB (Meng and Lowell, 1997). The higher resistance to endotoxic shock of Hck⁻/Fgr⁻ double knockout mice could be due to the requirement of Hck and Fgr for the action of endotoxin-induced mediators, such as cytokines.

One of the best-documented events in LPS-induced signal transduction is activation of MAP kinases. There are three families of MAP kinases (ERK1/2, p38, and JNK), and LPS strongly activates all these MAP kinases (Hambelton *et al.,* 1996; Dziarski *et al.,* 1996; Gupta *et al.,* 1995; Raingeaud *et al.,* 1995; Liu *et al.,* 1994; Han *et al.,* 1993, 1994; Weinstein *et al.,* 1992) (Fig. 5A). By contrast, PG strongly activates only ERK1/2 and JNK, and only marginally activates p38 (Dziarski *et al.,* 1996; Gupta *et al.,* 1996) (Fig. 5B).

MAP kinases are activated by dual phosphorylation on one threonine and one tyrosine by specialized kinases, called Mek or MKK (MAP kinase kinase), that show a high degree of specificity for individual MAP kinases. Thus, in several cells, including LPS-activated macrophages, Mek 1 and Mek 2 activate ERK 1 and ERK 2, Mek 3 and Mek 6 activate p38, and Mek 4 and Mek 7 activate JNK (although Mek 4 can also to some extent activate p38) (Yao *et al.,* 1997; Swantek *et al.,* 1997; Sanghera *et al.,* 1996; Buscher *et al.,* 1995; Raingeaud *et al.,* 1995; Reimann *et al.,* 1994; Geppert *et al.,* 1994) (Fig. 5A). It is not known which Meks activate which MAP kinases in PG-stimulated cells (Fig. 5B).

Meks are activated by a variety of different Mek kinases (Mekk or MKKK), whose specificity for substrates (Meks) is less stringent than the specificities of Meks for MAP kinases. Whereas various pathways leading to the activation of Meks have been identified in various cells stimulated by different stimuli, relatively little is known about the mechanisms of LPS- and PG-induced Mek activation. One pathway, Ras → Raf1 → Mek1/2 → ERK1/2, has been proposed to be activated by LPS (Hambelton *et al.,* 1995; Geppert *et al.,* 1994; Reimann *et al.,* 1994), although it was not determined what activates Ras. However, there also are some indications that the Ras → Raf is not the main activating pathway for ERK1/ 2 (Dziarski *et al.,* 1996; Buscher *et al.,* 1995) or that the Ras → Raf pathway is not activated at all by LPS (Guthridge *et al.,* 1997). PKC, whose phorbol-insensitive forms (ϵ and ζ) are activated by LPS (Shapira *et al.,* 1997; Herrera-Velit *et al.,* 1997), also could result in the activation of Mek1/2 → ERK1/2 pathway, although this connection has not yet been established for LPS.

Even though MAP kinases induce numerous transcription factors that can potentially induce numerous genes, the significance of the activation of each family of MAP kinases by LPS and PG is still not clear. For example, on one hand, inhibitors of ERK phosphorylation and activation, such as tyrphostins, block LPS-induced production of nitric oxide (NO) and TNF-α and prevent endotoxin lethality *in vivo* (Novogrodsky *et al.,* 1994). But on the other hand, strong activators of ERK, such as phorbol esters or colony-stimulating factor (CSF), do not mimic LPS effects, for example, they induce very little TNF-α production and are not toxic (Sweet and Hume, 1996; Gupta *et al.,* 1995). Similarly, selective activation of the Raf1 → ERK pathway by a chimeric Raf–estrogen receptor molecule induces, similarly to phorbol esters, only a small amount of TNF-α (Hambelton *et al.,* 1995). Also, activation of p38 MAP kinase alone cannot be responsible for or required for LPS- or PG-induced cytokine production, because PG, in contrast to LPS, does not effectively activate p38 (Dziarski *et al.,* 1996), yet both PG and LPS induce large amounts of TNF-α or IL-6 in macrophages (Gupta *et al.,* 1995; Weidemann *et al.,* 1994). Therefore, it appears that activation of individual MAP kinases is not sufficient, or in some cases may not be even necessary, for the full induction of LPS- or PG-stimulated genes.

Some other signal transduction pathways, for example, through the release or mimicking the action of ceramide (Wright and Kolesnick, 1995), through the action of G proteins (Daniel *et al.,* 1989; Dziarski, 1989; Jakway and DeFranco, 1986), or through interaction with microtubules (Ding *et al.,* 1992, 1996), also may be involved in LPS-induced cell activation, but more direct evidence is required to validate these proposals (Sweet and Hume, 1996).

In summary, it is likely that LPS and PG activate multiple signal transduction pathways that are mostly overlapping but partially different. The full spectrum of cell activation is likely to be the result of concerted action of all these pathways.

12. LPS- AND PEPTIDOGLYCAN-INDUCED TRANSCRIPTION FACTORS AND THEIR ROLE IN GENE ACTIVATION

LPS induces activation of a large number of transcription factors, including NF-κB, NF-IL-6, the CREB/ATF1 family, the AP-1 family, the Ets family (which includes Ets, Elk, Erg, and PU.1), and Egr (Fig. 5A) (Pan *et al.,* 1998; Groupp and Donovan-Peluso, 1996; Sweet and Hume, 1996). Some of these transcription factors are activated by phosphorylation by specific kinases (e.g., CREB/ATF1, or c-Jun, a member of the AP-1 family), and some other (e.g., JunB, c-Fos, or Egr-1) are synthesized *de novo* as a part of LPS-induced "early response genes." Activation of the NF-κB transcription factor involves phosphorylation of its inhibitor, IκB, which induces degradation of IκB and release and translocation of NF-κB into the nucleus. LPS-induced activation of several of these transcription factors has been shown to be CD14-dependent (Gupta *et al.,* 1999; Sweet and Hume, 1996), and activation of NF-κB is induced through TLR-2 (Kirschning *et al.,* 1998; Yang *et al.,* 1998).

Activated transcription factors bind to the specific sequences in the promoters of specific genes and induce transcription of these genes. Typically, activation of transcription of a given gene requires binding of several transcription factors. As can be seen in Fig. 5A, each transcription factor induced by LPS plays a role in regulation of transcription of several genes, and each gene is regulated by several transcription factors (Gupta *et al.,* 1999; Pan *et al.,* 1998; Groupp and Donovan-Peluso, 1996; Sweet and Hume, 1996).

Much less is known about transcription factors activated by PG. It has been shown that PG induces activation of NF-κB (Gupta et al, 1996), and recent results indicate that it also induces activation of CREB/ATF1, ATF2, AP-1, and EGR-1 (Gupta *et al.,* 1999; Gupta *et al.,* unpublished data) (Fig. 5B). NF-κB, CREB/ATF1, and Egr-1 are all involved in the activation of transcription of TNF-α gene (Gupta *et al.,* unpublished data). It is still not known which transcription factors are involved in the activation of other genes that are induced by PG.

ACKNOWLEDGMENTS

This work was supported by the National Institute of Health Grant AI28797 (to R. D. and D. G.), by a grant from the BMBF (No. 01KI9471) (to A. J. U.), and by the DFG (SFB 367, project C5) (to A. J. U.).

REFERENCES

Aliprantis, A. O., Yang, R.-B., Mark, M. R., Suggett, S., Devaux, B., Radolf, J. D., Klimpel, G. R., Godowski, P., Zychlinsky, A., 1999, Cell activation and apoptosis by bacterial lipoproteins through Toll-like receptor-2, *Science* **285:**736–739.
Alving, C. R., 1993, Lipopolysaccharide, lipid A, and liposomes containing lipid A as immunologic adjuvants, *Immunobiology* **187:**430–446.
Arditi, M., Zhou, J., Dorio, R., Rong, G. W., Goyert, S. M., and Kim, K. S., 1993, Endotoxin-mediated endothelial cell injury and activation: role of soluble CD14, *Infect. Immun.* **61:**3149–3156.
Baggiolini, M., Walz, A., and Kunkel, S. L., 1989, Neutrophil-activating peptide-1/interleukin 8, a novel cytokine that activates neutrophils, *J. Clin. Invest.* **84:**1045–1049.

Barclay, A. N., Brown, M. H., Law, S. K. A., McKnight, A. J., Tomlinson, M. G., and van der Merwe, P. A., 1997, *The Leukocyte Antigen Facts Book,* Academic Press, San Diego, CA.

Beaty, C. D., Franklin, T. L., Uehara, Y., and Wilson, C. B., 1994, Lipopolysaccharide-induced cytokine production in human monocytes: Role of tyrosine phosphorylation in transmembrane signal transduction, *Eur. J. Immunol.* **24:**1278–1284.

Bernardo, J., Billingslea, A. M., Blumenthal, R. L., Seetoo, K. F., Simons, E. R., and Fenton, M. J., 1998, Differential responses of human mononuclear phagocytes to mycobacterial lipoarabinomannans: Role of CD14 and the mannose receptor, *Infect. Immun.* **66:**28–35.

Beutler, B., and Cerami, A., 1988, Tumor necrosis, cachexia, shock, and inflammation: A common mediator, *Annu. Rev. Biochem.* **57:**505–518.

Beutler, B., and Grau, G. E., 1993, Tumor necrosis factor in the pathogenesis of infectious diseases, *Crit. Care Med.* **21:**S423–435.

Bevilacqua, M. P., 1993, Endothelial-leukocyte adhesion molecules, *Annu. Rev. Immunol.* **11:**767–804.

Biberstine, K. J., and Rosenthal, R. S., 1994, Peptidoglycan fragments decrease food intake and body weight gain in rats, *Infect. Immun.* **62:**3276–3281.

Blais, C., Couture, R., Drapeau, G., Colman, R. W., and Adam, A., 1997, Involvement of endogenous kinins in the pathogenesis of peptidoglycan-induced arthritis in the Lewis rat, *Arthrit. Rheumat.* **40:**1327–1333.

Bosco, M. C., Espinoza-Delgado, I., Rowe, T. K., Malabara, M. G., Longo, D. L., and Varesio, L., 1997, Functional role of the myeloid differentiation antigen CD14 in the activation of human monocytes by IL-2, *J. Immunol.* **159:**2922–2931.

Brightbill, H. D., Libraty, D. H., Krutzik, S. R., Yang, R.-B., Belisle, J. T., Bleharski, J. R., Maitland, M., Norgard, M. V., Plevy, S. E., Smale, S. T., Brennan, P. J., Bloom, B. R., Godowski, P. J., and Modlin, R. L., 1999, Host defense mechanisms triggered by microbial lipoproteins through Toll-like receptors, *Science* **285:**732–735.

Buscher, D., Hipskind, R. A., Krautwald, S., Reimann, T., and Baccarini, M., 1995, Ras-dependent and -independent pathways target the mitogen-activated protein kinase network in macrophages, *Mol. Cell. Biol.* **15:**466–475.

Camussi, G., Marino, F., Biancone, L., De Martino, A., Bussolati, B., Montrucchio, G., and Tobias, P. S., 1995, Lipopolysaccharide binding protein and CD14 modulate the synthesis of platelet-activating factor by human monocytes and mesangial and endothelial cells stimulated with lipopolysaccharide, *J. Immunol.* **155:**316–324.

Cavaillon, J. M., Marie, C., Caroff, M., Ledur, A., Godard. I., Poulain, D., Fitting, C., and Haeffner-Cavaillon, N., 1996, CD14/LPS receptor exhibits lectin-like properties, *J. Endotoxin Res.* **3:**471–480.

Chow, J. C., Young, D. W., Golenbock, D. T., Christ, W. J., and Gusovsky, F., 1999, Toll-like receptor-4 mediates lipopolysaccharide-induced signal transduction. *J. Biol. Chem.* **274:**10689–10692.

Christ, W. J., Asano, O., Robidoux, A. L. C., Perez, M., Wang, Y., Dubuc, G. R., Gavin, W. E., Hawkins, L. D., McGuinness, P. D., Mullarkey, M. A., Lewis, M. D., Kishi, Y., Kawata, T., Bristol, J. R., Rose, J. R., Rossignol, D. P., Kobayashi, S., Hishinuma, I., Kimura, A., Asakawa, N., Katayama, K., and Yamatsu, I., 1995, E5531, a pure endotoxin antagonist of high potency, *Science* **268:**80–83.

Cleveland, M. G., Gorham, J. D., Murphy, T. L., Tuomanen, E., and Murphy, K. M., 1996, Lipoteichoic acid preparations of gram-positive bacteria induce interleukin-12 through a CD14-dependent pathway, *Infect. Immun.* **64:**1906–1912.

Cooper, N. R., 1991, Complement evasion strategies of microorganisms, *Immunol. Today* **12:**327–331.

Cortran, R. S., Kumar, V., and Robbins, S. L., 1994, Inflammation and repair and hemodynamic disorders, thrombosis, and shock, in: *Robbins Pathologic Basis of Disease*, 5th Ed. (R. S. Cortran, V. Kumar, and S. L. Robbins, eds.), W. B. Saunders, Philadelphia, PA, pp. 51–121.

Crowley, M. T., Harmer, S. L., and DeFranco, A. L., 1996, Activation-induced association of a 145-kDa tyrosine-phosphorylated protein with Shc and Syk in B lymphocytes and macrophages, *J. Biol. Chem.* **271:**1145–1152.

Crowley, M. T., Costello, P. S., Fitzer-Attas, C. J., Turner, M., Meng, F., Lowell, C., Tybulewicz, V. L., and DeFranco, A. L., 1997, A critical role for Syk in signal transduction and phagocytosis by Fcγ receptors on macrophages, *J. Exp. Med.* **186:**1027–1039.

Daniel, I., Spiegel, S., and Strulovici, A. M., 1989, Lipopolysaccharide response is linked to the GTP binding protein, Gi2, in the promonocytic cell line U937, *J. Biol. Chem.* **264:**20240–20247.

De Kimpe, S. J., Kengatharan, M., Thiemermann, C, and Vane, J. R., 1995, The cell wall components peptidoglycan and lipoteichoic acid from *Staphylococcus aureus* act in synergy to cause shock and multiple organ failure, *Proc. Natl. Acad. Sic. USA* **92:** 10359–10363.

DeLa Cadena, R. A., Laskin, K. J., Pixley, R. A., Sartor, R. B., Schwab, J. H., Back, N., Bedi, G. S., Fisher, R. S., and Colman, R. W., 1991, Role of kallikrein–kinin system in pathogenesis of bacterial cell wall-induced inflammation, *Am. J. Physiol.* **260:** G213–219.

Delude, R. L., Savedra, R., Zhao, H., Thieringer, R., Yamamoto, S., Fenton, M. J., and Golenbock, D. T., 1995, CD14 enhances cellular responses to endotoxin without imparting ligand-specific recognition, *Proc. Natl. Acad. Sci. USA* **92:**9288–9292.

Devitt, A., Moffatt, O. D., Raykundalia, C., Capra, J. D., Simmons, D. L., and Gregory, C. D., 1998, Human CD14 mediates recognition and phagocytosis of apoptotic cells, *Nature* **392:**505–509.

de Vries, J. E., 1995, Immunosuppressive and anti-inflammatory properties of interleukin 10, *Ann. Med.* **27:**537–541.

Ding, A., Sanchez, E., Tancinco, M., and Nathan, C., 1992, Interactions of bacterial lipopolysaccharide with microtubule proteins, *J. Immunol.* **148:**2853–2858.

Ding, A., Chen, B., Fuortes, M., and Blum, E., 1996, Association of mitogen-activated protein kinases with microtubules in mouse macrophages, *J. Exp. Med.* **183:**1899–18904.

Dokter, W. H., Dijkstra, A. J., Koopmans, S. B., Mulder, A. B., Stulp, B. K., Halie, M. R., Keck, W., and Vellenga, E., 1994, G(AnH)MTetra, a naturally occurring 1,6-anhydro muramyl dipeptide, induces granulocyte colony-stimulating factor expression in human monocytes: A molecular analysis, *Infect. Immun.* **62:**2953–2957.

Dunn, P. E., Dai, W., Kanost, M. R., and Geng, C. X., 1985, Soluble peptidoglycan fragments stimulate antibacterial protein by fat body from larvae of *Manduca sexta, Dev. Comp. Immunol.* **9:**559–568.

Durum, S. K., Schmidt, J. A., and Oppenheim, J. J., 1985, Interleukin 1: An immunological perspective, *Ann. Rev. Immunol.* **3**:263–287.

Dziarski, R., 1986, Effects of peptidoglycan on the cellular components of the immune system, in: *Biological Properties of Peptidoglycan* (H. P. Seidl and K. H. Schleifer, eds.), Walter De Gruyter Co., Berlin, pp. 229–247.

Dziarski, R., 1989, Correlation between ribosylation of pertussis toxin substrates and inhibition of peptidoglycan-, muramyl didpeptide- and lipopolysaccharide-induced mitogenic stimulation in B lymphocytes, *Eur. J. Immunol.* **19**:125–130.

Dziarski, R., and Dziarski, A., 1979, Mitogenic activity of staphylococcal peptidoglycan. *Infect. Immun.* **23**:706–710.

Dziarski, R., Jin, Y., and Gupta, D., 1996, Differential activation of extracellular signal-regulated kinase (ERK) 1, ERK2, p38, and c-Jun NH$_2$-terminal kinase mitogen-activated protein kinases by bacterial peptidoglycan, *J. Infect. Dis.* **174**:777–785.

Dziarski, R., Tapping, R. I., and Tobias, P, 1998, Binding of bacterial peptidoglycan to CD14, *J. Biol. Chem.* **273**:8680–8690.

El-Samalouti, V. T., Schletter, J., Brade, H., Brade, L., Kusumoto, S., Rietschel, E. Th., Flad, H.-D., and Ulmer, A. J., 1997, Detection of LPS-binding membrane proteins by immunocoprecipitation with LPS and anti-LPS antibodies, *Eur. J. Biochem.* **250**:418–424.

English, B. K., Orlicek, S. L., Mei, Z., and Meals, E. A., 1997, Bacterial LPS and IFN-γ trigger the tyrosine phosphorylation of vav in macrophages: Evidence for involvement of the hck tyrosine kinase, *J. Leukocyt. Biol.* **62**:859–864.

Espevik, T., Otterlei, M., Skjåk-Bræk, G., Ryan, L., Wright, S. D., and Sundan, A., 1993, The involvement of CD14 in stimulation of cytokine production by uronic acid polymers, *Eur. J. Immunol.* **23**:255–261.

Frey, E. A., Miller, D. S., Jahr, T. G., Sundan, A., Bazil, V., Espevik, T., Finlay, B. B., and Wright, S. D., 1992, Soluble CD14 participates in the response of cells to lipopolysaccharide, *J. Exp. Med.* **176**:1665–1671.

Gegner, J. A., Ulevitch, R. J., and Tobias, P. S., 1995, Lipopolysaccharide (LPS) signal transduction and clearance: Dual roles for LPS binding protein and membrane CD14, *J. Biol. Chem.* **270**:5320–5325.

Geppert, T. D., Whitehurst, C. E., Thompson, P., and Beutler, B., 1994, Lipopolysaccharide signals activation of tumor necrosis factor biosynthesis through the Ras/Raf-1/MEK/MAPK pathway, *Mol. Med.* **1**:93–103.

Gold, M. R., Miller, C. L., and Mishell, R. I., 1985, Soluble non-cross-linked peptidoglycan polymers stimulate monocyte-macrophage inflammatory functions, *Infect. Immun.* **49**:731–741.

Golenbock, D. T., Hampton, R. Y., Qureshi, N., Takayama, K., and Raetz, C. R. H., 1991, Lipid A-like molecules that antagonize the effects of endotoxins on human monocytes, *J. Biol. Chem.* **266**:19490–19498.

Golenbock, D. T., Liu, Y., Millham, F. H., Freeman, M. W., and Zoeller, R. A., 1993, Surface expression of human CD14 in Chinese hamster ovary fibroblasts imparts macrophage-like responsiveness to bacterial endotoxin, *J. Biol. Chem.* **268**:22055–22059.

Goyert, S. M., Haziot, A., Jiao, D., Katz, I. R., Kruger, C., Ross, J., and Schütt, C., 1995, CD14 cluster workshop report, in: *Leukocyte Typing V: White Cell Differentiation*

Antigens (S. F. Schlossmann *et al.*, eds.), Oxford University Press, Oxford, pp. 778–782.

Groupp, E. R., and Donovan-Peluso, M., 1996, Lipopolysaccharide induction of THP-1 cells activates binding of c-Jun, Ets, and Egr-1 to the tissue factor promoter, *J. Biol. Chem.* **271:**12423–12430.

Guenounou, M., Goguel, A. F., and Nauciel, C., 1982, Study of adjuvant and mitogenic activities of bacterial with different structures, *Ann. Immunol.* **133D:**3–13.

Gupta, D., Jin, Y., and Dziarski, R., 1995, Peptidoglycan induces transcription and secretion of TNF-α and activation of Lyn, extracellular signal-regulated kinase, and Rsk signal transduction proteins in mouse macrophages, *J. Immunol.* **155:**2620–2630.

Gupta, D., Kirkland, T. N., Viriyakosol, S., and Dziarski, R., 1996, CD14 is a cell-activating receptor for bacterial peptidoglycan, *J. Biol. Chem.* **271:**23310–23316.

Gupta, D., Wang, Q., Vinson, C., and Dziarski, R., 1999, Bacterial peptidoglycan induces CD14-dependent activation of transcription factors CREB/ATF-1 and AP-1, *J. Biol. Chem.* **274:**14012–14020.

Guthridge, C. J., Eidlen, D., Arend, W. P., Gutierrez-Hartmann, A., and Smith, M. F., 1997, Lipopolysaccharide and Raf-1 kinase regulate secretory interleukin-1 receptor antagonistic gene expression by mutually antagonistic mechanisms, *Mol. Cell. Biol.* **17:**1118–1128.

Hailman, E., Lichenstein, H. S., Wurfel, M. M., Miller, D. S., Johnson, D. A., Kelley, M., Busse, L. A., Zukowski, M. M., and Wright, S. D., 1994, Lipopolysaccharide (LPS)-binding protein accelerates the binding of LPS to CD14, *J. Exp. Med.* **179:**269–277.

Hailman, E., Vasselon, T., Kelley, M., Busse, L. A., Hu, M. C. T., Lichenstein, H. S., Detmers, P. A., and Wright, S. D., 1996, Stimulation of macrophages and neutrophils by complexes of lipopolysaccharide and soluble CD14, *J. Immunol.* **156:**4384–4390.

Hamann, L., El-Samalouti, V. T., Schletter, J., Chyla, I., Lentschat, A., Flad, H. D., Rietschel, E. Th. and Ulmer, A. J., 1998, CD55, a new LPS-signaling element, in: *5th Conference of the International Endotoxin Society,* International Endotoxin Society Santa Fe, NM, p. 99.

Hambelton, J., McMahon, M., and DeFranco, A. L., 1995, Activation of Raf-1 and mitogen-activated protein kinase in murine macrophages partially mimics lipopolysaccharide-induced signaling events, *J. Exp. Med.* **182:**147–154.

Hambelton, J., Weinstein, S. L., Lem, L., and DeFranco, A. L., 1996, Activation of c-Jun N-terminal kinase in bacterial lipopolysaccharide-stimulated macrophages, *Proc. Natl. Acad. Sci. USA* **93:**2774–2778.

Han, J., Lee, J.-D., Tobias, P. S., and Ulevitch, R. J., 1993, Endotoxin induces rapid protein tyrosine phosphorylation in 70Z/3 cells expressing CD14, *J. Biol. Chem.* **268:**25009–25014.

Han, J., Lee, J.-D., Bibbs, L., and Ulevitch, R. J., 1994, A MAP kinase targeted by endotoxin and hyperosmolarity in mammalian cells, *Science* **265:**808–811.

Haskill, S., Peace, A., Morris, J., Sporn, S. A., Anisowicz, A., Lee, S. W., Smith, T., Martin, G., Ralph, P., and Sager, R., 1990, Identification of three related human GRO genes encoding cytokine functions, *Proc. Natl. Acad. Sci. USA* **87:**7732–7736.

Haworth, R., Platt, N., Keshav, S., Hughes, D., Darley, E., Suzuki, H., Kurihara, Y., Kodama, T., and Gordon, S., 1997, The macrophage scavenger receptor type A is expressed by macrophages and protects the host against lethal endotoxic shock, *J. Exp. Med.* **186:**1431–1439.

Haziot, A., Tsuberi, B.-Z., and Goyert, S. M., 1993a, Neutrophil CD14: Biochemical properties and role in the secretion of tumor necrosis factor-α in response to lipopolysaccharide, *J. Immunol.* **150:**5556–5565.

Haziot, A., Rong, G.-W., Silver, J., and Goyert, S. M., 1993b, Recombinant soluble CD14 mediates the activation of endothelial cells by lipopolysaccharide, *J. Immunol.* **151:**1500–1507.

Haziot, A., Ferrero, E., Kontgen, F., Hijiya, N., Yamamoto, S., Silver, J., Stewart, C. L., and Goyert, S. M., 1996, Resistance to endotoxin shock and reduced dissemination of gram-negative bacteria in CD14-deficient mice, *Immunity* **4:**407–414.

Haziot, A., Lin, X. Y., Zhang, F., and Goyert, S. M., 1998, The induction of acute phase proteins by lipopolysaccharide uses a novel pathway that is CD14-independent, *J. Immunol.* **160:**2570–2572.

Heine, H., Brade, H., Kusumoto, S., Kusama, T., Rietschel, E. Th., Flad, H.-D., and Ulmer, A. J., 1994, Inhibition of LPS-binding on human monocytes by phosphonooxyethyl analogs of lipid A, *J. Endotox. Res.* **1:**14–20.

Heine, H., Ulmer, A. J., Flad, H.-D., and Hauschildt, S., 1995, LPS-induced change of phosphorylation of two cytosolic proteins in human monocytes is prevented by inhibitors of ADP-ribosylation, *J. Immunol.* **155:**4899–4908.

Henricson, B. E., Carboni, J. M., Burkhardt, A. L., and Vogel, S. N., 1995, LPS and taxol activate *Lyn* kinase autophosphorylation in *LPS^n*, but not in *LPS^d*, macrophages, *Mol. Med.* **1:**428–435.

Herrera-Velit, P., and Reiner, N. E., 1996, Bacterial lipopolysaccharide induces the association and coordinate activation of p53/56^lyn and phosphatidylinositol 3-kinase in human monocytes, *J. Immunol.* **156:**1157–1165.

Herrera-Velit, P., Knutson, K. L., and Reiner, N. E., 1997, Phosphatidylinositol 3-kinase-dependent activation of protein kinase C-ζ in bacterial lipopolysaccharide-treated human monocytes, *J. Biol. Chem.* **272:**16445–16452.

Heymer, B., Seidl, P. H., and Schleifer, K. H., 1985, Immunochemistry and biological activity of peptidoglycan, in: *Immunology of the Bacterial Cell Envelope* (D. E. S. Stewart-Tull and M. Davis, eds.), J. Wiley & Sons, New York, pp. 11–46.

Hirschfeld, M., Kirschning C. J., Schwandner, R., Wesche, H., Weis, J. H., Wooten, R. M., and Weis, J. J., 1999, Inflammatory signaling by *Borrelia burgdorferi* lipoproteins is mediated by Toll-like receptor 2, *J. Immunol.* **163:**2382–2386.

Hoshino, K., Takeuchi, O., Kawai, T., Sanjo, H., Ogawa, T., Takeda, Y., Takeda, K., and Akira, S., 1999, Toll-like receptor 4 (TLR4)-deficient mice are hyporesponsive to lipopolysaccharide: evidence for TLR4 as the *Lps* gene product, *J. Immunol.* **162:**3749–3752.

Ingalls, R. R., and Golenbock, D. T., 1995, CD11c/CD18, a transmembrane signaling receptor for lipopolysaccharide, *J. Exp. Med.* **181:**1473–1479.

Ingalls, R. R., Arnaout, M. A., and Golenbock, D. T., 1997, Outside-in signaling by lipopolysaccharide through a tailless integrin, *J. Immunol.* **159:**433–438.

Jakway, J. P., and DeFranco, A. L., 1986, Pertussis toxin inhibition of B cell and macrophage responses to bacterial lipopolysaccharide, *Science* **234:**743–746.

Jarvis, B. W., Lichenstein, H., and Qureshi, N., 1997. Diphosphoryl lipid A from *Rhodobacter sphaeroides* inhibits complexes that form *in vitro* between lipopolysaccharide (LPS)-binding protein, soluble CD14, and spectrally pure LPS, *Infect. Immun.* **65:**3011–3016.

Jin, Y., Gupta, D., and Dziarski, R., 1998, Endothelial and epithelial cells do not respond to complexes of peptidoglycan with soluble CD14, but are activated indirectly by peptidoglycan-induced tumor necrosis factor-α and interleukin-1 from monocytes, *J. Infect. Dis.* **177:**1629–1638.

Johannsen, L., 1993, Biological properties of bacterial peptidoglycan, *APMIS* **101:**337–344.

Juan, T. S. C., Hailman, E., Kelley, M. J., Busse, L. A., Davy, E., Empig, C. J., Narhi, L. O., Wright, S. D., and Lichenstein, H. S., 1995a, Identification of a lipopolysaccharide binding domain in CD14 between amino acids 57 and 64, *J. Biol. Chem.* **270:**5219–5224.

Juan, T. S. C., Hailman, E., Kelley, M. J., Wright, S. D., and Lichenstein, H. S., 1995b, Identification of a domain in soluble CD14 essential for lipopolysaccharide (LPS) signaling but not LPS binding, *J. Biol. Chem.* **270:**17237–17242.

Juan, T. S., Kelley, M. J., Johnson, D. A., Busse, L. A., Hailman, E., Wright, S. D., and Lichenstein, H. S., 1995c, Soluble CD14 truncated at amino acid 152 binds lipopolysaccharide (LPS) and enables cellular response to LPS, *J. Biol. Chem.* **270:**1382–1387.

Kalter, E. S., van Dijk, W. C., Timmerman, A., Verhoef, J., and Bouma, B. N., 1983, Activation of purified human plasma prekallikrein triggered by cell wall fractions of *Escherichia coli* and *Staphylococcus aureus, J. Infect. Dis.* **148:**682–691.

Kasma, T., Strieter, R. M., Standiford, T. J., Burdick, M. D., and Kunkel, S. L., 1993, Expression and regulation of human neutrophil-derived macrophage inflammatory protein 1α, *J. Exp. Med.* **178:**63–72.

Kengatharan, K. M., De Kimpe, S., Robson, C., Foster, S. J., and Thiemermann, C., 1998, Mechanism of gram-positive shock: identification of peptidoglycan and lipoteichoic acid moieties essential in the induction of nitric oxide synthase, shock, and multiple organ failure, *J. Exp. Med.* **188:**305–315.

Kessler, C. M., Nussbaum, E., and Tuazon, C. U., 1991, Disseminated intravascular coagulation associated with *Staphylococcus aureus* septicemia is mediated by peptidoglycan-induced platelet aggregation, *J. Infect. Dis.* **1164:**101–107.

Kirkland, T. N., Finley, F., Leturq, D., Moiarty, A., Lee, J.-D., Ulevitch, R. J., and Tobias, P. S., 1993, Analysis of lipopolysaccharide binding by CD14, *J. Biol. Chem.* **268:**2418–2423.

Kirschning, C. J., Wesche, H., Ayres, T. M., and Rothe, M., 1998, Human Toll-like receptor 2 lipopolysaccharide confers responsiveness to bacterial LPS, *J. Exp. Med.* **188:**2091–2097.

Kissin, E., Tomasi, M., McCartney-Francis, N., Gibbs, C. L., and Smith, P. D., 1997, Age-related decline in murine macrophage production of nitric oxide, *J. Infect. Dis.* **175:**1004–1007.

Kitchens, R. L., and Munford, R. S., 1995, Enzymatically deacylated lipopolysaccharide (LPS) can antagonize LPS at multiple sites in the LPS recognition pathway, *J. Biol. Chem.* **270:**9904–9910.

Kitchens, R. J., and Munford, R. S., 1998, CD14-dependent internalization of bacterial lipopolysaccharide (LPS) is strongly influenced by LPS aggregation but not by cellular responses to LPS, *J. Immunol.* **160:**1920–1928.

Kitchens, R. L., Ulevitch, R. J., and Munford, R. S., 1992, Lipopolysaccharide (LPS) partial structures inhibit responses to LPS in a human macrophage cell line without inhibiting LPS uptake by a CD14-mediated pathway, *J. Exp. Med.* **176**:485–494.

Kruse-Elliott, K. T., Albert, D. H., Summers, J. B., Carter, G. W., Zimmerman, J. J., and Grossman, J. E., 1996, Attenuation of endotoxin-induced pathophysiology by a new potent PAF receptor antagonist, *Shock* **5**:265–273.

Krziwon, C., Zähringer, U., Kawahara, K., Weidemann, B., Kusumoto, S., Rietschel, E. Th., Flad, H. D., and Ulmer, A. J., 1995, Glycosphingolipids from *Sphingomonas paucimobilis* induce monokine production in human mononuclear cells, *Infect. Immun.* **63**:2899–2905.

Kusunoki, T., Hailman, E., Juan, T. S. C., Lichenstein, H. S., and Wright, S. D., 1995, Molecules from *Staphylococcus aureus* that bind CD14 and stimulate innate immune responses, *J. Exp. Med.* **182**:1673–1682.

LaRosa, G. J., Thomas, K. M., Kaufmann, M. E., Mark, R., White, M., Taylor, L., Gray, G., Witt, D., and Navarro, J., 1992, Amino terminus of the interleukin-8 receptor is a major determinant of receptor subtype specificity, *J. Biol. Chem.* **267**:25402–25406.

Lauener, R. P., Geha, R. S., and Vercelli, D., 1990, Engagement of the monocyte surface antigen CD14 induces lymphocyte function-associated antigen-1/intercellular adhesion molecule-1-dependent homotypic adhesion, *J. Immunol.* **145**:1390–1394.

Lawrence, C., and Nauciel, C., 1998, Production of interleukin-12 by murine macrophages in response to bacterial peptidoglycan, *Infect. Immun.* **66**:4947–4949.

Lee, J. D., Kato, K., Tobias, P. S., Kirkland, T. N., and Ulevitch, R. J., 1992, Transfection of CD14 into 70Z/3 cells dramatically enhances the sensitivity to complexes of lipopolysaccharide (LPS) and LPS binding protein, *J. Exp. Med.* **175**:1697–1705.

Levinson, A. I., Dziarski, A., Zweiman, B., and Dziarski, R., 1983, Staphylococcal peptidoglycan: T-cell-dependent mitogen and relatively T-cell-independent polyclonal B-cell activator of human lymphocytes, *Infect. Immun.* **39**:290–296.

Liu, M. K., Herrera-Velit, P., Brownsey, R. W., and Reiner, N. E., 1994, CD14-dependent activation of protein kinase C and mitogen-activated protein kinases (p42 and p44) in human monocytes treated with bacterial lipopolysaccharide, *J. Exp. Med.* **153**:2642–2652.

Loppnow, H., Libby, P., Freudenberg, M., Kraus, J. H., Weckesser, J., and Mayer, H., 1990, Cytokine induction by lipopolysaccharide (LPS) corresponds to the lethal toxicity and is inhibited by nontoxic *Rhodobacter capsulatus* LPS, *Infect. Immun.* **58**:3743–3750.

Loppnow, H., Stelter, F., Schonbeck, U., Schluter, C., Ernst, M., Schütt, C., and Flad, H.-D., 1995, Endotoxin activates human vascular smooth muscle endothelial cells despite lack of expression of CD14 mRNA or endogenous membrane CD14, *Infect. Immun.* **63**:1020–1026.

Lowell, C. A., and Berton, G., 1998, Resistance to endotoxic shock and reduced neutrophil migration in mice deficient for the Src-family kinases Hck and Fgr, *Proc. Natl. Acad. Sci. USA* **95**:7580–7584.

Luster, A. D., and Leder, P., 1993, IP-10, a -C-X-C- chemokine, elicits a potent thymus-dependent antitumor response *in vivo, J. Exp. Med.* **178**:1057–1065.

Mackman, N., Brand, K., and Edgington, T. S., 1991, Lipopolysaccharide-mediated transcriptional activation of the human tissue factor gene in THP-1 monocytic cells requires both activator protein 1 and nuclear factor kappa B binding sites, *J. Exp. Med.* **174**:1517–1526.

MacMicking, J., Xie, Q., and Nathan, C., 1997, Nitric oxide and macrophage function, *Annu. Rev. Immunol.* **15**:323–350.

Mannel, D. N., and Grau, G. E., 1997, Role of platelet adhesion in homeostasis and immunopathology, *Mol. Pathol.* **50**:175–85.

Mathison, J. C., Tobias, P. S., Wolfson, E., and Ulevitch, R. J., 1992, Plasma lipopolysaccharide (LPS)-binding protein: a key component in macrophage recognition of gram-negative LPS, *J. Immunol.* **149**:200–206.

Matsukawa, A., Ohkawara, S., Maeda, T., Takagi, K., and Yoshinaga, M., 1993, Production of IL-1 and IL-1 receptor antagonist and the pathological significance in lipopolysaccharide-induced arthritis in rabbits, *Clin. Exp. Immunol.* **93**:206–211.

Means, T. K., Wang, S., Lien, E., Yoshimura, A., Glolenbock, D. T., and Fenton, M. J., 1999, Human Toll-like receptors mediate cellular activation by *Mycobacterium tuberculosis, J. Immunol.* **163**:3920–3927.

Medvedev, A. E., Flo, T., Ingalls, R. R., Golenbock, D. T., Teti, G., Vogel, S. N., and Espevik, T., 1998, Involvement of CD14 and complement receptors CR3 and CR4 in nuclear factor-κB activation of TNF production induced by lipopolysaccharide and group B streptococcal cell walls, *J. Immunol.* **160**:4535–4542.

Meng, F., and Lowell, C. A., 1997, Lipopolysaccharide (LPS)-induced macrophage activation and signal transduction in the absence of Src-family kinases Hck, Fgr, and Lyn, *J. Exp. Med.* **185**:1661–1670.

Metcalf, D., 1991, Control of granulocytes and macrophages: molecular, cellular and clinical aspects, *Science* **254**:529–533.

Newman, S. L., Chaturvedi, S., and Klein, B. S., 1995, The WI-1 antigen of *Blastomyces dermatitidis* yeast mediates binding to human macrophage CD11/CD18 (CR3) and CD14, *J. Immunol.* **154**:753–761.

Novogrodsky, A., Vanichkin, A., Patya, M., Gazit, A., Osherov, N., and Levitzki, A., 1994, Prevention of lipopolysaccharide-induced lethal toxicity by tyrosine kinase inhibitors, *Science* **264**:1319–1322.

Noyori, K., Okamoto, R., Takagi, T., Hyodo, A., Suzuki, K., and Koshino, T., 1994, Experimental induction of arthritis in rats immunized with *Escherichia coli* O:14 lipopolysaccharide, *J. Rheumatol.* **21**:484–488.

Orr, S. L., and Tobias, P., 1998, Endothelial and epithelial cell lines can be stimulated by mycobacterial lipoarabinomannan via membrane-bound CD14 but fail to respond via soluble CD14, in: *5th Conference of the International Endotoxin Society,* International Endotoxin Society Santa Fe, NM, p. 135.

Otterlei, M., Varum, K. M., Ryan, L., and Espevik, T., 1994, Characterization of binding and TNF-alpha-inducing ability of chitosans on monocytes: the involvement of CD14, *Vaccine* **12**:825–832.

Pan, J., Xia, L., Yao, L., and McEver, R., 1998, Tumor necrosis factor-α or lipopolysaccharide-induced expression of the murine P-selectin gene in endothelial cells involves novel κB sites and a variant activating transcription factor/cAMP response element, *J. Biol. Chem.* **273**:10068–10077.

Perera, P.-Y., Vogel, S. N., Detore, G. R., Haziot, A., and Goyert, S. M., 1997, CD14-dependent and CD14-independent signaling pathways in murine macrophages from normal and CD14 knockout mice stimulated with lipopolysaccharide and taxol, *J. Immunol.* **158**:4422–4429.

Poltorak, A., He, X., Smirnova, I., Liu, M.-Y., Van Huffel, C., Du, X., Birdwell, D., Alejos, E., Silva, M., Galanos, C., Freudenberg, M., Riccardi-Castagnoli, P., Layton, B., and Beutler, B., 1998, Defective LPS signaling in C3H/HeJ and C57BL/10ScCr mice: mutations in *Tlr4* gene, *Science* **282**:2085–2088.

Proud, D., and Kaplan, A. P., 1988, Kinin formation: Mechanisms and role in inflammatory disorders, *Annu. Rev. Immunol.* **6**:49–83.

Pugin, J., Schurer-Maly, C.-C., Leturq, D., Moriarty, A., Ulevitch, R. J., and Tobias, P. S., 1993a, Lipopolysaccharide activation of human endothelial and epithelial cells is mediated by lipopolysaccharide-binding protein and soluble CD14, *Proc. Natl. Acad. Sci. USA* **90**:2744–2748.

Pugin, J., Ulevitch, R. J., and Tobias, P. S., 1993b, A critical role for monocytes and CD14 in endotoxin-induced endothelial cell activation, *J. Exp. Med.* **178**:2193–2200.

Pugin, J., Heumann, D., Tomasz, A., Kravchenko, V. V., Akamatsu, Y., Nishijima, M., Glauser, M. P., Tobias, P. S., and Ulevitch, R. J., 1994, CD14 is a pattern recognition receptor, *Immunity* **1**:509–516.

Pugin, J., Ulevitch, R. J., and Tobias, P. S., 1995, Tumor necrosis factor-α and interleukin-1β mediate human endothelial cell activation in blood at low endotoxin concentrations, *J. Inflammat.* **45**:49–55.

Qureshi, S. T., Lariviere, L., Leveque, G., Clermont, S., Moore, K. J., Gros, P., and Malo, D., 1999, Endotoxin-tolerant mice have mutations in Toll-like receptor 4 (*Tlr4*), *J. Exp. Med.* **189**:615–625.

Raingeaud, J., Gupta, S., Rogers, J. S., Dickens, M., Han, J., Ulevitch, R. J., and Davis, R. J., 1995, Pro-inflammatory cytokines and environmental stress cause p38 mitogen-activated protein kinase activation by dual phosphorylation on tyrosine and threonine, *J. Biol. Chem.* **270**:7420–7426.

Rasanen, L., and Arvilommi H., 1981, Cell walls, peptidoglycans, and teichoic acids of gram-positive bacteria as polyclonal inducers and immunomodulators of proliferative and lymphokine responses of human B and T lymphocytes, *Infect. Immun.* **34**:712–717.

Read, M. A., Cordle, S. R., Veach, R. A., Carlisle, C. D., and Hawiger, J., 1993, Cell-free pool of CD14 mediates activation of transcription factor NF-κB by lipopolysaccharide in human endothelial cells, *Proc. Natl. Acad. Sci. USA* **90**:9887–9891.

Redl, H., Schlag, G., Thurnher, M., Traber, L. D., and Traber, D. L., 1989, Cardiovascular reaction pattern during endotoxin or peptidoglycan application in awake sheep, *Circulatory Shock* **28**:101–108.

Reimann, T., Buscher, D., Hipskind, R. A., Krautwald, S., Lohmann-Matthes, M.-L., and Bacarini, M., 1994, Lipopolysaccharide induces activation of the Raf-1/MAP kinase pathway: A putative role for Raf-1 in the induction of the IL-1β and TNF-α genes, *J. Immunol.* **153**:5740–5749.

Rietschel, E. T., Kirikae, T., Schade, F. U., Mamat, U., Schmidt, G., Loppnow, H., Ulmer, A. J., Zahringer, U., Seydel, U., Di Padove, F., Schreier, M., and Brade, H., 1994, Bacterial endotoxin: Molecular relationships of structure to activity and function, *FASEB J.* **8**:217–225.

Rosenthal, R. S., and Dziarski, R., 1994, Isolation of peptidoglycan and soluble peptidoglycan fragments, *Methods Enzymol.* **235**:253–285.

Ryc, M., and Rotta, J., 1975, The thrombocytolytic activity of bacterial peptidoglycan, *Z. Immun. Forsch.* **149S**:265–272.

Saito-Taki, T., Tanabe, M. J., Mochizuki, H., Matsumoto, T., Nakano, M., Takada, H., Tsujimoto, M., Kotani, S., Kusumoto, S., Shiba, T., Yokogawa, K., and Kawata, S., 1980a, Polyclonal B cell activation by cell wall preparations of gram-positive bacteria. *In vitro* responses of spleen cells obtained from Balb/c, nu/nu, nu/+, C3H/He, C3H/HeJ and hybrid (DBA/N x Balb/c)F1 mice, *Microbiol. Immunol.* **24:**209–218.

Saito-Taki, T., Tanabe, M. J., Mochizuki, H., Nakano, M., Tsujimoto, M., Kotani, S., Yokogawa, K., and Kawata, S., 1980b, Mitogenicity of cell wall preparations of gram-positive bacteria on cultured spleen cells obtained from immunologically abnormal C3H/HeJ and CBA/N-defective mice, *Microbiol. Immunol.* **24:**249–254.

Sanghera, J. S., Weinstein, S. L., Aluwalia, M., Girn, J., and Pelech, S. L., 1996, Activation of multiple proline-directed kinases by bacterial lipopolysaccharide in murine macrophages, *J. Immunol.* **156:**4457–4465.

Savedra, R., Delude, R. L., Ingalls, R. R., Fenton, M. J., and Golenbock, D. T., 1996, Mycobacterial lipoarabinomannan recognition requires a receptor that shares components of the endotoxin signaling system, *J. Immunol.* **157:**2549–2554.

Schleifer, K. H., and Kandler, O, 1975, Peptidoglycan types in the bacterial cell walls and their taxonomic implications, *Bacteriol. Rev.* **36:**407–477.

Schroder, J. M., Persoon, N. L., and Christophers, E., 1990, Lipopolysaccharide-stimulated human monocytes secrete, apart from neutrophil-activating peptide 1/interleukin 8, a second neutrophil-activating protein: NH_2-terminal amino acid sequence identity with melanoma growth stimulatory activity, *J. Exp. Med.* **171:**1091–1100.

Schromm, A. B., Brandenburg, K., Rietschel, E. Th., Flad. H. D., Carroll. S. F., and Seydel, U., 1996, Lipopolysaccharide-binding protein mediates CD14-independent intercalation of lipopolysaccharide into phospholipid membranes, *FEBS Lett.* **399:**267–271.

Schumann, R. R., Leong, S. R., Flaggs, G. W., Gray, P. W., Wright, S. D., Mathison, J. C., Tobias, P. S., and Ulevitch, R. J., 1990, Structure and function of lipopolysaccharide binding protein, *Science* **249:**1429–1431.

Schütt, C., Ringel, B., Nausch, M., Bazil, V., Horejsi, V., Neels, P., Walzel, H., Jonas, L., Siegl, E., Friemel, H, and Plantikow, A., 1988, Human monocyte activation induced by an anti-CD14 monoclonal antibody, *Immunol. Lett.* **19:**321–328.

Schütt, C., Witt, U., Grunwald, U., Stelter, F., Schilling, T., Fan, X., Marquart, B.-P., Bassarab, S., and Krüger, C., 1995, Epitope mapping of CD14 glycoprotein, in: *Leukocyte Typing V: White Cell Differentiation antigens* (S. F. Schlossmann *et al.,* eds.), Oxford University Press, Oxford, pp. 784–788.

Schwandner, R., Dziarski, R., Wesche, H., Rothe, M., and Kirschning, C. J., 1999, Peptidoglycan- and lipoteichoic acid-induced cell activation is mediated by Toll-like receptor 2, *J. Biol. Chem.* **274:**17406–17409.

Sellati, T. J., Bouis, D. A., Kitchens, R. L., Darveau, R. P., Pugin, J., Ulevitch, R. J., Gangloff, S. C., Goyert, S. M., Norgard, M. V., and Radolf, J. D., 1998, *Treponema pallidum* and *Borrelia burgdorferi* lipoproteins and synthetic lipopeptides activate monocytic cells via a CD14-dependent pathway distinct from that used by lipopolysaccharide, *J. Immunol.* **160:**5455–5464.

Seydel, U., Labischinski, H., Kastowsky, M., and Brandenburg, K., 1993, Phase behavior, supramolecular structure, and molecular conformation of lipopolysaccharide, *Immunobiology* **187:**191–211.

Shapira, L., Sylvia, V. L., Halabi, A., Soskolne, W. A., Van Dyke, T. E., Dean, D. D., Boyan, B. D., and Schwartz, Z., 1997, Bacterial lipopolysaccharide induces early and late activation of protein kinase C in inflammatory macrophages by selective activation of PKC-ε, *Biochem. Biophys. Res. Commun.* **240**:629–634.

Shapiro, R. A., Cunningham, M. D., Ratcliffe, K., Seachord, C., Blake, J., Bajorath, J., Aruffo, A., and Darveau, R. P., 1997, Identification of CD14 residues involved in specific lipopolysaccharide recognition, *Infect. Immun.* **65**:293–297.

Shimazu, R., Akashi, S., Ogata, H., Nagai, Y., Fukudome, K., Miyake, K., and Kimoto, M., 1999, MD-2, a molecule that confers lipopolysaccharide responsiveness on Toll-like receptor 4, *J. Exp. Med.* **189**:1777–1782.

Soell, M., Lett, E., Holveck, F., Scholler, M., Wachsmann, D., and Klein, J. P., 1995, Activation of human monocytes by streptococcal rhamnose glucose polymers is mediated by CD14 antigen, and mannan binding protein inhibits TNF-alpha release, *J. Immunol.* **154**:851–860.

Spika, J. S., Peterson, P. K., Wilkinson, B. J., Hammerschmidt, D. E., Verbrugh, H. A., Verhoef, J., and Quie, P. G., 1982, Role of peptidoglycan from *Staphylococcus aureus* in leukopenia, thrombocytopenia, and complement activation associated with bacteriemia, *J. Infect. Dis.* **146**:227–234.

Stefanova, I., Horejsi, V., Ansotegui, I. J., Knapp, W., and Stockinger, H., 1991, GPI-anchored cell-surface molecules complexed to protein tyrosine kinases, *Science* **254**:1016–1019.

Stefanova, I., Corcoran, M. L., Horak, E. M., Wahl, L. M., Bolen, J. B., and Horak, I. D., 1993, Lipopolysaccharide induces activation of CD14-associated protein tyrosine kinase p53/56^{lyn}, *J. Biol. Chem.* **268**:20725–20728.

Stelter, F., Bernheiden, M., Menzel, R., Jack, R. S., Witt, S., Fan, X., Pfister, M., and Schutt, C., 1997, Mutation of amino acids 39–44 of human CD14 abrogates binding of lipopolysaccharide and *Escherichia coli*, *Eur. J. Biochem.* **243**:100–109.

Suda, Y., Tochio, H., Kawano, K., Takada, H., Yoshida, T., Kotani, S., and Kusumoto, S., 1995, Cytokine-inducing glycolipids in the lipoteichoic acid fraction from *Enterococcus hirae* ATCC 9790, *FEMS Immunol. Med. Microbiol.* **12**:97–112.

Swantek, J. L., Cobb, M. H., and Geppert, T. D., 1997, Jun N-terminal kinase/stress-activated protein kinase (JNK/SAPK) is required for lipopolysaccharide stimulation of tumor necrosis factor-α (TNF-α) translation: Glucocorticoids inhibit TNF-α translation by blocking JNK/SAPK, *Mol. Cell. Biol.* **17**:6274–6282.

Sweet, M. J., and Hume, D. A., 1996, Endotoxin signal transduction in macrophages, *J. Leukoc. Biol.* **60**:8–26.

Takada, H., Kawabata, Y., Arakaki, R., Kusumoto, S., Fukase, K., Suda, Y., Yoshimura, T., Kokeguchi, S., Kato, K., Komuro, T., Tanaka, N., Saito, M., Yoshida, T., Sato, M., and Kotani, S., 1995, Molecular and structural requirements of a lipoteichoic acid from *Enterococcus hirae* ATCC 9790 for cytokine-inducing, antitumor, and antigenic activities, *Infect. Immun.* **63**:57–65.

Takayama, K., Quereshi, N., Beutler, B., and Kirkland, T. N., 1989, Diphosphoryl lipid A from *Rhodopseudomonas sphaeroides* ATCC 17023 blocks induction of cachectin in macrophages by lipopolysaccharide, *Infect. Immun.* **57**:1336–1338.

Takayama, K., Mitchell, D. H., Din, Z. Z., Mukerjee, P., Li, C., and Coleman DL., 1994, Monomeric Re lipopolysaccharide from *Escherichia coli* is more active than the aggregated form in the *Limulus* amebocyte lysate assay and in inducing Egr-1 mRNA in murine peritoneal macrophages, *J. Biol. Chem.* **269**:2241–2244.

Tapping, R. I., and Tobias, P. S., 1997, Cellular binding of soluble CD14 requires lipopolysaccharide (LPS) and LPS-binding protein, *J. Biol. Chem.* **272**:23157–23164.

Timmerman, C. P., Mattsson, E., Martinez-Martinez, L., De Graaf, L., Van Strijp, J. A. G., Verbrugh, H. A., Verhoef, J., and Fleer, A., 1993, Induction of release of tumor necrosis factor from human monocytes by staphylococci and staphylococcal peptidoglycans, *Infect. Immun.* **61**:4167–4172.

Tobias, P. S., Soldau, K., and Ulevitch, R. J., 1986, Isolation of a lipopolysaccharide-binding acute phase reactant from rabbit serum, *J. Exp. Med.* **164**:777–793.

Tobias, P. S., Soldau, K., and Ulevitch, R. J., 1989, Identification of a lipid A binding site in the acute phase reactant lipopolysaccharide binding protein, *J. Biol. Chem.* **264**:10867–10871.

Tobias, P. S., Soldau, K., Kline, L., Lee, J. D., Kato, K., Martin, T. P., and Ulevitch, R. J., 1993, Cross-linking of lipopolysaccharide (LPS) to CD14 on THP-1 cells mediated by LPS-binding protein, *J. Immunol.* **150**:3011–3021.

Tobias, P. S., Soldau, K., Gegner, J. A., Mintz, D., and Ulevitch, R. J., 1995, Lipopolysaccharide binding protein-mediated complexation of lipopolysaccharide with soluble CD14, *J. Biol. Chem.* **270**:10482–10488.

Trinchieri, G., 1995, Interleukin-12: A proinflammatory cytokine with immunoregulatory functions that bridge innate resistance and antigen-specific adaptive immunity, *Annu. Rev. Immunol.* **13**:251–276.

Ulevitch, R. J., and Tobias, P. S., 1995, Receptor-dependent mechanisms of cell stimulation by bacterial endotoxin, *Annu. Rev. Immunol.* **13**:437–457.

Ulmer, A. J., Feist, W., Heine, H., Kirikae, T., Kirikae, F., Kusumoto, S., Kusama, T., Brade, H., Schade, U., Rietschel, E. Th., and Flad, H.-D., 1992, Modulation of endotoxin-induced monokine release in human monocytes by lipid A partial structures inhibiting the binding of ^{125}I-LPS, *Infect. Immun.* **60**:5145–5152.

Ulmer, A. J., Dziarski, R., El-Samalouti, V., Rietschel, E. T., and Flad, H. D., 1999, CD14, an innate immune receptor for various bacterial cell wall components, in: *Endotoxin in Health and Disease* (D. Morrison, ed.), Marcel Dekker, New York, pp. 463–472.

Vacheron, F., Guenounou, M., and Nauciel, C., 1983, Induction of interleukin 1 secretion by adjuvant-active peptidoglycans, *Infect. Immun.* **42**:1049–1053.

Vacheron, F., Guenounou, M., Zinbi, H., and Nauciel, C., 1986, Release of a cytotoxic factor by macrophages stimulated with adjuvant-active peptidoglycans, *J. Natl. Cancer Inst.* **77**:549–553.

Van Snick, J., 1990, Interleukin-6: An overview, *Annu. Rev. Immunol.* **8**:253–278.

Vassalli, P., 1992, The pathophysiology of tumor necrosis factors, *Annu. Rev. Immunol.* **10**:411–452.

Verhoef, J., and Kalter, E., 1985, Endotoxic effects of peptidoglycan, *Prog. Clin. Biol. Res.* **189**:101–113.

Viriyakosol, S., and Kirkland, T. N., 1995, A region of human CD14 required for lipopolysaccharide binding, *J. Biol. Chem.* **270**:361–368.

Viriyakosol, S., and Kirkland, T. N., 1996, The N-terminal half of membrane CD14 is a functional cellular lipopolysaccharide receptor, *Infect. Immun.* **64**:653–656.

Vita, N., Lefort, S., Sozzani, P., Reeb, R., Richards, S., Borysiewicz, L., Ferrara, P., and La-

beta, M., 1997, Detection and biochemical characteristics of the receptor for complexes of soluble CD14 and bacterial lipopolysaccharide, *J. Immunol.* **158:**3457–
3462.

Vowels, B. R., Yang, S., and Leyden, J. J., 1995, Induction of proinflammatory cytokines
by a soluble factor of *Propionibacterium acnes:* Implications for chronic inflammatory acne, *Infect. Immun.* **63:**3158–3165.

Wahl, S. M., Feldman, G. M., and McCarthy, J. B., 1996, Regulation of leukocyte adhesion
and signaling in inflammation and disease, *J. Leukoc. Biol.* **59:**789–796.

Weidemann, B., Brade, H., Rietschel, E. T., Dziarski, R., Bazil, V., Kusumoto, S., Flad,
H.-D., and Ulmer, A. J., 1994, Soluble peptidoglycan-induced monokine production
can be blocked by anti-CD14 monoclonal antibodies and by lipid A partial structures,
Infect. Immun. **62:**4709–4715.

Weidemann, B., Schletter, J., Dziarski, R., Kusumoto, S., Stelter, F., Rietschel, E. T., Flad,
H. D., and Ulmer, A. J, 1997, Specific binding of soluble peptidoglycan and mu-
ramyldipeptide to CD14 on human monocytes, *Infect. Immun.* **65:**858–864.

Weinstein, S. L., Sanghera, J. S., Lemke, K., DeFranco, A. L., and Pelech, S. L., 1992, Bacterial lipopolysaccharide induces tyrosine phosphorylation and activation of mitogen-
activated protein kinases in macrophages, *J. Biol. Chem.* **267:**14955–14962.

Wildfeuer, A., Heymer, B., Spilker, D., Schleifer, K.-H., Vanek, E., and Haferkamp, O.,
1975, Use of *Limulus* assay to compare the biological activity of peptidoglycan and
endotoxin, *Z. Immun. Foersch.* **149S:**258–264.

Wolpe, S. D., and Cerami A., 1989, Macrophage inflammatory proteins 1 and 2: Members
of a novel family of cytokines, *FASEB J.* **3:**2565–2573.

Wooten, M., Morrison, T. B., Weis, J. H., Wright, S. D., Thieringer, R., and Weis, J. J., 1998,
The role of CD14 in signaling mediated by outer membrane lipoproteins of *Borrelia
burgdorferi, J. Immunol.* **160:**5485–5492.

Wright, S. D., 1995, CD14 and innate recognition of bacteria, *J. Immunol.* **155:**6–8.

Wright, S. D., and Kolesnick, R. N., 1995, Does endotoxin stimulate cells by mimicking
ceramide? *Immunol. Today.* **16:**297–302.

Wright, S. D., Ramos, R. A., Tobias, P. S., Ulevitch, R. J., and Mathison, J. C., 1990, CD14,
a receptor for complexes of lipopolysaccharide (LPS) and LPS binding protein, *Science* **249:**1431–1433.

Wurfel, M. M., and Wright, S. D., 1997, Lipopolysaccharide-binding protein and soluble
CD14 transfer lipopolysaccharide to phospholipid bilayers: Preferential interaction
with particular classes of lipid, *J. Immunol.* **158:**3925–3934.

Wurfel, M. M., Kunitake, S. T., Lichenstein, H., Kane, J. P., and Wright, S. D., 1994,
Lipopolysaccharide (LPS)-binding protein is carried on lipoproteins and acts as a co-
factor in the neutralization of LPS, *J. Exp. Med.* **180:**1025–1035.

Yang, R.-B., Mark, M. R., Gray, A., Huang, A., Xie, M. H., Zhang, M., Goddard, A., Wood,
W. I., Gurney, A. L., and Godowski, P. J., 1998, Toll-like receptor-2 mediates
lipopolysaccharide-induced cellular signaling, *Nature* **395:**284–288.

Yang, R.-B., Mark, M. R., Gurney, A. L., and Godowski, P. J., 1999, Signaling events induced by lipopolysaccharide-activated Toll-like receptor 2, *J. Immunol.* **163:**639–643.

Yao, Z., Diener, K., Wang, X. S., Zukowski, M., Matsumoto, G., Zhou, G., Mo, R., Sasaki,
T., Nishina, H., Hui, C. C., Tan, T. H., Woodgett, J. P., and Penninger, J. M., 1997, Activation of stress-activated protein kinase/c-Jun N-terminal protein kinase (SAPKs/

JNKs) by a novel mitogen-activated protein kinase kinase, *J. Biol. Chem.* **272:**32378–32383.

Yoshida, H., Kinoshita, K., and Ashida, M., 1996, Purification of a peptidoglycan recognition protein from hemolymph of the silkworm, *Bombyx mori, J. Biol. Chem.* **271:** 13854–13860.

Yoshimura, A., Lien, E., Ingalls, R. R., Tuomanen, E., Dziarski, R., and Golenbock, D., 1999, Recognition of Gram-positive bacterial cell wall components by the innate immune system occurs via Toll-like receptor 2, *J. Immunol.* **163:**1–5.

Young, L. S., 1995, Sepsis syndrome, in: *Principles and Practice of Infectious Diseases* (G. L. Mandell, J. E. Bennett, and R. Dolin, eds.), Churchill Livingstone, New York, pp. 690–705.

Yu, B., and Wright, S. D., 1996, Catalytic properties of lipopolysaccharide (LPS) binding protein. Transfer of LPS to soluble CD14, *J. Biol. Chem.* **271:**4100–4105.

Yu, B., Hailman, E., and Wright, S. D., 1997, Lipopolysaccharide binding protein and soluble CD14 catalyze exchange of phospholipids, *J. Clin. Invest.* **99:**315–324.

Zhang, Y., Doerfler, M., Lee, T. C., Guillemin, B., and Rom, W. N., 1993, Mechanisms of stimulation of interleukin-1 beta and tumor necrosis factor-alpha by *Mycobacterium tuberculosis* components, *J. Clin. Invest.* **91:**2076–2083.

Ziegler-Heitbrock, H. W., 1995, Molecular mechanism in tolerance to lipopolysaccharide, *J. Inflammat.* **45:**13–26.

7

Pathways for the *O*-Acetylation of Bacterial Cell Wall Polysaccharides

Anthony J. Clarke, Hendrik Strating, and Neil T. Blackburn

1. COMPOSITION AND STRUCTURE OF BACTERIAL CELL WALL POLYSACCHARIDES

1.1. Peptidoglycan

Peptidoglycan is a heteropolymer of distinctive composition and structure, associated uniquely with bacterial cell walls. Its chemistry and structure have been reviewed (Höltje, 1998; Schleifer and Kandler, 1972), but briefly, peptidoglycan is composed of β 1 → 4 linked *N*-acetylglucosaminyl and *N*-acetylmuramyl residues (Fig. 1A). In the mature polymer, the latter amino sugar is modified by a tetrapeptide (stem peptide) composed of alternating L and D amino acids. The amino group of the muramyl residues is generally acetylated, but it also may be free, acylated with glycolic acid, or form an internal amide with an adjacent carboxyl group. Muramyl residues also may exist at the nonreducing ends of glycan chains with intramolecular 1,6-anhydro linkages as a product of lytic transglycosylase activities (Höltje, 1998), or be replaced by its isomer with the D-manno configuration. The stem peptide is linked to the carboxyl group of muramic acid and is initially syn-

Anthony J. Clarke, Hendrik Strating, and Neil T. Blackburn • Department of Microbiology, University of Guelph, Guelph, Ontario, N1G 2W1, Canada.

Glycomicrobiology, edited by Doyle.
Kluwer Academic/Plenum Publishers, New York, 2000.

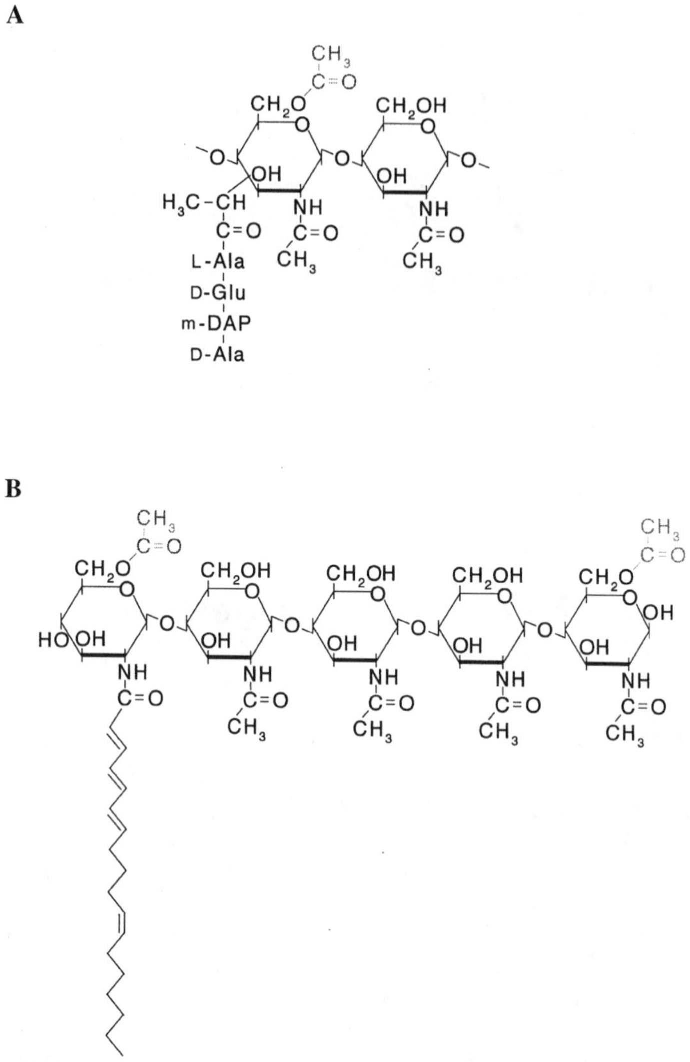

A

B

Figure 1.   Structure of (A) peptidoglycan from gram-negative bacteria (type AIγ) and (B) nodulation factor from *Rhizobium leguminosarum* biovar *viciae.*

thesized in the periplasm as a pentapeptide. Although there may be considerable variation in its composition, particularly amongst gram-positive bacteria, the pentapeptides are synthesized with a terminal D-Ala-D-Ala dipeptide. The D-Ala-D-Ala bond is used to form a cross-link between the penultimate D-Ala residue and a neighboring stem peptide, either directly or through a bridging peptide, usually to the third diamino acid residue of the acceptor peptide.

The overall complexity of peptidoglycan that exists within a single species of bacteria is exemplified in a study by Glauner *et al.* (1988). From a digest of *Escherichia coli* peptidoglycan, 80 different types of muropeptides were isolated by high-performance liquid chromatography (HPLC) and characterized. The diversity of structures is due to the free combination of seven different types of side chains with two types of cross-bridges between the peptides of adjacent strands.

In gram-positive bacteria, the cross-linked peptidoglycan chains are arranged in a thick three-dimensional concentric array, whereas that of gram-negative bacteria generally has been accepted to exist as a monolayer of two-dimensionally arranged, cross-linked chains (Shockman and Barrett, 1983). However, some reports suggest that the peptidoglycan sacculus of gram-negative bacteria may be up to three layers thick (Labischinski *et al.*, 1991). Regardless of its dimensions, both types of peptidoglycan cover the inner cytoplasmic membrane of bacterial cells to maintain the structural integrity of the cell. This fact has made the enzymatic machinery responsible for the biosynthesis of peptidoglycan a popular and prime target for antimicrobial therapy.

1.2. Lipopolysaccharides

Lipopolysaccharides are uniquely found in gram-negative bacteria and are localized to the outer leaflet of the outer membrane. They are characteristically a tripartite molecule comprised of lipid A, core oligosaccharide, and O-antigen (or O-chain) (recently reviewed by Raetz, 1996). Lipid A (also known as endotoxin) is the innermost domain of the entire lipopolysaccharide molecule and serves to anchor it to the lipid phase of the outer leaflet. *E. coli* lipid A, which shares most features of those characterized from other gram-negative bacteria, consists of $\beta 1 \rightarrow$ 6-linked disaccharide of D-glucosamine that is acylated at positions 2, 3, 2', and 3', involving six fatty acid chains. This acylated disaccharide is also phosphorylated at positions 1 and 4' (Raetz, 1990).

The core oligosaccharide regions of lipopolysaccharide, which connect the lipid A anchor to O-antigen, can be divided into inner and outer core regions (Osborn, 1979). The inner core region, which is proximal to lipid A, is composed of the unusual monosaccharides 2-keto-3-deoxyoctonic acid (KDO), L-glycero-D-*manno*-heptose, often together with phosphate or sulfate groups and ethanolamine (Rivera and McGroarty, 1989; Wilkinson, 1983). KDO attaches the core oligosaccharide to lipid A through an acid-labile ketosidic linkage, and three to six more monosaccharides comprise the rest of the inner core region. The outer core is attached to the O-antigen and is generally referred to as the hexose region, as its composition is predominantly neutral and basic hexoses. For *Pseudomonas aeruginosa* lipopolysaccharide, these hexoses form a branched hexasaccharide comprised of D-glucose, L-rhamnose, and D-galactosamine, with the latter being modified by an L-alanine residue (Kropinski *et al.,* 1985; Wilkinson, 1983). The lipopolysaccharide of *Salmonella* spp. also contains one core oligosaccharide type, a trisaccharide of D-glucosyl and a *N*-acetyl-D-glucosaminyl residue (Raetz, 1990), whereas *E. coli* produces five distinct core structures, designated R1 to R-4, and R-12.

The O-antigen is a repeating polysaccharide that extends from the core oligosaccharide out into the cell's environment. The chemical composition and structure of O-antigens is strain specific, and hence its antigenicity forms the basis of O-serotyping for species of gram-negative bacteria. The composition of the repeat units of O-antigens ranges from two to seven acidic, neutral, and/or basic monosaccharides, many of which are rarely found elsewhere in nature, such as abequose, colitose, fucosamine, *N*-acetylquinovosamine, tyvelose, 2,3-diamino-2,3-dideoxyuronic acid, and 5,7-diamino-3,5,7,9-tetradeoxynonulsonic acid, to name a few (recently reviewed by Knirel and Kochetkov, 1994). These repeat units may be either linear or branched. In addition, many of the amino sugars can be acylated with substitutions that are also very uncommon, such as formyl, hydroxybutyryl, and acetamidoly groups. With the availability of so many different unmodified and modified monosaccharides that may be polymerized by a variety of glycosidic linkages, the structural diversity of the O-antigens is enormous. Indeed, *Salmonella* spp. produces more than 1000 chemically and structurally distinct variants (Lindberg and LeMinor, 1984), whereas 173 O-antigens have been reported for *E. coli* (Ørskov and Ørskov, 1992). Although the majority of O-antigens are heteropolymers, some have been found to consist of a single monosaccharide. Thus, the O-antigens of *Klebsiella pneumoniae* serotypes 03 and 05, and *E. coli* serotypes 08, 09 and 09a are homopolymers of mannose that are glycosidically linked in different patterns. Only one of these O-antigens, that of *E. coli* serotype 08, is modified; it possesses a 3-*O*-methyl group on the nonreducing terminal mannose of the repeat unit (Knirel and Kochetkov, 1994). Another subset of *Klebsiella* O-antigens consists of homopolymers of galactose. Two discrete repeat unit structures, designated D-galactan I and D-galactan II, have been observed (Whitfield *et al.*, 1991). For the prototype member of this family, serotype 01, D-galactan II is attached to the distal end of D-galactan I to form the O-antigen, with D-galactan I antigens being linked directly to the core oligosaccharide (Kol *et al.*, 1992).

The length of O-antigens (that is, number of repeat units) will vary within a strain. For example, the lipopolysaccharide of *P. aeruginosa* PAO1 is a mixture of molecules composed of lipid A-core without O-antigen units (rough or R-form), lipid A core with one repeat unit (smooth–rough, or SR-form), and lipid A core with with 2 to 50 repeat O-antigen units (smooth, or S-form) (Lam *et al.*, 1992; Kropinski *et al.*, 1985). The mole percent, or capping frequency, of O-antigen has been shown to vary *in vitro* with environmental conditions. With *P. aeruginosa* lipopolysaccharides, the capping frequency will increase from 19.3% to 37.6% in cells grown at 15°C and 45°C, respectively (Kropinski *et al.*, 1987), which results from an increase in the proportion of SR-form rather and a concomitant decrease in the concentration of S-form (McGroarty and Rivera, 1990; Kropinski *et al.,* 1987). Likewise, high concentrations of NaCl, $MgCl_2$, glycerol, or sucrose result in an observed decrease in proportions of long-chain S-form lipopolysaccharide relative to SR-form (McGroaty and Rivera, 1990).

1.3. Lipo-oligosaccharides

The lipopolysaccharides produced by species of *Haemophilus* and *Neisseria* are quite distinct from those produced by the *Enterobacteriaceae* and other gram-negative bacteria in that they lack O-antigens. These molecules are termed lipooligosaccharides because the oligosaccharide chain attached to the lipid A moiety is equivalent in size and location to the core region of enterobacterial lipopolysaccharide (Flesher and Insel, 1978).

The secreted lipo-oligosaccharides produced by the rhizobia (species of *Rhizobium, Bradyrhizobium,* and *Azorhizobium*), although chemically distinct from each other, are composed of a backbone of three to five β 1 → 4 linked *N*-acetyl-glucosaminyl residues with an *N*-acyl group (fatty acid) attached to the nonreducing terminal residue (Fig. 1B). Specificity is conferred by variations of the acyl moiety or by modifications to the *N*-acetylglucosamine backbone. For example, the acyl group anchoring the lipo-oligosaccharide produced by *R. melilotii* to its plasma membrane is hexadecadienoic acid ($C_{16:2}$) (Lerouge *et al.,* 1990), whereas that of the *B. japonicum* and *Rhizobium* sp. NGR234 lipo-oligosaccharides is oleic acid ($C_{18:1}$) (Price *et al.,* 1992; Sanjaun *et al.,* 1992). Modifications to the backbone oligosaccharide include: *O*-sulfation of the reducing end residue (Price *et al.,* 1992; Roche *et al.,* 1991; Lerouge *et al,.,* 1990), carbamoylation of the nonreducing terminal *N*-acetylglucosamine (Price *et al.,* 1992), and the addition of 2-*O*-methylfucose to the reducing *N*-acetylglucosaminyl residue (Price *et al.,* 1992).

1.4. Exopolysaccharides

The exopolysaccharides are a group of high-molecular polysaccharides that are exported to comprise the capsular or slime layers of bacteria. Capsule polysaccharides are a form of extracellular polysaccharide that remains attached to the cell. In view of this tight association with the bacterial cell surface, capsular polysaccharides are thought to be linked to the underlying rigid peptidoglycan, but a covalent association has yet to be identified. Slime, on the other hand, is chemically and structurally similar to capsular polysaccharides but is not held tightly by

Alginate

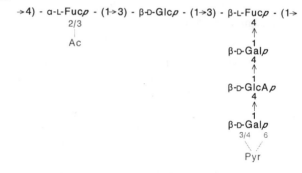

Colanic acid

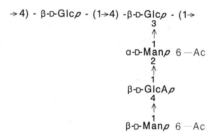

Xanthan

→4) - β-D-Glc*p* - (1→4) -β-D-Glc*p* - (1→
3
↑
1
α-D-Man*p* 6 — Ac
2
↑
1
β-D-GlcA*p*
4
↑
1
β-D-Man*p* 6 — Ac

Figure 2.   Structures of repeating units of capsular polysaccharides.

the cell, and hence it easily sloughs off. Both capsules and slimes typically consist of high-molecular weight polymers composed of units of repeating oligosaccharides. The composition, structure, and number of the repeating oligosaccharides will vary with genera, species, and strain and will define K-antigen or K-serogroup specificity. Such diversity arises through the combination of different monosaccharide composition, the positioning of their linkages, and further modifications. For example, 11 capsular serotypes have been reported for *Staphylococcus aureus* (Arbeit *et al.,* 1984; Sompolinsky *et al.* 1985), whereas 80 unique capsular antigens (K antigens) to date have been detected for *E. coli.* The latter have been divided into four groups (IA, IB, II, and III) based on a number of criteria (Jann and Jann, 1990).

Individual capsular polysaccharide chains are considerably larger than the O-antigens of lipopolysaccharide. Although it has been difficult to accurately determine, the degree of polymerization of the K1 capsule of *E. coli* is thought to be as high as 230, resulting in a molecular weight of approximately 700 kDa (Jann and Jann, 1990; Pelkonen *et al.,* 1988). The capsular polysaccharide produced by *K. pneumoniae* is much larger, being estimated at over 2000 kDa in size (Sutherland, 1985).

In addition to the production of serotype-specific capsules, many species of either closely related or totally unrelated bacteria may produce structurally and immunologically identical polysaccharides. These common capsular polysaccharides include alginate, colanic acid, enterobacterial common antigen, and xanthan.

1.4.1. ALGINATE

Alginate is a high-molecular weight, linear, nonrepeating polymer of β 1 → 4 linked D-mannuronic acid residues and variable amounts of its C-5 epimer, L-guluronic acid (Gacesa, 1988; Evans and Linker, 1973) (Fig. 2). It is produced as an exopolysaccharide by *Azobacter vinelandii* (Larsen and Jones, 1971), several species of *Pseudomonas,* including *P. aeruginosa* (Linker and Jones, 1966), and *P. syringae* (Fett *et al.,* 1986), as well as by seaweeds. A notable difference between the alginate produced by *A. vinelandii* and those of the pseudomonads is the complete lack of consecutive L-guluronyl residues in the latter (Sherbroack-Cox *et al.,* 1984; Skjåk-Bræk *et al.,* 1986).

1.4.2. COLANIC ACID

E. coli, like other species of *Enterobacteriaceae,* also can produce colanic acid (M antigen), a high-molecular weight, acidic polymer resembling the group IA capsular polysaccharides. The structure of colanic acid involves a repeating trisaccharide of β 1 → 3 linked fucopyranose residues on either side of a glucopyranose with a trisaccharide side chain of two galactopyranosyl residues and glucuronic acid attached to the reducing end main chain fucopyranose (Fig. 2) (Garegg *et al.,* 1971). Each hexasaccharide repeat unit is linked by an α 1 → 4 glycosidic bond.

1.4.3. XANTHAN

The exopolysaccharide produced by species of *Xanthomonas* is xanthan, a high-molecular weight polymer of repeating cellobiosyl units with α 1 → 3-linked branches of a tetrasaccharide composed of α- and β-D-mannopyranoses, pyruvate, and β-D-glucuronic acid (Fig. 2) (Stankowski *et al.,* 1993).

2. LOCALIZATION AND EXTENT OF *O*-ACETYLATION

The *O*-acetylation of saccharides involves the attachment of acetate to a free hydroxyl group on the residue through a base-labile ester linkage. Despite the enormous amount of information gathered regarding the composition and structure of the various bacterial cell wall polysaccharides, less is known about the frequency of *O*-acetylation. In many cases, it may have escaped detection due to the lability of the ester linkage or alternatively, it simply has not been examined for. Nevertheless, it is known that *O*-acetylation occurs on some peptidoglycans, lipopolysaccharides, the exopolysaccharide capsules including alginate, and colanic acid, and lipo-oligosaccharides. In each case, the modification is specific to a given monosaccharide residue, but without exception naturally occurring acetylation on bacterial cell wall polysaccharides is not stoichiometric.

2.1. Peptidoglycan

O-Acetylation of peptidoglycan occurs exclusively at the C-6 hydroxyl group of *N*-acetylmuramyl residues resulting in the generation of the *N*-2,*O*-6-diacetylmuramyl derivative (Fig. 1A). It was first detected independently 40 years ago in *Streptococcus faecalis* (*Enterococcus faecalis*) by Abrams (1958) and in *Micrococcus lysodeikticus* (*M. luteus*) by Brumfitt and co-workers (1958). Since then, this modification to peptidoglycan has been observed in a total of 69 strains of 19 species of both gram-positive and gram-negative bacteria, including a variety of important human pathogens (Table I). Interestingly, the bacteria listed in Table I include all species of the three genera comprising the *Proteeae* (*Proteus, Providencia,* and *Morganella*). These three genera constitute a medically important group of bacteria, and a number of their species together are responsible for approximately 10% of all nosocomial infections in North America (Brenner, 1992). The full extent of peptidoglycan *O*-acetylation among the bacteria is unknown, but it is very likely that many others do perform this modification. However, the peptidoglycan produced by *E. coli, Serratia marcescens,* and *P. aeruginosa* has been shown to be devoid of the modification (Clarke, 1993; Dupont and Clarke, 1991a).

The natural levels of peptidoglycan *O*-acetylation range from approximately 20 to 70% relative to muramic acid content (Clarke, 1993; Clarke and Dupont, 1992), with only one exception; the peptidoglycan of *N. gonorrhoeae* RD_5 is reported to contain between 10 and 15% ester-linked acetate (Rosenthal *et al.*, 1982; Swim *et al.*, 1983). An *M. luteus* mutant cultured on plates containing hen egg-white lysozyme was claimed to have a molar ratio of *N*-acetylmuramic acid–*O*-acetyl groups of 1–1 (that is, 100%) (Brumfitt *et al.*, 1958), but it is possible that this value may be unreliable, considering the simple nature of the assay used to de-

Table I

Bacteria Reported to Possess O-Acetylated Peptidoglycan

Species	Strain	% O-Acetyl[a]	Reference
Gram positive			
Lactobacillus acidophilus	63 AM Gasser	60–70	Coyette and Ghuysen (1970)
Lactobacillus fermentum	ATCC 9338	n.r.[b]	Logardt and Neujahr (1975)
Micrococcus luteus	n.r.	100	Brumfitt et al. (1958)
Staphylococcus aureus	Copenhagen	60	Ghuysen and Strominger (1963)
	SG511Berlin	35–90	Burghaus et al. (1983)
	H(NCIB 6571)	60	Snowden et al. (1989)
Streptococcus faecalis	ATCC 9790	n.r.	Abrams (1958)
Gram negative			
Moraxella glucidolytica	n.r.	n.r.	Martin et al. (1973)
Morganella morganii	UGM 92	50	Clarke (1993)
	ATCC 25830	46	Clarke (1993)
	UGM 326	43	Clarke (1993)
Neisseria gonorrhoeae	FA140	52	Swim et al. (1983)
	CL1	52	Swim et al. (1983)
	BR87	51	Swim et al. (1983)
	1342	50	Swim et al. (1983)
	FA171	50	Swim et al. (1983)
	I1260	40–50	Blundell et al. (1980)
	FA19	40–50	Blundell et al. (1980)
	FA136	49	Swim et al. (1983)
	7502	49	Swim et al. (1983)
	2686	48	Swim et al. (1983)
	7405	47	Swim et al. (1983)
	CS7	43	Swim et al. (1983)
	609	43	Swim et al. (1983)
	F62	43	Swim et al. (1983)
	624	40	Swim et al. (1983)
	8035	40	Swim et al. (1983)
	FA102	39	Swim et al. (1983)
	8038	37	Swim et al. (1983)
	1291	35	Swim et al. (1983)
	JW31	34	Swim et al. (1983)
	RD$_5$	10–15	Swim et al. (1983)
Neisseria perflava	n.r.	n.r.	Martin et al. (1973)
Proteus mirabilis	19	66	Martin and Gmeiner (1979)
	19	53	Dupont and Clarke (1991a)
	Perkins	41	Blundell and Perkins (1981)
	Perkins	40	Dupont and Clarke (1991a)
	GB8	36	Dupont and Clarke (1991a)
	ATCC 7002	44	Dupont and Clarke (1991a)
	ATCC 29906	42	Clarke (1993)
	ATCC 29245	32	Dupont and Clarke (1991a)
	ATCC 33583	32	Dupont and Clarke (1991a)

(*continued*)

Table I (*Continued*)

Species	Strain	% O-Acetyl[a]	Reference
	ATCC 43071	27	Dupont and Clarke (1991a)
	ATCC 25933	26	Dupont and Clarke (1991a)
	ATCC 33659	25	Dupont and Clarke (1991a)
Proteus mirabilis	ATCC 12453	24	Dupont and Clarke (1991a)
	ATCC 14273	20	Dupont and Clarke (1991a)
	TGH 9041	30	Dupont and Clarke (1991a)
	TGH 7341	27	Dupont and Clarke (1991a)
	F16	31	Dupont and Clarke (1991a)
	F491	28	Dupont and Clarke (1991a)
Proteus myxofaciens	ATCC 19692	53	Clarke (1993)
Proteus penneri	ATCC 33519	36	Clarke (1993)
Proteus vulgaris	P18	50	Fleck *et al.* (1971)
	ATCC 33420	35	Clarke (1993)
	ATCC 6380	32	Clarke (1993)
	ATCC 13315	31	Clarke (1993)
	ATCC 8427	29	Clarke (1993)
Providencia alcalifaciens	ATCC 9886	42	Clarke (1993)
Providencia heinbachae	ATCC 35613	34	Clarke (1993)
Providencia rettgeri	UGM 565	42	Clarke (1993)
	ATCC 29944	37	Clarke (1993)
Providencia rustigianii	ATCC 33673	41	Clarke (1993)
Providencia stuartii	ATCC 35031	57	Clarke (1993)
	ATCC 29914	54	Clarke (1993)
	PR50	54	Payie *et al.* (1995)
	ATCC 33672	40	Clarke (1993)
	UGM 603	39	Clarke (1993)
Pseudomonas alcaligenes	n.r.	n.r.	Martin *et al.* (1973)

[a]Expressed relative to concentration of muramic acid.
[b]n.r., Not recorded.

tect the acetylation. Strains of *S. aureus* appear to produce the highest levels of *O*-acetylation, averaging 60%, which can be further increased to above 85% by treatment of cells with the bacteriostatic antibiotic chloramphenicol (Johannsen *et al.*, 1983). Similarly, the presence of chloramphenicol enhances the degree of *O*-acetylation of the peptidoglycan from *N. gonorrhoeae*, raising it from 46 to 70% (Rosenthal *et al.*, 1985). Generally however, the majority of the strains of species producing *O*-acetylated peptidoglycan maintain levels above 30%. It also should be noted that of those species known to *O*-acetylate their peptidoglycan, no strain completely devoid of the modification has been detected. This would suggest that *O*-acetylation may play a role in the metabolism of peptidoglycan in those species that perform it (discussed in Section 3.1).

At present, it is not clear whether the peptidoglycan at the poles and/or in the cylindrical wall serves as the substrates for acetylation. Attempts to address this issue have been made by applying the immunogold-labeling technique on both mixed populations (Gyorffy and Clarke, 1992) and synchronized cells (Clarke *et al.*, 1993) of *P. mirabilis* using a monoclonal antibody specific for *O*-acetylated peptidoglycan. These investigations failed to detect any pattern of the modification on the murein sacculus, which may suggest that the modification simply occurs evenly over the entire peptidoglycan sacculus.

2.2. Lipopolysaccharides

In view of the number of lipopolysaccharides characterized to date, an exhaustive review of all that bear *O*-acetylation would be of little value. However, it would appear that, in general, *O*-acetylation is restricted to the O-antigens. One noted exception concerns the extensive *O*-acetylation of each of the monosaccharides (rhamnopyranose, *N*-acetylglucosamine, and *N*-acetylquivalosamine) comprising the core oligosaccharide of *Legionella pneumophila* (Knirel *et al.*, 1996). As with peptidoglycan, the presence of *O*-acetylation on the polysaccharides was largely overlooked until techniques involving high-field nuclear magnetic resonance (NMR) were applied. Also, the situation with lipopolysaccharides and capsules may be further complicated by the phenomenon of *O*-acetyl group migration during storage in solution, a situation noted with the *Neisseria meningitidis* capsular polysaccharides (Lemercinier and Jones, 1996).

The *O*-Acetylation is a relatively common feature of O-antigens, but usually occurring nonstoichiometrically. For example, of the 31 different O-antigen chemotypes produced by strains of *P. aeruginosa,* 14 bear at least one *O*-acetyl moiety, with the extent of the modification ranging from approximately 30 to 90% (Stanislavsky and Lam, 1997). In some cases, *O*-acetylation of a specific residue will represent the only chemical difference between the O-antigens of different serotypes. This is the case for the D-galactan-type O-antigens of *K. pneumoniae* serotypes 01 and 08, which are structurally identical with the exception that approximately 40% of the galactofuranosyl residues of the D-galactan I subunit are *O*-acetylated at the C-2 (20%) or C-6 (20%) hydroxyl group (Kelly *et al.*, 1993). It is also possible that localization of *O*-acetyl groups serves to distinguish the otherwise identical O-antigens of serotypes 09, 02(2a,2e) and 02(2a,2e,2H) (Kelly *et al.*, 1995).

The *O*-acetylation of O-antigens does not seem to be limited to specific types of monosaccharides. As indicated by the examples cited above, *O*-acetylation may occur at basic or neutral monosaccharides. These include some of the more unusual monosaccharides found in O-antigens, such as the abequose residue of the *Salmonella enterica* serovar *typhi* serotype O5 (Slauch *et al.*, 1995). Other unusual monosaccharides that are naturally *O*-acetylated include, L-rhamnose, *N*-acetylfucosamine, *N*-acetyl-7-*N*-formylpseudaminic acid (5-acetamido-7-formamido-3,5,7,9-tetradeoxy-L-*glycero*-L-*manno*-nonulosonic acid), and 6-deoxy-D-talose (Perry *et al.*, 1996; Knirel and Kochetkov, 1994).

2.3. Lipo-oligosaccharides

The lipo-oligosaccharides produced by the rhizobial bacteria *Rhizobium leguminosarum* biovar *viciae* and *R. meliloti* are *O*-acetylated, with the modification specifically occurring at the C-6 hydroxyl group of the acylated, nonreducing terminal *N*-acetylglucosamine residue (Price *et al.*, 1992; Roche *et al.*, 1991; Lerouge *et al.*, 1990). With the *R. leguminosarum* bv. *viciae* oligosaccharide, a second *O*-acetyl group is present and it is located on the C-6 hydroxyl of the reducing *N*-acetylglucosamine residues (Firmin *et al.*, 1993). Presumably, as with other saccharide polymers, the extent of these *O*-acetylations is not stoichiometric.

2.4. Exopolysaccharides

Of the few capsular polysaccharides that have been characterized structurally, those produced by *S. aureus* have received the most attention. As noted above, *S. aureus* produces 11 different capsular polysaccharides but not all have been characterized biochemically. However, most isolates of *S. aureus* belong to only two capsule serotypes, 5 and 8 (Arbeit *et al.*, 1984). The structures of these two capsules are very similar and both are *O*-acetylated on their mannosaminuronic acid backbone residues. In type 5 capsule, the C-3 hydroxyl group of approximately 50% of the mannosaminuronyl residues is *O*-acetylated (Moreau *et al.*, 1990), whereas the C-4 hydroxyl carries the *O*-acetylation in type 8 capsules (Fournier *et al.*, 1984). In addition to these two, type 1 and type 2 capsules are also *O*-acetylated, with the modification occurring on the type 2 polysaccharide at the C-3 hydroxyl of the glucosaminuronyl backbone residues (Hanessian and Haskell, 1964). The site of *O*-acetylation of type 1 capsule remains unknown.

2.4.1. ALGINATE

A conspicuous difference between the alginates produced by bacteria and seaweed is that the bacterial polymer is *O*-acetylated (Skjåk-Broeke *et al.*, 1986; Davidson *et al.*, 1977; Linker and Jones, 1966). Acetylation of alginate occurs exclusively on its mannuronyl residues (Fig. 2). The primary site of acetylation involves the C-2 hydroxyl group but the C-3 hydroxyl or both can be acetylated with the latter giving 2,3-di-*O*-acetylmannuronyl residues (Skjåk-Bræke *et al.*, 1986).

The extent of alginate *O*-acetylation is strain specific and has been reported in the range of 2–57%, being dependent on the amount of mannuronyl residues present in the heteropolymer. There does not appear to be any relationship between the level of alginate production and extent of *O*-acetylation; high-alginate-producing strains of pseudomonads do not necessarily highly *O*-acetylate the polysaccharide (Lee and Day, 1998). However, the extent of *O*-acetylation has been shown to be affected by nutrient availability. In particular, nitrogen limitation causes an apparent increase in the level of *O*-acetylation of pseudomonad alginates (Lee and Day, 1998).

2.4.2. COLANIC ACID

As with alginate, *O*-acetylation of colanic acid occurs at the C-2 and/or C-3 hydroxyl group of one specific residue, the fucopyranose at the nonreducing end of the main chain trisaccharide (Fig. 2). However, the extent of this *O*-acetylation is unknown.

2.4.3. XANTHANS

The side chain of xanthan begins and terminates with mannopyranose residues; as observed with *Xanthomonas campestris,* both may be *O*-acetylated to varying degrees at their C-6 hydroxyl groups (Stankowski *et al.,* 1993) (Fig. 2).

3. PHYSIOLOGICAL ROLES OF *O*-ACETYLATION

3.1. Peptidoglycan

The physiological role of peptidoglycan *O*-acetylation remains largely unknown, but it has been speculated to control the action of autolysins (Clarke, 1993), enzymes that hydrolyse peptidoglycan during its biosynthesis and turnover, and there is a growing body of data to support this postulate. The availability of mutants of *Providencia stuartii* underexpressing and overexpressing *aac(2′)*-Ia, a gene encoding a gentamicin acetyltransferase that is also responsible for peptidoglycan *O*-acetylation (discussed in Section 4.1), has provided the opportunity to evaluate the morphological consequences of changes to the levels of *O*-acetylation in peptidoglycan. Underexpression of *aac(2′)*-Ia in *P. stuartii* PR100 results in a low level of *O*-acetylation and a concomitant change in cell morphology from short to long, distorted rods (Payie *et al.,* 1995, 1996) (Table II). Consistent with this trend, the cells of mutant strains PR50.LM3 and PR51, which overexpress

Table II
Characteristics of Wild-Type and Mutant Strains of *Providencia stuartii*[a]

Strain	MIC (μg/ml^{-1}) gentamicin	Peptidoglycan O-acetylation (%)[b]	Cell morphology
PR50	4.0	54 ± 4.3	Short rods
PR100	0.5	42 ± 4.1	Irregular rod
PR50.LM3	>70	65 ± 2.1	Spherical
PR51	0.5	63 ± 3.9	Spherical, chains

[a]Data obtained from Payie *et al.* (1995).
[b]Expressed relative to concentration of muramic acid.

aac(2′)-Ia and consequently have increased O-acetylation levels, adopt a coccobacilli morphology. Given the role of autolysins in the growth of bacterial cells and that mutant strains of various gram-positive bacteria expressing deficient autolysins have altered cellular morphologies with defects in cell septation and/ or filamentation (Wuenscher *et al.*, 1993; Sanchez-Puelles *et al.*, 1986; Fein and Rogers, 1976), it would appear that O-acetylation of peptidoglycan could provide a level of control of these enzymes. This view is supported by experiments conducted by Labischinski and co-workers (Sidow *et al.*, 1990) showing that the onset of penicillin G-induced autolysis of *S. aureus* cells correlates with a decrease in O-acetylation. The fact that, with only one exception, all characterized strains of the various species of bacteria modifying their peptidoglycan in this manner do so at levels greater than 20% provides further evidence, albeit indirect, for a true physiological role for O-acetylation in the metabolism of peptidoglycan.

Unfortunately, very little is currently known concerning the autolytic enzymes expressed by *P. stuartii* (or for that matter, by any of the species of the *Proteeae*), but an endopeptidase from *N. gonorrhoeae* with specificity for O-acetylated peptidoglycan has been described (Blundell and Perkins, 1985). Nonetheless, we have obtained preliminary evidence based on zymogram analyses of *P. mirabilis* extracts that O-acetylation does affect the activity of different autolysins (Payie *et al.*, 1996). At least eight separate and distinct activity bands, ranging in apparent molecular weights from between 15 kDa and 69 kDa, were detected in the zymogram gels. By comparing the relative intensities of each band within a given lane, two enzymes with apparent molecular weights of 69 and 32 kDa were clearly more active on the peptidoglycan devoid of the acetyl modification. This suggested that the activity of at least two autolysins from *P. mirabilis* are controlled at the enzymatic level by the O-acetylation of their substrate. Two other enzymes with apparent molecular weights of 42 and 31 kDa appeared to be slightly more active on the native, O-acetylated peptidoglycan of *P. mirabilis* compared to the chemically de-O-acetylated material. The efficacy of the other apparent autolytic enzymes is presumably unaffected by the presence of the acetate modification, a

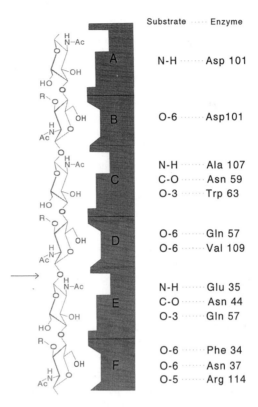

Substrate ····· Enzyme

N-H ······ Asp 101	
O-6 ······ Asp101	
N-H ······ Ala 107	
C-O ····· Asn 59	
O-3 ····· Trp 63	
O-6 ····· Gln 57	
O-6 ····· Val 109	
N-H ····· Glu 35	
C-O ······ Asn 44	
O-3 ····· Gln 57	
O-6 ····· Phe 34	
O-6 ····· Asn 37	
O-5 ····· Arg 114	

Figure 3. Schematic of the active site cleft of hen egg-white lysozyme complexed with peptidoglycan. The six subsites, labeled A–F, bind the glycosyl residues of substrate with hydrolysis occurring between residues in subsites D and E (indicated by arrow). Alternate sites A, C, and E interact with the acetamido groups of *N*-acetylglucosaminyl residues, while subsites B, D, and F accommodate the C-6 hydroxyl groups of *N*-acetylmuramyl residues. Some of the hydrogen bonds observed in the X-ray crystal structure (Blake *et al.*, 1967) of the individual subsites are noted on the right. The figure is adapted from Zubay (1993).

situation analogous to that with mutanolysin and the *N,O*-diacetylmuramidase of *Streptomyces globisporus* (Hamada *et al.*, 1978) and the fungus *Chalaropsis* (Hash and Rothlauf, 1967), respectively.

With the determination of the three-dimensional structure of hen egg-white lysozyme by X-ray crystallography (the first enzyme structure to be determined by this technique) by Phillips and colleagues in 1967 (Blake *et al.*, 1967), it is clear how the *O*-acetylation of peptidoglycan would sterically inhibit the action of a muramidase. As depicted in Fig. 3, the substrate-binding cleft of lysozyme is organized into six subsites (labeled A–F), each of which accommodates a glycosyl unit of peptidoglycan. The catalytic acidic residues Glu35 and Asp52 oppose each other between subsites D and E, allowing for the hydrolysis of the β 1 → 4 linkage between the *N*-acetylmuramyl and *N*-acetylglucosaminyl residues bound to these subsites, respectively. The C-6 hydroxyl groups of the *N*-acetylmuramyl residues bound in subsites B, D, and F make extensive contacts with amino acid residues lining the active site cleft, and each of these contacts would be blocked by the presence of O-acetyl groups. Indeed, this weaker affinity of lysozyme for substrate has been demonstrated by correlating the extent of hydrolysis of *P. mirabilis* peptidoglycan with the level of O-acetylation, which revealed that the overall change in standard Gibbs free energy of activation ($\Delta(\Delta G)$) of the enzyme increased with increasing O-acetylation (Dupont and Clarke, 1991a). Likewise, the extent of lysozyme resistance to the solubilization of gonococcal peptidoglycan by both hen egg-white and human enzymes is dependent on the degree of O-acetylation (Swim *et al.*, 1983; Rosenthal *et al.*, 1982, 1983; Blundell *et al.*, 1980).

The protection from hydrolysis conferred by O-acetylation of peptidoglycan is consistent with that observed elsewhere in nature. For example, stability and preservation of cellulose in plant cell walls is maintained by the overlaying of polysaccharides, primarily xylan, which is in turn protected by O-acetylation (Clarke, 1997). The presence of numerous O-acetyl groups on the xylan backbone, a linear polymer of xylosyl residues, protects it from the hydrolytic action of xylanases. Thus, the efficient biodegradation of cellulose and xylan requires the action of O-acetylesterases.

Based on the preliminary observations made with both *P. stuartii* and *P. mirabilis,* a working model for the apparent essential role of peptidoglycan O-acetylation, at least for the *Proteeae,* can be formulated. We propose that a class of autolysins with specificity for non-O-acetylated peptidoglycan are secreted and catalyze the hydrolysis of bonds at the growing, immature sites within the cylindrical wall of the sacculus to provide new sites for chain insertion. A second class of autolysins that either have specificity for N,O-diacetylmuramic acid or are active at other residues of peptidoglycan monomers that are not sterically hindered by the O-acetyl moieties, such as peptide cross-links, catalyze the formation of septa and subsequent release of daughter cells. Overexpression of the O-acetyl transferase activity would serve to inhibit the activity of the former class of enzymes and result in the observed production of cocco-bacilli shaped cells. Similarly, decreased O-acetylation of peptidoglycan would permit the continued production of insertion sites and extension of the cylindrical wall. This model is consistent with that proposed for the pathway of peptidoglycan O-acetylation in *P. mirabilis* that invokes the incorporation of non-O-acetylated subunits to a growing peptidoglycan chain, followed by the assembly and cross-linking of a second peptidoglycan chain prior to the O-acetylation of the former (Gmeiner and Sarnow, 1987; Gmeiner and Kroll, 1981). It also would account for the observed increase in O-acetyl content associated with the peptidoglycan isolated from intact cells that were obtained from either stationary cultures or treated with chloramphenicol (Rosenthal *et al.*, 1985; Johannsen *et al.*, 1983). The continued O-acetylation of peptidoglycan at the growing regions of the sacculus under these bacteriostatic conditions would prevent the continued activity of the potentially lethal autolytic activities of the enzymes associated with this region. Finally, preliminary analysis of the peptidoglycan isolated from swarmer cells of *P. mirabilis* indicates that their peptidoglycan sacculus is characterized by a decreased level of *O*-acetylation compared to the normal, shorter vegetative cells (unpublished data). It is also conceivable that this model applies to *S. aureus.* The typical levels of *O*-acetylation associated with these coccal-shaped cells are the highest among the bacteria known to modify peptidoglycan in this manner, and it could thus preclude the possibility for the formation of cylindrical regions.

3.2. Lipopolysaccharides and Exopolysaccharides

It is unlikely that there is any physiological function for the *O*-acetylation of lipopolysaccharides and serospecific capsules because it is thought that in most cases the modification arises through lysogenic phage conversions. These phages recognize the polysaccharide chains as receptors for adsorption and infection of the host bacterium, which once inside cause the serospecific conversion through the expression of an *O*-acetyltransferase. This has been directly demonstrated for the invasion of *Shigella flexneri* by bacteriophage SF6, which upon invasion, converts the group 3,4 lipopolysaccharide O-antigen to the group 6 O-antigen by *O*-acetylation (Gemski *et al.,* 1975). This *O*-acetyltransferase also has been demonstrated to convert the Y strain O-antigen (group 3,4) to a hybrid, type 3b, that expresses O-antigens 6,3,4 (Verma *et al.,* 1991). That *O*-acetylation can alter binding sites clearly has been demonstrated for the recognition of lipopolysaccharides by monoclonal antibodies. Detailed immunological studies with *S. typhimurium* lipopolysaccharides revealed that the *O*-acetylation affects the three-dimensional structure of the O-antigens thereby creating or destroying a series of conformational antigenic determinants (Michetti *et al.,* 1992; Slauch *et al.,* 1995).

The *O*-acetylation of alginate has been suggested to protect the modified mannuronate residues from being converted to guluronate residues (Narbad *et al.,* 1990). The presence of the *O*-acetyl groups on alginate are also thought to control the activity of alginate lyase (Ertesvåg *et al.,* 1998), an enzyme required for the production of this exopolysaccharide (Monday and Schiller, 1996; Boyd and Chakrabarty, 1994). This would thus be analogous to the proposed role of the *O*-acetylation of peptidoglycan in controlling the levels of autolysis. Also, it has been demonstrated that the *O*-acetylation of *P. aeruginosa* alginate results in a large increase in hydration leading to an overall increase in gel volume (Skjåk-Broek *et al.,* 1989), which may protect cells from periods of dehydration. This response to a detrimental environmental change is consistent with the observation that nitrogen limitation also causes an increase in *O*-acetylation (Lee and Day, 1998).

The lipo-oligosaccharides produced by the rhizobia have been demonstrated to serve as signaling factors for the successful formation of nitrogen-fixing nodules on leguminous plants (Dénairié *et al.,* 1992). These nodulation factors are host specific and such specificity is conferred in part by *O*-acetylation. For example, *O*-

acetylation of the nonreducing terminal *N*-acetylglucosaminyl residue of the nodulation factor produced by *Rhizobium leguminosarum* bv. *viciae* is required for the efficient nodulation of peas (*Pisum sativum*), lentil (*Lens* sp.), and sweet pea (*Lathyrus* spp.) but not for nodulation of vetch (*Vicia* spp.) (Surin and Downie, 1988). One cultivar of peas (Afghanistan) is resistant to nodulation by most strains of *R. leguminosarum* bv. *viciae.* but this resistance is overcome by the *O*-acetylation of the reducing terminal *N*-acetylglucosaminyl residue in addition to that at the nonreducing terminus (Firmin *et al.,* 1993).

4. PATHWAY FOR *O*-ACETYLATION

4.1. Peptidoglycan

Despite considerable effort over the past 15 years, little is still known about the pathway for the *O*-acetylation of peptidoglycan. Substantial evidence does exist, however, to indicate that it follows both the linkage of nascent peptidoglycan strands to the existing sacculus and their transpeptidation to generate the peptide cross-links (reviewed by Clarke and Dupont, 1991). The results of pulse-chase experiments with *P. mirabilis* led Gmeiner and co-workers (1981, 1987) to propose a model for the pathway in which newly inserted subunits are attached to "old" *O*-acetylated subunits by peptide cross-linkage and the new chain is extended by non-*O*-acetylated subunits. A second non-*O*-acetylated chain is then assembled and linked to the previously incorporated chain. This would imply that both *O*-acetylated and non-*O*-acetylated subunits may serve as acceptors in the transpeptidation reactions. The new peptidoglycan region subsequently becomes *O*-acetylated chain by chain. Support for this model was provided by others investigating *N. gonorrhoeae* (Lear and Perkins, 1983, 1986, 1987; Dougherty, 1985), where the *O*-acetylation of gonococcal peptidoglycan was also observed to be a slower process than cross-linking, indicating that subunits already incorporated into the preexisting sacculus must then undergo *O*-acetylation. These data indicate that *O*-acetylation of peptidoglycan must occur in the periplasm, which was confirmed by the inability to isolate lipid (bactoprenyl)-linked *N*-acetylglucosamine-*N,O*-diacetylmuramyl-pentapeptide precursors in either the cytoplasm or the cytoplasmic membrane of both *N. gonorrhoeae* (Lear and Perkins, 1986) and *S. aureus* (Snowden *et al.,* 1989). Thus, both the acetyltransferase and the acetyl donor must be present and available in the periplasm. This is consistent with the observation that *O*-acetylation continues in an *in vitro,* cell-free biosynthetic system of *P. mirabilis* peptidoglycan (Dupont and Clarke, 1991b; Martin, 1984). The fact that the transferase(s) and cosubstrate are not lost to the soluble fraction during the preparation of the biosynthetic components indicates that they are not free in the periplasm but remain associated with either the cytoplasmic membrane or pepti-doglycan.

Both *in vivo* and *in vitro* radiotracer studies with *P. mirabilis* supported the proposal that an $N \rightarrow O$ acetyltransferase catalyzes the transfer of acetyl groups from the N-2 position of either glucosaminyl or muramyl residues to the C-6 po-sition of the latter (Dupont and Clarke, 1991b,c). Similar results recently have been obtained with *Providencia stuartii* (Payie and Clarke, 1997). The activity of this putative peptidoglycan $N \rightarrow O$-acetyltransferase would thus be similar to *N*-arylhydroxamic acid N,O-acetyltransferase (EC 2.3.1.56), which transfers acetate from aromatic *N*-acethydroxamates to the *O*-position of aromatic hydroxylamines (Smith and Hanna, 1986). This enzymatic system would account for the lack of readily available activated acetate (e.g., acetyl-CoA or acetyl phosphate) in the periplasm and instead utilize the conserved bond energies stored within peptido-glycan to achieve *O*-acetylation in a manner analogous to the action of the peni-cillin-binding proteins during the latter stages of peptidoglycan biosynthesis. These enzymes form new bonds in the periplasm where ATP (or other such ener-gy source) is unavailable by catalyzing *trans*glycosylation and *trans*peptidation reactions involving bonds created in the cytoplasm and carried out with the disac-charide-pentapeptide peptidoglycan precursor. Unacetylated amino sugars, the ex-pected by-product of the putative N,O-acetyltransferase, were not detected on the peptidoglycan sacculus, but instead were observed in the spent culture medium (Dupont and Clarke, 1991c), suggesting that the processes of peptidoglycan *O*-acetylation and turnover would have to be linked. Moreover, recognition that au-tolytic (Li *et al.*, 1996; Blundell and Perkins, 1985; Chapman and Perkins, 1983) and $N \rightarrow O$ acetyl transfer activities are retained in a cell-free system led to the further speculation that, by analogy to the higher-molecular-weight penicillin-binding proteins (PBPs), the two activities may be catalyzed by a single bifunc-tional enzyme (Dupont, 1991). Interestingly, both a "lysozymelike" and trans-acetylase activity have been observed with Vi phage particles as they degrade the peptidoglycan of *Citrobacter* sp. O-serogroup Ci23Vi+ (Jastrzemski and Kwiatkowski, 1984).

Recent developments from our laboratory relating to the enzymatic process of peptidoglycan *O*-acetylation pertain to a relationship between the acetylation of aminoglycosides and peptidoglycan in *Providencia stuartii*. A direct correlation exists between the expression of gentamicin 2'-*N*-acetyltransferase (EC 3.2.1.59; AAC(2')-Ia) and peptidoglycan *O*-acetylation (Payie *et al.*, 1995, 1996). The *aac(2')-Ia* gene coding for this aminoglycoside resistance factor is located on the *P. stuartii* chromosome, which is quite unusual for this class of enzyme. The ap-parent cross-reactivity of AAC(2')-Ia thus would suggest that its primary physio-logical role concerns peptidoglycan metabolism, while conferring resistance to aminoglycoside antibiotics is a fortuitous (to the bacterium, not the clinician!) side reaction. This view is supported by the observation that peptidoglycan serves as a source of acetate for the acetylation of aminoglycosides (Payie and Clarke, 1997). However, recent studies involving Southern hybridizations using *aac(2′)*-Ia to probe each species of the *Proteeae* have revealed that homologous gene sequences are only found in *Proteus penneri* and *Providencia rettgeri* (Clarke *et al.*, 1996). As this enzyme does not appear to be common to all the *Proteeae*, each species of which produces *O*-acetylated peptidoglycan (Clarke, 1993), together with the finding that a frameshift mutation of *aac(2′)-Ia* does not lead to the complete loss of *O*-acetylation (Payie *et al.*, 1995) (Table II), it would appear that AAC(2′)-Ia contributes to but is not solely responsible for the modification of peptidoglycan.

4.2. Lipopolysaccharides and Capsules

There is a paucity of information regarding the pathway for the *O*-acetylation of lipopolysaccharides and capsular polysaccharides. As might be expected, much of what is known stems from studies pertaining to the more common surface polysaccharides, alginate, and nodulation factors.

It has been assumed that acetyl-CoA is the direct source of acetate for the modification of alginates, and that the process occurs intracellularly (Skjåk-Bræk *et al.*, 1986). This was largely based on a postulate that *O*-acetylation controls the level of epimerization of mannuronyl residues to its C-5 epimer, L-guluronic acid, which also was assumed to occur in the cytoplasm. However, addition of seaweed alginates into cultures of *Pseudomonas syringae* at any point in their growth cycle leads to their *O*-acetylation (Day and Lee, 1994). This would imply that, at least for *P. syringae*, both the acetyltransferase and source of acetate must be present extracellularly, and that the biosynthesis and subsequent *O*-acetylation of the polysaccharide are separated (Lee and Day, 1998). Analysis of the gene cluster responsible for the production of alginate in *P. aeruginosa* does indeed support the proposal for an enzymatic system to catalyze the extracellular *O*-acetylation of the polysaccharide. In fact, everything else known regarding the pathway for the *O*-acetylation of all other cell wall polysaccharides has arisen from homology searches within gene clusters encoding proteins responsible for their biosynthesis. Such analyses has permitted the placement of the various putative *O*-acetyltransferases into distinct families.

4.3. Families of *O*-Acetyltransferases

With the recent emergence and development of techniques in the sequencing of DNA and the initiation of a number of genome sequencing projects, a plethora of deduced amino acid sequences for a variety of enzymes from a broad range of

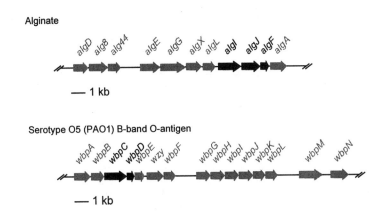

Figure 4. Organization of the gene clusters for the biosynthesis of alginate (Chitnis and Ohman, 1993) and O-antigen (Burrows *et al.*, 1996) in *Pseudomonas aeruginosa*. The genes encoding hypothetical acetyltransferases are shown in black.

organisms has accumulated in the literature and data banks. Comparison and alignment of these amino acid sequences together with complementation and mutation studies have served to identify a number of enzymes involved in the *O*-acetylation of different bacterial polysaccharides. Thus, four different families of *O*-acetyltransferases with common sequence motifs are now recognized, two of which have previously been reported (Bhasin *et al.*, 1998; Slauch *et al.*, 1996). These analyses also have permitted speculation on the pathway in which the modification arises.

Many of the genes required for the biosynthesis of alginate in *P. aeruginosa* are clustered in an 18-kb operon at 35 min on the chromosome (Fig. 4), and this gene cluster appears to have an operonic structure (Chitnis and Ohman, 1993). Three of the genes, *algF, algI,* and *algJ,* have been shown by transposon mutagenesis to encode enzymes responsible for *O*-acetylation (Franklin and Ohman, 1993, 1996; Shinabarger *et al.*, 1993). The 24.5-kDa protein encoded by *algF* contains an *N*-terminal signal sequence of 28 amino acids, and expression of this gene in *P. aeruginosa* results in the production of a processed 19.5-kDa protein (Shinabarger *et al.*, 1993). This strongly suggests that Alg F is processed and exported to participate in the extracellular *O*-acetylation of alginate. Furthermore, the deduced amino acid sequence of Alg I is homologous to a family of membrane-bound Dlt B proteins expressed by different gram-positive bacteria (Franklin and Ohman, 1996) (Fig. 5). These latter proteins are thought to be involved in the transport of activated alanine through the cytoplasmic membrane for its incorporation into teichoic and lipoteichoic acids (Heaton and Neuhaus, 1992). Thus, by analogy, Alg I has been proposed to transfer acetate from pools of cytoplasmic acetyl-CoA through the cytoplasmic membrane for its subsequent addition to alginate, cat-

```
Pa AlgI     235  YTAQ-LYFDFSGYSDMAIGLGLMMGFRFMENFNQPYISQSITEFWRRWHISLSTWLRDYLYISL
Az AlgI     235  YTAQ-LYFDFSGYSDMAIGLGLMIGFRFMENFNQPYISQSITEFWRRWHISLSTWLRDYLYISL
Hp HP0855   276  YSFQ-LYFDFSGYCDMAIGIGLFFNIKLPINFNSPYKALNIQDFWRRWHITLSRFLKEYLYIPL
Tp TP0566   254  YCDFSGYSDLAI----AVGLL-F-GFETPANFKRPYISQSVTEFWRRWHISFSQWLKEYLYFSL
Lc DltB     242  YSGY-LFFDFAGYSLFAVAISYLMGIETPMNFNKPW-SHITSRLLNRWQLSLSFWFRDYIYMRF
Sa DltB     247  YSLY-LFFDFAGYSLFAIAFSYLFGIKTPPNFDKPFKAKNIKDFWNRWHMTLSFWFRDCIYMRS
Sm DltB     243  YGLD-LFFDFAGYSMFAIAISNFMGIKSPTNFNQPFKSQDLKEFWNRWHMSLSFWFRDFVFMRL
Bs DltB     234  YSMY-LFFDFAGYTMFAVGVSYIMGIKSPENFNKPFISKNIKDFWNRWHMSLSFWFRDYVFMRF
```

Figure 5. Amino acid sequence alignment of a highly homologous region of the integral membrane, family 1 *O*-acetyltransferases. Highly conserved amino acids are shaded, while those totally conserved are underlined. The sequences and their accession numbers (in parenthesis) are: alginate *O*-acetyltransferases *Pa* Alg I (U50202) and *Az* Alg I (AF027499) from *Pseudomonas aeruginosa* and *Azotobacter vinelandii*, respectively; hypothetical proteins *Hp* HP0855 (AE000596) and *Tp* TP0566 (AE001232) from *Helicobacter pylori* and *Treponema pallidum*, respectively; and membrane-bound D-alanine transporter (Dlt B) proteins *Lc* Dlt B (P35855), *Sa* Dlt B (D86240), *Sm* Dlt B (AF049357), and *Bs* Dlt B (P39580) from *Lactobacillus casei*, *Staphylococcus aureus*, *Streptococcus mutans*, and *Bacillus subtilis*, respectively.

alyzed by Alg F and/or Alg I (Franklin and Ohman, 1996). Homologues of *algI* have been found on the chromosomes of another alginate-producing bacterium, *Azotobacter vinelandii* and two other bacteria, *Helicobacter pylori* and *Treponema pallidum* (Fig. 5). Both the latter bacteria are serious human pathogens but they do not synthesize alginate. However, *H. pylori* does produce a ganglioside that terminates with a residue of 5-acetylsialic acid (Saitoh *et al.*, 1991).

Considerable amino acid sequence homology is shared by each of the two enzyme types within this family of *O*-acetyltransferases (that is, Alg I and Dlt B). In addition to being integral membrane proteins, all the enzymes are characterized by a highly conserved amino acid sequence motif involving aromatic (Phe, Tyr, Trp), charged (Asp, Arg, His), and proline residues. In particular, a common F-W-R/N-R-W-H motif exists in the central region of each enzyme.

A second distinct family of relatively large integral membrane proteins involved in the *O*-acetylation of bacterial polysaccharides has been recognized (Slauch *et al.*, 1996). This family originally comprised 21 *O*-acetyltransferases acting on capsules, O-antigens, xanthan, and nodulation factors and involved two regions of significant sequence homology. A recent search of the various genome sequencing projects has permitted the addition of a further six hypothetical bacterial proteins and a closer examination of the entire family of sequences has revealed a third region of significant homology (Fig. 6). It is interesting to note that one of these homologous regions, near the N-terminus of the proteins, shares some similarities to the N-terminal region of the acetyl-CoA transporter of humans. Also, this second family of membrane-associated *O*-acetyltransferases includes four distinct *O*-acetyltransferases with activity toward macrolide antibiotics. It is conceivable that these enzymes produced by *Streptomyces thermotolerans* and *S. mycarofaciens* have been detected as antibiotic resistance factors, but like the

```
SF6  Oac       40  GGIAVIIFFSISGYLISKSAIRSDS----------------FIDFMAKRARRIFP   78
St   OafA      34  GFIGVDVFFVISGFLMTGIVLERV-----------------DHDFYIARFLRIVP   71
Pa   WpbC      49  GFVGVDVFFVISGFIITALLVERGVKVDLVE----------------FYAGRIKRIFP   90
Pa   WpbC2     18  ILSVVDVFLVISGYLITSIIRRDRQAGRQAG--------RFSFVDFWARRARRILP   65
Lp   Lag1      56  QSLAVNAFFWLSGFLITYHCITKKPYT---------------FAEYMIDRFCRIYV   96
Xc   GumF      60  YSFHVPLFFLVSGWLAAGYASRTTSLL--------------QTITKQARGLLL-P   99
Aa   Imat      43  GPTGVDIFFVISGFVVTLVAPKDSNSNDC----------CRSAFVFFVKRVIRTYP   89
At   PicA      40  YLA-VDLFFALSGFVLAHAYGKKLYEGTITP-----------GFFLKARFARLYP   83
Hi   0392       1  MDIFFVISGFLITGIIITEIQQNSFSL------------KQFYTRRIKRIYP   40
Ng   Orf       34  GFLGVDIFFVISGFLITNIILSEIQNGSFS-----------FRDFYTRRIKRIYP   77
Bs   Yrhl      38  GFIGVDIFFVLSGYLITSILLPAYGNDINLD----------FRDFWVRRIRRLLP   82
Mt   RVi565c   56  VSGGVDVFLALSGFFFGGKILRAALNPD----------LSLSPIAEVIRLIRRLLP  101
Rl   Nodx      53  AP-GVAIFFLTSGFLVTDSYIRSSSAAS---------------FFVKRSLRIFP   90
Rm   ExoZ      17  GAAGVDVFFVISGFIMWVISDRRSVT--------------PVEFIADRARRIVP   56
St   CarE      49  GPLTVSFFFMLSGFVLTWAGLPDKSKVN--------------FWRRRTVRAYS   87
St   AcyA      44  GSLAVSLFFVLSGYVLTWSARDGDSVRS---------------FWQRRFAKIYP   87
Sm   Mpt       49  GPVAV-FFFMLSGFVLTWAGMPDPSKPA--------------PWRRRWVRVYS   86
Sm   MdmB      49  GSIAVSVFFLLSGFVLAWSARDKDSVTT--------------FWRRRFAKIYP   87
Ce   RO3H4.1   29  GFLGVDIFFVISGFLMAKILTKSSLRSV----------QDITAFYFRRFRRILT   72
Ce   RO3H4.5   29  GFLGVDIFFVISGFLMANNLTNLNLLNV-----------HDFLLFYYKRFRRILP   72
Ce   RO3H4.6   29  GFLGVDIFFVISGFLMAQNLSKSKLVTV-----------QDFFIFYYRRFRRILP   72
Ce   F09B9.1  305  AFVSVDTFFVLSGLVLTYMFFKTTPKKKMI-------VNPVTWIMFYVHRYLRLTP  353
Ce   C06B3.2  316  AVFSVDTFFLVSGITVAYSFFRLKPTTKT-------LKSPATWILFYNHRYVRLTP  364
Ce   C08B11.4 379  APLAVDSFFFLSGMLAAFSFFKKTMKADPNHPPKLSAFNWQTWPYYYKRYIRIIT  434
Ce   C08H10.4  28  GYIGVDMFFVLSGFLMAMIISSKPITWN-----------SVYQFYFRRSKRILP   70

Hs   AcCoAT   104  VSYTDQAFF--S-FVFWPFSLKLLWAPLVD--------AVYVKNFG-RRKSWLVP  146
```

```
SF6  Oac       140  WTLPLEFLCYIITGVAVALLKNGK  163    251  VGDPLVKGRFDYSYGVYIYAFP  383
St   OafA      136  WSLSVEWQFYILYPLLVIIV--KK  157    268  TSNRIAQWVGKISYSVYLWHWP  289
Pa   WpbC      151  CSIANEMQFYLFYPVFMCLPCRWR  174    283  LASRPMVWIGGISYSLYLWHWP  304
Pa   WpbC2     139  WSLSVEEGYYILFPLLLLAAISGQR  162   277  LGWQPLVWFGLISYSLYLWHWP  298
Lp   Lag-1     145  WSIAVEWWLYTLFGIAFFFH---K  165    273  KIELISAFLAFISYTLYLSHEP  294
Xc   GumF      199  WGL--DVLPVSLCFYALGALLI-H  129    269  LSAVAGSLMVICAARMVQEWTW  290
Aa   Imat      135  WTLAYEMFFYLVLTLAIFF--RKK  168    267  RLPGILEFFGNISFSLYLWH-Q  287
At   PicA      149  WSLFNELVVNAVYARWGARAT-MK  181
Hi   0392      101  WSLAVEGQYYLIFPLILILA-YKK  123    287  LSASPIVFVGKISYSLYLYHWI  308
Ng   Orf       138  WSLAVEEQIFC----------YKK  151    244  IRNKAIVFIGKISYSLYLYHWI  265
Bs   Yrhl      143  WSLAIEEQFYIIWPMFLVVGMYIM  166    294  LSWRPLRWLGTRSYGIYLWHYP  315
Mt   RV1565c   163  WSMSVQGQFYLAFLLLVAGCAYLL  186    322  LATAPLVALGAMAYSWYLWHWP  343
Rl   NodX      157  WTLTVELTFYLTLPMLLEIWRRWK  190    292  LPRPNLLRRQDLSYGIYLYHML  313
Rm   ExoZ      180  WTLNFEMLFYAVFAGSLFMPRNWR  141    226  RALSLPGLLGDASYSIYLWHTF  247
St   CarE      141  WSLSCEMFFYLFPLFAFFT--K  172     290  LGTRTMVLLGELTFAFVLVWYL  290
St   AcyA      142  WSLSCEMAFYLTFPLWYRLLLRIR  173    289  LRAAVLVRLGEWSYAFYLIHFL  310
Sm   Mpt       140  WSLSCEMLFYAAFPFLFAFFSKMR  173    289  LGTRTMVLLGELTFAFVIHYL   310
Sm   MdmB      142  WSLSCEAFYLTFPLWYRLVR--K  173     289  LRAAVLVRLGEWSFAFYLVWFH  310
Ce   RO3H4.1   138  WSLSVEMQFYILAPIVFFGLQFLK  171    282  LKSKILGYIGDISYMYLVHWP   303
Ce   RO3H4.5   138  WSLSVEMQFYLLVPFIFLGIQFLK  171    281  LNSKVLVYIGDISYVVYLVHWP  302
Ce   RO3H4.6   138  WSLGVEMQFYLLVPFIFLGIQFLK  171    281  LKSKTLCYIGDISYVIYLVHWP  302
Ce   F09B9.1   415  WYLAVDTQLYLVAPIVLIGLYFSF  448    584  MSHPIWQPFGRLSYCAYIVHWV  605
Ce   C06B3.2   425  WYLAVDTQLYLVAPILLVALTWTP  458    594  MSHPIWQPLGRLSYSAYIVHLM  615
Ce   C08B11.4  491  WYLANDMQFHIFLMPLLVIV-FLK  523    688  LSWRLFVPLSKITFCAYLLH-P  708
```

Figure 6. Three regions of strong amino acid sequence homology amongst the family 2 integral membrane acetyltransferases. Highly conserved amino acids are shaded, while those totally conserved are underlined. The sequences and their accession numbers (in parethesis) are: O-antigen O-acetyltransferases F6 Oac, bacteriophage SF6 (X56800); St OafA, Salmonella typhimurium (U65941); Pa Wbp C and Pa Wbp C2, Pseudomonas aeruginosa (U50396); Lp Lag 1, Legionella pneumophila (U32118); and Xc Gum F, Xanthomonas campestris (S47286); integral membrane acetyltransferase Aa Imat, Actinobacillus actinomycetemcomitans (AB010415); hypothetical proteins At Pic A, Agrobacterium tumefaciens (M62814); Hi 0392, Haemophilus influenzae Rd (U32723); Ng Orf, Neisseria gonorrhoeae; Bs Yrh l, Bacillus subtilis (U93874); and Mt RV1565c, Mycobacterium tuberculosis (G70539); nodulation factor O-acetyltransferases Rl Nod X (X07990) and Rm Exo Z (X58126) from Rhizobium leguminosarum and R. meliloti, respectively; macrolide O-acetyltransferases St Car E (D32821), St Acy A (D30759), Sm Mpt ((D63662), and Sm Mdm B (M93958) from Streptomyces thermotolerans and S. mycarofaciens, respectively; hypothetical proteins Ce R03H4.1 (U50300), Ce R03H4.5 (U50300), Ce R03H4.6 (U50300), Ce F09B9.1 (Z49887), Ce C06B3.2 (Z77652), Ce C08B11.4 (Z46676), Ce C08H10.4 (U55368), from Caenorhabditis elegans; and acetyl-coenzyme A transporter Hs AcCoA T from Homo sapiens (D88152).

Providencia stuartii gentamicin 2′-*N*-acetyltransferase, which *O*-acetylates peptidoglycan, they may also have a physiological function.

One of the family 2 hypothetical enzymes produced by *P. aeruginosa*, Wbp C, is encoded within the *wbp* cluster involved with the synthesis of the serotype 05 B-band O-antigen (Burrows *et al.*, 1996) (Fig. 4). Located immediately downstream from *wbpC* is *wbpD*, which encodes a second hypothetical *O*-acetyltransferase. The gene product, Wbp D, would be considerably smaller than Wbp C (17.4 kDa compared to 69.9 kDa, respectively) and homologous to a large family of soluble *O*-acetyltransferases. This third family, currently comprising 39 sequences but first identified by Vuorio *et al.* (1991), is characterized by a C-terminal re-

```
Sa Cap5H  118 SRTTIKNDVWIGANVIIMDGLTINTGAVIAAGSVVTKNVGAYEVVGGVPAKVIK 171
Sa Cap1G   80 HRIFIGNNVFIGINSIILPGVTIGNNVVVGAGSVVTKDVPDNVIIGGNPAKKIK 133
Ec Wcab   108 ACPHIGNGVELGANVIILGDITLGNNVTVGAGSVVLKSVPDNALVVGEKAR-VK 160
Ec WcaF   123 TPIVIGEKCWLATDVPVAPGVTIGDGTVVGARSSVFKSLPANVVCRGNPA-VIR 175
Va rfb        HRIFIGNNVGIGINSIILPGVTIGNNVVVGAGSVVTKDVPNDVIIGGNPAKKIK
Pa WbpD   100 RNTLVKKGATLGANCTIVCGVTIGEYAFLGAGAVINKNVPSYALMVGVPARGIG 153
Bp BplB    99 RDTLVRQGATLGANCTIVCGATVGRYAFVGAGAVVNKDVPDFALVVGVPARQIG 152
Ec WbbJ   131 SAVVIGQR-WLGEN-TVLPGTIIGNGVVVGANSVVRGSIPENTVIAGVPAKIIK 182
Rl NodL   130 RPVSIGRHAWIGGGAIILPGVTIGDHAVIGAGSVVTRDVPAGSTAMGNPAR-VK 182
Sm NodL   130 RPVRIGKHVWIGGGAIILPGVTIGDHAVVGAGSVVTRDVPPGAKVMGSPAR-RG 182
Bs yvoF   105 GKVLIGDEVMIGANTTILPGVKIGDGAVVSAGTLVHKDVPDGAFVGGNPMIIYT 158
Ll THGA   132 KKVYIEENVWLGAGVIVLPGVRIGKNSVIGAGSLVTKDIPDNVVAFGTPC-MVK 184
Ec LacA   130 FPITIGNNVWIGSHVVINPGVTIGDNSVIGAGSIVTKDIPPNVVAAGVPCRVIR 183
Ec MAA    128 KPVTIGNNVWIGGRAVINPGVTIGDNVVASGAVVTKDVPDNVVGGNPARIIK 181
Bs MAA    129 KPVTIGDQVWIGGRAVINPGVTIGDNAVIAGSVVTKDVPANTVVGGNPARILK 182
Ec Cat    109 GDTVIGSDVWIGSEAMIMPGIKIGHGAVIGSRALVAKDVEPYTIVGGNPAKSIR 162
St Cat    109 GNTVIGNDVWIGSEAMVMPGIKIGHGAVIGSRSLVTKDVEPYAIVGSNPAKKIK 162
Mm Cat    109 GDTAIGMDVWIGSEAMIMPGIKIGHGAVIGSRSLVTKDVVPYAIIGGSPAKQIK 162
Ea Cat    109 GNTVIGNDVWIGSEAMVMPGIKIGHGAVIGSRSLVTKDVEPYAIVGGNPAKKIK 162
Pa Cat    109 GDTLIGHEVWIGTEAMFMPGVRIGHGAIIGSRALVTGDVEPYAIVGGNPARTIR 162
At Cat    108 GDTVIGNDVWIGSEAIIMPGITVGDAGVIGTRALVTKDVEPYAIVGGNPAKTIR 161
Bs cat        GDTVIGNDVWIGQNVTIMPGVIIGDGAIIAANSTVVKSVEPYSIYSGNPAKFIK
Rs cat    109 LSVVIGSDVWIGRDTIIQAGVRIGHGAVIGTRALVTSDIEPYTIAAGIPAKPLR 162
Sa Vat    112 GDIEIGNDVWIGRDVTIMPGVKIGDGAIIAAEAVVTKNNAPYSIVGGNPLKFIR 165
Sa VatB   115 GDTVVGNDVWIGQNVTVMPGIQIGDGAIVAANSVVTKDVPPYRI-GGNPSRIIK 167
Ef SatA   116 GDTVVGNDVWIGQNVTVMPGIQIGDGAIVAANSVVTKDVPPYRIIGGNPSRIIK 169
Ba CysE   196 ---IIRKNVTIGAGAKILGNIEVGQGVKVGAGSIVLKNIPPFVTVVGVPAKIIK 246
```

Figure 7. Amino acid sequence alignment of a highly homologous region of the soluble, family 3 *O*-acetyltransferases. Highly conserved amino acids are shaded. The sequences and their accession numbers (in parenthesis) for the *O*-acetyltransferases are: capsules, *Sa* Cap5 H (U81973), *Sa* Cap1 G (U10927), *Ec* Wca B (U38473), and *Ec* Wca F (U38473) from *Staphylococcus aureus* and *Escherichia coli*, respectively; O-antigen, *Va* rfb (AF025396), *Pa* Wbp D (U50396), *Bp* Bpl B (X90711), and *Ec* Wbb J (P37750) from *Vibrio anguillarum*, *Pseudomonas aeruginosa*, *Bordetella purtusis*, and *E. coli*, respectively; nodulation factors *Rl* Nod L (Y00548) and *Sm* Nod L (P28266) from *Rhizobium meliloti* and *Sinorhizobium meliloti*, respectively; *Bs* Yvo F, *Bacillus subtilis* *O*-acetyltransferase (Z99121); thiogalactoside *O*-acetyltransferases *Ll* THGA (P52984) and *Ec* Lac A (U73857) from *Lactococcus lactis* and *E. coli*, respectively; maltose transacetylases *Ec* MAA (P77791) and *Bs* MAA (P37515) from *E. coli* and *B. subtilis*, respectively; xenobiotic acetyltransferases *Ec* cat (P26838), *St* cat (AJ009818), *Mm* cat (P50869), *Ea* cat (P50868), *Pa* cat (G3318874), *At* cat (P23374), *Bs* cat (P00486), *Rs* cat (AF010496), *Sa* Vat (L07778), and *Ef* SatA (P50870) from *E. coli*, *Salmonella typhimurium*, *Morganella morgani*, *Enterobacter aerogenes*, *P. aeruginosa*, *Agrobacterium tumefaciens*, *B. subtilis*, *Rhodobacter capsulatus*, *S. aureus*, and *Enterococcus faecium*, respectively; and one representative of the serine acetyltransferases, *Ba* Cys E, from *Buchnera aphidicola* (P32003).

```
Ps AAC2'   75 GYVEAMVVEQSYRRQGIGRQLMLQTNK-IIASCYQLG--LLSA-SDD---GQKLYHSVGW 127
Mt AAC2'   79 GYVEGVAVRADWRGQRLVSALLD-AVEQVMRGAYQLG--ALSS-SAR---ARRLYASRGW 131
Mf AAC2'   89 GYVEAVAVREDWRGQGLATAVMD-AVEQVLRGAYQLG--ALSA-SDT---ARGMYLSRGW 141
Ms AAC2'  103 GYVEAVAVREDRRGDGLGTAVLD-ALEQVIRGAYQIG--ALSA-SDI---ARPMYIARGW 155
Ml orf        GYLEGVAVRKDCRGRGLVHALLD-AIEQVIRGAYQFG--ALSS-SDR---ARRVYMSRGW
Sp orf        -FVQDLIVLPSYQRQGIGSSLMKEALG-NFKEAYQV---QL-ATEET-EKNVGFYRSMGW
Ng orf        YWLGDVFVLPEYRGKGIGRRLVAHCIG-AARS---LGIKFL--YLYTPD-VQIFYESFGW
Pa orf        I-LEDMVVDRHARGQGVGRELIGRAVER-ARSWGCYK-LALSSHQDR-ETAQRFYAALGF
Pa orf        --VAKMLVHRRARRRGIGEALMGEL-DRLARACGKSL-LVLDTVSGS--AAERLYLKTGW
Ps TTR     91 AEVQKLMVLPSARGRGLGRQLMDE-VEQVAVKHKR-G--LLHLDTEAGSVAEAFYSALAY 147
              -                          -                    -         -
```

Figure 8.   Amino acid sequence alignment of a highly homologous region of the soluble, family 4 *O*-acetyltransferases. Highly conserved amino acids are shaded, while those totally conserved are underlined. The sequences and their accession numbers (in parenthesis) are: aminoglycoside 2'-*N*-acetyltransferases *Ps* AAC2' (L061560), *Mt* AAC2' (U72714), *Mf* AAC2' (U41471), and *Ms* AAC2' (U72743) from *Providencia stuartii, Mycobacterium tuberculosis, M. fortuitum,* and *M. smegmatis*, respectively; hypothetical proteins *Ml* orf (L78819), *Sp* orf (STP—4 39), *Ng* orf (contig182), *Pa* orf (contig292), and *Pa* orf (contig230) from *Mycobacterium leprae, Streptococcus pneumoniae, Neisseria gonorrhoeae,* and *Pseudomonas aeruginosa*, respectively; and *Ps* TTR, tabtoxin acetyltransferase (X17150).

peating hexapeptide motif of hydrophobic residues and glycine (Fig. 7). Members of the family 3 *O*-acetyltransferases include those acting on capsular polysaccharides and O-antigens, in addition to cysteine *O*-acetyltransferases and a class of xenobiotic (chloramphenicol, streptogramin A) acetyltransferases. A second nodulation factor *O*-acetyltransferase, Nod L, is also a member of the family 3 enzymes (Downie, 1989). Nod L represents the only polysaccharide *O*-acetyltransferase that has been studied in homogeneous form. Purified Nod L was shown to *O*-acetylate lipooligosaccharides, chitin fragments, and *N*-acetylglucosamine *in vitro* using acetyl-CoA as the acetyl donor (Bloemberg *et al.,* 1994). With nodulation factor as substrate, Nod L was later shown to be highly specific, adding acetyl groups only to the C-6 hydroxyl group of the nonreducing terminal *N*-acetylglucosaminyl residue (Bloemberg *et al.,* 1995). Thus, as with alginate and the *P. aeruginosa* serotype 05 O-antigen, both an integral membrane protein and a smaller soluble enzyme appear to be involved with the *O*-acetylation of nodulation factor.

The fourth family of *O*-acetyltransferases is composed of gentamicin 2'-*N*-acetyltransferases from *Providencia stuartii* and species of *Mycobacterium, Pseudomonas syringae* tabtoxin acetyltransferase, and several unidentified proteins (Fig. 8). Although initially characterized as an aminoglycoside acetyltransferase, the *P. stuartii* enzyme has subsequently been shown to catalyze the *O*-acetylation of peptidoglycan (Payie and Clarke, 1997; Payie *et al.,* 1995, 1996). Moreover, in addition to its use of acetyl-CoA as acetate source, this peripheral membrane protein is apparently capable of catalyzing an $N \rightarrow O$ transfer of acetate on amino sugars in a manner analogous to arylamine acetyltransferases (Smith and Hanna, 1986).

4.4. Model for the *O*-Acetylation of Bacterial Cell Wall Polysaccharides

There is considerable evidence to indicate that the *O*-acetylation of peptido-glycan, and likely alginate, occurs within the periplasm of bacteria. Unfortunate-ly, nothing is presently known about the location for the *O*-acetylation of the other bacterial cell wall polymers. With each of these polysaccharides, however, *O*-acetylation has been shown to be nonstoichiometric, suggesting that the precursor units are synthesized and polymerized prior to modification. Indeed, the biosynthesis and secretion of nodulation factor (Surin and Downie, 1988), O-antigen (Verma *et al.*, 1991; Kuzio and Kropinski, 1983; Gemski *et al.*, 1975), and capsular polysaccharides (Sau *et al.*, 1997) is clearly not affected by the pre-vention of *O*-acetylation even though the modification alters conformation (Slauch *et al.*, 1995). These observations thus provide indirect evidence to suggest that the *O*-acetylation of all bacterial cell wall polysaccharides is a maturation event oc-curring in the periplasm. If this is the case, then a pathway for the modification would have to involve the provision of a source of exported acetate.

Based on the observations reviewed above, two separate pathways can be proposed for the periplasmic *O*-acetylation of bacterial cell wall polysaccharides. The first would involve at least two proteins and is modeled on the pathway for the *O*-acetylation of glycoconjugate-bound sialic acids in the golgi of rat liver (Higa *et al.*, 1989). An integral membrane protein would serve to translocate ac-etate from the cytoplasmic pools of acetyl-CoA to the outside surface of the cyto-plasmic membrane and present it to the second protein, a peripheral membrane protein that acts as the true *O*-acetyltransferase (Fig. 9A). With the acetylation of sialic acids in the human golgi, the integral membrane acetyl-CoA transporter that has limited homology to the family 2 enzymes (Fig. 6) serves this function (Kanamori *et al.*, 1997). The prototype for this pathway involving bacterial cell wall polysaccharides would be the *O*-acetylation of alginate (Franklin and Ohman, 1996). Thus, membrane-bound Alg I (family 1) would translocate acetate for its transfer to substrate by the secreted Alg F (unclassified) (Fig. 9A). In this model, specificity for the *O*-acetylation would be conferred by the smaller, soluble fami-ly 3 or 4 enzymes, and it is conceivable that the integral membrane proteins of fam-ilies 1 and 2 serve as more generic translocators of acetate. This would be analo-gous to the use of the common lipid carrier bactoprenol (undecaprenyl phosphate) for the translocation of peptidoglycan, lipopolysaccharide, or capsular polysac-charide precursors across the cytoplasmic membrane. The acetate may remain bound to the family 1 or 2 translocating protein for subsequent transfer by a fam-ily 3 or 4 enzyme. The concept that a acetyltransferase accepts acetate from a translocating protein rather than directly from acetyl-CoA is supported by the in-ability of a purified preparation of the family 3 capsule *O*-acetyltransferase from *S. aureus* to function in isolation (Bhasin *et al.*, 1998). Nevertheless, it is equally

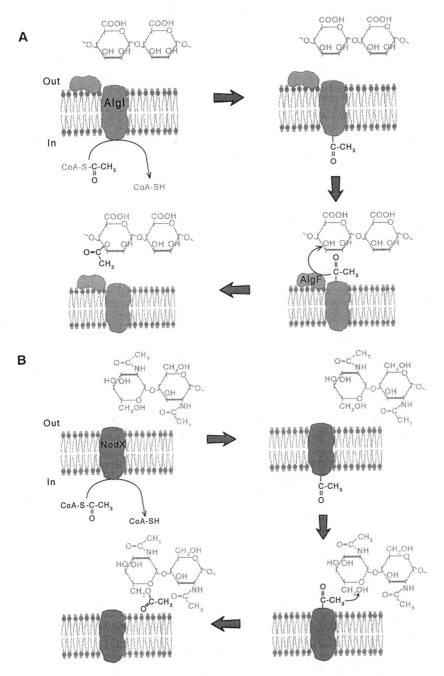

Figure 9. Proposed pathways for the *O*-acetylation of bacterial cell wall polysaccharides. (A) Two protein system involving an integral membrane acetate transporter (Alg I for alginate biosynthesis) and an extracellular peripheral membrane acetyltransferase (Alg F?) that accepts the translocated acetate and catalyzes its specific transfer to substrate, alginate. (B) A single-enzyme system, possibly exemplified by the acetylation of nodulation factor involving Nod X, whereby the integral membrane protein accepts acetate from intracellular pools of acetyl-CoA and both translocates and transfers it to a specific substrate.

possible that for other systems, acetyl-CoA could be translocated to the periplasm with the family 1 or 2 proteins serving as an acetyl-CoA–CoA antiporter.

It is tempting to apply this model to the hypothetical proteins encoded by *wbpC* (family 2) and *wbpD* (family 3) of the *P. aeruginosa* serotype O5 O-antigen gene cluster, O5, but the synthesized polysaccharide does not bear *O*-acetylation. This could be interpreted as either the assignment of function to the genes is incorrect, one of the two is not expressed, or one of the two expressed proteins is inactive. Support for the latter two explanations is provided by the observation that the genome of bacteriophage D3 encodes a homologue of *wbpC* (A. Kropinsky, personal communication) and its infection of *P. aeruginosa* O5 results in an O-antigen conversion to serotype O20, which is identical to O5 except for *O*-acetylation of its *N*-acetylfucosaminyl residues (Stanislavsky and Lam, 1997; Kuzion and Kropinski, 1983). Hence, bacteriophage D3 may complement an inactive Wbp C to permit the *O*-acetylation of the O-antigen by Wbp D.

It also is speculated that the *O*-acetylation of peptidoglycan follows this pathway, at least for *Providencia stuartii*. In this case, the family 4 gentamicin 2'-*N*-acetyltransferase would accept acetate from a currently unidentified acetate translocator and transfer it to the C-6 hydroxyl group of muramyl residues.

As an alternative to this first model, it is possible that in some cases the integral membrane protein of families 1 and 2 may act as both translocator and transacetylase (Fig. 9B). This second model is based on the apparent pathway for the acetylation of glucosamine by the lysosomal membrane enzyme acetyl-CoA–α-glucosaminide *N*-acetyltransferase in the production of heparan sulfate (Bame and Rome, 1985). Hence, the enzyme would acquire acetate from cytoplasmic pools of acetyl-CoA and add it directly to substrates on the other side of the cytoplasmic membrane. The family 2 enzyme Nod X may act in this manner, as it alone appears to be responsible for the specific *O*-acetylation of nodulation factor at the reducing terminal *N*-acetylglucosaminyl residue (Firmin *et al.*, 1993). However, it is equally possible that for this relatively small lipooligosaccharide, the *O*-acetylation occurs at the cytoplasmic face of the membrane prior to the export of the mature nodulation factor.

Investigations are currently in progress to test the validity of the two-component model for the *O*-acetylation of peptidoglycan. The experimental strategy involves the transformation of *E. coli* with both the family 4 peptidoglycan *O*-acetyltransferase, aminoglycoside 2'-*N*-acetyltransferase, and the *P. aeruginosa* bacteriophage D3 homologue of the family 2 Wbp C. Expression of these two proteins will have to be tightly regulated, because *E. coli* does not naturally perform the *O*-acetylation of peptidoglycan, and it is quite possible that such a transformation could be lethal. Such studies thus should reflect the physiological role of peptidoglycan *O*-acetylation and its potential as a new target for antimicrobial therapy.

ACKNOWLEDGMENTS

Our studies on the *O*-acetylation of peptidoglycan have continually been supported by operating grants to A. J. C. from the Medical Research Council of Canada. N. B. is the recipient of a Natural Sciences and Engineering Research Council Post-Graduate Scholarship (PGSB).

REFERENCES

Abrams, A., 1958, *O*-Acetyl groups in the cell wall of *Streptococcus faecalis, J. Biol. Chem.* **230:**949–959.

Arbeit, R. D., Karakawa, W. W., Vann, W. F., and Robbins, J. B., 1984, Predominance of two newly described capsular polysaccharide types among clinical isolates of *Staphylococcus aureus, Diagn. Microbiol. Infect. Dis.* **2:**85–91.

Bame, K. J., and Rome, L. H., 1985, Acetyl coenzyme A:α-glucosaminide *N*-acetyltransferase: Evidence for a transmembrane acetylation mechanism, *J. Biol. Chem.* **260:**11293–11299.

Bhasin, N., Albus, A., Michon, F., Livolsi, P. J., Park, J.-S., and Lee, J. C., 1998, Identification of a gene essential for O-acetylation of the *Staphylococcus aureus* type 5 capsular polysaccharide, *Mol. Microbiol.* **271:**9–21.

Blake, C. C. F., Johnson, L. N., Mair, G. A., North, A. T. C., Phillips, D. C., and Sarma, V. R., 1967, Crystallographic studies of the activity of hen egg-white lysozyme, *Proc. R. Soc. Lond. B. Biol. Sci.* **167:**378–388.

Bloemberg, G. V., Thomas-Oates, J. E., Lugtenberg, J. J., and Spaink, H. P., 1994, Nodulation protein NodL of *Rhizobium leguminosarum O*-acetylates lipo-oligosaccharides, chitin fragments and *N*-acetylglucosamine *in vitro, Mol. Microbiol.* **11:**793–804.

Bloemberg, G. V., Lagas, R. M., van Leeuwen, S., Van der Marel, G. A., Van Boom, J. H., Lugtenberg, B. J. J., and Spanik, H. P., 1995, Substrate specificity and kinetic studies of nodulation protein NodL of *Rhizobium leguminosarum, Biochemistry* **34:**12717–12720.

Blundell, J. K., and Perkins, H. R., 1981, Effects of β-lactam antibiotics on peptidoglycan synthesis in growing *Neisseria gonorrhoeae,* including changes in the degree of *O*-acetylation, *J. Bacteriol.* **147:**633–641.

Blundell, J. K., and Perkins, H. R., 1985, Selectivity for *O*-acetylated peptidoglycan during endopeptidase action by permeabilized *Neisseria gonorrhoeae, FEMS Microbiol Lett.* **30:**67–69.

Blundell, J. K., Smith, G. J., and Perkins, H. R., 1980, The peptidoglycan of *Neisseria gonorrhoeae: O*-acetyl groups and lysozyme sensitivity, *FEMS Microbiol. Lett.* **9:**259–261.

Boyd, A., and Chakrabarty, A. M., 1994, Role of alginate lyase in cell detachment of *Pseudomonas aeruginosa, Appl. Env. Microbiol.* **60:**2355–2359.

Brenner, D. J., 1992, Introduction to the *Enterobacteriaceae,* in: *The Prokaryotes,* 2nd Ed.,

Vol. 3 (A. Balows, H. G. Truper, M. Dworkin, W. Harder, and K.-H. Schleifer, eds.), Springer-Verlag, New York, pp. 2673–2695.

Brumfitt, W., 1959, The mechanism of development of resistance to lysozyme by some gram-positive bacteria and its results, *Br. J. Exp. Pathol.* **40**:441–451.

Brumfitt, W., Wardlaw, A. C., and Park, J. T., 1958, Development of lysozyme resistance in *Micrococcus lysodeikticus* and its association with an increased O-acetyl content in the cell wall, *Nature* **181**:1783–1784.

Burghaus, P., Johannsen, L., Naumann, D., Labischinski, H., Bradaczek, H., and Giesbrecht, P., 1983, The influence of different antibiotics on the degree of O-acetylation of staphylococcal cell walls, in: *The Target of Penicillin. The Murein Sacculus of Bacterial Cell Walls Architecture and Growth* (R. Hakenbeck, J.-V. Höltje, and H. Labischinski, eds.), Walter de Gruyter, New York, pp. 317–322.

Burrows, L. L., Charter, D. F., and Lam, J. S., 1996, Molecular characterization of the *Pseudomonas aeruginosa* serotype 05 (PAO1) B-band lipopolysaccharide gene cluster, *Mol. Microbiol.* **22**:481–495.

Chapman, S. J., and Perkins, H. R., 1983, Peptidoglycan degrading enzymes in ether-treated cells of *Neisseria gonorrhoeae. J. Gen. Microbiol.* **129**:877–883.

Chitnis, C. E., and Ohman, D. E., 1993, Genetic analysis of the alginate biosynthetic gene cluster of *Pseudomonas aeruginosa* shows evidence for an operonic structure, *Mol. Microbiol.* **8**:583–590.

Clarke, A. J., 1993, Extent of peptidoglycan O-acetylation in the tribe *Proteeae, J. Bacteriol.* **175**:4550–4553.

Clarke, A. J., 1997, *Biodegradation of Cellulose. Enzymology and Biotechnology.* Technomic Publishing Co., Lancaster, PA.

Clarke, A. J., and Dupont, C., 1992, O-Acetylated peptidoglycan: Its occurrence, pathobiological significance and biosynthesis, *Can. J. Microbiol.* **38**:85–91.

Clarke, A. J., Gyorffy, S., and Chase, J., 1993, Distribution of O-acetylation in the peptidoglycan from *Proteus mirabilis* 19, in: *Bacterial Growth and Lysis: Metabolism and Structure of the Bacterial Sacculus* (M. A. dePedro, J.-V. Holtje, W. Loffelhardt, eds.), Plenum Publishing, New York, pp. 83–89.

Clarke, A. J., Francis, D., and Keenleyside, W. J., 1996, The prevalence of gentamicin 2'-N-acetyltransferase in the *Proteeae* and its role in the O-acetylation of peptidoglycan, *FEMS Microbiol. Lett.* **145**:201–207.

Coyette, J., and Ghuysen, J. M., 1970, Structure of the walls of *Lactobacillus acidophilus* strain 63 AM Gasser. *Biochemistry* **9**:2935–2943.

Davidson, I. W., Sutherland, I. W., and Lawson, C. J., 1977, Localization of O-acetyl groups in bacterial alginate, *J. Gen. Microbiol.* **98**:603–606.

Day, F., and Lee, J. W., 1994, Process for acetylating seaweed alginate with *Pseudomonas syringae* subsp. *phaseolicola,* US Patent 5308761.

Dénairié, J., Debellé, F., and Rosenburg, C., 1992, Signalling and host range variation in nodulation, *Annu. Rev. Microbiol.* **46**:497–531.

Dougherty, T. J., 1985, Involvement of a change in penicillin target and peptidoglycan structure in low-level resistance to β-lactam antibiotics in *Neisseria gonorrhoeae, Antimicrob. Agents Chemother.* **28**:90–95.

Downie, J. A., 1989, The *nodL* gene from *Rhizobium leguminosarum* is homologous to the acetyl transferases encoded by *lacA* and *cysE, Mol. Microbiol.* **3**:1049–1051.

Dupont, C., 1991, *Characterization of Peptidoglycan O-Acetylation in Proteus mirabilis,* PhD thesis, University of Guelph.

Dupont, C., and Clarke, A. J., 1991a, Dependence of lysozyme-catalysed solubilization of *Proteus mirabilis* peptidoglycan on the extent of *O*-acetylation, *Eur. J. Biochem.* **195:**763–769.

Dupont, C., and Clarke, A. J., 1991b, Evidence for N-O acetyl migration as the mechanism for the O-acetylation of peptidoglycan in *Proteus mirabilis, J. Bacteriol.* **173:**4318–4324.

Dupont, C., and Clarke, A. J., 1991c, *In vitro* synthesis and O-acetylation of peptidoglycan by permeabilized cells of *Proteus mirabilis, J. Bacteriol.* **173:**4618–4624.

Ertesvåg, H., Erlien, F., Skjåk-Broek, G., Rehm, B. H., and Valla, S., 1998, Biochemical properties and substrate specificities of a recombinantly produced *Azotobacter vinelandii* alginate lyase, *J. Bacteriol.* **180:**3779–3784.

Evans, L. R., and Linker, A., 1973, Production and characterization of the slime polysaccharide of *Pseudomonas aeruginosa, J. Bacteriol.* **116:**915–924.

Fein, J. E., and Rogers, H. J., 1976, Autolytic enzyme-deficient mutants of *Bacillus subtilis* 168, *J. Bacteriol.* **127:**1427–1442.

Fett, W. F., Osman, S., Fishman, M. L., and Siebles, T. S. III, 1986, Alginate production by plant-pathogenic pseudomonads, *Appl. Env. Microbiol.* **52:**466–473.

Firmin, J. L., Wilson, K. E., Carlson, R. W., Davies, A. E., and Downie, J. A., 1993, Resistance to nodulation of cv. Afghanistan peas is overcome by *nodX,* which mediates an *O*-acetylation of the *Rhizobium leguminosarum* lipo-oligosaccharide nodulation factor, *Mol. Microbiol.* **10:**351–360.

Fleck, J., Mock, M., Minck, R. and Ghuysen, J.-M., 1971, The cell envelope of *P. vulgaris* P18: Isolation and characterization of the peptidoglycan component. *Biochim. Biophys. Acta* **233:**489–503.

Flesher, A. R., and Insel, R. A., 1978, Characterization of lipopolysaccharide of *Haemophilus influenzae, J. Infect. Dis.* **138:**719–730.

Fournier, J. M., Vann, W. F., and Karakawa, W. W., 1984, Purification and characterization of *Staphylococcus aureus* type 8 capsular polysaccharide, *Infect. Immun.* **45:**87–93.

Franklin, M. J., and Ohman, D. E., 1993, Identification of *algF* in the alginate biosynthetic gene cluster of *Pseudomonas aeruginosa* which is required for alginate acetylation, *J. Bacteriol.* **175:**5057–5065.

Franklin, M. J., and Ohman, D. E., 1996, Identification of *algI* and *algJ* in the *Pseudomonas aeruginosa* alginate biosynthetic gene cluster which are required for alginate *O*-acetylation, *J. Bacteriol.* **178:**2186–2195.

Gacęsa, P., 1988, Alginates, *Carbohydr. Polym.* **8:**161–182.

Garegg, P. J., Lindberg, B., Onn, T., and Sutherland, I. W., 1971, Comparative structural studies on the M-antigen from *Salmonella typhimurium, Escherichia coli* and *Enterobacter cloacae, Acta Chem. Scand.* **25:**2103–2108.

Gemski, P., Jr., Koeltzow, D. E., and Formal, S. B., 1975, Phage conversion of *Shigella flexneri* group antigens, *Infect. Immun.* **11:**685–691.

Ghuysen, J. M., and Strominger, J. L., 1963, Structure of the cell wall of *S. aureus* strain Copenhagen. II. Separation and structure of disaccharides, *Biochemistry* **2:**1119–1125.

Glauner, B., Holjte, J. V., and Schwarz, U., 1988, The composition of the murein of *Escherichia coli, J. Biol. Chem.* **263:**10088–10095.

Gmeiner, J., and Kroll, H. P., 1981, Murein biosynthesis and *O*-acetylation of *N*-acetylmuramic acid during the cell division cycle of *Proteus mirabilis, Eur. J. Biochem.* **177:**171–177.

Gmeiner, J., and Sarnow, E., 1987, Murein biosynthesis in synchronized cells of *Proteus mirabilis.* Quantitative analysis of *O*-acetylated murein subunits and of chain terminators incorporated into the sacculus during the cell cycle, *Eur. J. Biochem.* **163:**389–395.

Gyorffy, S., and Clarke, A. J., 1992, Production and characterization of a monoclonal antibody to the *O*-acetylated peptidoglycan of *Proteus mirabilis, J. Bacteriol.* **174:**5043–5050.

Hamada, S., Torii, M., Kotani, S., Masuda, N., Oushima, T., Yokogawa, K., and Kawata, S., 1978, Lysis of *Streptococcus mutans* cells with mutanolysin a lytic enzyme prepared from a culture liquor of *Streptomyces globisporous* 1829, *Arch. Oral Biol.* **23:**543–549.

Hanessian, S., and Haskell, T., 1964, Structural studies on staphylococcal polysaccharide antigen, *J. Biol. Chem.* **239:**2758–2764.

Hash, J. H., and Rothlauf, M. V., 1967, The *N,O*-diacetylmuramidase of *Chalaropsis* species. 1. Purification and crystallization, *J. Biol. Chem.* **242:**5586–5590.

Heaton, M. P., and Neuhaus, F. C., 1992, Biosynthesis of D-alanyl-lipoteichoic acid: Cloning, nucleotide sequence, and expression of the *Lactobacillus casei* gene for the D-alanine activating enzyme, *J. Bacteriol.* **174:**4707–4717.

Higa, H. H., Butor, C., Diaz, S., and Varki, A., 1989, *O*-Acetylation and de-*O*-acetylation of sialic acids. *O*-Acetylation of sialic acids in rat liver Golgi apparatus involves an acetyl intermediate and essential histidine and lysine residues: A transmembrane reaction, *J. Biol. Chem.* **264:**19427–19434.

Höltje, J.-V., 1998, Growth of the stress-bearing and shape-determining murein sacculus of *Escherichia coli, Microbiol. Mol. Biol. Rev.* **62:**181–203.

Jann, B., and Jann, K., 1990, Structure and biosynthesis of the capsular antigens of *Escherichia coli, Curr. Top. Microbiol.* **150:**19–42.

Jastrzemski, K. B., and Kwiatkowski, B., 1984, Degradation of *Citrobacter* O-serogroup Ci23Vi+ murein by Vi phages, *Zentrabl. Bakteriol. Mikrobiol. Hyg. A* **258:**32–37.

Johannsen, L., Labischinski, H., Burghaus, P., and Giesbrecht, P., 1983, Acetylation in different phases of growth of staphylococci and their relation to cell wall degradability by lysozyme, in: *The Target of Penicillin: The Murein Sacculus of Bacterial Cell Walls Architecture and Growth* (R. Hakenbeck, J.-V. Höltje, and H. Labischinski, eds.), Walter de Gruyter, Berlin, pp. 261–266.

Kanamori, A., Nakayama, J., Fukuda, M. N., Stallcup, W. B., Sasaki, K., Fukada, M., and Hirabayashi, Y., 1997, Expression, cloning, and characterization of a cDNA encoding a novel membrane protein required for the formation of *O*-acetylated ganglioside: A putative acetyl-CoA transporter, *Proc. Natl. Acad. Sci. USA* **94:**2897–2902.

Kelly, R. F., Severn, W. B., Richards, J. C., Perry, M. B., MacLean, L. L., Tomás, J. M., Merion, S., and Whitfield, C., 1993, Structural variation in the O-specific polysaccharides of *Klebsiella pneumoniae* serotype O1 and O8 lipopolysaccharide: Evidence for clonal diversity in *rfb* genes, *Mol. Microbiol.* **10:**615–625.

Kelly, R. F., MacLean, L. L., Perry, M. B., and Whitfield, C., 1995, Structures of the O-antigens of *Klebsiella* serotypes O(2a,2e), O2(2a,2e,2h), and O2(2a,2f,2g), members of a family of related D-galactan O-antigens in *Klebsiella* spp., *J. Endotoxin Res.* 2:131–140.

Knirel, Y. A., and Kochetkov, N. K., 1994, The structure of lipopolysaccharides of gram-negative bacteria. III. The structure of O-antigens: A review, *Biochemistry (Moscow)* 59:1325–1383.

Knirel, Y. A., Moll, H., and Zahringer, U., 1996, Structural study of a highly *O*-acetylated core of *Legionella pneumophila* serogroup 1 lipopolysaccharide, *Carbohydr. Res.* 293:223–234.

Kol, O., Wieruszeski, J.-M., Strecker, G., Fournet, B., Zalisz, R., and Smets, P., 1992, Structure of the O-specific polysaccharide chain of *Klebsiella pneumoniae* O1:K2 (NCTC 5055) lipopolysaccharide. A complementary elucidation, *Carbohydr. Res.* 236:339–344.

Kropinski, A. M., Jewell, B., Kuzio, J., Milazzo, F., and Berry, D., 1985, Structure and functions of *Pseudomonas aeruginosa* lipopolysaccharide, *Antibiot. Chemother.* 36:58–73.

Kropinski, A. M., Lewis, V., and Berry, D., 1987, Effect of growth temperature on the lipids, outer membrane proteins, and lipopolysaccharides of *Pseudomonas aeruginosa* PAO, *J. Bacteriol.* 169:1960–1966.

Kuzio, J., and Kropinski, A., 1983, O-Antigen conversion in *Pseudomonas aeruginosa* PAO1 by bacteriophage D3, *J. Bacteriol.* 155:203–212.

Labischinski, H., Goodell, E. W., Goodell, A., and Hochberg, M. L., 1991, Direct proof of a "more-than-single-layered" peptidoglycan architecture of *Escherichia coli* W7: A neutron small-angle scattering study, *J. Bacteriol.* 173:751–756.

Lam, J. S., Handelsman, M. Y. C., Chivers, T. R., and MacDonald, L. A., 1992, Monoclonal antibodies as probes to examine serotype-specific and cross-reactive epitopes of lipopolysaccharides from serotypes O2, O5, and O16 of *Pseudomonas aeruginosa*, *J. Bacteriol.* 174:2178–2184.

Larsen, B., and Haug, A., 1971, Biosynthesis of alginate. I. Composition and structure of alginate produced by *Azotobacter vinelandii* (Lipman), *Carbohydr. Res.* 17:287–296.

Lear, A. L. and Perkins, H. R., 1983, Degrees of *O*-acetylation and cross-linking of the peptidoglycan of *Neisseria gonorrhoeae* during growth, *J. Gen. Microbiol.* 129:885–888.

Lear, A. L., and Perkins, H. R., 1986, *O*-Acetylation of peptidoglycan in *Neisseria gonorrhoeae*. Investigation of lipid-linked intermediates and glycan chains newly incorporated into the cell wall, *J. Gen. Microbiol.* 132:2413–2420.

Lear, A. L., and Perkins, H. R., 1987, Progress of *O*-acetylation and cross-linking of peptidoglycan in *Neisseria gonorrhoeae* grown in the presence of penicillin, *J. Gen. Microbiol.* 133:1743–1750.

Lee, J. W., and Day, F., 1998, The separation of alginate biosynthesis and acetylation in *Pseudomonas syringae, Can. J. Microbiol.* 44:394–398.

Lemercinier, X., and Jones, C., 1996, Full ^{1}H NMR assignment and detailed *O*-acetylation patterns of capsular polysaccharide from *Neisseria meningitidis* used in vaccine production, *Carbohydr. Res.* 296:83–96.

Lerouge, P. Roche, P, Faucher, C., Maillet, F., Truchet, G., Promé, J.-C., and Dénairié, J., 1990, Symbiotic host-specificity of *Rhizobium meliloti* is determined by a sulphated and acylated glucosamine oligosaccharide signal, *Nature* 344:781–784.

Li, Z. S., Clarke, A. J., and Beveridge, T. J., 1996, A major autolysin of *Pseudomonas aeru-*

ginosa, its subcellular distribution, its potential role in cell growth and division, and its secretion in surface membrane vesicles, *J. Bacteriol.* **178:**2479–2488.

Lindberg, A. A., and LeMinor, L., 1984, Serology of *Salmonella, Methods Microbiol.* **15:**1–141.

Linker, A., and Jones, R. S., 1966, A new polysaccharide resembling alginic acid isolated from pseudomonads, *J. Biol. Chem.* **241:**3845–3851.

Logardt, I. M., and Neujahr, H. Y., 1975, Lysis of modified walls from *Lactobacillus fermentum, J. Bacteriol.* **124:**73–77.

Martin, H. H., 1984, *In vitro* synthesis of peptidoglycan by spheroplasts of *Proteus mirabilis* grown in the presence of penicillin, *Arch. Microbiol.* **139:**371–375.

Martin, H. H., and Gmeiner, J., 1979, Modification of peptidoglycan structure by penicillin action in cell walls of *Proteus mirabilis, Eur. J. Biochem.* **95:**487–495.

Martin, J.-P., Fleck, J., Mock, M. and Ghuysen, J.-M., 1973, The wall peptidoglycans of *Neisseria perflava, Moraxella glucidolytica, Pseudomonas alcaligenes* and *Proteus vulgaris* strain P18, *Eur. J. Biochem.* **38:**301–306.

McGroarty, E. J., and Rivera, M., 1990, Growth-dependent alterations in production of serotype-specific and common antigenic lipopolysaccharides in *Pseudomonas aeruginosa* PAO1, *Infect. Immun.* **58:**1030–1037.

Michetti, P., Mahan, M. J., Slauch, J. M., Mekalanos, J. J., and Neutra, M. R., 1992, Monoclonal secretory immunoglobulin A protects mice against oral challenge with the invasive pathogen *Salmonella typhimurium, Infect. Immun.* **60:**1786–1792.

Monday, S. R., and Schiller, N. L., 1996, Alginate synthesis in *Pseudomonas aeruginosa:* The role of AlgL (alginate lyase) and AlgX, *J. Bacteriol.* **178:**625–632.

Moreau, M., Richards, J. C., Fournier, J. M., Byrd, R. A., Karakawa, W. W., and Vann, W. F., 1990, Structure of the type-5 capsular polysaccharide of *Staphylococcus aureus, Carbohydr. Res.* **201:**285–297.

Narbad, A., Gacesa, P., and Russell, N. J., 1990, Biosynthesis of alginate, in: *Pseudomonas Infection and Alginates* (N. J. Russell and P. Gacesa, eds.), Chapman and Hall, London, pp. 181–205.

Ørskov, F., and Ørskov, I., 1992, *Escherichia coli* serotyping and disease in man and animals, *Can. J. Microbiol.* **38:**699–704.

Osborn, M. J., 1979, Biosynthesis and assembly of the lipopolysaccharide of the outer membrane, in: *Bacterial Outer Membranes* (M. Inouye, ed), John Wiley and Sons, New York, pp. 15–34.

Payie, K. G., and Clarke, A. J., 1997, Characterization of gentamicin 2'-N-acetyltransferase from *Providencia stuartii:* Its use of peptidoglycan metabolites for acetylation of both aminoglycosides and peptidoglycan, *J. Bacteriol.* **179:**4106–4114.

Payie, K., Rather, P. N., and Clarke, A. J., 1995, Contribution of gentamicin 2'-N-acetyltransferase to the O-acetylation of peptidoglycan in *Providencia stuartii, J. Bacteriol.* **177:**4303–4310.

Payie, K. G., Strating, H., and Clarke, A. J., 1996, The role of O-acetylation in metabolism of peptidoglycan in *Providencia stuartii, Microb. Drug Resist.* **2:**135–140.

Pelkonen, S., Hayrinen, J., and Finne, J., 1988, Polyacrylamide gel electrophoresis of the capsular polysaccharides of *Escherichia coli* K1 and other bacteria, *J. Bacteriol.* **170:**2646–2653.

Perry, M. B., MacLean, L. M., Brisson, J. R., and Wilson, M. E., 1996, Structures of the antigenic *O*-polysaccharides of lipopolysaccharides produced by *Actinobacillus actin-omycetemcomitans* serotypes a, c, d and e, *Eur. J. Biochem.* **242**:682–688.

Price, N. P. J., Relic, B., Talmont, F., Lewin, A., Promé, D., Pueppke, S. G., Maillet, F., Dé-narié, J., Promé, J.-C., and Broughton, W. J., 1992, Broad-host range *Rhizobium* species strain NGR234 secretes a family of carbamoylated, and fucosylated, nodula-tion signals that are *O*-acetylated or sulphated, *Mol. Microbiol.* **6**:3575–3584.

Raetz, C. R. H., 1990, Biochemistry of endotoxins, *Annu. Rev. Biochem.* **59**:129–170.

Raetz, C. R. H., 1996, Bacterial lipopolysaccharides: A remarkable family of bioactive macroamphiphiles, in: *Escherichia coli and Salmonella. Cellular and Molecular Bi-ology* (F. C. Neidhart *et al.,* eds.), American Society for Microbiology Press, Wash-ington DC, pp. 1035–1063.

Rivera, M., and McGroarty, E. J., 1989, Analysis of a common antigen lipopolysaccharide from *Pseudomonas aeruginosa, J. Bacteriol.* **171**:2244–2248.

Roche, P., Lerouge, P., Ponthus, C., and Promé, J.-C., 1991, Structural determination of bac-terial nodulation factors involved in the *Rhizobium meliloti*-alfalfa symbiosis, *J. Biol. Chem.* **266**:10933–10940.

Rosenthal, R. S., Blundell, J. K. and Perkins, H. R., 1982, Strain-related differences in lysozyme sensitivity and extent of *O*-acetylation of gonococcal peptidoglycan, *Infect. Immun.* **37**:826–829.

Rosenthal, R. S., Folkening, W. J., Miller, D. R. and Swim, S. C., 1983, Resistance of *O*-acetylated gonococcal peptidoglycan to human peptidoglycan-degrading enzymes, *Infect. Immun.* **40**:826–829.

Rosenthal, R. S., Gfell, M. A., and Folkening, W. J., 1985, Influence of protein synthesis inhibitors on regulation of extent of *O*-acetylation of gonococcal peptidoglycan, *In-fect. Immun.* **49**:7–13.

Saitoh, T., Natomi, H., Zhao, W., Okuzumi, K., Sugano, K., Iwamori, M., and Nagai, Y., 1991, Indentification of glycolipid receptors for *Helicobacter pylori* by TLC-immunostaining, *FEBS Lett.* **282**:385–387.

Sanchez-Puelles, J. M., Ronda, C., Garcia, J. L., Lopez, R., and Garcia, E., 1986, Search-ing for autolysin functions. Characterization of a pneumoncoccal mutant deleted in the *lytA* gene, *Eur. J. Biochem.* **175**:289–293.

Sanjuan, J., Carlson, R. W., Spaink, H. P., Bhat, U. R., Barbour, W. M., Glushka, J., and Stacey, G., 1992, A 2-*O*-methylfucose moiety is present in the lipo-oligosaccharide nodulation signal of *Bradyrhizobium japonicum, Proc. Natl. Acad. Sci. USA* **89**:8789–8793.

Sau, S., Sun, J., and Lee, C. Y., 1997, Molecular characterization and transcriptional ana-lysis of type 8 capsule genes in *Staphylococcus aureus, J. Bacteriol.* **179**:1614–1621.

Schleifer, K. H., and Kandler, O., 1972, Peptidoglycan types of bacterial cell walls and their taxonomic implications, *Bacteriol. Rev.* **36**:407–477.

Sherbroack-Cox, V., Russell, N., and Gacesa, P., 1984, The purification and chemical char-acterization of the alginate present in extracellular material produced by mucoid strains of *Pseudomonas aeruginosa, Carbohydr. Res.* **135**:147–154.

Shinabarger, D., May, T. B., Boyd, A., Ghosh, M., and Chakrabarty, A. M., 1993, Nucleotide sequence and expression of the *Pseudomonas aeruginosa algF* gene controlling acety-lation of alginate, *Mol. Microbiol.* **9**:1027–1035.

Shockman, G. D., and Barrett, J. F., 1983, Structure, function and assembly of cell walls of gram-positive bacteria, *Annu. Rev. Microbiol.* **37:**501–527.

Sidow, T., Johannsen, L., and Labischinski, H., 1990, Penicillin-induced changes in the cell wall composition of *Staphylococcus aureus* before the onset of bacteriolysis, *Arch. Microbiol.* **154:**73–81.

Skjåk-Broek, G., Grasdalen, H., and Larsen, B., 1986, Monomer sequence and acetylation pattern in some bacterial alginates, *Carbohydr. Res.* **154:**239–250.

Skjåk-Broek, G., Paoletti, S., and Gianferrara, T., 1989, Selective acetylation of mannuronic acid residues in calcium alginate gels, *Carbohydr. Res.* **185:**119–129.

Slauch, J. M., Mahan, M. J., Michette, P., Neutra, M. R., and Mekalanos, J. J., 1995, Acelation (O-factor 5) affects structural and immunological properties of *Salmonella typhimurium* lipopolysaccharide O antigen, *Infect. Immun.* **63:**437–441.

Slauch, J. M., Lee, A. A., Mahan, M. J., and Mekalanos, J. J., 1996, Molecular characterization of the *oafA* locus resposible for acetylation of *Salmonella typhimurium* O-antigen: OafA is a member of a family of integral membrane trans-acylases, *J. Bacteriol.* **178:**5904–5909.

Smith, T. J., and Hanna, P. E., 1986, *N*-Acetyltransferase multiplicity and the bioactivation of *N*-arylhydroxamic acids by hamster hepatic and intestinal enzymes, *Carcinogenesis* **7:**697–702.

Snowden, M. A., Perkins, H. R., Wyke, A. W., Hayes, M. V., and Ward, J. B., 1989, Crosslinking and O-acetylation of newly synthesized peptidoglycan in *Staphylococcus aureus* H, *J. Gen. Microbiol.* **135:**3015–3022.

Sompolinsky, D., Samra, Z., Karakawa, W. W., Vann, W. F., Schneerson, R., and Malik, Z., 1985, Encapsulation and capsular types in isolates of *Staphylococcus aureus* from different sources and relationship to phage types, *J. Clin. Microbiol.* **22:**828–834.

Stanislavsky, E. S., and Lam, J. L., 1997, *Pseudomonas aeruginosa* antigens as potential vaccines, *FEMS Microbiol. Rev.* **21:**243–277.

Stankowski, J. D., Mueller, B. E., and Zeller, S. G., 1993, Location of a second O-acetyl group in xanthan gum by the reductive-cleavage method, *Carbohydr. Res.* **241:**321–326.

Surin, B. P., and Downie, J. A., 1988, Characterization of the *Rhizobium leguminosarum* genes *nodLMN* involved in efficient host-specific nodulation, *Mol. Microbiol.* **2:**173–183.

Sutherland, I. W., 1985, Biosynthesis and composition of gram-negative bacterial extracellular and wall polysaccharides, *Annu. Rev. Microbiol.* **39:**243–270.

Swim, S. C., Gfell, M. A., Wilde, C. E. III, and Rosenthal, R. S. 1983, Strain distribution in extents of lysozyme resistance and O-acetylation of gonococcal peptidoglycan determined by high-performance liquid chromatography. *Infect. Immun.* **42:**446–52.

Verma, N. K., Brandt, J. M., and Lindberg, A. A., 1991, Molecular characterization of the O-acetyl transferase gene of converting bacteriophage SF6 that adds group antigen 6 to *Shigella flexneri, Mol. Microbiol.* **5:**71–75.

Vuorio, R., Hirvas, L., and Vaara, M., 1991, The Ssc protein of enteric bacteria has significant homology to the acyltransferase Lpxa of lipid A biosynthesis, and to three acetyltransferases, *FEBS Lett.* **292:**90–94.

Whitfield, C., Richards, J. C., Perry, M. B., Clarke, B. R., and MacLean, L. L., 1991, Expression of two structurally distinct galactan O-antigens in the lipopolysaccharide of *Klebsiella pneumoniae* serotype 01, *J. Bacteriol.* **173:**1420–1431.

Wilkinson, S. G., 1983, Composition and structure of lipopolysaccharides from *Pseudomonas aeruginosa, Rev. Infect. Dis.* **5:**S941–S949.

Wuenscher, M. D., Kohler, S., Bubert, A., Gerike, U., and Goebel, W., 1993, The *iap* gene of *Listeria monocytogenes* is essential for cell viability, and its gene product, p60, has bacteriolytic activity, *J. Bacteriol.* **175:**3491–3501.

Zubay, G., 1993, *Biochemistry,* 3rd Ed. Wm. C. Brown Publishers, Dubuque, IA.

8

Glycobiology of the Mycobacterial Surface

Structures and Biological Activities of the Cell Envelope Glycoconjugates

Mamadou Daffé and Anne Lemassu

1. INTRODUCTION

The genus *Mycobacterium* is composed of obligate aerobes that grow more slowly than most other bacteria, generation times between a few hours to several days, and includes *Mycobacterium tuberculosis,* the causative agent of tuberculosis. Of all the infectious diseases that have plagued man, tuberculosis probably has been responsible for the greatest morbidity and mortality. Even today, when the incidence of tuberculosis in the Western nations has markedly decreased, more people died from tuberculosis in 1995—about 3.1 million—than in any other year in history, according to the 1996 report of the World Health Organization (Moran, 1996). Almost 2 billion people (one third of the world's population) have been infected by *M. tuberculosis,* of whom 5–10% will develop the active disease. Although most tuberculous patients currently are living primarily in developing countries, tuberculosis is becoming a major health problem in industrialized coun-

Mamadou Daffé • Institut de Pharmacologie et de Biologie Structurale du Centre National de la Recherche Scientifique, 31077 Toulouse Cedex, France. *Anne Lemassu* • Université Paul Sabatier, 31062 Toulouse Cedex, France.

Glycomicrobiology, edited by Doyle.
Kluwer Academic/Plenum Publishers, New York, 2000.

tries, due to the emergence of drug-resistant strains and the coincidence of tuberculosis and human immunodeficiency virus (HIV) infection (Bloom and Murray, 1992). However, it should be borne in mind that the genus *Mycobacterium* comprises more than 50 recognized bacterial species whose properties range from those of saprophytes (organisms that ordinarily do not produce disease) to those of the apparently obligate intracellular parasite *M. leprae* (the causative agent of leprosy), which is still recovered only from tissues of leprosy patients or of experimentally infected animals. In between these two extremes are mycobacterial species that are opportunistic pathogens occurring naturally in the environment but may cause serious disease in humans occasionally, especially in immunosuppressed individuals. The best examples of this group are the members of the so-called *M. avium–intracellulare* complex that commonly infect HIV patients. Finally, animal pathogens also exist in the mycobacterial world: *M. bovis,* bovine tubercle bacillus, and *M. paratuberculosis,* which may cause economically important diseases.

The ability of mycobacterial pathogens to multiply inside phagocytic cells, in spite of the bactericidal properties of such cells, and also the general resistance of mycobacteria to strong chemicals such as acids, alkalis, and hypochlorite have been associated with the uncommon composition of their cell envelopes. The high lipid contents of mycobacterial cells have been recognized for a long time and much effort has been devoted to the identification of the various types of lipids present, many of which are glycolipids, unique to mycobacteria (Brennan, 1988). Most of this lipid is associated with the mycobacterial envelope and is believed to form an outer layer. Indeed, the distinctive property of acid fastness (the failure of dilute acids to decolorize bacteria stained with various basic dyes) may be explained by the presence of this outer lipid permeability barrier, which would obstruct access by hydrophilic substances, and could also explain their limited permeability, their tendency to grow in large clumps, and their rather general insusceptibility to toxic substances. Certainly the envelope contains much lipid, but it is now becoming clear that the mycobacterial envelope has a complex and unusual structure that is not a simple lipid coat (Daffé and Draper, 1998) and that the protection of the organism from its environment is due not only to the mere presence of a large amount of it. The lipid itself is covered by an outermost capsular layer of polysaccharide and protein (Daffé and Etienne, 1999).

Based on the most recent developments in knowledge of the ultrastructure and chemistry of mycobacteria, the mycobacterial cell envelope (Fig. 1) possesses three structural components: plasma membrane, wall, and outer layer or capsule (Daffé and Draper, 1998). The plasma membrane appears to be a typical bacterial membrane that contains a class of phospholipids whose distribution is restricted to mycobacteria and related genera, the phosphatidylinositol mannosides. The complex cell wall that surrounds this membrane partly resembles a

Figure 1. Proposed model for the mycobacterial cell envelope showing its three structural three entities: plasma membrane, cell wall, and outer layer or capsule. The plasma membrane appears to be a typical bacterial membrane. In principle, a compartment analogous to the periplasmic space in gram-negative bacteria could exist in mycobacteria, between the membrane

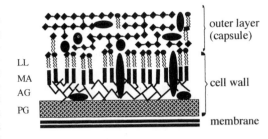

and the peptidoglycan, but this has not been directly demonstrated. The complex cell wall that surrounds this membrane is formed by a (1) peptidoglycan (PG) composed of oligosaccharides formed from disaccharide units of N-acetylglucosamine and N-glycolylmuramic acid cross-linked by short peptides; (2) D-arabino-D-galactan (AG) composed of β-furanosyl units and esterified by α-branched, β-hydroxylated long-chain (up to 90 carbon atoms) fatty acids and mycolic acids (MA); and (3) a great variety of other lipids (LL) associated with the mycoloyl-arabinogalactan but not covalently attached to it that are probably arranged to form with the cell wall mycolate monolayer an asymmetric bilayer that represents a permeability barrier to polar molecules. The outer layer consists primarily of a protein–carbohydrate matrix loosely bound to the cell wall. Ovals and lozenges represent proteins and polysaccharides, respectively.

gram-positive wall, but is unusual in having a layer of D-arabino-D-galactan, composed of β-furanosyl units, esterified by α-branched, β-hydroxylated long-chain (up to 90 carbon atoms) fatty acids, mycolic acids. This huge glycoconjugate is covalently linked to peptidoglycan, composed of oligosaccharides formed from disaccharide units of N-acetylglucosamine and N-glycolylmuramic acid cross-linked by short peptides, to form the "cell wall skeleton" (Daffé and Draper, 1998). A great variety of other lipids, mainly glycolipids, which are probably arranged to form an asymmetric bilayer with the cell wall mycolate monolayer, but not covalently attached to it (Minnikin, 1982), presenting a permeability barrier to polar molecules (Brennan and Nikaido, 1995). In principle, a compartment analogous to the periplasmic space in gram-negative bacteria could exist in mycobacteria between the membrane and the peptidoglycan, but this has not been directly demonstrated (Daffé and Draper, 1998). The outer layer consists primarily of a protein–carbohydrate matrix loosely bound to the cell wall. It follows then that glycoconjugates are both structurally and functionally important compounds of the mycobacterial cell envelope. In addition, several mycobacterial pathogens produce species-specific glycolipids of unusual structures that have been shown to modulate the host immune system by interacting with cell membranes (Vergne and Daffé, 1998). This chapter will summarize the structural work conducted on the three classes of mycobacterial glycoconjugates and the relevant biological functions of the molecules.

2. GLYCOLIPIDS

Mycobacteria elaborate several types of glycolipids that may be divided into two major groups: ubiquitous glycolipids, such as phosphoglycolipids found in all the species examined so far, and species- or type-specific glycolipids, such as glycopeptidolipids and phenolic glycolipids produced by a restricted number of species or strains.

2.1. Ubiquitous Glycolipids

2.1.1. TREHALOSE MYCOLATES

No other group of substances isolated from mycobacteria has stimulated so much work as mycolic acids and mycolate-containing glycolipids. Early morphological studies revealed that virulent tubercle bacilli grew in the form of "sepentine cords" (Middlebrook *et al.*, 1947), whereas avirulent and attenuated tubercle bacilli as well as saprophytes did not (Bloch, 1950). This growth pattern arose from the division of bacilli inside a lipid matrix, because disruption of the cords by washing cultures of virulent tubercle bacilli with petroleum ether removed a presumably peripherally located lipid that might be responsible for cord formation and related to virulence. Interestingly, the material, coated onto dried *Bacillus subtilis* or avirulent tubercle bacilli and ingested by leukocytes, inhibited their migration. On the assumption that the substance causing the migration inhibition was also responsible for the formation of cords, he named it "cord factor." The chemical structure of cord factor was established by Noll *et al.* (1956) through conventional chemical analyses, who demonstrated that cord factor corresponds to a family of 6,6'-dimycolate of trehalose (Fig. 2), differing one from the other by the chemical groups present in the mycolic acyl substituents. Briefly, alkaline saponification of purified cord factor demonstrated that it contains two moles of mycolic acids, a mixture of structurally related molecules, and one molecule of trehalose. Per-*O*-methylation of the glycolipid in conditions that block free hydroxyl groups, followed by alkaline saponification and acid hydrolysis, yielded only 2,3,4-tri-*O*-methyl glucose, revealing that in the intact cord factor trehalose is symmetrically substituted with mycolic acid residues in 6,6' positions. The separation of the different classes of cord factor on the basis of the types of mycolates they contain may be achieved by thin-layer chromatography (TLC) of their ether derivatives (Strain *et al.*, 1977; Promé *et al.*, 1976). Indeed, cord factor and structurally related analogues have been characterized from noncording saprophytes and nontuberculous mycobacteria examined so far and from the related *Nocardia* and *Corynebacterium* genera (Asselineau and Asselineau, 1978). A notable exception

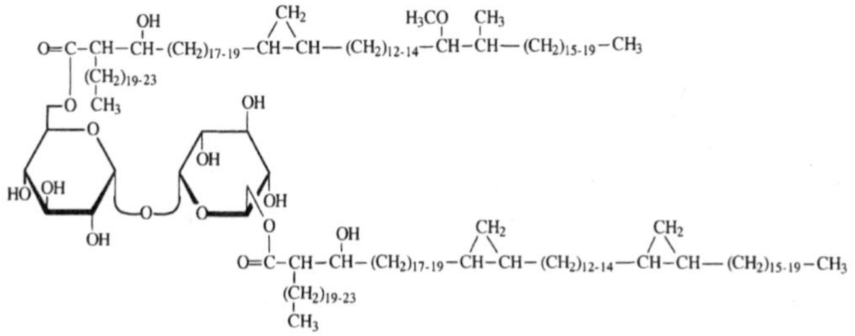

Figure 2. Structure of a 6,6′-dimycolate of trehalose (cord factor); members of this family differ from one another by the chemical groups present in the mycolic acyl substituents.

is *M. leprae* in which no cord factor was detected; rather, in the noncultivable leprosy bacillus, a related compound, 6-monomycolate trehalose, another mycolate-containing glycolipid widely distributed in the genus *Mycobacterium,* was the only trehalose ester characterized (Dhariwal *et al.,* 1987). Thus, the substance responsible for cord formation remains to be discovered. Consistent with this fact is the recent study that demonstrated that in most mycobacterial species, including *M. tuberculosis,* cord factor was not present on the bacterial surface but in deeper peripheral layers of the cell envelope (Ortalo-Magné *et al.,* 1996b). Nevertheless, the early hypothesis that cord factor might be related to virulence has stimulated the investigation of the biological effects of the compound and has yielded a rich harvest in terms of relationships of structure to biological functions. Cord factor has been shown to be highly toxic for mice by altering mitochondrial phosphorylation and respiration. The inhibition is site II-specific and is widely sensitive to the configuration of both sugar units and the nature of the acyl substituents and to the organization state of suspensions. The molecule also has been shown to exhibit granulomagenic and antitumor activities, to stimulate macrophages to secrete cachectin, to inhibit fusion between phospholipid vesicles, and to induce apoptosis (see Vergne and Daffé, 1998; Goren, 1990; Asselineau and Asselineau, 1978).

2.1.2. PHOSPHATIDYLINOSITOLMANNOSIDES

Phosphatidylinositolmannosides (PIM) and phosphatidylethanolamine are the major phospholipid constituents of the mycobacteria cell envelope, except in *M. vaccae* (Brennan, 1988). The structural features of the phosphoglycolipids were established in 1930, by Anderson, who isolated phosphatide fractions from *M. tuberculosis, M. bovis,* and *M. avium* (formerly human, bovine, and avian tubercle bacilli, respectively), which on complete hydrolysis yielded inositol and an aldose, identified as mannose. By saponification of the phosphatide, Anderson and Roberts obtained fatty acids, an organophosphoric acid, and a phosphorylated glycan, manninositose, which upon acid hydrolysis yielded roughly 2 moles of mannose and 1 mole of inositol (see Asselineau, 1966; Anderson, 1939). Later, Lederer's and Ballou's groups established the structure of PIM_2; they showed that the glycerol phosphate group was attached to the L-1 position of the *myo*-inositol (Ballou *et al.*, 1963). They also established by methylation analysis the glycosyl linkage composition (Vilkas, 1960) and showed by nuclear magnetic resonance (NMR) analysis that the mannosyl units were glycosidically linked to positions 2 and 6 of the *myo*-inositol (Lee and Ballou, 1964). Lee and Ballou (1965) also determined the structure of PIM_5 in which a tetramannoside of sequence α-D-mannopyranosyl-(1 → 2)-α-D-mannopyranosyl-(1 → 6)-α-D-mannopyranosyl-(1 → 6)-α-manno-pyranosyl-(1 → was shown to be attached to position 6 of the *myo*-inositol ring (Fig. 3). PIM_3 and PIM_4 differ from the latter in lacking, respectively, one and two terminal mannosyl residues in the tetramannosyl chain. PIM_5 may be further substituted at position 2 by an α-D-mannosyl, leading to PIM_6 (Goren and Brennan, 1979). Pangborn and McKinney (1966) isolated from *M. tuberculosis* a series of PIM_2 containing a total of two, three, and four acyl residues and described a presumed PIM_5, which potentially contained three and four fatty acyl residues. Conventional PIM already was known to contain two acyl groups substituting the glycerol moiety. Using fast atom bombardment–mass spectrometry analyses of PIM derivatives, Khoo *et al.* (1995b) established the exact fatty acyl compositions of

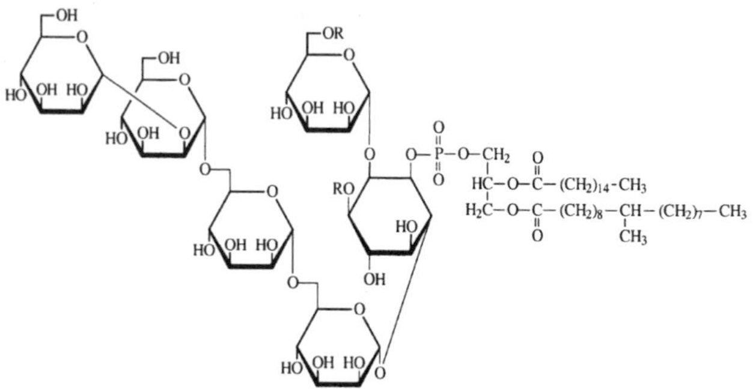

Figure 3. Structure of the phosphatidylinositolpentamannoside (PIM_5); mannosyl units were glycosidically linked to positions 2 and 6 of the *myo*-inositol to yield the common core of all PIM (PIM_2) and PIM-containing glycoconjugates (lipomannan and lipoarabinomannan). In PIM_5, a tetramannoside of sequence α-D-mannopyranosyl-(1 → 2)-α-D-mannopyranosyl-(1 → 6)-α-D-mannopyranosyl-(1 → 6)-α-D-mannopyranosol-(1→ is attached to position 6 of the *myo*-inositol ring. Di-*O*-acylated PIM_2 and relatives contain either two C_{16} or one C_{16} and one C_{19} (presumably 10-methyl-octadecanoyl, the so-called tuberculostearoyl) substituents. Tri- and tetra-*O*-acylated PIM and related compounds also exist and consist of various combinations of C_{16}, C_{18}, and C_{19} molecules. In tri-*O*-acylated PIM and relatives, the extra acyl group is a C_{16} fatty acyl substituent and is located on position 6 of the mannosyl unit linked to C-2 of *myo*-inositol; the fourth fatty acyl substituent in the tetra-*O*-acylated PIM and relatives is situated on position 3 of the *myo*-inositol ring.

the multiacylated heterogeneous PIM families of *M. tuberculosis* and *M. leprae*. The di-*O*-acylated PIM_2 and PIM_6 contain either two C_{16} or one C_{16} and one C_{19} (presumably 10-methyl-octadecanoyl, the so-called tuberculostearoyl) substituents. The fatty acid substituents of the tri- and tetra-*O*-acylated PIM consisted of various combinations of C_{16}, C_{18}, and C_{19} molecules. In the tri-*O*-acylated PIM the "extra" acyl group is located on position 6 of the mannosyl unit linked to C-2 of *myo*-inositol (Khoo *et al.*, 1995b). Recent NMR data (Gilleron *et al.*, 1999; Nigou *et al.*, 1999) showed that the same location of fatty acyl residues occurs in the structurally related lipoarabinomannan (LAM) and lipomannan (LM) and demonstrated that the fourth fatty acyl substituent in the tetra-*O*-acylated PIM, LM, and LAM is situated on position 3 of the *myo*-inositol ring. Thus, PIMs occur generally in mycobacteria as a mixture of compounds differing one from another by the numbers of mannosyl and fatty acyl residues.

A recent study showed that PIM, which were known for a long time to be present in the plasma membrane, are also found on the bacterial surface of both pathogenic and nonpathogenic mycobacteria (Ortalo-Magné *et al.*, 1996b). This would explain why polar PIM (PIM_5 or PIM_6) mediate the binding of *M. tuberculosis* (strain H37Rv) and *M. smegmatis* to Chinese hamster ovary (CHO) fibroblasts and porcine aortic endothelial cells, either directly or after opsonisation with serum proteins, such as the mannose-binding protein (Hoppe *et al.*, 1997). Although the mode of association of PIM with these nonphagocytic mammalian cells is not yet determined, the cell specificity and the inhibiting effect of periodate treatment strongly suggest the participation of a lectinlike receptor in these interactions (Hoppe *et al.*, 1997). Nonopsonic binding experiments using murine macrophages also have shown that PIM, as well as the mycobacterial lipopolysaccharide, LAM, inhibit the binding of *M. tuberculosis* and that the phosphatidylinositol moiety was important in the abrogation of the binding (Stokes and Speert, 1995). This work also suggested that a competitive inhibition of a receptor–ligand interaction was probably not the cause of the observed phenomenon. In that connection, Barratt *et al.* (1986) have previously shown that PIM-containing liposomes inhibit the uptake of mannosylated bovine serum albumin by mouse inflammatory macrophages, suggesting an interaction with the mannose receptor. PIM stimulate B cells to synthesize antibodies, since the molecules readily react with tuberculous sera (Reggiardo and Middlebrook, 1974). PIM have been reported to protect guinea pigs and mice against infection by *M. tuberculosis* (Mehta and Khuller, 1988; Khuller *et al.*, 1983) and to induce the release of tumor necrosis factor-α (TNF-α), interleukin (IL)-6, -8, and -10 (Zhang *et al.*, 1995; Barnes *et al.*, 1992), and nitrite oxide synthase activity in mouse peritoneal macrophages primed by interferon (INF)-γ or cord factor (Tenu *et al.*, 1995). Finally, PIMs are capable of suppressing antigen-induced T-cell lymphoproliferation (Zhang *et al.*, 1995). The interference of PIM with cell signaling mechanisms may be explained by the anchorage of the mycobacterial phospholipids in lymphomonocytic cell plasma membranes *via* the glycosylphosphatidylinositol (GPI) moiety, and especially into glycosylinositol-rich domains, as do the structurally related LAM (Ilangumaran *et al.*, 1995), resulting in the modulation of the signal transduction used by GPI-linked proteins of host cells.

2.2. Species-Specific Glycolipids

In addition to ubiquitous lipids, mycobacteria elaborate species-specific lipids whose existence was revealed by Smith and colleagues (1960a,b) in their efforts to chemically characterize chromatographically fractionated ethanol–diethyl ether extracts of *M. tuberculosis, M. bovis,* and *M. avium* (formerly the human, bovine, and avium tubercle bacilli) and other mycobacterial species by infrared spectroscopy. They used the term "mycosides" to typify species-specific glycolipids of mycobacterial origin (Smith *et al.*, 1960a,b) and found several substances fulfilling this requirement. For instance, the terms "mycosides" A, B, and G were used to define characteristic substances from *M. kansasii, M. bovis,* and *M. marinum,* respectively. Although the subsequent detailed analyses of the purified compounds demonstrated that the distribution of these substances is not restricted to a single species (mycosides A, B, and G are closely related phenolic glycolipids), they also reinforced the original notion that each mycobacterial species is endowed with characteristic lipids, mainly glycolipids, which can be used to distinguish it from all the other species. The species-specific mycobacterial glycolipids described to date may be grouped into families, based on the nature of their constituents.

2.2.1. PHENOLIC GLYCOLIPIDS

Phenolic glycolipids (PGLs) share a common backbone composed of phenol-phthiocerol and relatives (long-chain C_{33}-C_{41} β-diols) whose two aliphatic hydroxyl groups are esterified by two multimethyl-branched C_{27}-C_{34} acids. The structure of this backbone has been established largely by Gastambide-Odier and colleagues (Gastambide-Odier and Sarda, 1970; Gastambide-Odier *et al.*, 1965, 1967). PGLs are found in several obligate and opportunistic mycobacterial pathogens. These include members of the *M. tuberculosis* complex (*M. tuberculosis, M. bovis, M. microti*), *M. leprae, M. kansasii, M. marinum* (Puzo, 1990; Brennan, 1988; Daffé and Lanéelle, 1988), *M. haemophilum* (Besra *et al.*, 1991), and *M. ulcerans* (Daffé *et al.*, 1992). The nonpathogenic species *M. gastri,* a species closely related to *M. kansasii,* also contains PGL (Daffé and Lanéelle, 1988). It has to be noted that not all the strains of the same species elaborate PGL; this is particularly true for *M. tuberculosis* (Watanabe *et al.*, 1994; Daffé *et al.*,

Table I
Structures of the Oligosaccharidyl Residues of the Major Phenolic Glycolipids
from Mycobacteria

Species	Structure of the oligosaccharide moietie branched on the phenol-dimycocerosyl phthiocerol
M. tuberculosis	2,3,4-tri-O-Me-α-L-Fucp-(1→3)-α-L-Rhap-(1→3)-2-O-Me-α-L-Rhap-(1→
M. bovis	2-O-Me-α-L-Rhap-(1→
M. leprae	3,6-di-O-Me-β-D-Glcp-(1→4)-2,3-di-O-Me-α-L-Rhap-(1→2)-3-O-Me-α-L-Rhap-(1→
M. marinum	3-O-Me-α-L-Rhap-(1→
M. ulcerans	3-O-Me-α-L-Rhap-(1→
M. kansasii	2,6-dideoxy-4-O-Me-α-D-arabino-hexp-(1→3)-4-O-Ac-2-O-Me-α-L-Fucp-(1→3)-2-O-Me-α-L-Rhap-(1→3)-2,4-di-O-Me-α-L-Rhap-(1→
M. gastri	2,6-dideoxy-4-O-Me-α-D-arabino-hexp-(1→3)-4-O-Ac-2-O-Me-α-L-Fucp-(1→3)-2-O-Me-α-L-Rhap-(1→3)-2,4-di-O-Me-α-L-Rhap-(1→
M. haemophilum	2,3-di-O-Me-α-L-Rhap-(1→2)-3-O-Me-α-L-Rhap-(1→4)-2,3-di-O-Me-α-L-Rhap-(1→

1987, 1991a), but similar observations have been made for *M. marinum* and *M. bovis* (Daffé and Lanéelle, 1988). According to the species (Daffé and Lanéelle, 1988), the chiral centers bearing the methyl branch in the fatty acids are called mycocerosic and phthioceranic acids, respectively, but the major difference between the various PGLs identified so far is in the nature of the oligosaccharidyl units glycosically linked to the aromatic hydroxyl–phenol group (Table I). Generally, a given mycobacterial strain contains a major type of PGL, but also several minor variants of PGL differing from the major type by the nature of the oligosaccharide. For instance, the strain Canetti of *M. tuberculosis* elaborates, in addition to the major triglycosyl phenol–phthiocerol (Fig. 4), a minor triglycosyl variant (Daffé, 1989) and two minor monoglycosyl types that include a PGL structurally identical to the major PGL of *M. bovis* (Daffé *et al.,* 1988b). The structures of the oligosaccharides, consisting of one to four *O*-methylated sugars, principally desoxysugars, were first established through conventional analyses following complete and partial acid hydrolyses, Smith degradation and identification of the resulting products by paper or gas chromatography, and later by mass spectrometry and NMR spectroscopy.

PGLs are surface located on the mycobacterial species that contain these glycolipids (Ortalo-Magné *et al.,* 1996b; Gilleron *et al.,* 1990). In agreement with their surface exposure, PGL of *M. leprae* has been reported to activate the human complement system through both the classical and the alternative pathways (Ramanathan *et al.,* 1990) and to mediate the phagocytosis of the leprosy bacillus by macrophages (Schlesinger and Horwitz, 1991). Phenolic glycolipids constitute up to 2% of the mass of the leprosy bacillus and represent a major component of the

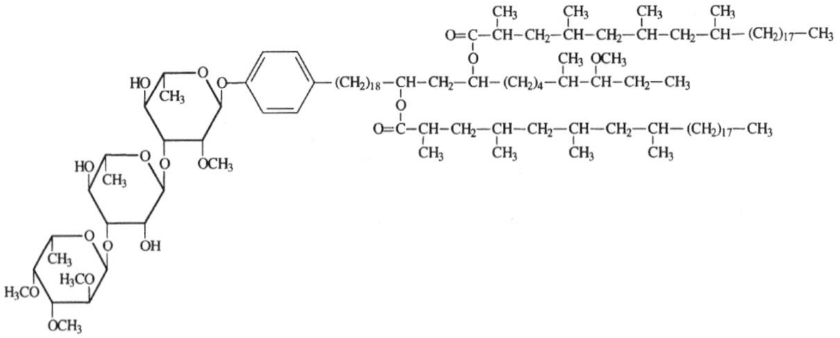

Figure 4. Structure of the major phenolic glycolipid from the Canetti strain of *Mycobacterium tuberculosis*. The glycolipid is composed of phenol–phthiocerol (long-chain C_{37} β-diols) whose two aliphatic hydroxyl groups are esterified by two multimethyl-branched C_{30} mycocerosic acids. The trisaccharidyl unit of *M. tuberculosis* consists of 2,3,4-tri-*O*-methyl-α-L-fucopyranosyl-(1 → 3)-α-L-rhamnopyranosyl-(1 → 3)-2-*O*-methyl-α-L-rhamnopyranosyl-(1→.

ultrastructurally observed electron-transparent zone surrounding the leprosy bacillus (Hunter and Brennan, 1981), and thus may protect this bacterium from host attack. However, it is clear that accumulation of PGL is not a general explanation for this zone in pathogenic species, because some strains of *M. tuberculosis* do not synthesize significant amounts of PGL (Cho *et al.*, 1992; Daffé *et al.*, 1987, 1991a), while other species, such as *M. kansasii* and *M. bovis,* do produce PGL but only in small amounts compared to *M. leprae* (Daffé and Draper, 1998). Similarly, although the PGL of *M. leprae* may represent a virulence factor through scavenging oxygen radicals (Chan *et al.*, 1989; Vachula *et al.*, 1989; Neill and Klebanoff, 1988), this activity seems specific to this glycolipid because PGL from other mycobacterial sources lack such activity (Launois *et al.*, 1989; Vachula *et al.*, 1989). In addition, PGL from *M. leprae* but not those from *M. bovis* and *M. kansasii* inhibit the concanavalin A (ConA) stimulation of lymphocytes from patients with a lepromatous leprosy, the most severe form of the disease. Finally, PGLs are serologically active and have been used for the diagnosis of both leprosy (see Gaylord and Brennan, 1987) and tuberculosis.

2.2.2. GLYCOPEPTIDOLIPIDS

In the course of characterizing species-specific lipids in mycobacteria, Smith *et al.* (1960a,b) described infrared characteristics of lipid fractions (called "J substances") common to some nontuberculous mycobacterial species. The J substances (Smith *et al.*, 1960b), renamed "mycoside C" (Smith *et al.*, 1960a), showed absorption bands attributable to peptide bonds and ester linkages. Hydrolysis of mycoside C released an unknown fatty acid, three amino acids, alanine, threonine and phenylalanine, and monosaccharides, the migrations of which were similar to those of *O*-methylated 6-deoxyhexoses. Subsequent structural investigations by several groups established the structures of mycoside C, which are now merely called glycopeptidolipids (see Brennan, 1988); however, in view of the recent description of a novel family of alkali-labile serine-containing glycopeptidolipids in *M. xenopi* (Besra *et al.*, 1993b; Rivière and Puzo, 1992), whose structures differ greatly from those of the alkali-stable mycoside C, we will name the former family mycoside C-type glycopeptidolipids (C-type GPL). These molecules have been characterized so far from saprophytic (*M. smegmatis*), opportunistic pathogens for human (*M. avium, M. intracellulare, M. scrofulaceum, M. peregrinum, M. chelonae, M. abscessus*), and for animals (*M. lepraemurium, M. paratuberculosis, M. porcinum, M. senegalense*), but remarkably never from species producing PGL. C-type GPL share a same lipopeptidyl core consisting of a mixture of 3-hydroxy and 3-methoxy long chain (C_{26}-C_{34}) (Daffé *et al.*, 1983) amidated by a tripeptide D-Phe-D-*allo*-Thr-D-Ala and terminated by L-alaninol (from alanine). They differ from one another by the number and the nature of the saccharidyl units linked to the hydroxyl group of *allo*-threonine and/or alaninol and occur generally as a mixture of compounds. In the most abundant molecular species, the apolar C-type GPL, the hydroxyl groups are substituted by a mono- or a disaccharidyl unit composed of deoxysugar residue(s) that are usually *O*-methylated and/or *O*-acylated. In members of the *M. avium–intracellulare* complex an oligosaccharidyl of unusual composition is linked to the 6-deoxytalosyl unit, which in turn substitutes the *allo*-threonine residue, leading to polar C-type GPL.

In a series of elegant structural studies combining classical carbohydrate chemistry, NMR spectroscopy, and mass spectrometry, Brennan and colleagues (1988) established the molecular bases of both the variability and the specificity of the antigenically distinct serovariants within the *M. avium–intracellulare* complex and several other C-type GPL-containing mycobacteria. The Schaefer typing antigens used in seroagglutination assays for purposes of identification and classification of several types of nontuberculous mycobacteria were thus resolved (Brennan, 1988). They convincingly demonstrated that the glycosyl units of polar C-type GPL contain a nether region unique to the particular serovariant that was responsible for antigenicity and chromatographic distinctiveness. The oligosaccharide haptens from several polar C-type GPL contain "exotic" sugars as exemplified in Fig. 5: glucuronic acid and variants, acetamido-dideoxy-hexosyl residues and other branched sugars (Table II). In addition, several species other than members of the *M. avium–intracellulare* complex and related species elaborate C-type GPL in which a monoglycosyl residue or a sulfate group substitutes the *O*-methylated rhamnosyl unit linked to alaninol, whereas a monosaccharidyl residue, usually an *O*-methylated rhamnosyl unit that may be acylated, replaces the 6-deoxytalosyl-containing oligosaccharide (López-Marín *et al.*, 1991, 1992, 1993, 1994a) (Table II).

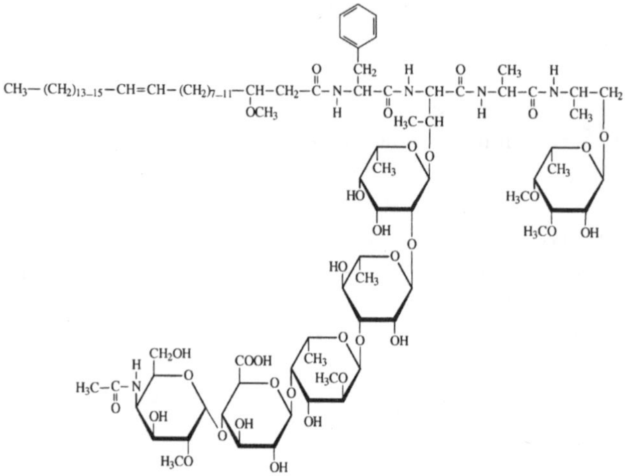

Figure 5. Structure of a mycoside C-type glycopeptidolipid (C-type GPL) from *Mycobacterium avium* (serovariant 25). C-type GPL possesses a lipopeptidyl core consisting of a mixture of 3-hydroxy and 3-methoxy long chain (C_{26}-C_{34}) amidated by a tripeptide (D-Phe-D-*allo*-Thr-D-Ala) and terminated by L-alaninol from alanine. The structure of the oligosaccharidyl unit is generally species- or type-specific and contains "exotic" sugars such as *O*-methylated deoxyhexosyl, glucuronic acid, and acetamido-dideoxy-hexosyl residues. That of *M. avium* (serovariant 25) consists of 4-acetamido-4,6-dideoxy-2-*O*-methyl-α-L-Galp-(1 → 4)-β-glucuronyl-(1 → 4)-2-*O*-methyl-α-L-fucopyranosyl-(1 → 3)-α-L-rhammnopyranosyl-(1 → 2)-6-deoxy-L-talosyl-(1→.

Localization studies demonstrated that C-type GPL are exposed on the bacterial surface (Ortalo-Magné *et al.,* 1996b), in agreement with these compounds being the Schaefer typing antigens (Brennan, 1988) and their identification as the receptor of mycobacteriophage D4 (Furuchi and Tokunaga, 1972; Goren *et al.,* 1972). The association with possession of C-type GPL and the smooth morphology of mycobacteria (Belisle *et al.,* 1993; Barrow and Brennan, 1982) also is consistent with the surface-exposure. In addition, as stated above, polar C-type GPL correspond to Schaefer typing antigens, which implies that these compounds occur on the peripheral bacterial surface, since antisera to them react strongly with intact mycobacterial cells. *In vivo,* C-type GPL, which are composed of amino acids of the unusual D-series, are poorly degraded after phagocytosis of *M. avium* by macrophages (Hooper and Barrow, 1988) and accumulate into the phagosome during bacterial growth, contributing to the formation of a capsule surrounding the bacteria (Rulong *et al.,* 1991; Tereletsky and Barrow, 1983; Draper, 1974). However, strains of *M. avium–intracellulare,* devoid of C-type GPL, also are sur-

Table II

Table II

Structures of the Oligosaccharide Moieties of Mycobacterial C-type Glycopeptidolipids

⟩ecies		Oligosaccharidyl unit linked to the *allo* Thr residue	Saccharidyl unit linked to the AlaOH residue
, avium	sv 1[a]	α-L-Rha*p*-(1→2)-6-deoxy-α-L-Tal	3,4-di-*O*-Me-α-L-Rha*p*
	sv 2	2,3-di-*O*-Me-α-L-Fuc*p*-(1→3)-α-L-Rha*p*-(1→2)-6-deoxy-α-L-Tal	3,4-di-*O*-Me-α-L-Rha*p*
	sv 4	4-*O*-Me-α-L-Rha*p*-(1→4)-2-*O*-Me-α-L-Fuc*p*-(1→3)- α-L-Rha*p*-(1→2)-6-deoxy-α-L-Tal	
	sv 8	4,6-(1′-carboxyethylidene)-3-*O*-Me-β-D-Glc*p*-(1→3)-α-L-Rha*p*-(1→2)-6-deoxy-α-L-Tal	3,4-di-*O*-Me-α-L-Rha*p*
	sv 9	4-*O*-Ac-2,3-di-*O*-Me-α-L-Fuc*p*-(1→4)-β-D-Glc*p*A-(1→4)-2,3-di-*O*-Me-α-Fuc*p*-(1→3)-α-L-Rha*p*-(1→2)-6-deoxy-α-L-Tal	3,4-di-*O*-Me-α-L-Rha*p*
	sv 12	4-(2′-hydroxy)propionamido-3-*O*-Me-4,6-dideoxy-β-Glc-(1→3)-4-*O*-Me-α-L-Rha*p*-(1→3)-α-L-Rha*p*-(1→3)-α-L-Rha*p*-(1→2)-6-deoxy-α-L-Tal	3,4-di-*O*-Me-α-L-Rha*p*
	sv 14	4-formamido-4,6,-dideoxy-2-*O*-Me-3-C-Me-α-L-Man*p*-(1→3)-2-*O*-Me-α-D-Rha*p*-(1→3)-2-*O*-Me-α-L-Fuc*p*-(1→3)-α-L-Rha-(1→2)-6-deoxy-α-L-Tal	3,4-di-*O*-Me-α-L-Rha*p*
	sv 17	4-(2′-Methyl-3′-hydroxybutyramido)-4-6-dideoxy-Glc*p*-(1→3)-4-*O*-Me-α-L-Rha*p*-(1→3)-α-L-Rha*p*-(1→3)-α-L-Rha*p*-(1→2)-6-deoxy-α-L-Tal	3,4-di-*O*-Me-α-L-Rha*p*
	sv 19	3,4-di-*O*-Me-β-D-Glc*p*A-(1→3)-2,4-di-*O*-Me-3-C-Me-6-deoxy-α-Hex-(1→3)-α-L-Rha*p*-(1→3)-α-L-Rha*p*-(1→2)-6-deoxy-α-L-Tal	3,4-di-*O*-Me-α-L-Rha*p*
	sv 20	2-*O*-Me-α-D-Rha*p*-(1→3)-2-*O*-Me-α-Fuc*p*-(1→3)-α-L-Rha*p*-(1→2)-6-deoxy-α-L-Tal	3,4-di-*O*-Me-α-L-Rha*p*
	sv 21	4,6-(1′-carboxyethylidene)-β-D-Glc*p*-(1→3)-α-L-Rha*p*-(1→2)-6-deoxy-α-L-Tal	3,4-di-*O*-Me-α-L-Rha*p*
	sv 25	4-acetamido-4,6-dideoxy-2-*O*-Me-α-D-Gal*p*-(1→4)-β-D-Glc*p*A-(1→4)-2-*O*-Me-α-L-Fuc*p*-(1→3)-α-L-Rha*p*-(1→2)-6-deoxy-α-L-Tal	3,4-di-*O*-Me-α-L-Rha*p*
	sv 26	2,4-di-*O*-Me-α-L-Fuc*p*-(1→4)-β-D-GlcA-(1→4)-2-*O*-Me-α-L-Fuc*p*-(1→3)-α-L-Rha*p*-(1→2)-6-deoxy-α-L-Tal	3,4-di-*O*-Me-α-L-Rha*p*
⟨. paratuberculosis		2,3-di-*O*-Me-α-L-Fuc*p*-(1→3)-L-Rha-(1→2)-6-deoxy-α-L-Tal	3,4-di-*O*-Me-α-L-Rha*p*
⟨. peregrinum		3-*O*-Me-α-L-Rha*p*	3,4-di-*O*-Me-α-L-Rha*p*-(1→2)-3,4-di-*O*-Me-α-L-Rha*p*
⟨. senegalense		2-*O*-Ac-3-*O*-Me-α-L-Rha*p*	3,4-di-*O*-Me-α-L-Rha*p*-(1→2)-3,4-di-*O*-Me-α-L-Rha*p*
⟨. porcinum		3-*O*-Me-α-L-Rha*p*	3,4-di-*O*-Me-α-L-Rha*p*-(1→2)-3,4-di-*O*-Me-α-L-Rha*p*
⟨. abscessus		3,4-di-*O*-Ac-6-deoxy-α-L-Tal	α-L-Rha*p*-(1→2)-3,4-di-*O*-Me-α-L-Rha*p*
⟨. chelonae		3,4-di-*O*-Ac-6-deoxy-α-L-Tal	α-L-Rha*p*-(1→2)-3,4-di-*O*-Me-α-L-Rha*p*

⟩v, Serovariant.

rounded by a capsule (Rastogi *et al.,* 1989), indicating that these glycolipids represent only part of the capsular constituents.

Mice with severe *M. lepraemurium* infection (Brown and Draper, 1976), as well as tissues of immunosuppressed people with *M. avium* infections (Klatt *et al.,* 1987), contain heavily infected but apparently undamaged macrophages, suggesting that C-type GPL elaborated by these species are not toxic. Evidence that purified glycolipids can affect cells, however, has been published. An intraperitoneal injection of C-type GPL from *M. avium–intracellulare* serovars 4 and 20 causes the inhibition of the blastogenic response of murine splenic lymphocytes to nonspecific mitogens [Con A, phytohemagglutinin, and lipopolysaccharide, (LPS)] (Brownback and Barrow, 1988; Hooper *et al.,* 1986). As the same effects were not observed *in vitro,* these data suggested a production of active GPL metabolites *in vivo* (Brownback and Barrow, 1988). Because the responsiveness of lymphocytes was down-regulated to a greater extent by the lipid moiety obtained by β-elimination of C-type GPL, the putative metabolites may be structurally related to the lipid moiety (Pourshafie *et al.,* 1993). The lymphoproliferative response of human peripheral blood mononuclear cells to the stimulation by phytohemagglutinin was also affected in the same way by lipid moiety (Barrow *et al.,* 1993). This suppression of the lymphoproliferation seems to be mediated by soluble factors, different from prostaglandin E2 released by macrophages treated by the lipid part of the C-type GPL (Pourshafie *et al.,* 1993). The lipid moiety obtained by chemical means from C-type GPL also affects the ability of human peripheral blood macrophages to control the growth of *M. avium* serovar 2. The lipid also can induce the release of high levels of prostaglandin E2 by the cells and some changes in the membrane ultrastructure (Pourshafie *et al.,* 1993), resulting in the alteration of macrophage functions (Barrow *et al.,* 1993). It is noteworthy that polar GPL from serovar 8 but not those from serovars 4 and 20 exhibit similar properties, pointing out the probable role of the glycosyl composition of the sugar moiety in the biological activity (Barrow *et al.,* 1995). It also has been shown that the surface-exposed diglycosylpeptidolipids, apolar C-type GPL, from *M. smegmatis* (Daffé *et al.,* 1983) cause a decrease of the phosphorylation efficiency (ADP–O ratio) of mitochondria isolated from rat liver without modifying the active respiration (Sut *et al.,* 1990). In addition, these molecules were shown to increase the permeability of liposomes to carboxyfluorescein, suggesting that GPL could act on mitochondria by enhancing the passive permeability of the inner membrane to protons (López-Marín *et al.,* 1994b). Among the C-type GPL tested, the monoglycosylated lipopeptides, resulting from the β-elimination of the apolar diglycosylpeptidolipids, were more active than the native apolar diglycosylpeptidolipids, which in turn were more active than the relatively more polar triglycosylated. The aglycosylated substances (GPLO) were poorly active (López-Marín *et al.,* 1994b).

2.2.3. LIPO-OLIGOSACCHARIDES

Historically, lipo-oligosaccharides (LOS) are the newest species- or type-species mycobacterial glycolipids (Brennan, 1988). The recognition of LOS was based on the realization that the mycobacterial glycolipids may differ greatly in terms of stability to chemicals as exemplified by the *M. fortuitum* complex. Members of this complex, like most mycobacterial species, elaborate species-specific glycolipids and may be divided into two groups according to their lipid patterns on TLC plates, the stability of these patterns and of the seroreactivity of the lipid extracts on treatment with alkalis (Tsang *et al.*, 1984). Structural studies demonstrated that the first group contained alkali-stable C-type GPL-containing glycolipids (Lanéelle *et al.*, 1996; López-Marín *et al.*, 1991, 1992, 1993, 1994a; Tsang *et al.*, 1984), which conserved most of their seroactivity following treatment with alkali, a treatment that would eliminate the *O*-acetyl groups commonly occurring on these glycolipids. Loss of *O*-acetyl would induce a slight modification of their migration on TLC (López-Marín *et al.*, 1994a). In contrast, both the seroreactivity and the observation that glycolipid spots on TLC are lost on deacylation of the lipid extracts of the second group (Lanéelle *et al.*, 1996; Tsang *et al.*, 1984), point to the occurrence of a class of alkali-labile species-specific glycolipids, the LOS in mycobacteria. These glycolipids have been structurally characterized in several mycobacteria, including PGL-containing species (e.g., *M. kansasii, M. gastri, M. tuberculosis*) and the C-type GPL-containing *M. smegmatis* (Table III). LOS consist of a common core of poly-*O*-acylated trehalose, which may be *O*-methylated (Besra *et al.*, 1993a; Daffé *et al.*, 1991b), further glycosylated by a mono- or more frequently an oligosaccharidyl unit (Table III). The structures of these alkali-labile glycolipids have been fully defined by acetolysis, partial acid hydrolysis, NMR spectroscopy, and mass spectrometry and in many cases have proved to be specific for the mycobacterial species due to the presence of unusual sugars such as 4,6-dideoxy-2-*O*-methyl-3-*C*-methyl-4-(2'-methoxypropionamido)-α-L-*manno*-hexopyranose. In the same mycobacterial species they generally occur as a mixture of compounds, differing one from another by both the composition of the oligosaccharide moiety and the number of fatty acyl substituents, straight chain and methyl-branched acyl substituents (see Brennan, 1988). The different species- and type-specific LOS differ mainly by their glycosyl composition (Table III). An example of mycobacterial LOS (Gilleron and Puzo, 1995) is illustrated in Fig. 6.

LOS are exposed on the surface of the mycobacterial species that synthesize them (Ortalo-Magné *et al.*, 1996b; Belisle and Brennan, 1989) and their presence has been associated with the smooth morphology of *M. kansasii* (Belisle and Brennan, 1989) and *M. mucogenicum* (Muñoz *et al.*, 1998). This correlation was not established, however, for strains of *M. tuberculosis* (Lemassu *et al.*, 1992) and of

Table III

Illustration of the Structural Diversity of Mycobacterial Lipo-oligosaccharides

Species	Structure of the oligosaccharide	Positions of the acyl residues on the terminal trehalose
M. tuberculosis	2-*O*-Me-α-L-Fuc*p*-(1→3)-β-D-Glc*p*-(1→3)-2-*O*-Me-α-L-Rha*p*-(1→3)-2-*O*-Me-α-L-Rha*p*-(1→3)-β-D-Glc*p*-(1→3)-4-*O*-Me-α-L-Rha*p*-(1→3)-6-*O*-Me-α-D-Glc*p*-(1↔1)-α-D-Glc*p*	3,4,6 or 2,3,6 hydroxyls of terminal Glc*p* unit
M. kansasii	4,6-dideoxy-2-*O*-Me-3-*C*-Me-4-(2′-methoxypropionamido)-L-*manno*-hex*p*-(1→3)-Fuc*p*-(1→4)-[β-L-Xyl*p*-(1→4)]₆-2-*O*-Ac-3-*O*-Me-Rha*p*-(1→3)α-D-β-D-Glc*p*-(1→4)-β-D-Glc*p*-(1→4)-α-D-Glc*p*-(1↔1)-α-D-Glc*p*	4,6 hydroxyls of terminal Glc*p* unit and 2 hydrox of the other Glc*p* unit
M. gastri	3,6-dideoxy-4-*C*-(1,3-dimethoxy-4,5,6,7-tetrahydroxy-heptyl)-α-xylo-hex*p*-(1→3)-[β-L-Xyl*p*-(1→4)]₆₋₇-3-*O*-Me-Rha*p*-(1→3)-β-D-Gal*p*-(1→3)-β-D-Glc*p*-(1→4)-α-D-Glc*p*-(1↔1)-α-D-Glc*p*	4,6 hydroxyls of terminal Glc*p* unit and 2 hydrox of the other Glc*p* unit
M. fortuitum (3rd biovariant)	β-D-Glc*p*-(1→6)-α-D-Glc*p*-(1↔1)-α-D-Glc*p*	3,4,6 hydroxyls of termin Glc*p* unit and 2 hydrox of the other Glc*p* unit
M. malmoense	α-D-Man*p*-(1→3)-α-D-Man*p*-(1→2)-α-L-Rha*p*-(1→2)-[α-L-3-*O*-Me-Rha*p*-(1→2)]₂-α-L-Rha*p*-(1→3)-α-D-Glc*p*-(1↔1)-α-D-Glc	3,4,6 hydroxyls of termin Glc*p* unit
M. szulgai	α-L-2-*O*-Me-Fuc*p*-(1→3)-α-L-Rha*p*-(1→3)-α-L-Rha*p*-(1→3)-β-D-Glc*p*-(1→6)-α-D-Glc*p*-(1↔1)-α-D-2-O-Me-Glc	3,4,6 hydroxyls of termin Glc*p* unit
M. smegmatis	4,6-(1′-carboxyethylidene-3-*O*-Me-β-D-Glc*p*-(1→3)-4,6-(1-carboxyehtylidene-β-D-Glc*p*-(1→4)-β-D-Glc*p*-(1→4)-β-D-Glc*p*-(1→6)-α-D-Glc*p*-(1↔1)-α-D-Glc*p*	4 and 6 hydroxyls of both terminal Glc*p* units
"*M. linda*"[a]	β-D-Glc*p*-(1→3)-α-L-Rha*p*-(1→3)-α-D-Glc*p*-(1↔1)-α-D-Glc*p*	3,4,6 hydroxyls of termin Glc*p* unit

[a]Unrecognized species.

those belonging to the third biovariant complex of *M. fortuitum* (Lanéelle *et al.*, 1996). To our best knowledge, apart from their high antigenicity when LOS-containing lipid extracts or purified LOS mixed with the appropriate adjuvant are injected to mice or rabbits to make specific antibodies that recognize the terminal species- or type-specific terminal mono- or di-saccharidyl unit, no other biological activity has been published for LOS.

2.2.4. ESTERS OF TREHALOSE

In addition to PGL, C-type GPL, and LOS, mycobacteria are known to contain other glycolipids whose distribution is restricted to few species.

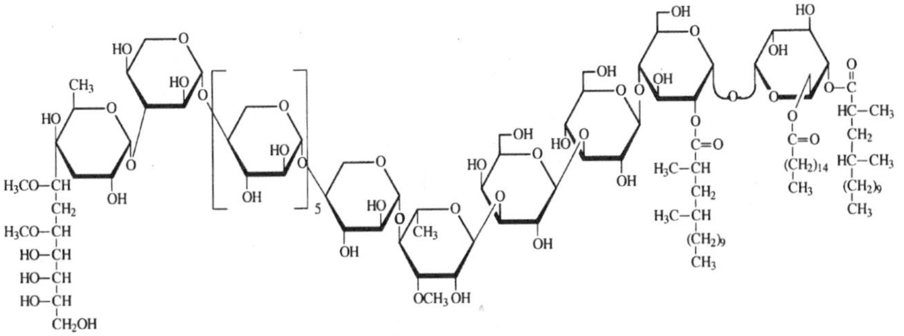

Figure 6. Structure of a lipo-oligosaccharide (LOS) isolated from *Mycobacterium gastri*. LOS consist of a common core of poly-*O*-acylated trehalose, which may be *O*-methylated, and further glycosylated by a mono- and more frequently an antigenic oligosaccharidyl unit. In the LOS of *M. gastri* presented, the oligosaccharidyl unit is composed of 3,6-dideoxy-4-*C*-(1,3-dimethoxy-4,5,6,7-tetrahydroxy-heptyl)-α-*xylo*hexopyranosyl-(1 → 3)-[β-L-xylopyranosyl-(1 → 4)]$_7$-3-*O*-methyl-rhamnopyranosyl-(1 → 3)-β-D-galactopyranosyl-(1 → 3)-β-D-glucopyranosyl-(1 → 4)-α-D-glucopyranosyl-(1 ↔ 1)-α-D-glucopyranosyl.

2.2.4a. Sulfatides

Virulent strains of *M. tuberculosis* differ from avirulent ones and from saprophytic mycobacteria by their growth pattern, in the form of serpentine cords, and also by their coloration with the cationic phenazine dye, neutral red (Dubos and Middlebrook, 1948). While virulent tubercle bacilli fix neutral red to become red, the other strains remain yellow; a mild and brief extraction with hexane is sufficient to remove the compounds responsible for the coloration, suggesting the presence of a peripherally located substance. The search for such a factor led Goren (1970a) to characterize a family of closely related sulfate esters (sulfatides). These consist of five glycolipids typified by a sulfate substituent on position 2′ of trehalose. They differ from one another by the numbers and types of acyl substituents and by their positions on trehalose. The fatty acyl substituents consist of straight chain ($C_{16:0}$ and $C_{18:0}$) and characteristic multimethyl-branched fatty acid residues, the phthioceranic and hydroxyphthioceranoyl substituents. The principal sulfatide, SL-I (Fig. 7), is a 2,3,6,6′-tetra-*O*-acyl-α,α′-D-trehalose-2′-sulfate (Goren, 1970b). Sulfatides have been characterized so far only from *M. tuberculosis*.

Sulfatides are believed to be peripherally located substances because of their ease of extraction with hexane and their presumed reactivity with neutral red in the context of intact bacilli. Further data are needed to determine their precise location, since many components of the tubercle bacillus may potentially react with this dye. Although Goren and colleagues (1974) found a correlation between virulence in a guinea pig model and the content of 40 tubercle bacilli in strongly acidic

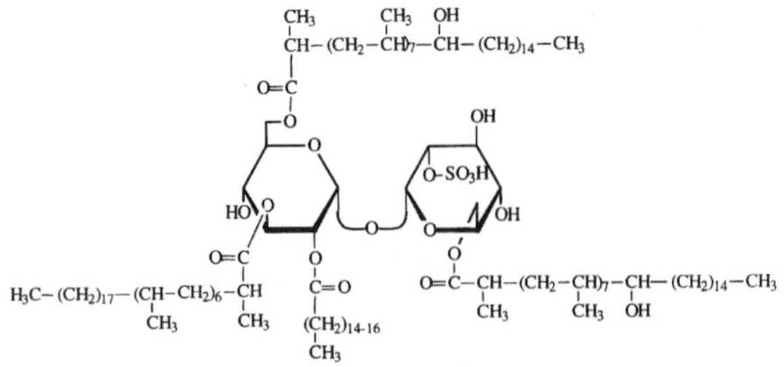

Figure 7. Structure of the major sulfatide (SL-I) of *Mycobacterium tuberculosis*. Sulfatides consist of five glycolipids typified by a sulfate substituent on position 2' of trehalose. They differ by the numbers and types of acyl substituents and by their positions on trehalose. The fatty acyl substituents consist of straight chain ($C_{16:0}$ and $C_{18:0}$) and characteristic multimethyl-branched fatty acid residues and the phthioceranic and hydroxyphthioceranoyl substituents. SL-I is a 2,3,6,6'-tetra-O-acyl-α,α'-D-trehalose-2'-sulfate.

lipids (SAL), sulfatides, and phospholipids, the high content of SAL of a few attenuated strains led the authors to conclude that the content of SAL is not sufficient for expression of virulence. Nevertheless, the initial observation that has led to their discovery has simulated much work with the goal of characterizing a potential virulence factor. The first of these, and probably the most important effect of sulfatides in terms of pathogenicity, was the initial report of Goren *et al.* (1976) who found that sulfatides prevented phagosome–lysosome fusion in macrophages following the phagocytosis of virulent tubercle bacilli. Unfortunately, Goren and colleagues showed years later that the observed inhibition was questionable, because the method they used introduced artifacts (Goren *et al.*, 1987). Thus, although the inhibition of the fusion of lysosomes with phagosomes containing intact tubercle bacilli and other pathogenic species is well established (Clemens and Horwitz, 1995; de Chastellier *et al.*, 1995; Sturgill-Koszycki *et al.*, 1994; Crowle *et al.*, 1991; Armstrong and Hart, 1975), the most recent studies demonstrate it is the maturation of the phagosomes containing pathogenic mycobacteria that is inhibited. Early endosomes but not late endosomes (lysosomes) fuse with the phagosomes, but the nature of the mycobacterial substances involved in this fusion phenomenon remains to be discovered. Thus, in contrast to what is quoted in most secondary literature, this property of SL-I is still unproven. If further studies should demonstrate this property for SL-I in *M. tuberculosis,* in other mycobacterial species the potential candidates are necessarily different from sulfatides, since these latter compounds so far have been characterized only from *M. tuberculosis.*

It remains, however, that SL-I has been shown to exhibit numerous biological activities, such as an *in vitro* intrinsic toxicity on mitochondria membrane and *in vivo* synergistic effect with cord factor, trehalose dimycolate (see Goren, 1990;

Asselineau and Asselineau, 1978), and an inhibition of macrophage priming (Pabst *et al.,* 1988). SL-I also blocked the effect of several priming agents such as INF-γ, LPS, IL-1β, TNF-α, and muramyl dipeptide on macrophages, resulting in the abolishment of the release of O_2-induced by phorbol 12-myristate 13-acetate (PMA) or formyl-methionyl-leucyl-phenylalanine (FMLP). This property was shown to be specific for sulfatide, because dextran sulfate was inactive and trehalose dimycolate promoted macrophage priming. No effect of SL-I was observed on unprimed cells, suggesting that sulfatide altered the priming rather than the O_2-release. Inhibition of the macrophage priming seems to occur indirectly by the inactivation of protein kinase C (Pabst *et al.,* 1988; Brozna *et al.,* 1991). In contrast, SL-I activates human neutrophils (Zhang *et al.,* 1991), apparently by a mechanism involving guanine nucleotide-binding proteins, which implies that the molecules interfere with cell-signaling mechanisms.

2.2.4b. Miscellaneous Trehalose Esters

During the search for specific probes for the serological detection of tuberculosis, it was realized (Simmoney *et al.,* 1995; Cruaud *et al.,* 1989, 1990; Martin Casabona *et al.,* 1989; Papa *et al.,* 1989) that *M. tuberculosis* contains trehalose esters reactive with tuberculous sera with a high specificity (97%) and a good sensitivity (86%). These consist of family of 2,3-di-*O*-acyl (Fig. 8) (Besra *et al.,* 1992; Lemassu *et al.,* 1991) and 2,3,6-tri-*O*-acyl trehalose (Muñoz *et al.,* 1997). Structurally similar molecules have been characterized from *M. fortuitum* (Gautier *et al.,* 1992), which also elaborates the 2,3,4-tri-*O*-acyl trehalose antigens, and have been shown to be equally effective in terms of serodiagnosis of tuberculosis (Muñoz *et al.,* 1997; Escamilla *et al.,* 1996).

Some mycobacteria also elaborate species- or type-specific poly-*O*-acylated trehalose esters that are unlikely to be antigenic, due to the number of the fatty acyl substituents. These include the fully *O*-acylated trehalose esters with highly unusual polyunsaturated acyl substituents. The hexatriaconta-4,8,12,16,20-pentaenoyl homologue is the principal homologue of these phleic acids produced by the saprophytes *M. phlei* and *M. smegmatis* (Asselineau *et al.,* 1972), and the polyphthienoyl (also called mycolipenoyl) (Minnikin *et al.,* 1985) trehalose typi-

Figure 8. Proposed structure of the antigenic di-*O*-acyl trehalose (DAT) from *Mycobacterium tuberculosis.* DAT is a mixture of closely related trehalose esters differing by the type of the acyl substituents on trehalose. Although the fatty acyl substituents shown are the major constituents of the saponification products of DAT, the relative positions of the residues on trehalose are unknown.

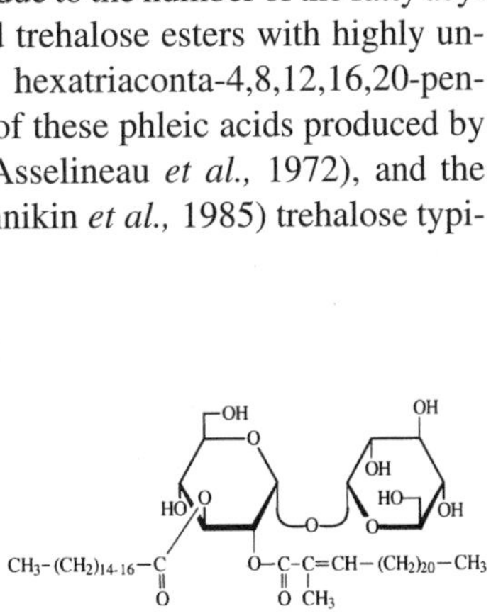

fying *M. tuberculosis* (Daffé *et al.,* 1988a). These glycolipids are easily extractable from the cells with petroleum ether, a procedure that does not affect the viability of the bacteria, and thus are assumed to be peripherally located compounds.

3. POLYSACCHARIDES AND LIPOPOLYSACCHARIDES

3.1. Glucan

The mycobacterial glucan is the major carbohydrate of the surface-exposed and extracellular constituents of *M. tuberculosis* and some other slow-growing mycobacteria (Lemassu *et al.,* 1996; Ortalo-Magné *et al.,* 1995; Lemassu and Daffé, 1994). This polysaccharide probably corresponds to the highly branched glycogen-type glucan found associated with cell wall preparations of *M. tuberculosis* (Amar-Nacasch and Vilkas, 1970) and *M. bovis* bacille Calmette-Guérin (BCG) (Misaki and Yukawa, 1966) and to the so-called polysaccharide-II of Seibert (1949), a 100-kDa glucan, despite the different structure proposed earlier (Kent, 1951). It has an apparent molecular mass of 100 kDa, 1000-fold less than that of the cytosolic glycogen (Antoine and Tepper, 1969) and is composed of repeating units of five or six →4-α-D-glucosyl residues substituted at position 6 with a mono- or diglucosyl residue (Fig. 9).

The mycobacterial glucan is poorly soluble in water and easily recoverable at the interface of a chloroform–methanol–water partition (Lemassu and Daffé, 1994), a property that may explain the hydrophobicity of mycobacteria when

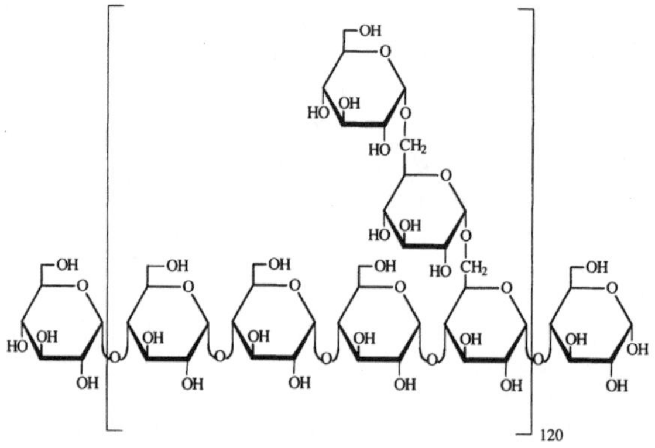

Figure 9.   The structural motif of the mycobacterial glycogen-like D-glucan. This consists of →4-α-D-glucopyranosyl-1→ residues substituted at position 6 with a mono- or diglucosyl residue.

grown as pellicles on liquid media. In phagocytic cells the outermost glucan-rich layer would separate the mycobacterial cell from host components, especially macromolecular ones, and would impede diffusion of smaller molecules (Daffé and Etienne, 1999). The surface-exposed glucan has been shown to be involved in the nonopsonizing binding of M. tuberculosis to mammalian cells through the complement receptor 3 (CR3), a proposed privileged receptor for the entry of bacterial pathogens (Ehlers and Daffé, 1998). The binding of M. tuberculosis to CR3-transfected CHO cells (Cywes et al., 1996) was shown to occur through the CR3 lectin site and was inhibited by both the purified glucan and the treatment of bacilli with amyloglucosidase, an enzyme capable of hydrolyzing glycogen and structurally related polysaccharides (Cywes et al., 1997).

3.2. Arabinomannan, Lipoarabinomannan, and Related Compounds

3.2.1. ARABINOMANNAN AND LIPOARABINOMANNAN

Mycobacteria are known to contain a heteropolysaccharide composed of arabinosyl residues of the unusual D series and of mannopyranosyl units [D-arabino-D-mannan (AM)] (Misaki et al., 1977). Subsequently, Hunter et al. (1986) isolated a lipopolysaccharide composed of a phosphatidylinositol group similar to that of PIM covalently linked to arabinomannan, originally called lipoarabinomannan-B (LAM-B). Further analysis showed that LAM-A, which contaminates LAM-B preparations, was a lipomannan (Hunter and Brennan, 1990) and led to the recognition of a single family of arabinose and mannose-containing mycobacterial lipopolysaccharides called LAM. On gel permeation columns AM and LAM have an apparent molecular mass of 13 kDa (Lemassu and Daffé, 1994) and 1000 kDa (Venisse et al., 1993). On sodium dodecyl sulfate–polyacrylamide gel electrophoresis (SDS-PAGE), LAM exhibits an apparent molecular mass of 30–40 kDa (Hunter et al., 1986), suggesting that its behavior in gel permeation chromatography is probably due to the formation of micelles in the aqueous buffer used. Using matrix-assisted laser-desorption ionization mass spectrometry, Venisse et al. (1993) determined that the molecular mass of LAM is around 17 kDa.

3.2.1a. Structural Features of Arabinomannan and Lipoarabinomannan

The mycobacterial AM and the polysaccharide moiety of LAM are composed of arabinan and mannan chains. The D-arabinan segment of the AM and LAM (Fig. 10A), whose structure has been established through the application of the conventional per-O-alkylation method of the polysaccharide and by NMR analyses (Lemassu et al., 1996; Ortalo-Magné et al., 1995, 1996a; Lemassu and Daffé,

A

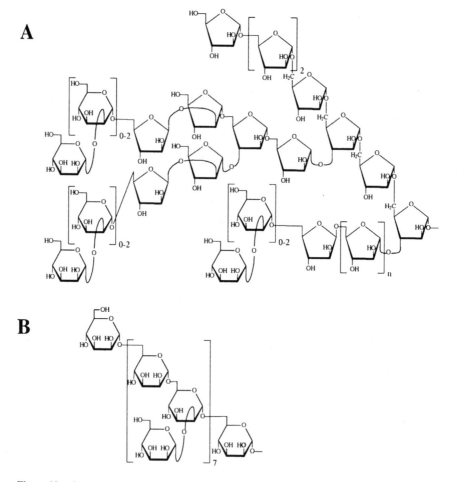

B

Figure 10. Structural motifs of the D-arabino-D-mannan (AM) and the structurally related lipoarabinomannan (LAM). LAM differs from AM by the terminus of the mannan segment: AM is terminated by a reducing mannosyl unit, whereas that of LAM consists of PIM (see Fig. 3). (A) The D-arabinan segment of the AM and LAM consists of linear →5-α-D-arabinofuranosyl-1→ residues with branching produced by 3,5-linked-α-D-arabinofuranosyl units substituted at both positions by a β-D-arabinofuranosyl-(1→2)-α-D-arabinofuranosyl, leading to a penta-arabinoside at the nonreducing terminus of the arabinan chains. In addition, AM and LAM also contain several other types of branching in the arabinan segment, including a linear nonreducing terminus of β-D-arabinofuranosyl-(1 → 2)-α-D-arabinofuranosyl-(1→5)-α-D-arabinofuranosyl-(1→. In slowly-growing species, most of the nonreducing termini of the arabinan chains of AM and LAM are capped by small oligomannosides, and accordingly are called ManAM and ManLAM. In contrast, the AM-like molecules, called "phosphoinositol-glyceroarabinomannans" (PI-GAM) of the rapidly-growing species *M. smegmatis* and the LAM of a rapid grower the nonreducing β-D-arabinofuranosyl termini of the arabinan segment was found capped by alkali-labile inositolphosphate groups and the native molecules are referred as AraAM and AraLAM. Variants with succinyl and lactyl substituents also occur in both AM and LAM but are not shown; these acidic short chain acyl residues esterify some hydroxyl groups of the arabinan segment. (B) The mannan chain of AM and LAM is composed of a →6-α-D-mannopyranosyl-1→ core substituted at some positions 2 with an α-D-mannosyl unit; a phosphate group may occur on the mannan segment of quantitatively minor ManLAM. A neutral mannan, exhibiting the same structural features as the mannan segment of AM and LAM, is also recoverable from the extracellular and surface-exposed materials of *M. tuberculosis* and other slowly-growing mycobacteria.

1994; Venisse *et al.*, 1993; Chatterjee *et al.*, 1991, 1992c), consists of linear →5-α-D-arabinofuranosyl-1→ residues with branching produced by 3,5-linked-α-D-arabinofuranosyl units substituted at both positions by a β-D-arabinofuranosyl-(1→2)-α-D-arabinofuranosyl-(1→, leading to a penta-arabinoside at the nonreducing terminus of the arabinan chains; this pentasaccharide was originally found terminating the arabinan segment of the cell wall AG (Daffé *et al.*, 1990). In addition, AM and LAM but not AG also contain several other types of branching in the arabinan segment, including a linear nonreducing terminus consisting of β-D-arabinofuranosyl-(1→2)-α-D-arabinofuranosyl-(1→5)-α-D-arabinofuranosyl-(1→ (Prinzis *et al.*, 1993; Chatterjee *et al.*, 1991). In slow-growing species, most of the nonreducing termini of the arabinan chains of AM and LAM (50–90% of the molecules) are capped by small oligomannosides, and accordingly are called ManAM and ManLAM (Fig. 9) (Lemassu *et al.*, 1996; Ortalo-Magné *et al.*, 1995, 1996a; Lemassu and Daffé, 1994; Venisse *et al.*, 1993; Chatterjee *et al.*, 1992c), a feature not found in the wall AG (Daffé *et al.*, 1990, 1993). In contrast, in the AM-like molecules, called "phosphoinositol-glyceroarabinomannans" (PI-GAM) (Gilleron *et al.*, 1997) of the rapid-growing species *M. smegmatis* and the LAM of a rapid grower, originally but incorrectly identified as *M. tuberculosis* (Khoo *et al.*, 1995a), the nonreducing β-D-arabinofuranosyl termini of the arabinan segment was found capped by alkali-labile inositolphosphate groups, and the native molecules are referred as AraAM and AraLAM.

   Most of the AM molecules (90%) are neutral polysaccharides, but variants with succinyl and lactyl substituents also occur (Ortalo-Magné *et al.*, 1996a; Weber and Gray, 1979). These acidic short-chain acyl residues are known to be present on LAM (Hunter *et al.*, 1986) and to esterify some hydroxyl groups of the arabinan segment (Delmas *et al.*, 1997). This location is consistent with the fact that no acidic mannan was isolated from the extracellular and surface-exposed materials of mycobacteria (Lemassu *et al.*, 1996; Ortalo-Magné *et al.*, 1995; Lemassu and Daffé, 1994) and may explain the migration of LAM and acidic arabinomannan on SDS-PAGE as a broad band with an apparent mass of 30–40 kDa (Ortalo-Magné *et al.*, 1996a; Hunter *et al.*, 1986).

   The mannan chain of AM and LAM is composed of a →6-α-D-mannopyranosyl-1→ core substituted at some positions 2 with an α-D-mannosyl unit (Fig. 10B) (Ortalo-Magné *et al.*, 1995, 1996a; Lemassu and Daffé, 1994; Venisse *et al.*, 1993; Chatterjee *et al.*, 1992a; Misaki *et al.*, 1977); a phosphate group may occur on the mannan segment of quantitatively minor ManLAM (Venisse *et al.*, 1995a). A neutral mannan, with an apparent molecular mass of 4 kDa and exhibiting the same structural features as the mannan segment of AM and LAM, is also recoverable from the extracellular and surface-exposed materials of *M. tuberculosis* and other slowly-growing mycobacteria (Lemassu *et al.*, 1996; Ortalo-Magné *et al.*, 1995; Lemassu and Daffé, 1994).

   LAM differs from AM by the terminus of the mannan segment: AM is terminated by a reducing mannosyl unit, whereas that of LAM consists of PIM (see Fig. 3). It has to be noted, however, that in the AM-like molecules isolated from *M. smegmatis,* the so-called PI-GAM, the mannan segment is terminated by an alkali-stable phosphoinositol glycerol (Gilleron *et al.,* 1997). LAM is highly heterogenous, as revealed by mass spectrometry analysis (Venisse *et al.,* 1993). Fractionation of the LAM molecules showed that they differ in terms of capping with mannosyl residues (Nigou *et al.,* 1997), numbers of fatty acyl residues occurring in the PIM moiety of LAM (Nigou *et al.,* 1997; Khoo *et al.,* 1995b; Leopold and Fischer, 1993), and the nature of these residues (Nigou *et al.,* 1997). In *M. bovis* BCG, an unusual fatty acid, a 12-*O*-(methoxypropanoyl)-12-hydroxystearic acid, was identified as the only fatty acid esterifying C-1 of the glycerol residue of a class of ManLAM, called "parietal" ManLAM, whereas palmitoyl and tuberculostearoyl residues were identified in the so-called "cellular" ManLAM (Nigou *et al.,* 1997); these two latter fatty acyl residues were known to occur in LAM (Hunter *et al.,* 1986) as found in PIM. Parietal and cellular ManLAM were defined as the families of LAM extractable from delipidated cells with ethanol–water prior and after the disruption of these cell residues, respectively (Nigou *et al.,* 1997). Further analyses of cellular ManLAM of *M. bovis* BCG, using NMR spectroscopy, showed that it consists of four types of lipopolysaccharides that differ both in the number and the location of fatty acids esterifying the PIM moiety of ManLAM (Nigou *et al.,* 1999). In addition to C-1 and C-2 of the glycerol moiety occurring in most ManLAM of *M. bovis* BCG, fatty acyl residues may be found on position 6 of the mannosyl unit linked to C-2 of *myo*-inositol and C-3 of this latter residue in the PIM moiety (see Fig. 3).

3.2.1b. Localization and Biological Activities of Arabinomannan

AM is the second major carbohydrate constituent of the extracellular and capsular materials of the tubercle bacillus and other slowly-growing mycobacteria (Lemassu *et al.,* 1996; Ortalo-Magné *et al.,* 1995, 1996a; Lemassu and Daffé, 1994); only trace amounts of AM, if any at all, exist in the extracellular fluids and surface-exposed materials of the rapidly-growing species examined (Lemassu *et al.,* 1996). The AM recoverable from the culture filtrates of mycobacteria is serologically active (Miller *et al.,* 1984), and that from *M. tuberculosis* has been reported to be immunosuppressive (Moreno *et al.,* 1988; Ellner and Daniel, 1979).

3.2.1c. Putative Localization and Biological Activities of Lipoarabinomannan

The localization of LAM in the mycobacterial envelope is unknown. Analysis of the outermost constituents of the tubercle bacillus failed to detect LAM in the capsule (Ortalo-Magné *et al.*, 1996a). As the anchor possessed by lipoteichoic acid, a known membrane component of some gram-positive bacteria, is similar to the phosphatidylinositol moiety of LAM (Chatterjee *et al.*, 1992a; Khoo *et al.*, 1995b), it seems likely that LAM is similarly situated. Indeed, both LAM and lipomannan have been found apparently associated with the plasma membrane fraction (Hunter *et al.*, 1986). Due to technical problems involved in obtaining purified plasma membrane, however, it is difficult to demonstrate the occurrence of LAM, since high-molecular-weight glycans may sediment with $100,000 \times g$ membrane fractions. Both the mycobacterial glucan and arabinomannan are recoverable from such fractions (M. Daffé and A. Lemassu, unpublished data), so it is possible that capsular polysaccharides shed from the cells due to the mechanical stress and also intracellular glycogen may contaminate the membrane preparations. LAM is shown in some schematic models of the mycobacterial envelope (Gaylord and Brennan, 1987; McNeil and Brennan, 1991) spanning the cell wall, including the mycolate monolayer, but there is no direct evidence for this, and the reported serological detection of LAM on the bacterial cell surface (Hunter and Brennan, 1990) is compromised by cross-reactivities between LAM and the capsular AM (Ortalo-Magné *et al.*, 1995). LAM may serve to link the plasma membrane and the cell wall. A third possibility also exists: the PIM anchor of LAM may be located in the membrane and the remaining AM moiety in the hypothetical mycobacterial periplasmic space (Daffé and Draper, 1998). This latter hypothesis may explain in part the asymmetrical appearance of the mycobacterial plasma membrane in thin sections of viable bacterial cells attributable to the presence of excess glycoconjugates in the thicker electron-dense outer layer compared with the inner one (Daffé *et al.*, 1989; Silva and Macedo, 1983a, 1984). However, it has been noted that the asymmetrical appearance of the mycobacterial plasma membrane and the accumulation of material in its outer leaflet depend on conditions of fixation and are not seen in sections of bacteria that were dead before fixation or have been subjected to membrane-damaging treatments (Silva and Macedo, 1983b). Thus, the symmetrical appearance of the membrane on dead bacteria is hard to explain in terms of the static presence of a chemical substance.

As stated above, LAM is likely to be a membrane component and in general one would expect little interaction between the components of the mycobacterial membrane and its animal host. One of the primary functions of the outer two compartments of the envelope is presumably to prevent such interaction and to protect the vital functions of the membrane, and only dead or damaged mycobacteria would expose membrane components to the host. Paradoxically, purified LAM exhibits a remarkable range of biological activities that seem to suit it ideally to the role of immunomodulator in mycobacterial disease and LAM has been included in a general class of bacterial virulence factors named modulins, which operate by inducing synthesis of host cytokines (Henderson *et al.*, 1996).

AraLAM from the rapidly-growing *Mycobacterium* species used by Chatterjee *et al.* (1992b) and several other groups was not in fact from *M. tuberculosis* H37Ra but from an unidentified rapidly-growing species (see Khoo *et al.*, 1995a), and like bacterial LPS—but not ManLAM—stimulates murine macrophages to produce TNF-α, IL-1 and IL-6, and several interleukins (Dahl *et al.*, 1996; Adams and Czuprynski, 1994; Bradbury and Moreno, 1993; Barnes *et al.*, 1992; Chatterjee *et al.*, 1992b). AraLAM but not ManLAM also induces several potent early genes (c-fos, KC, JE, and TNF-α) involved in the activation of macrophages (Roach *et al.*, 1993, 1994). Consistent with these findings is the capacity of AraLAM to rapidly activate the critical component in the regulation of genes central to immune function, the transcription factor NF-κB, in both bone marrow-derived macrophages and murine macrophage-like cell lines (Brown and Taffet, 1995). AraLAM but not ManLAM stimulates the production of inducible nitric oxide synthase (iNOS), a microbiocidal substance, synergistically with IFN-γ (Anthony *et al.*, 1994; Schuller-Levis *et al.*, 1994). It is necessary to mention that the potential contribution of contaminating (gram-negative) LPS in these activities (Molloy *et al.*, 1990) has been ruled out by appropriate control experiments (Adams *et al.*, 1993). In addition, data showing that Man LAM was significantly less potent than AraLAM in induction of these factors (Dahl *et al.*, 1996; Roach *et al.*, 1994) further tend to make the LPS hypothesis unlikely to be true. Deacylation of AraLAM reduces (Chatterjee *et al.*, 1992b) or eliminates (Barnes *et al.*, 1992) these biological effects. It should be noted that the treatment with alkali needed for deacylation also removes the alkali-labile inositolphosphate groups that cap AraLAM and PI-GAM of certain rapid-growing mycobacteria such as *M. smegmatis* (Gilleron *et al.*, 1997). Since this treatment abrogates the secretion of TNF-α induced by PI-GAM (Gilleron *et al.*, 1997), the alkali-labile phosphoinositides are likely to be the major epitope involved in this process. Parietal ManLAM from *M. bovis* BCG, which contains the unusual 12-*O*-(methoxypropanol)-12-hydroxystearoyl substituent, but not cellular ManLAM—devoid of this fatty acyl residue—stimulates both IL-8 and TNF-α secretion from dendritic cells (Nigou *et al.*, 1997).

Although LAM cannot be detected in the mycobacterial capsule (Ortalo-Magné *et al.*, 1996a), it has been implicated in phagocytosis of mycobacteria. Purified ManLAM binds to mouse macrophages (Schlesinger *et al.*, 1994; Venisse *et al.*, 1995b) and mannose seems to be important in this binding (Schlesinger *et al.*, 1994). LAM and also PIM and phosphatidylinositol inhibit the nonopsonic binding of *M. tuberculosis* to mouse macrophages (Stokes and Speert, 1995). Further, it is chemotactic for T lymphocytes (Berman *et al.*, 1996), an activity lost after deacylation but shared by ManLAM and AraLAM, though curiously not by LAM from *M. bovis* BCG.

Like PIM and AM, LAM is serologically active and as expected from their chemical structural resemblances cross-reacts with the two former molecules, so that identification of LAM by serological means needs caution. Finally, it is a free-radical scavenger (Chan *et al.,* 1991), though this may be a nonspecific property of carbohydrates.

3.2.2. LIPOMANNAN

In addition to LAM, a phosphorylated lipopolysaccharide structurally related to LAM, originally called LAM-A (Hunter *et al.,* 1986), but lacking the arabinan segment also has been characterized in mycobacteria (Hunter and Brennan, 1990). This lipomannan (LM) corresponds to a multiglycosylated form of PIM (Hunter and Brennan, 1990), and consequently is likely located in the plasma membrane. Using mass spectrometric analyses of time course acetolysates of LM and LAM, Khoo *et al.* (1995b) showed evidence for tri-*O*-acylated mannophosphoinositide, which also occurs in PIM, as a common anchor for both LM and LAM; more recent NMR studies (Gilleron *et al.,* 1999) revealed the location of the three fatty acyl substituents of LM as being identical to that of PIM, that is, position 1 and 2 of glycerol and position 6 of the mannosyl residue linked to C-2 of the *myo*-inositol ring (see Fig. 3) and the occurrence of quantitatively minor variants containing four fatty acyl substituents. As expected from the structural analogies between PIM, LM, and LAM, the fourth fatty acyl substituent in the tetra-*O*-acylated LM is again located on C-3 of *myo*-inositol (Gilleron *et al.,* 1999). The mannan moiety of LM shares a common structure with that of AM and LAM (Fig. 10B), consisting of a →6-α-D-mannopyranosyl-1→ core substituted at some positions 2 with an α-D-mannosyl unit. As expected from the location of the succinyl, lactyl residues, and alkali-labile inositolphosphate groups on the arabinan segment of AM-containing polysaccharides (Delmas *et al.,* 1997, Gilleron *et al.,* 1997; Khoo *et al.,* 1995a), the presence of these acidic substituents has not been noted in LM. Nevertheless, LM migrates on SDS-PAGE as a broad band with an apparent mass of 16 kDa (Hunter *et al.,* 1986). As the phosphate group is not responsible for the migration of these polysaccharides, LM may contain some acidic substituents whose chemical nature remains to be determined. In contrast to LAM, LM is not recognized by either whole human leprosy serum or monoclonal antibodies raised against mycobacteria (Hunter *et al.,* 1986), indicating that the epitopes of the AM and LAM molecules reside on the arabinan segment.

3.3. Arabinogalactan

The cell wall skeleton of the mycobacterial cell envelope, that is, the material remaining after removal of all noncovalently linked substances (Kotani *et al.,* 1959), consists of a lipopolysaccharide covalently bound to peptidoglycan. Digestion of this material with alkali yields mycolic acids and solubilizes a polysaccharide composed of arabinosyl and galactosyl residues [arabinogalactan (AG)]. The molecular mass of this heteropolysaccharide was estimated to 31 kDa by gradient density (Misaki and Yukawa, 1966) and 15 kDa by glycosyl composition analysis (Daffé et al., 1990) and by gel filtration (M. Daffé and A. Lemassu, unpublished data).

3.3.1. STRUCTURAL FEATURES OF ARABINOGALACTAN

The AG is composed of D-arabinofuranosyl and D-galactofuranosyl residues, and a tentative structural formula was proposed some 25 years ago by Misaki et al. (1974). Subsequently, this model has been extensively revised using data obtained by applying selective ion monitoring gas chromatography–mass spectrometry, high-resolution NMR to the partially depolymerized per-O-alkylated AG of M. tuberculosis (Daffé et al., 1990). The same structure occurs in all the rapid- and slow-growing mycobacterial species examined and also in the noncultivable M. leprae (Daffé et al., 1993).

The structural motifs of the mycobacterial AG are illustrated in Fig. 11. The polysaccharide is composed of arabinan chains (composed of roughly 27 D-arabinofuranosyl units each) attached to the homogalactan core (consisting of 32 D-galactofuranosyl units) of linear alternating 5- and 6-linked β-D-galactofuranosyl residues. The homoarabinan chains are composed of linear α-D-arabinofuranosyl residues with branching produced by 3,5-linked-α-D-arabinofuranosyl units substituted at both positions by α-D-arabinofuranosyl residues. The nonreducing termini of the arabinan chains of mycobacterial AG consist exclusively of pentaarabinosyl units. In contrast to AM and LAM, the nonreducing penta-arabinosyl termini contain all the 2-linked arabinofuranosyl residues and are devoid of mannosyl or phosphoinositol caps. Characterization of larger per-O-alkylated oligosaccharides by fast-atom bombardment mass spectrometry (Besra et al., 1995) demonstrated that the arabinan chains are attached to the homogalactan core in a region near the reducing end of the molecule.

3.3.2. THE LINK BETWEEN ARABINOGALACTAN
AND PEPTIDOGLYCAN

The mycobacterial AG is attached to the peptidoglycan through a linker oligosaccharide the structure of which was defined precisely by McNeil et al. (1990) who used mild acid hydrolysis of lysozyme-degraded cell wall skeleton to obtain fragments of AG still attached to peptidoglycan oligomers, followed by [31P]-NMR. The linker oligosaccharide consists of a galactofuranosyl unit at-

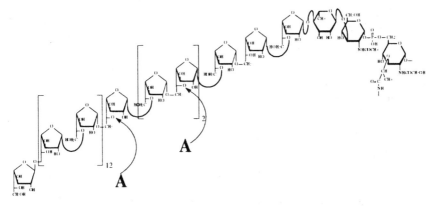

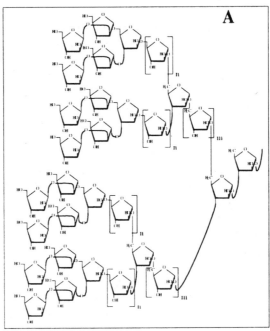

Figure 11. Structural motifs of the mycobacterial arabinogalactan. The polysaccharide is composed of 2 or 3 arabinan chains attached to the homogalactan core (consisting of roughly 32 D-galactofuranosyl units) of linear alternating 5- and 6-linked β-D-galactofuranosyl residues (upper part of the figure). (A) The homoarabinan chains are composed of linear α-D-arabinofuranosyl residues with branching produced by 3,5-linked-α-D-arabinofuranosyl units substituted at both positions by α-D-arabinofuranosyl residues. The nonreducing termini of the arabinan chains of mycobacterial arabinogalactan consist exclusively of penta-arabinosyl units. This oligosaccharidyl unit is formed by branching produced by 3,5-linked-α-D-arabinofuranosyl units substituted at both positions by a β-D-arabinofuranosyl-(1→2)-α-D-arabinofuranosyl. In the intact cell wall skeleton, arabinogalactan is linked to peptidoglycan and is esterified by mycolic acids. The linker oligosaccharide consists of a galactofuranosyl unit attached at position 4 of a L-rhamnopyranosyl residue that substitutes position 3 of an N-acetyl-D-glucosaminyl unit; the latter sugar is in turn connected to position 6 of a muramic acid residue of the peptidoglycan through a phosphodiester link. The mycoloyl residues are clustered in groups of four on the nonreducing penta-arabinosyl unit. In *M. tuberculosis,* about two thirds of the available penta-arabinosyl units are so substituted; the remaining one third is mycolate free. In *M. leprae, M. bovis* BCG, and *M. smegmatis,* however, the cell wall skeleton is less mycoloylated than that of the tubercle bacillus.

tached at position 4 of a L-rhamnopyranosyl residue that substitutes position 3 of an N-acetyl-D-glucosaminyl unit (Fig. 11); the latter sugar is in turn connected to position 6 of a muramic acid residue of the peptidoglycan through a phosphodiester link (McNeil *et al.*, 1990).

3.3.3. THE LINK BETWEEN ARABINOGALACTAN AND MYCOLIC ACIDS

The arabinan segment of the AG has long been known to be the site of esterification of mycolic acids, the exclusive lipids of the mycobacterial cell skeleton, since Azuma and Yamamura (1962, 1963) isolated a glycolipid and identified a 5-mycoloyl arabinose and Amar-Nacash and Vilkas (1970) were able to obtain a mycoloyl arabinobiose from walls of *M. tuberculosis*. However, in light of the new structural model for the AG (Daffé *et al.*, 1990), several possibilities arose for the location of mycoloyl residues on the 5-OH functions of the nonreducing penta-arabinofuranosyl termini of the arabinan segment of the polysaccharide. The exact location of the mycoloyl residues was determined by McNeil *et al.* (1991) through per-*O*-methylation of the cell wall skeleton with methyl trifluoromethanesulfonate, an acid-catalyzed methylation procedure that does not result in deacylation. Following the replacement of mycoloyl residues by ethyl groups using a conventional alkali-catalyzed reaction, depolymerization of the per-*O*-alkylated polysaccharide and Smith degradation of the native AG demonstrated that the mycoloyl units are clustered in groups of four on the nonreducing penta-arabinosyl unit. In *M. tuberculosis*, about two thirds of the available penta-arabinosyl units are so substituted; the remaining one third is mycolate free (McNeil *et al.*, 1991). In *M. leprae*, *M. bovis* BCG, and *M. smegmatis*, however, the cell wall skeleton is less mycoloylated than that of tubercle bacillus. The clustering of the highly hydrophobic mycolic acids may explain in part the exceptional impenetrability of the mycobacterial cell envelope, with permeability coefficients two to five orders of magnitude lower than that of *Escherichia coli* (see Connell and Nikaido, 1994).

3.3.4. BIOLOGICAL ACTIVITIES OF ARABINOGALACTAN

The mycobacterial AG is a serologically active polysaccharide that reacts not only with antisera against mycobacterial walls but also with antisera against related bacterial genera, showing the structural analogy between the antigens (Misaki *et al.*, 1974). Indeed, detailed structural analyses of the AG from representative strains of the phylogenically close *Rhodococcus* and *Nocardia* (Daffé *et al.*, 1993) confirmed the structural analogies between the various AG but also demonstrated the occurrence of specific arabinan and galactan motifs in each AG. Using various D-arabino-oligosaccharides generated from the heteropolysaccharides, the same authors demonstrated that the 2-linked arabinosyl-containing oligosaccha-rides, that is, the nonreducing termini of the arabinan segments, are responsible for the serological activity (Misaki *et al.*, 1974). In addition, the mycobacterial AG has been shown to inhibit antigen-induced T-cell proliferation (Moreno *et al.*, 1988; Kleinhenz *et al.*, 1981).

3.4. Peptidoglycan

Mycobacterial peptidoglycan, which belongs to a family of structures pos-sessed by almost all eubacteria, consists of chains of a glycan formed from alter-nating units of *N*-acetylglucosaminyl linked β-1→4 to muramic acid (Adam *et al.*, 1969). Tetrapeptide chains attached to the muramic acid residues cross-link the glycan chains. In the mycobacterial species examined, except *M. leprae,* the pep-tide consists of L-alanyl-D-isoglutaminyl-*meso*-diaminopimelyl-D-alanine and the diaminopimelic acid is amidated (Wietzerbin-Falszpan *et al.*, 1970). In the puri-fied peptidoglycan of *M. leprae,* however, L-alanine is specifically replaced by glycine (Draper *et al.*, 1987), a specific feature probably unrelated to the intracel-lular growth, still the only way in which *M. leprae* can be produced, since the con-ventional peptidoglycan tetrapeptide constituents occur in walls of *M. leprae-murium* prepared from bacteria grown in mice (Draper, 1971). This substitution of L-alanine by glycine occurs in at least one other microorganism, *Micromonospo-ra olivastereospora* (Nara *et al.*, 1977).

The mycobacterial peptidoglycan is similar to one of the most common types found, for example, in *E. coli*. However, its structure, deduced from partial acid and enzymic hydrolyses and mass spectrometry of the resulting fragments (Petit *et al.*, 1969; Wietzerbin-Falszpan *et al.*, 1970), differs slightly from the common type. One difference is that the muramic acid is *N*-acylated with a glycolyl residue, rather than the usual acetyl residue (Adam *et al.*, 1969). The second difference con-cerns the occurrence of a substantial number of unusual cross-links between two chains of peptidoglycan. In addition to the usual D-alanyl-diaminopimelate link-ages, a proportion of bonds involving two residues of diaminopimelic acid has been characterized in mycobacteria (Wietzerbin *et al.*, 1974).

In contrast to structurally close walls, which occur in numerous species of bacteria, the mycobacterial walls are powerful adjuvants; thus, although the adju-vant activity of the mycobacterial wall has been narrowed down to a small frag-ment of the peptidoglycan muramyl-L-alanyl-D-isoglutamine [muramyl dipeptide (MDP)], the whole of the adjuvant activity of the wall may not reside in this struc-ture. MDP and various derivatives have been used experimentally as adjuvants, but have not been widely adopted for use in human vaccines. An extraordinary sec-ond biological activity of MDP is its ability to alter sleep patterns and its possible role as a normal animal "hormone" (Johannsen *et al.*, 1989). However, MDP can arise from many sorts of bacteria in the animal, and this aspect of its activities is unlikely to have any connexion with mycobacterial disease.

4. GLYCOPROTEINS

Mycobacteria are among the prokaryotes that have been shown to modify the proteins they synthesize by posttranslational addition of carbohydrate residues (see Chapters 1 and 4, this volume). Although various proteins among the numerous proteins found in the extracellular fluid of *M. tuberculosis* [more than 200 polypeptides are observed on two-dimensional SDS-PAGE (Sonnenberg and Belisle, 1997)] are suspected to be glycosylated, primarily based on their ability to bind ConA (Espitia *et al.*, 1995; Garbe *et al.*, 1993; Espitia and Mancilla, 1989), only two glycoproteins have been well characterized to date, namely, the 45/47-kDa antigen complex (Romain *et al.*, 1993) and the 19-kDa lipoprotein (Young and Garbe, 1991).

4.1. The 45/47-kDa-Antigen Complex

In the search of antigens selectively reacting with antibodies present in sera of animals immunized with living *M. bovis* (BCG), Marchal and collaborators (Romain *et al.*, 1993) discovered a family of immunodominant antigens with an apparent molecular mass of 45/47-kDa on SDS-PAGE; the same group cloned the corresponding gene of *M. tuberculosis* in the rapidly-growing *M. smegmatis* (Laqueyrerie *et al.*, 1995). The gene was referred to as *apa* because of the high percentages of proline (22%) and alanine (19%) in the purified protein, which has a predicted molecular mass of 28,779 Da; the high percentage of proline probably explains the apparently higher molecular mass as determined by SDS-PAGE, resulting from the increased rigidity of molecules due to proline residues (Laqueyrerie *et al.*, 1995). In the meantime, Belisle and colleagues (Dobos *et al.*, 1995) have performed structural analyses of the protein complex through proteolytic digestion, followed by reaction of the resulting peptides with ConA and analysis of putative glycopeptides by liquid chromatography and electrospray mass spectrometry, and demonstrated the presence of *O*-glycosylated peptides, one of which was composed of a dihexosyl unit. Further analyses defined the full extent and the nature of glycosylation as well as the location of glycosylated amino acid residues (Dobos *et al.*, 1996). In addition to the presumably 50 monoglycosylated peptides, all of which produced only the loss of 162 mass units, fast atom bombardment–mass spectrometry, N-terminal amino acid sequencing, and α-mannosidase digestion demonstrated *O*-glycosylation of threoninyl residues with mannosyl, mannobiose, and mannotriose units possessing α-1 $\rightarrow$ 2 linkages in the oligomannosides (Dobos *et al.*, 1996). This was the first demonstration of the glycosylation of mycobacterial proteins.

Fibronectin-binding proteins are known to be present in the culture filtrates and on the surface of several mycobacteria (Abou-Zeid *et al.*, 1988, 1991); purification and sequence comparison of a class of fibronectin-binding proteins, the so-called fibronectin attachment proteins (FAP) (Schorey *et al.*, 1995), demonstrated that FAB-A of *M. avium* (Schorey *et al.*, 1996) has an unusually large number of proline and alanine residues (40% overall) and is 50% identical to the 45/47-kDa proline-rich antigen complex of *M. tuberculosis* (Laqueyrerie *et al.*, 1995) and FAB-L of *M. leprae* (Schorey *et al.*, 1995). Furthermore, a FAB-like protein has been previously recognized in *M. vaccae* (Ratliff *et al.*, 1993) and the antibodies raised against this molecule cross-react with a cell wall component in *M. bovis* (BCG) (Kuroda *et al.*, 1993) and other mycobacterial species including *M. tuberculosis, M. avium,* and *M. kansasii.* The glycosylation of FAB remains to be established, however, in mycobacterial species other than *M. tuberculosis.*

Fibronectin significantly enhances both attachment and ingestion of *M. leprae* by epithelial and Schwann cell lines and anti-FAB-L antibodies significantly block *M. leprae* attachment and internalization by both cell lines (Schorey *et al.*, 1995). These data strongly suggest that FAP may play an important role in the adhesion of this mycobacterium to fibronectin for recognition and uptake of bacteria by nonphagocytic cells. In the case of *M. tuberculosis,* however, the clinical significance of the infection of nonphagocytic cells is not well established. It remains that the recognition of the 45/47 kDa by sera from animals immunized by living *M. bovis* but not sera from animals immunized by dead bacilli (Romain *et al.*, 1993) is an important step in the development of sensitive and specific detection of active tuberculosis.

4.2. The 19-kDa Lipoprotein

The 19-kDa protein was originally identified by using a set of monoclonal antibodies binding to *M. tuberculosis* and a limited number of slowly-growing mycobacterial species (Andersen *et al.*, 1986). Data from nucleotide sequence analysis (Ashbridge *et al.*, 1989) and biochemical characterization (Young and Garbe, 1991) suggested that the mature protein is a secreted lipoprotein. Analysis of the purified 19-kDa protein from *M. bovis,* as well as the recombinant protein of *M. tuberculosis* expressed in *M. smegmatis,* has provided evidence for glycosylation, since the proteins bound strongly to ConA (Garbe *et al.*, 1993; Fifis *et al.*, 1991). Furthermore, the recombinant protein of *M. tuberculosis* expressed in *E. coli* was not stainable with ConA and exhibited an apparent molecular mass approximately 4 kDa lower than the recombinant protein expressed in *M. smegmatis* (Garbe *et al.*, 1993). By production of a set of alkaline phosphatase hybrid proteins in a mycobacterial expression system, the peptide region required for glycosylation of the 19 kDa of *M. tuberculosis* was defined (Herrmann *et al.,* 1996). Mutagenesis of two threonine clusters within this region abolished lectin binding by hybrids and by the 19 kDa itself. Substitution of the threoninyl residues also resulted in generation of truncated polypeptides, probably as a result of proteolysis (Herrmann *et al.,* 1996). Although further structural evidence of a covalent carbohydrate–protein linkage is obviously needed, altogether the above data provide strong support for the glycosylation of the 19 kDa of *M. tuberculosis.* The 19-kDa lipoprotein is a target of humoral and cellular immune responses to *M. tuberculosis* (Garbe *et al.,* 1993; Andersen *et al.,* 1986) and may play a role in the virulence of *M. tuberculosis,* since tubercle bacilli that do not produce this protein are of low virulence for mice, whereas recombinant cells producing it exhibit enhanced virulence (Lathigra *et al.,* 1996).

4.3. Putative Glycoproteins

Among the putative mycobacterial glycoproteins is the 38-kDa lipoprotein, a phosphate-binding protein with features very similar to those of well-characterized periplasmic proteins of *E. coli* (Chang *et al.,* 1994). The deduced amino acid sequence of the cloned gene shows a relatively high homology with Pho S (Pst S) of *E. coli* (Andersen and Hansen, 1989), and expression of the protein is enhanced in phosphate-starved cultures of *M. tuberculosis* (Espitia *et al.,* 1992). This protein possesses a threonine-rich *O*-glycosylation site present in the N-terminal region and similar to that of the 19-kDa lipoprotein (Herrmann *et al.,* 1996). Furthermore, alkaline phosphatase hybrid protein containing the N-terminal amino acids from the 38 kDa is positive in a ConA binding assay (data cited in Herrmann *et al.,* 1996).

Finally, a mycobacterial heparin-binding hemagglutinin (HBHA), a protein involved in attachment of mycobacteria to epithelial cells and induction of bacterial autoaggregation (Menozzi *et al.,* 1996), recently has been characterized (Menozzi *et al.,* 1998). The calculated molecular mass of HBHA (21,331 Da) is substantially lower than its apparent molecular mass estimated by SDS-PAGE (28 kDa), probably due to its high content in lysine and proline residues occurring at the C-terminal region but also to the presence of a putative carbohydrate moiety. The recombinant HBHA in *E. coli* has a slightly lower molecular mass (27 kDa) than the native protein on SDS-PAGE, suggesting a posttranslational modification of the protein in mycobacteria. Indeed, analysis of HBHA purified from *M. bovis* (BCG) culture fluids showed that carbohydrate (glucose, xylose, mannose, arabinose, and an unidentified component) represents 2.8% of the mass of the protein; however, some of the major sugar constituents, notably glucose, mannose, and arabinose, may be contaminants derived from the extracellular polysaccharides (Lemassu and Daffé, 1994). Further structural studies are clearly needed to confirm this glycosyl composition and to establish the glycoprotein nature of HBHA.

5. FINAL REMARKS

From the data presented in this chapter it is obvious that glycoconjugates are both structurally and functionally important compounds of the mycobacterial cell envelope. The cell wall arabinogalactan and the capsular polysaccharides are certainly major structural components of the mycobacterial envelope. The biological activities of some purified mycobacterial glycoconjugates, notably lipoarabinomannan, are likely to be important in diseases produced by mycobacteria. Thus, the enzymes involved in the biosynthesis of mycobacterial cell surface glycoconjugates represent potential targets for new drugs. With the development of molecular biology of mycobacteria and the recent availability of the entire genome sequence of *M. tuberculosis* it is hoped that the isolation of mutants devoid of a specific glycoconjugate and the study of their fate in infected animals will establish the importance of these fascinating molecules in the pathogenicity of the organisms producing them.

REFERENCES

Abou-Zeid, C., Ratliff, T. L., Wiker, H. G., Harboe, M., Bennedsen, J., and Rook, G. A., 1988, Characterization of fibronectin-binding antigens released by *Mycobacterium tuberculosis* and *Mycobacterium bovis* BCG, *Infect. Immun.* **56**:3046–3051.

Abou-Zeid, C., Garbe, T., Lathigra, R., Wiker, H. G., Harboe, M., Rook, G. A. W., and Young, D. B., 1991, Genetic and immunological analysis of *Mycobacterium tuberculosis* fibronectin-binding proteins, *Infect. Immun.* **59**:2712–2718.

Adam, A., Petit, J. F., Wietzerbin-Falszpan, J., Sinay, P., Thomas, D. W., and Lederer, E., 1969, L'acide *N*-glycolyl-muramique, constituant des parois de *Mycobacterium smegmatis;* identification par spectrométrie de masse, *FEBS Lett.* **4**:87–92.

Adams, J. L., and Czuprynski, C. J., 1994, Mycobacterial cell wall components induce the production of TNF-α, IL-1, and IL-6 by bovine monocytes and the murine macrophage cell line RAW 264.7, *Microb. Pathog.* **16**:410–411.

Adams, L. B., Fukutomi, Y., and Krahenbuhl, J. L., 1993, Regulation of murine macrophage effector functions by lipoarabinomannan from mycobacterial strains with different degrees of virulence, *Infect. Immun.* **61**:4173–4181.

Amar-Nacasch, C., and Vilkas, E., 1970, Etude des parois de *Mycobacterium tuberculosis*. II. Mise en évidence d'un mycolate d'arabinobiose et d'un glucane dans les parois de *M. tuberculosis* H_{37}Ra, *Bull. Soc. Chim. Biol.* **52**:145–151.

Andersen, Å. B., and Hansen, E. B., 1989, Structure and mapping of antigenic domains of protein antigen b, a 38,000-molecular-weight protein of *Mycobacterium tuberculosis,* *Infect. Immun.* **57**:2481–2488.

Andersen, Å. B., Yuan, Z. L., Hasløv, K., Vergmann, B., and Bennedsen, J., 1986, Inter- species reactivity of five monoclonal antibodies to *Mycobacterium tuberculosis* as ex- amined by immunoblotting and enzyme-linked immunosorbent assay, *J. Clin. Micro- biol.* **23**:446–451.

Anderson, R. J., 1939, The chemistry of the lipids of the tubercle bacillus and certain oth- er microorganisms, *Prog. Chem. Org. Nat. Prod.* **3**:145–202.

Anthony, L. S. D., Chatterjee, D., Brennan, P. J., and Nano, F. E., 1994, Lipoarabinoman- nan from *Mycobacterium tuberculosis* modulates the generation of reactive nitrogen intermediates by gamma interferon-activated macrophages, *FEMS Immunol. Med. Mi- crobiol.* **8**:299–306.

Antoine, A. D., and Tepper, B. S., 1969, Characterization of glycogen from mycobacteria, *Arch. Biochem. Biophys.* **134**:207–213.

Armstrong, J. A., and Hart, P. d'Arcy, 1975, Phagosome-lysosome interactions in cultured macrophages infected with virulent tubercle bacilli, *J. Exp. Med.* **142**:1–15.

Ashbridge, K. R., Booth, R. J., Watson, J. D., and Lathigra, R. B., 1989, Nucleotide se- quence of the 19 kDa antigen gene from *Mycobacterium tuberculosis, Nucleic Acids Res.* **17**:1249.

Asselineau, C., and Asselineau, J., 1978, Trehalose-containing glycolipids, *Prog. Chem. Fats Other Lipids,* 16:59–99.

Asselineau, C., Montrozier, H. L., Promé, J. C., Savagnac, A. M., and Welby, M., 1972, Étude d'un glycolipide polyinsaturé synthétisé par *Mycobacterium phlei, Eur. J. Biochem.* **28**:102–109.

Asselineau, J., 1966, *The Bacterial Lipids,* Hermann, Paris.

Azuma, I., and Yamamura, Y., 1962, Studies on the firmly bound lipids of human tubercle bacillus. I. Isolation of arabinose mycolate, *J. Biochem.* **52**:200–206.

Azuma, I., and Yamamura, Y., 1963, Studies on the firmly bound lipids of human tubercle bacillus. II. Isolation of arabinose mycolate and identification of its chemical structure, *J. Biochem.* **53**:275–281.

Ballou, C. E., Vilkas, E., and Lederer, E., 1963, Structural studies on the myo-inositol phos- pholipids of *Mycobacterium tuberculosis, J. Biol. Chem.* **238**:69–76.

Barnes, P. F., Chatterjee, D., Abrams, J. S., Lu, S., Wang, E., Yamamura, M., Brennan, P. J., and Orme, I. M., 1992, Cytokine production induced by *Mycobacterium tuberculosis* lipoarabinomannan. Relationship to chemical structure, *J. Immunol.* **149**:541–547.

Barratt, G., Tenu, J-P., Yapo, A., and J-F. Petit, 1986, Preparation and characterisation of li- posomes containing mannosylated phospholipids capable of targeting drugs to macrophages, *Biochim. Biophys. Acta* **862**:153–164.

Barrow, W. W., and Brennan, P. J., 1982, Isolation in high frequency of rough variants of *Mycobacterium intracellulare* lacking C-mycoside glycopeptidolipid antigens, *J. Bac- teriol.* **150**:381–384.

Barrow, W. W., Carvalho de Sousa, J. P., Davis, T. L., Wright, E. L., Bachelet, M., and Ras- togi, N., 1993, Immunomodulation of human peripheral blood mononuclear cell func- tions by defined lipid fractions of *Mycobacterium avium, Infect. Immun.* **61**:5286– 5293.

Barrow, W. W., Davis, T., Wright, E. L., Labrousse, V., Bachelet, M., and Rastogi, N., 1995,

Immunomodulatory spectrum of lipids associated with *Mycobacterium avium* serovar 8, *Infect. Immun.* **63**:126–133.

Belisle, J. T., and Brennan, P. J., 1989, Chemical basis of rough and smooth variation in mycobacteria, *J. Bacteriol.* **171**:3465–3470.

Belisle, J. T., McNeil, M., Chatterjee, D., Inamine, J. M., and Brennan, P. J., 1993, Expression of the core lipopeptide of the glycopeptidolipid surface antigens in rough mutants of *Mycobacterium avium, J. Biol. Chem.* **268**:10510–10516.

Berman, J. S., Blumenthal, R. L., Kornfeld, H., Cook, J. A., Cruikshank, W. W., Vermeulen, M. W., Chatterjee, D., Belisle, J. T., and Fenton, M. J., 1996, Chemotactic activity of mycobacterial lipoarabinomannans for human blood T lymphocytes *in vitro, J. Immunol.* **156**:3828–3835.

Besra, G. S., McNeil, M. R., Minnikin, D. E., Portaels, F., Ridell, M., and Brennan, P. J., 1991, Structural elucidation and antigenicity of a novel phenolic glycolipid antigen from *Mycobacterium haemophilum, Biochemistry* **30**:7772–7777.

Besra, G. S., Bolron, R. C., McNeil, M. R., Ridell, M., Simpson, K. E., Glushka, J., van Halbeek, H., Brennan, P. J., and Minnikin, D. E., 1992, Structural elucidation of a novel family of acyltrehaloses from *Mycobacterium tuberculosis, Biochemistry* **31**:9832–9837.

Besra, G. S., McNeil, M. R., Khoo, K-H., Dell, A., Morris, H. R., and Brennan, P. J., 1993a, Trehalose-containing lipooligosaccharides of *Mycobacterium gordonae:* Presence of a mono-*O*-methyltetra-*O*-acyltrehalose "core" and branching in the oligosaccharide backbone, *Biochemistry* **32**:12705–12714.

Besra, G. S., Mc Neil, M. R., Rivoire, B., Khoo, K-H., Morris, H. R., Dell, A., and Brennan, P. J., 1993b, Further structural definition of a new family of glycopeptidolipids from *Mycobacterium xenopi, Biochemistry* **32**:347–355.

Besra, G. S., Khoo, K-H., McNeil, M. R., Dell, A., Morris, H. R., and Brennan, P. J., 1995, A new interpretation of the structure of the mycolyl-arabinogalactan complex of *Mycobacterium tuberculosis* as revealed through characterization of oligoglycosylalditol fragments by fast-atom bombardment mass spectrometry and ^{1}H nuclear magnetic resonance spectroscopy, *Biochemistry* **34**:4257–4266.

Bloch, H., 1950, Studies on the virulence of tubercle bacilli. Isolation and biological properties of a constituent of virulent organisms, *J. Exp. Med.* **91**:197–218.

Bloom, B. R., and Murray, C. J. L., 1992, Tuberculosis: Commentary on a reemergent killer, *Science* **257**:1055–1064.

Bradbury, M. G., and Moreno, C., 1993, Effect of lipoarabinomannan and mycobacteria on tumor necrosis factor production by different populations of murine macrophages, *Clin. Exp. Immunol.* **94**:57–63.

Brennan, P. J., 1988, *Mycobacterium* and other actinomycetes, in: *Microbial Lipid,* Vol. 1 (C. Ratledge and S. G. Wilkinson, eds.), Academic Press, London, pp. 203–298.

Brennan, P. J., and Nikaido, H., 1995, The envelope of mycobacteria, *Annu. Rev. Microbiol.* **64**:29–63.

Brown, C. B., and Taffet, S. M., 1995, Lipoarabinomannans derived from different strains of *Mycobacterium tuberculosis* differentially stimulate the activation of NF-κB and KBF1 in murine macrophages, *Infect. Immun.* **63**:1960–1968.

Brown, I. N., and Draper, P., 1976, Growth of *Mycobacterium lepraemurium* in the mouse bone marrow: An ultrastructural study, *Infect. Immun.* **13**:1199–1204.

Brownback, P. E., and Barrow, W. W., 1988, Modified lymphocyte response to mitogens after intraperitoneal injection of glycopeptidolipid antigens from *Mycobacterium avium* complex, *Infect. Immun.* **56:**1044–1050.

Brozna, J. P., Horan, M., Rademacher, J. M., Pabst, K. M., and Pabst, M. J., 1991, Monocyte responses to sulfatide from *Mycobacterium tuberculosis:* Inhibition of priming for enhanced release of superoxide, associated with increased secretion of interleukin-1 and tumor necrosis factor alpha, and altered protein phosphorylation, *Infect. Immun.* **59:**2542–2548.

Chan, J., Fujiwara, T., Brennan, P., McNeil, M., Turco, S. J., Sibille, J-C., Snapper, M., Aisen, P., and Bloom, B. R., 1989, Microbial glycolipids: Possible virulence factors that scavenge oxygen radicals, *Proc. Natl. Acad. Sci. USA* **86:**2453–2457.

Chan, J., Fan, X., Hunter, S. W., Brennan, P. J., and Bloom, B. R., 1991, Lipoarabinomannan, a possible virulence factor involved in persistence of *Mycobacterium tuberculosis* within macrophages, *Infect. Immun.* **59:**1755–1761.

Chang, Z., Choudhary, A., Lathigra, R., and Quiocho, F. A., 1994, The immunodominant 38-kDa lipoprotein antigen of *Mycobacterium tuberculosis* is a phosphate-binding protein, *J. Biol. Chem.* **269:**1956–1958.

Chatterjee, D., Bozic, C., McNeil, M., and Brennan, P. J., 1991, Structural features of the lipoarabinomannan of *Mycobacterium tuberculosis, J. Biol. Chem.* **266:**9652–9660.

Chatterjee, D., Hunter, S. W., McNeil, M., and Brennan, P. J., 1992a, Lipoarabinomannan. Multiglycosylated form of the mycobacterial mannosylphosphatidylinositols, *J. Biol. Chem.* **267:**6228–6233.

Chatterjee, D., Roberts, A. D., Lowell, K., Brennan, P. J., and Orme, I. M., 1992b, Structural basis of capacity of lipoarabinomannan to induce secretion of tumor necrosis factor, *Infect. Immun.* **60:**1249–1253.

Chatterjee, D., Lowell, K., Rivoire, B., McNeil, M. R., and Brennan, P. J., 1992c, Lipoarabinomannan of *Mycobacterium tuberculosis.* Capping with mannosyl residues in some strains, *J. Biol. Chem.* **267:**6234–6239.

Cho, S-N., Shin, J-S., Daffé, M., Chong, Y., Kim, S-K., and Kim, J-D., 1992, Production of monoclonal antibody to phenolic glycolipid of *Mycobacterium tuberculosis* and use in the detection of the antigen in clinical isolates, *J. Clin. Microbiol.* **30:**3065–3069.

Clemens, D. L., and Horwitz, M. A., 1995, Characterization of the *Mycobacterium tuberculosis* phagosome and evidence that phagosomal maturation is inhibited, *J. Exp. Med.* **181:**257–270.

Connell, N. D., and Nikaido, N., 1994, Membrane permeability and transport in *Mycobacterium tuberculosis,* in: *Tuberculosis* (B. R. Bloom, ed.), American Society for Microbiology, Washington, DC, pp. 333–352.

Crowle, A. J., Dahl, R., Ross, E., and May, M. H., 1991, Evidence that vesicles containing living, virulent *Mycobacterium tuberculosis* or *Mycobacterium avium* in cultured human macrophages are not acidic, *Infect. Immun.* **59:**1823–1831.

Cruaud, P., Torgal Garcia, J., Papa, F., and David, H. L., 1989, Human IgG antibodies immunoreacting with specific sulfolipids from *M. tuberculosis, Zbl. Bakt* **271:**481–485.

Cruaud, P., Yamashita, J. T., Martin Casabona, N., Papa, F., and David, H. L., 1990, Evaluation of a novel 2,3-diacyl trehalose 2'-sulphate (SL-IV) antigen for case finding and diagnosis of leprosy and tuberculosis, *Res. Microbiol.* **141:**679–694.

Cywes, C., Godenir, N. L., Hoppe, H. C., Scholle, R. R., Steyn, L. M., Kirsch, R. E., and

Ehlers, M. R. W., 1996, Nonopsonic binding of *Mycobacterium tuberculosis* to human complement receptor type 3 expressed in Chinese hamster ovary cells, *Infect. Immun.* **64:**5373–5383.

Cywes, C., Hoppe, H. C., Daffé, M., and Ehlers, M. R. W., 1997, Nonopsonic binding of *Mycobacterium tuberculosis* to complement receptor type 3 is mediated by capsular polysaccharides and is strain dependent, *Infect. Immun.* **65:**4258–4266.

Daffé, M., 1989, Further specific triglycosylphenol phthiocerol diester from *Mycobacterium tuberculosis, Biochim. Biophys. Acta* **1002:**257–260.

Daffé, M., and Draper, P., 1998, The envelope layers of mycobacteria with reference to their pathogenicity, *Adv. Microbial Physiol.* **39:**131–203.

Daffé, M., and Etienne, G., 1999, The capsule of *Mycobacterium tuberculosis* and its implications for pathogenicity, *Tuberc. Lung Dis.* **79:**153–169.

Daffé, M., and Lanéelle, M.-A., 1988, Distribution of phthiocerol diesters phenolic mycosides and related compounds in mycobacteria, *J. Gen. Microbiol.* **134:**2049–2055.

Daffé, M., Lanéelle, M.-A., and Puzo, G., 1983, Structural elucidation by field desorption and electron-impact mass spectrometry of the C-mycosides isolated from *Mycobacterium smegmatis, Biochim. Biophys. Acta* **751:**439–443.

Daffé, M., Lacave, C., Lanéelle, M.-A., and Lanéelle, G., 1987, Structure of the major triglycosyl phenol-phthiocerol of *Mycobacterium tuberculosis* (strain Canetti), *Eur. J. Biochem.* **167:**155–160.

Daffé, M., Lacave, C., Lanéelle, M.-A., Gillois, M., and Lanéelle, G., 1988a, Polyphthienoyl trehalose, glycolipids specific for virulent strains of the tubercle bacillus, *Eur. J. Biochem.* **172:**579–584.

Daffé, M., Lanéelle, M.-A., Lacave, C., and Lanéelle, G., 1988b, Monoglycosyl diacyl phenol-phthiocerol of *Mycobacterium tuberculosis* and *Mycobacterium bovis, Biochim. Biophys. Acta* **958:**443–449.

Daffé, M., Dupont, M.-A., and Gas, N., 1989, The cell envelope of *Mycobacterium smegmatis:* Cytochemistry and architectural implications, *FEMS Microbiol. Lett.* **61:**89–94.

Daffé, M., Brennan, P. J., and McNeil, M., 1990, Predominant structural features of the cell wall arabinogalactan of *Mycobacterium tuberculosis* as revealed through characterization of oligoglycosyl alditol fragments by gas chromatography/mass spectrometry and by ^{1}H- and ^{13}C-NMR analyses, *J. Biol. Chem.* **265:**6734–6743.

Daffé, M., Cho, S-N., Chatterjee, D., and Brennan, P. J., 1991a, Chemical synthesis and seroactivity of a neoantigen containing the oligosaccharide hapten of the *Mycobacterium tuberculosis* specific phenolic glycolipid, *J. Infect. Dis.* **163:**161–168.

Daffé, M., McNeil, M., and Brennan, P. J., 1991b, Novel type-specific lipooligosaccharides from *Mycobacterium tuberculosis, Biochemistry* **30:**378–388.

Daffé, M., Varnerot, A., and Vincent Lévy-Frébault, V., 1992, The phenolic mycoside of *Mycobacterium ulcerans:* Structure and taxonomic implications, *J. Gen. Microbiol.* **138:**131–137.

Daffé, M., Brennan, P. J., and McNeil, M., 1993, Major structural features of the cell wall arabinogalactans of *Mycobacterium, Rhodococcus,* and *Nocardia* spp., *Carbohydr. Res.* **249:**383–398.

Dahl, K. E., Shiratsuchi, H., Hamilton, B. D., Ellner, J. J., and Toosi, Z., 1996, Selective induction of transforming growth factor b in human monocytes by lipoarabinomannan of *Mycobactererium tuberculosis, Infect. Immun.* **64:**399–405.

de Chastellier, C., Lang, T., and Thilo, L., 1995, Phagocytic processing of the macrophage endoparasite, *Mycobacterium avium,* in comparison to phagosomes which contain *Bacillus subtilis* or latex beads, *Eur. J. Cell. Biol.* **68:**167–182.

Delmas, C., Gilleron, M., Brando, T., Vercellone, A., Gheorghui, M., Rivière, M., and Puzo, G., 1997, Comparative structural study of the mannosylated-lipoarabinomannans from *Mycobacterium bovis* BCG vaccine strains: characterization and localization of suc- cinates, *Glycobiology* **7:**811–817.

Dhariwal, K. R., Yang, Y. M., Fales, H. M., and Goren, M. B., 1987, Detection of trehalose monomycolate in *Mycobacterium leprae* grown in armadillo tissues, *J. Gen. Microbi- ol.* **133:**201–209.

Dobos, K. M., Swiderek, K., Khoo, K-H., Brennan, P. J., and Belisle, J. T., 1995, Evidence for glycosylation sites on the 45-kilodalton glycoprotein of *Mycobacterium tubercu- losis, Infect. Immun.* **63:**2846–2853.

Dobos, K. M., Khoo, K-H., Swiderek, K. M., Brennan, P. J., and Belisle, J. T., 1996, Defi- nition of the full extent of glycosylation of the 45-kilodalton glycoprotein of *My- cobacterium tuberculosis, J. Bacteriol.* **178:**2498–2506.

Draper, P., 1971, The walls of *Mycobacterium lepraemurium:* Chemistry and ultrastructure, *J. Gen. Microbiol.* **69:**313–324.

Draper, P., 1974, The mycoside capsule of *Mycobacterium avium* 357, *J. Gen. Microbiol.* **83:**431–433.

Draper, P., Kandler, O., and Darbre, A., 1987, Peptidoglycan and arabinogalactan of *My- cobacterium leprae, J. Gen. Microbiol.* **133:**1187–1194.

Dubos, R. J., and Middlebrook, G., 1948, Cytochemical reaction of virulent tubercle bacil- li, *Am. Rev. Tuberc.* **58:**698–699.

Ehlers, M. R. W., and Daffé, M., 1998, Interactions between *Mycobacterium tuberculosis* and host cells: Are mycobacterial sugars the key: *Trends Microbiol.* **3:**328–335.

Ellner, J. J., and Daniel, T. M., 1979, Immunosuppression by mycobacterial arabinoman- nan, *Clin. Exp. Immunol.* **35:**250–257.

Escamilla, L., Mancilla, R., Glender, W., and López-Marín, L. M., 1996, *Mycobacterium fortuitum* glycolipids for the serodiagnosis of pulmonary tuberculosis, *Am. J. Respir. Crit. Care Med.* **154:**1864–1867.

Espitia, C., and Mancilla, R., 1989, Identification, isolation and partial characterization of *Mycobacterium tuberculosis* glycoprotein antigens, *Clin. Exp. Immunol.* **77:**378–383.

Espitia, C., Elinos, M., Hernandez-Pando, R., and Mancilla, R. A., 1992, Phosphate star- vation enhances expression of the immunodominant 38-kilodalton protein antigen of *Mycobacterium tuberculosis:* Demonstration by immunogold electron microscopy, *In- fect. Immun.* **60:**2998–3001.

Espitia, C., Espinosa, R., Saavedra, R., Mancilla, R., Romain, F., Laqueyrerie, A., and Moreno, C., 1995, Antigenic and structural similarities between *Mycobacterium tu- berculosis* 50- to 55-kilodalton and *Mycobacterium bovis* BCG 45- to 47-kilodalton antigens, *Infect. Immun.* **63:**580–884.

Fifis, T., Costopoulos, C., Radford, A. J., Bacic, A., and Wood, P. R., 1991, Purification and characterization of major antigens from *Mycobacterium bovis* culture filtrate, *Infect. Immun.* **59:**800–807.

Furuchi, A., and Tokunaga, T., 1972, Nature of the receptor substance of *Mycobacterium smegmatis* for D4 bacteriophage adsorption, *J. Bacteriol.* **111:**404–411.

Garbe, T., Harris, D., Vordermeier, M., Lathigra, R., Ivanyi, J., and Young, D., 1993, Expression of the *Mycobacterium tuberculosis* 19-kilodalton antigen in *Mycobacterium smegmatis:* Immunological analysis and evidence of glycosylation, *Infect. Immun.* **61:**260–267.

Gastambide-Odier, M., and Sarda, P., 1970, Contributions à l'étude de la structure et de la biosynthèse de glycolipides spécifiques isolés de mycobactéries, les mycosides A et B, *Pneumologie* **142:**241–255.

Gastambide-Odier, M., Sarda, P., and Lederer, E., 1965, Structure des aglycones des mycosides A et B, *Tetrahedron Lett.* **35:**3135–3143.

Gastambide-Odier, M., Sarda, P., and Lederer, E., 1967, Biosynthèse des aglycones des mycosides A et B, *Bull. Soc. Chim. Biol.* **49:**849–864.

Gautier, N., López-Marín, L. M., Lanéelle, M.-A., and Daffé, M., 1992, Structure of mycoside F, a family of trehalose-containing glycolipids of *Mycobacterium fortuitum, FEMS Microbiol. Lett.* **98:**81–88.

Gaylord, H., and Brennan, P. J., 1987, Leprosy and the leprosy bacillus: Recent developments in characterization of antigens and immunology of disease, *Annu. Rev. Microbiol.* **41:**645–675.

Gilleron, M., and Puzo, G., 1995, Lipooligosaccharidic antigens from *Mycobacterium kansasii* and *Mycobacterium gastri, Glyconjugate J.* **12:**298–308.

Gilleron, M., Venisse, A., Fournié, J.-J., Rivière, M., Dupont, M.-A., and Puzo, G., 1990, Structural and immunological properties of the phenolic glycolipids from *Mycobacterium gastri* and *Mycobacterium kansasii, Eur. J. Biochem.* **186:**167–173.

Gilleron, M., Himoudi, N., Adam, O., Constant, P., Venisse, A., Rivière, M., and Puzo, G., 1997, *Mycobacterium smegmatis* phosphoinositols-glyceroarabinomannans: Structure and localization of alkali-labile and alkali-stable phosphoinositides, *J. Biol. Chem.* **272:**117–124.

Gilleron, M., Nigou, J., Cahuzac, B., and Puzo, G., 1999, Structural study of the lipomannans from *Mycobacterium bovis* BCG: Characterization of multiacylated forms of the phosphatidyl-*myo*-inositol anchor, *J. Mol. Biol.* **285:**2147–2160.

Goren, M. B., 1970a, Sulfolipid I of *Mycobacterium tuberculosis,* strain H37Rv.I. Purification and properties, *Biochim. Biophys. Acta* **210:**116–126.

Goren, M. B., 1970b, Sulfolipid I of *Mycobacterium tuberculosis,* strain H37Rv.II. Structural studies, *Biochim. Biophys. Acta* **210:**127–138.

Goren, M. B., 1990, Mycobacterial fatty acid esters of sugars and sulfosugars, *Handbook Lip. Res.* **6:**363–461.

Goren, M. B., and Brennan, P. J., 1979, Mycobacterial lipids: Chemistry and biologic activities, in: *Tuberculosis* (G. P. Youmans, ed.), W B Saunders, Philadelphia, pp. 63–193.

Goren, M. B., McClatchy, J. K., Martens, B., and Brokl, O., 1972, Mycosides C: Behavior as receptor site substance for mycobacteriophage D4, *J. Virol.* **9:**999–1003.

Goren, M. B., Brokl, O., and Schaefer, W. B., 1974, Lipids of putative relevance to virulence in *Mycobacterium tuberculosis:* Correlation of virulence and elaboration of sulfatides and strongly acidic lipids, *Infect. Immun.* **9:**142–149.

Goren, M. B., Hart, P. d'Arcy, Young, M. R., and Armstrong, J. A., 1976, Prevention of phagosome–lysosome fusion in cultured macrophages by sulphatide of *Mycobacterium tuberculosis, Proc. Natl. Acad. Sci. USA* **73:**2510–1514.

Goren, M. B., Vatter, A. E., and Fiscus, J., 1987, Polyanionic agents as inhibitors of phago-some–lysosome fusion in cultured macrophages: Evolution of an alternative interpretation, *J. Leukocyte Biol.* **41:**111–121.

Henderson, B., Poole, S., and Wilson, M., 1996, Bacterial modulins: A novel class of virulence factors which cause host tissue pathology by inducing cytokine synthesis, *Microbiol. Rev.* **60:**316–341.

Herrmann, J. L., O'Gaora, P., Gallagher, A., Thole, J. E. R., and Young, D. B., 1996, Bacterial glycoproteins: A link between glycosylation and proteolytic cleavage of a 19 kDa antigen from *Mycobacterium tuberculosis, EMBO J.* **15:**3547–3554.

Hooper, L. C., and Barrow, W. W., 1988, Decreased mitogenic response of murine spleen cells following intraperitoneal injection of serovar-specific glycopeptidolipid antigens from the *Mycobacterium avium* complex, *Adv. Exp. Med. Biol.* **239:**309–325.

Hooper, L. C., Johnson, M. M., Khera, V. R., and Barrow, W. W., 1986, Macrophage uptake and retention of radiolabeled glycopeptidolipid antigens associated with the superficial L₁ layer of *Mycobacterium intracellulare* serovar 20, *Infect. Immun.* **54:**133–141.

Hoppe, H., Barend, C., de Wet, J. M., Cywes, C., Daffé, M., and Ehlers, M. R. W., 1997, Identification of phosphatidylinositol mannoside as a mycobacterial adhesin mediating both direct and opsonic binding to nonphagocytic mammalian cells, *Infect. Immun.* **65:**3896–3905.

Hunter, S. W., and Brennan, P. J., 1981, A novel phenolic glycolipid from *Mycobacterium leprae* possibly involved in immunogenicity and pathogenicity, *J. Bacteriol.* **147:**728–735.

Hunter, S. W., and Brennan, P. J., 1990, Evidence for the presence of a phosphatidylinositol anchor on the lipoarabinomannan and lipomannan of *Mycobacterium tuberculosis, J. Biol. Chem.* **265:**9272–9279.

Hunter, S. W., Gaylord, H., and Brennan, P. J., 1986, Structure and antigenicity of the phosphorylated lipopolysaccharides from the leprosy and tubercle bacilli, *J. Biol. Chem.* **261:**12345–12351.

Ilangumaran, S., Arni, S., Poincelet, M., Theler, J. M., Brennan, P. J., Nasir-ud-Din, and Hoessli, D. C., 1995, Integration of mycobacterial lipoarabinomannans into glycosylphosphatidylinositol-rich domains of lymphomonocytic cell plasma membranes, *J. Immunol.* **155:**1334–1342.

Johannsen, L., Rosenthal, R. S., Martin, S. A., Cady, A. B., Obal, F., Guinand, M., and Krueger, J. M., 1989, Somnogenic activity of *O*-acetylated and dimeric muramyl peptides, *Infect. Immun.* **57:**2726–2732.

Kent, P. W., 1951, Structure of an antigenic polysaccharide isolated from tuberculin, *J. Chem. Soc.* **1:**364–368.

Khoo, K-H., Dell, A., Morris, H. R., Brennan, P. J., and Chatterjee, D., 1995a, Inositol phosphate capping of the nonreducing termini of lipoarabinomannan from rapidly growing strains of *Mycobacterium, J. Biol. Chem.* **270:**12380–12389.

Khoo, K-H., Dell, A., Morris, H. R., Brennan, P. J., and Chatterjee, D., 1995b, Structural definition of acylated phosphatidylinositol mannosides from *Mycobacterium tuberculosis:* Definition of a common anchor for lipomannan and lipoarabinomannan, *Glycobiology* **5:**117–127.

Khuller, G. K., Penumarti, N., and Subrahmanyam, D., 1983, Induction of resistance to tuberculosis in guinea pigs with mannophosphoinositides of mycobacteria, *IRCS Med. Sci.* **11**:288–289.

Klatt, E. C., Jensen, D. F., and Meyer, P. R., 1987, Pathology of *Mycobacterium avium-intracellulare* infection in acquired immunodeficiency syndrome, *Hum. Pathol.* **18**:709–714.

Kleinhenz, M. E., Ellner, J. J., Spagnulo, P. J., and Daniel, T. M., 1981, Suppression of lymphocyte response by tuberculous plasma and mycobacterial arabinogalactan. Monocyte dependence and indomethacin reversibility, *J. Clin. Invest.* **68**:153–162.

Kotani, S., Kitaura, T., Hirano, T., and Tanaka, A., 1959, Isolation and chemical composition of the cell walls of BCG, *Bikens J.* **2**:129–141.

Kuroda, K., Brown, E. J., Telle, W. B., Russell, D. G., and Ratliff, T. L., 1993, Characterization of the internalization of bacillus Calmette-Guérin by human bladder tumor cells, *J. Clin. Invest.* **91**:69–76.

Lanéelle, M.-A., Silve, G., López-Marín, L. M., and Daffé, M., 1996, Structures of the glycolipid antigens of members of the third biovariant complex of *Mycobacterium fortuitum, Eur. J. Biochem.* **238**:270–279.

Laqueyrerie, A., Militzer, P., Romain, F., Eiglmeier, K., Cole, S., and Marchal, G., 1995, Cloning, sequencing, and expression of the *apa* gene coding for the *Mycobacterium tuberculosis* 45/47-kilodalton secreted antigen complex, *Infect. Immun.* **63**:4003–4010.

Lathigra, R., Zhang, Y., Hill, M., Garcia, M. J., Jackett, P. S., and Ivanyi, J., 1996, Lack of production of the 19-kDa glycolipoprotein in certain strains of *Mycobacterium tuberculosis, Res. Microbiol.* **147**:237–249.

Launois, P., Blum, L., Dieye, A., Millan, J. J., Sarthou, L., and Bach, M. A., 1989, Phenolic glycolipid-1 from *M. leprae* inhibits oxygen free radical production by human mononuclear cells, *Res. Immunol.* **140**:847–855.

Lee, Y. C., and Ballou, C. E., 1964, Structural studies on the myo-inositol mannosides from the glycolipids of *Mycobacterium tuberculosis* and *Mycobacterium phlei, J. Biol. Chem.* **239**:1316–1327.

Lee, Y. C., and Ballou, C. E., 1965, Complete structures of the glycophospholipids of mycobacteria, *Biochemistry* **4**:1395–1404.

Lemassu, A., and Daffé, M., 1994, Structural features of the exocellular polysaccharides of *Mycobacterium tuberculosis, Biochem. J.* **297**:351–357.

Lemassu, A., Lanéelle, M.-A., and Daffé, M., 1991, Revised structure of a trehalose-containing immunoreactive glycolipid of *Mycobacterium tuberculosis, FEMS Microbiol. Lett.* **62**:171–175.

Lemassu, A., Vincent Lévy-Frébault, V., Lanéelle, M.-A., and Daffé, M., 1992, Lack of correlation between colony morphology and lipooligosaccharide content in the *Mycobacterium tuberculosis* complex, *J. Gen. Microbiol.* **138**:1535–1541.

Lemassu, A., Ortalo-Magné, A., Bardou, F., Silve, G., Lanéelle, M.-A., and Daffé, M., 1996, Extracellular and surface-exposed polysaccharides of non-tuberculous mycobacteria, *Microbiology* **142**:1513–1520.

Leopold, K., and Fisher, W., 1993, Molecular analysis of the lipoglycans of *Mycobacterium tuberculosis, Anal. Biochem.* **208**:57–64.

López-Marín, L. M., Lanéelle, M.-A., Promé, D., Daffé, M., Lanéelle, G., and Promé, J-C., 1991, Glycopeptidolipids from *Mycobacterium fortuitum:* A variant in the structure of C-mycoside, *Biochemistry* **30**:10536–10542.

López-Marín, L. M., Lanéelle, M.-A., Promé, D., Lanéelle, G., Promé, J-C., and Daffé, M., 1992, Structure of a novel sulfate-containing mycobacterial glycolipid, *Biochemistry* **31**:11106–11111.

López-Marín, L. M., Lanéelle, M.-A., Promé, D., and Daffé, M., 1993, Structures of the glycopeptidolipid antigens of two animal pathogens: *Mycobacterium senegalense* and *Mycobacterium porcinum, Eur. J. Biochem.* **215**:859–866.

López-Marín, L. M., Gautier, N., Lanéelle, M.-A., Silve, G., and Daffé, M., 1994a, Structures of the glycopeptidolipid antigens of *Mycobacterium abscessus* and *Mycobacterium chelonae* and possible chemical basis of the serological cross-reactions in the *Mycobacterium fortuitum* complex, *Microbiology* **140**:1109–1118.

López-Marín, L. M., Quesada, D., Lakhdar-Ghazal, F., Tocanne, J-F., and Lanéelle, G., 1994b, Interactions of mycobacterial glycopeptidolipids with membranes: Influence of carbohydrate on induced alterations, *Biochemistry* **33**:7056–7061.

Martin Casabona, N., Gonzalez Fuente, T., Arcalis Arce, L., Ortal Entraigos, J., and Vidal Pla, R., 1989, Evaluation of a phenolglycolipid antigen (PGL-Tb1) from *M. tuberculosis* in the serodiagnosis of tuberculosis: comparison with PPD antigen, *Acta Leprol.* **7**(suppl. 1):S89–S93.

McNeil, M. R., and Brennan, P. J., 1991, Structure, function and biogenesis of the cell envelope of mycobacteria in relation to bacterial physiology, pathogenesis and drug resistance; Some thoughts and possibilities arising from recent structural information, *Res. Microbiol.* **142**:451–463.

McNeil, M., Daffé, M., and Brennan, P. J., 1990, Evidence for the nature of the link between the arabinogalactan and peptidoglycan of mycobacterial cell walls, *J. Biol. Chem.* **265**:18200–18206.

McNeil, M., Daffé, M., and Brennan, P. J., 1991, Location of the mycoloyl ester substituents in the cell walls of mycobacteria, *J. Biol. Chem.* **266**:13217–13223.

Mehta, P. K., and Khuller, G. K., 1988, Protective immunity to experimental tuberculosis by mannophosphoinositides of mycobacteria, *Med. Microbiol. Immunol.* **177**:265–284.

Menozzi, F. D., Rouse, J. H., Alavi, M., Laude-Sharp, M., Muller, J., Bischoff, R., Brennan, M. J., and Locht, C., 1996, Identification of a heparin-binding hemagglutinin present in mycobacteria, *J. Exp. Med.* **184**:993–1001.

Menozzi, F. D., Bischoff, R., Fort, E., Brennan, M., and Locht, C., 1998, Molecular characterization of the mycobacterial heparin-binding hemagglutinin, a mycobacterial adhesin, *Proc. Natl. Acad. Sci. USA* **95**:12625–12630.

Middlebrook, G., Dubos, R. J., and Pierce, C., 1947, Virulence and morphological characteristics of mammalian tubercle bacilli, *J. Exp. Med.* **86**:175–184.

Miller, R. A., Harnisch, J. P., and Buchanan, T. M., 1984, Antibodies to mycobacterial arabinomannan in leprosy: Correlation with reactional states and variation during treatment, *Int. J. Lepr.* **52**:133–139.

Minnikin, D. E., 1982, Lipids: Complex lipids, their chemistry, biosynthesis and roles, in: *The Biology of the Mycobacteria,* Vol. 1 (C. Ratledge and J. Stanford, eds.), Academic Press, London, pp. 95–184.

Minnikin, D. E., Dobson, G., Sesardie, D., and Ridell, M., 1985, Mycolipenates of trehalose from *Mycobacterium tuberculosis, J. Gen. Microbiol.* **131**:1369–1374.

Misaki, A., and Yukawa, S., 1966, Studies on cell walls of Mycobacteria. II. Constitution of polysaccharides from BCG cell walls, *J. Biochem.* **59**:511–520.

Misaki, A., Seto, N., and Azuma, I., 1974, Structure and immunological properties of D-arabino-D-galactan isolated from cell walls of *Mycobacterium* species, *J. Biochem.* **76**:15–27.

Misaki, A., Azuma, I., and Yamamura, Y., 1977, Structural and immunochemical studies on D-arabino-D-mannans and D-mannans of *Mycobacterium tuberculosis* and other *Mycobacterium* species, *J. Biochem.* **82**:1759–1770.

Molloy, A., Gaudernack, G., Levis, W. R., Cohn, Z. A., and Kaplan, G., 1990, Suppression of T-cell proliferation by *Mycobacterium leprae* and its products: The role of lipopolysaccharide, *Proc. Natl. Acad. Sci. USA* **87**:973–977.

Moran, N., 1996, WHO issues another gloomy tuberculosis report, *Nature Med.* **2**:377.

Moreno, C., Mehlert, A., and Lamb, J., 1988, The inhibitory effects of mycobacterial lipoarabinomannan and polysaccharides upon polyclonal and monoclonal human T-cell proliferation, *Clin. Exp. Immunol.* **74**:206–210.

Muñoz, M., Lanéelle, M.-A., Luquin, M., Torrelles, J., Julián, E., Ausina, V., and Daffé, M., 1997, Occurrence of an antigenic triacyl trehalose in clinical isolates and reference strains of *Mycobacterium tuberculosis, FEMS Microbiol. Lett.* **157**:251–259.

Muñoz, M., Raynaud, C., Lanéelle, M.-A., Julián, E., López-Marín, L., Silve, G., Ausina, V., Daffé, M., and Luquin, M., 1998, Seroreactive species-specific lipooligosaccharides of *Mycobacterium mucogenicum* sp. nov. (formerly *Mycobacterium chelonae*-like organisms): Identification and chemical characterization, *Microbiology* **144**:137–148.

Nara, T., Kawamoto, I., Okachi, R., and Oka, T., 1977, Source of antibiotics other than *Streptomyces, Jpn. J. Antibiot.* **30**(suppl.):S174–S189.

Neill, M. A., and Klebanoff, S. J., 1988, The effect of phenolic glycolipid-1 from *Mycobacterium leprae* on the antimicrobial activity of human macrophages, *J. Exp. Med.* **167**:30–42.

Nigou, J., Gilleron, M., Cahuzac, B., Bounéry, J-D., Herold, M., Thurnher, M., and Puzo, G., 1997, The phosphatidyl-*myo*-inositol anchor of the lipoarabinomannans from *Mycobacterium bovis* Bacillus Calmette Guérin, *J. Biol. Chem.* **272**:23094–23103.

Nigou, J., Gilleron, M., and Puzo, G., 1999, Lipoarabinomannans: characterization of multiacylated forms of the phosphatidyl-*myo*-inositol anchor by nuclear magnetic resonance spectroscopy, *Biochem. J.* **337**:453–460.

Noll, H., Bloch, H., Asselineau, J., and Lederer, E., 1956, The chemical structure of the cord factor of *Mycobacterium tuberculosis, Biochim. Biophys. Acta* **20**:299–309.

Ortalo-Magné, A., Dupont, M-A., Lemassu, A., Andersen, A. B., Gounon, P., and Daffé, M., 1995, Molecular composition of the outermost capsular material of the tubercle bacillus, *Microbiology* **141**:1609–1620.

Ortalo-Magné, A., Andersen, A. B., and Daffé, M., 1996a, The outermost capsular arabinomannans and other mannoconjugates of virulent and avirulent tubercle bacilli, *Microbiology* **142**:927–935.

Ortalo-Magné, A., Lemassu, A., Lanéelle, M.-A., Bardou, F., Silve, G., Gounon, P., Marchal, G., and Daffé, M., 1996b, Identification of the surface-exposed lipids on the cell envelopes of *Mycobacterium tuberculosis* and other mycobacterial species, *J. Bacteriol.* **178**:456–461.

Pabst, M. J., Gross, J. M., Brozna, J. P., and Goren, M. B., 1988, Inhibition of macrophage priming by sulfatide from *Mycobacterium tuberculosis, J. Immunol.* **140**:634–640.

Pangborn, M. C., and McKinney, J. A., 1966, Purification of serologically active phosphoinositides of *Mycobacterium tuberculosis, J. Lipid Res.* **7**:627–637.

Papa, F., Cruaud, P., and David, H. L., 1989, Antigenicity and specificity of selected glycolipid fractions from *Mycobacterium tuberculosis, Res. Microbiol.* **140:**569–578.

Petit, J. F., Adam, A., Wietzerbin-Falszpan, J., Lederer, E., and Ghuysen, J. M., 1969, Chemical structure of the cell wall of *Mycobacterium smegmatis.* I—Isolation and partial
characterization of the peptidoglycan, *Biochem. Biophys. Res. Commun.* **35:**478–485.

Pourshafie, M., Ayub, Q., and Barrow, W. W., 1993, Comparative effects of *Mycobacterium avium* glycopeptidolipid and lipopeptide fragment on the function and ultrastructure of mononuclear cells, *Clin. Exp. Immunol.* **93:**72–79.

Prinzis, S., Chatterjee, D., and Brennan, P. J., 1993, Structure and antigenicity of lipoarabinomannan from *Mycobacterium bovis* BCG, *J. Gen. Microbiol.* **139:**2649–2658.

Promé, J-C., Lacave, C., Ahibo-Coffy, A., and Savagnac, A., 1976, Séparation et étude
structurale des espèces moléculaires de monomycolates et de dimycolates de α-D-
tréhalose présents chez *Mycobacterium phlei, Eur. J. Biochem.* **63:**543–552.

Puzo, G., 1990, The carbohydrate- and lipid-containing cell wall of mycobacteria, phenolic glycolipids: Structure and immunological properties, *Crit. Rev. Microbiol.* **17:**305–
327.

Ramanathan, V. D., Parkash, O., Tyagi, P., Sengupta, U., and Ramu, G., 1990, Activation
of the human complement system by phenolic glycolipid 1 of *Mycobacterium leprae,
Microb. Pathog.* **8:**403–410.

Rastogi, N., Lévy-Frebault, V., Blom-Potar, M-C., and David, H. L., 1989, Ability of
smooth and rough variants of *Mycobacterium avium* and *M. intracellulare* to multiply
and survive intracellularly: Role of C-mycosides, *Zbl. Bakt. Hyg.* **270:**345–360.

Ratliff, T. L., McCarthy, R., Telle, W. B., and Brown, E. J., 1993, Purification of a mycobacterial adhesin for fibronectin, *Infect. Immun.* **61:**1889–1894.

Reggiardo, Z., and Middlebrook, G., 1974, Serologically active glycolipid families from
Mycobacterium bovis BCG. II. Serologic studies on human sera, *Am. J. Epidemiol.*
100:477–486.

Rivière, M., and Puzo, G., 1992, Use of ^{1}H NMR ROESY for structural determination of
O-glycosylated amino acids from a serine-containing glycopeptidolipid antigen, *Biochemistry* **31:**3757–3580.

Roach, T. I. A., Barton, C. H., Chatterjee, D., and Blackwell, J. M., 1993, Macrophage activation: Lipoarabinomannan from avirulent and virulent strains of *Mycobacterium tuberculosis* differentially induces the early genes c-fos, KC, JE, and tumor necrosis factor-α, *J. Immunol.* **150:**1886–1896.

Roach, T. I. A., Chatterjee, D., and Blackwell, J. M., 1994, Induction of early-response
genes KC and JE by mycobacterial lipoarabinomannan: Regulation of KC expression
in murine macrophages by *Lsh/Ity/Bcg* (candidate *Nramp*), *Infect. Immun.* **62:**1176–
1184.

Romain, F., Laqueyrerie, A., Militzer, P., Pescher, P., Chavarot, P., Lagranderie, M., Aure-
gan, G., Gheorghiu, M., and Marchal, G., 1993, Identification of a *Mycobacterium bovis* BCG 45/47-kilodalton antigen complex, an immunodominant target for antibody
response after immunization with living bacteria, *Infect. Immun.* **61:**742–750.

Rulong, S., Aguas, A. P., da Silva, P. P., and Silva, M. T., 1991, Intramacrophagic *Mycobacterium avium bacilli* are coated by multiple lamella structure: freeze fracture
analysis of infected mouse liver, *Infect. Immun.* **59:**3895–3902.

Schlesinger, L. S., and Horwitz, M. A., 1991, Phenolic glycolipid-1 of *Mycobacterium lep-*

rae binds complement component C3 in serum and mediates phagocytosis by human monocytes, *J. Exp. Med.* **174:**1031–1038.

Schlesinger, L. S., Hull, S. R., and Kaufman, T. M., 1994, Binding of terminal mannosyl units of lipoarabinomannan from a virulent strain of *Mycobacterium tuberculosis* to human macrophages, *J. Immunol.* **152:**4070–4079.

Schorey, J. S., Li, Q., McCourt, D. W., Bong-Mastek, M., Clark-Curtiss, J. E., Ratliff, T. L., and Brown, E. J., 1995, A *Mycobacterium leprae* gene encoding a fibronectin binding protein is used for efficient invasion of epithelial cells and Schwann cells, *Infect. Immun.* **63:**2652–2657.

Schorey, J. S., Holsti, M-A., Ratliff, T. L., Allen, P. M., and Brown, E. J., 1996, Characterization of the fibronectin-attachment protein of *Mycobacterium avium* reveals a fibronectin-binding motif conserved among mycobacteria, *Mol. Microbiol.* **21:**321–329.

Schuller-Levis, G. B., Levis, W. R., Ammazzalorso, M., Nostrati, A., and Park, E., 1994, Mycobacterial lipoarabinomannan induces nitric oxide and tumor necrosis factor alpha production in a macrophage cell line: down regulation by taurine chloramine, *Infect. Immun.* **62:**4671–4674.

Seibert, F. B., 1949, The isolation of three different proteins and two polysaccharides from tuberculin by alcohol fractionation. Their chemical and biological properties, *Am. Rev. Tuberc.* **59:**86–101.

Silva, M. T., and Macedo, P. M., 1983a, A comparative ultrastructural study of the membranes of *Mycobacterium leprae* and of cultivable mycobacteria, *Biol. Cell.* **47:**383–386.

Silva, M. T., and Macedo, P. M., 1983b, The interpretation of the ultrastructure of mycobacterial cells in transmission electron microscopy in ultrathin sections, *Int. J. Lepr.* **51:**225–234.

Silva, M. T., and Macedo, P. M., 1984, Ultrastructural characterization of normal and damaged membranes of *Mycobacterium leprae* and cultivable mycobacteria, *J. Gen. Microbiol.* **130:**369–380.

Simonney, N., Molina, J. M., Molimard, M., Oksenhendler, E., Peronne, C., and Lagrange, P., 1995, Analysis of the immunological humoral response to *Mycobacterium tuberculosis* glycolipid antigens (DAT, PGLTb1) for diagnosis of tuberculosis in HIV-seropositive and -seronegative patients, *Eur. J. Clin. Microbiol. Infect. Dis.* **14:**883–891.

Smith, D. W., Randall, H. M., MacLennan, A. P., and Lederer, E., 1960a, Mycosides: A new class of type-specific glycolipids of mycobacteria, *Nature* **186:**887–888.

Smith, D. W., Randall, H. M., MacLennan, A. P., Putney, R. K., and Rao, S. V., 1960b, Detection of specific lipids in mycobacteria by infrared spectroscopy, *J. Bacteriol.* **79:**217–229.

Sonnenberg, M. G., and Belisle, J. T., 1997, Definition of *Mycobacterium tuberculosis* culture filtrate proteins by two-dimensional polyacrylamide gel electrophoresis, N-terminal amino-acid sequencing and electrospray mass spectrometry, *Infect. Immun.* **65:**4515–4524.

Stokes, R. W., and Speert, D. P., 1995, Lipoarabinomannan inhibits nonopsonic binding of *Mycobacterium tuberculosis* to murine macrophages, *J. Immunol.* **155:**1361–1369.

Strain, S. M., Toubiana, R., Ribi, E., and Parker, R., 1977, Separation of the mixture of trehalose 6,6'-dimycolates comprising the mycobacterial glycolipid fraction, "P3," *Biochem. Biophys. Res. Commun.* **77**:449–456.

Sturgill-Koszycki, S., Schlesinger, P. H., Chakraborty, P., Haddix, P. L., Collins, H. L., Fok, A. K., Allen, R. D., Gluck, S. L., Heuser, J., and Russel, D. G., 1994, Lack of acidification in *Mycobacterium* produced by exclusion of the vesicular proton-ATPase, *Science* **263**:678–681.

Sut, A., Sirugue, S., Sixou, S., Lakhdar-Ghazal, F., Tocanne, J-F., and Laneelle, G., 1990, Mycobacteria glycolipids as potential pathogenicity effectors: Alteration of model and natural membranes, *Biochemistry* **29**:8498–8502.

Tenu, J-P., Sekkaï, D., Yapo, A., Petit, J-F., and Lemaire, G., 1995, Phosphatidylinositol-mannoside-based liposomes induce NO synthase in primed mouse peritoneal macrophages, *Biochem. Biophys. Res. Commun.* **208**:295–301.

Tereletsky, M. J., and Barrow, W. W., 1983, Postphagocytic detection of glycopeptidolipids associated with superficial L1 layer of *Mycobacterium intracellulare*, *Infect. Immun.* **41**:1312–1321.

Tsang, A. Y., Barr, V. L., McClatchy, J. K., Goldberg, M., Drupa, I., and Brennan, P. J., 1984, Antigenic relationships of the *Mycobacterium fortuitum, Mycobacterium chelonae* complex, *Int. J. Syst. Bacteriol.* **34**:35–44.

Vachula, M. T., Holzer, J., and Andersen, B. R., 1989, Suppression of monocyte oxidative response by phenolic glycolipid I of *Mycobacterium leprae*, *J. Immunol.* **60**:203–206.

Venisse, A., Berjeaud, J.-M., Chaurand, P., Gilleron, M., and Puzo, G., 1993, Structural features of lipoarabinomannan from *Mycobacterium bovis* BCG. Determination of molecular mass by laser desorption mass spectrometry, *J. Biol. Chem.* **268**:12401–12411.

Venisse, A., Rivière, M., Vercauteren, J., and Puzo, G., 1995a, Structural analysis of the mannan region of lipoarabinomannan from *Mycobacterium bovis* BCG. Heterogeneity in phosphorylation state, *J. Biol. Chem.* **270**:15012–15021.

Venisse, A., Fournié, J-J., and Puzo, G., 1995b, Mannosylated lipoarabinomannan interacts with phagocytes, *Eur. J. Biochem.* **231**:440–447.

Vergne, I., and Daffé, M., 1998, Interaction of mycobacterial glycolipids with host cells, *Front. Biosci.* **3**:d865–d876.

Vilkas, E., 1960, Sur divers types de phospholipides présents dans le bacille Calmette-Guèrin, *Bull. Soc. Chim. Biol.* **42**:1005–1011.

Watanabe, M., Yamada, Y., Iguchi, K., and Minnikin, D. E., 1994, Structural elucidation of new phenolic glycolipids from *Mycobacterium tuberculosis*, *Biochim. Biophys. Acta* **1210**:174–180.

Weber, P. L., and Gray, G. R., 1979, Structural and immunochemical characterization of the acidic arabinomannan of *Mycobacterium smegmatis*, *Carbohydr. Res.* **74**:259–278.

Wietzerbin-Falszpan, J., Das, B. C., Azuma, I., Adam, A., Petit, J. F., and Lederer, E., 1970, Isolation and mass spectrometric identification of the peptide subunits of mycobacterial cell walls, *Biochem. Biophys. Res. Commun.* **40**:57–63.

Wietzerbin, J., Das, B. C., Petit, J-F., Lederer, E., Leyh-Bouille, M., and Ghuysen, J-M., 1974, Occurrence of D-alanyl-(D)-*meso*-diaminopimelic acid and *meso*-diaminopime-lyl-*meso*-diaminopimelic acid interpeptide linkages in the peptidoglycan of mycobacteria, *Biochemistry* **13**:3471–3476.

Young, D. B., and Garbe, T. R., 1991, Lipoprotein antigens of *Mycobacterium tuberculosis*, *Res. Microbiol.* **142**:55–65.

Zhang, L., English, D., and Andersen, B. R., 1991, Activation of human neutrophils by *Mycobacterium tuberculosis*-derived sulfolipid-1, *J. Immunol.* **146:**2730–2736.
Zhang, Y., Broser, M., Cohen, H., Bodkin, M., Law, K., Reibman, J., and Rom, W. N., 1995, Enhanced interleukin-8 release and gene expression in macrophages after exposure to *Mycobacterium tuberculosis* and its components, *J. Clin. Invest.* **95:**586–592.

9

Biosynthesis and Regulation of Expression of Group 1 Capsules in *Escherichia coli* and Related Extracellular Polysaccharides in Other Bacteria

Chris Whitfield, Jolyne Drummelsmith,
Andrea Rahn, and Thomas Wugeditsch

1. INTRODUCTION: DIVERSITY OF CELL SURFACE POLYSACCHARIDES IN *ESCHERICHIA COLI*

Strains of *Escherichia coli* produce an extensive range of different cell surface polysaccharides (reviewed in Hull, 1997; Jann and Jann, 1997; Whitfield *et al.,* 1994). All strains produce the enterobacterial common antigen, although this polymer currently has no established role in the biology of *E. coli* (Rick and Silver, 1996). Many (but not all) strains of *E. coli* can produce a loosely cell-associated exopolysaccharide called colanic acid. This polymer is generally not produced at growth temperatures above 30°C, but large amounts can be synthesized in response to specific mutations (Gottesman, 1995). As might be expected given its absence at 37°C, colanic acid has no obvious role in the pathogenesis of *E. coli* and it most

Chris Whitfield, Jolyne Drummelsmith, Andrea Rahn, and Thomas Wugeditsch • Department of Microbiology, University of Guelph, Guelph, Ontario, N1G 2W1, Canada.

Glycomicrobiology, edited by Doyle.
Kluwer Academic/Plenum Publishers, New York, 2000.

likely plays a role in survival of this bacterium outside the host. In *E. coli*, the capsular polysaccharides and the side chain polysaccharides of the lipopolysaccharide (LPS) molecules form the major surface polysaccharides expressed at 37°C. These polymers are serotype-specific, giving rise to the K- and O-antigens, respectively. Variations in sugar composition, linkage specificity, as well as substitution with noncarbohydrate residues results in 167 different O-serogroups and more than 70 polysaccharide K-antigens in *E. coli*. The primary structures of many of these antigens have been determined (see the Complex Carbohydrate Structure Database; www.ccrc.uga.edu/). The O- and K-antigens provide recognized virulence determinants (Hull, 1997; Jann and Jann, 1997; Whitfield *et al.*, 1994). Generally, the O-antigens are important for resistance to complement-mediated serum killing, whereas the capsular K-antigens are responsible for resistance against phagocytosis. However, as is usual with such generalizations, there are some notable exceptions.

The significant advances that have been made in our understanding of synthesis and expression of *E. coli* capsules have allowed their classification into four different groups (Whitfield and Roberts, 1999). In this chapter, we will describe the group 1 capsules in *E. coli*. The other major capsule group (group 2) provides the subject of Chapter 15, this volume. *Escherichia coli* capsular polysaccharides (CPSs) have provided important model systems for studies in other bacteria. As will become apparent, a variety of bacteria produce cell surface polysaccharides that share biochemical and genetic properties with group 1 capsules. These bacteria include pathogens of humans and livestock, as well as plant-associated bacteria.

2. STRUCTURE AND SURFACE ORGANIZATION OF GROUP 1 CAPSULES AND RELATED POLYMERS

The group 1 capsules of *E. coli* provide the K-antigen identified in serological tests. However, the group 1 K-antigen itself is found in two different forms on the cell surface. High-molecular-weight K-antigen forms the capsule structure that can be seen in electron microscopy. There is no known terminal molecule on these polymers that might serve as an anchor to the cell surface, yet the capsules are firmly adherent. The high-molecular-weight form of the K-antigen protects the cell against phagocytosis and masks underlying surface antigens in serological tests. This latter property is one that defines the *E. coli* capsule. Also present on the cell surface are oligosaccharides consisting of one or a few K-antigen repeat units, in a form that is linked to the lipid A core of LPS. These molecules are termed K_{LPS} (MacLachlan *et al.*, 1993). The impact of the expression of the two K-antigen forms on the biology of *E. coli* has yet to be established.

In terms of structure, the *E. coli* group 1 capsules resemble those of *Klebsiella pneumoniae* (Fig. 1). In fact, there are examples where an identical polysaccharide structure is produced by isolates of *E. coli* and *K. pneumoniae*. Both bacteria

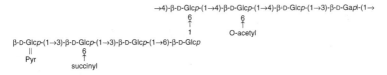

K30 capsule - *Escherichia coli*

→2)-α-D-Man*p*-(1→3)-β-D-Gal*p*-(1→
3
↑
1
β-D-Glc*p*A-(1→3)-α-D-Gal*p*

K2 capsule - *Klebsiella pneumoniae*

→3)-β-D-Glc*p*-(1→4)-β-D-Man*p*-(1→4)-α-D-Glc*p*-(1→
3
↑
1
α-D-Glc*p*A

colanic acid - *Escherichia coli*

→4)-α-L-Fuc*p*-(1→3)-β-D-Glc*p*-(1→3)-β-L-Fuc*p*-(1→
4
↑
1
β-D-Gal*p*-(1→4)-β-D-Glc*p*A-(1→3)-β-D-Gal*p*
‖
Pyr

amylovoran - *Erwinia amylovora*

→4)-β-D-Glc*p*A-(1→3)-α-D-Gal*p*-(1→3)-β-D-Gal*p*-(1→6)-α-D-Gal-(1→
4
↑
1
β-D-Gal*p*
‖
Pyr

xanthan gum - *Xanthomonas campestris*

→4)-β-D-Glc*p*-(1→4)-β-D-Glc*p*-(1→
3
↑
1
β-D-Man*p*-(1→4)-β-D-Glc*p*A-(1→2)-α-D-Man*p*
‖   6
Pyr   ↑
                                    O-acetyl

sphingan S-88 - *Sphingomonas sp. S88*

→4)-α-L-Rha*p*-(1→3)-β-D-Glc*p*-(1→4)-β-D-Glc*p*A-(1→4)-β-D-Glc*p*-(1→
(or Man)   ↑   3
                            ?   ↑
                        O-acetyl   1
                                                            α-L-Rha*p*

succinoglycan - *Sinorhizobium meliloti*

→4)-β-D-Glc*p*-(1→4)-β-D-Glc*p*-(1→4)-β-D-Glc*p*-(1→3)-β-D-Ga*p*l-(1→
6   6
↑   ↑
1   O-acetyl
β-D-Glc*p*-(1→3)-β-D-Glc*p*-(1→3)-β-D-Glc*p*-(1→6)-β-D-Glc*p*
‖   6
Pyr   ↑
                  succinyl

Figure 1. Structure of representative group 1 capsular polysaccharides from *E. coli* and *K. pneumoniae* and related polymers from other bacteria.

require their capsules for virulence (Jann and Jann, 1997; Podschun and Ullmann, 1998; Whitfield *et al.,* 1994). Despite their similarities, *K. pneumoniae* does not appear to make the K_{LPS} form. This is presumably due to differences in the LPS core structures and perhaps in the ligase enzyme (Heinrichs *et al.,* 1998) that attaches polysaccharides to lipid A core acceptors. The molecular basis for the structural relationships in the group 1 capsules of *E. coli* and *K. pneumoniae* will be discussed, but from the perspective of this review, they are indistinguishable and both are included under the generic "group 1" name. The colanic acid polymer elaborated by many *E. coli* strains also has a structure that resembles group 1 capsules (Fig. 1), but due to differences in patterns of expression, we consider colanic acid separately from group 1 capsules. Additional extracellular polysaccharides whose structures resemble the group 1 capsules are particularly prevalent in plant-associated bacteria (Fig. 1). These polysaccharides all contain either glucose or galactose (or both) in their repeat unit structures and uronic acids and ketal groups provide negative charges. Most have side branch substitutions. These structural features place limitations on the systems involved in their synthesis, and indeed all are assembled by a common mechanism.

3. CHROMOSOMAL ORGANIZATION OF THE REGION RESPONSIBLE FOR EXPRESSION OF GROUP 1 CAPSULES

In *E. coli* K-12, the colanic acid biosynthesis genes map near the chromosomal *his* locus. The observation that genes for *E. coli* group 1 capsules are also located near *his* led to speculation that the group 1 capsule and colanic acid biosynthesis genes are allelic (Keenleyside *et al.*, 1992). As a result, the group 1 capsule locus was designated *cps*, following the terminology established for colanic acid. The chromosomal region responsible for colanic acid is now known in detail resulting from studies by P. Reeves and colleagues (Stevenson *et al.*, 1996) and from the *E. coli* K-12 genome project. The corresponding regions have also been described in group 1 capsule prototypes for *E. coli* (serotype K30) (Rahn *et al.*, 1999; Drummelsmith and Whitfield, 1999, and references therein) and *K. pneumoniae* (serotype K2) (Arakawa *et al.*, 1995). Closer examination of these regions indicates that colanic acid is not simply a widespread allelic variant of group 1 capsule and there are important differences in the equivalent regions of the chromosomes (Fig. 2). These organizational differences explain the inability of strains with group 1 capsules to produce colanic acid (Jayaratne *et al.*, 1993). A particularly important difference between *E. coli* K-12 and group 1 capsule producers is the lack of the *wzz* (LPS O-antigen chain-length regulator) gene in the latter. This has a dramatic bearing on the structure of K_{LPS} (see Section 4.1).

The gene clusters responsible for *E. coli* and *K. pneumoniae* group 1 capsules show a common organization (Fig. 2). Located immediately upstream of the *gnd* (6-phosphogluconate dehydrogenase) gene is a contiguous block of genes required for synthesis and polymerization of the K-antigen repeat units. The genes in this region vary in each gene cluster because their products determine the serotype-specific repeat unit structure. In terms of genetic content, this region of the *cps* clusters is indistinguishable from some *rfb* (O-antigen biosynthesis) gene clusters.

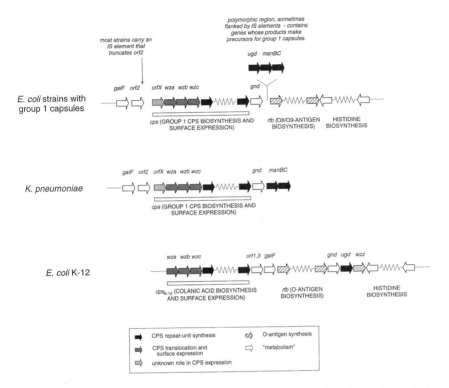

Figure 2. Organizations of chromosomal regions responsible for expression of group 1 capsules in *E. coli* and *K. pneumoniae* and comparison with the colanic acid synthesis locus from *E. coli* K-12. Individual gene functions are described in the text and in Fig. 3. Modified from Whitfield and Roberts (1999).

However, located immediately upstream of this are three genes, *wza–wzb–wzc*, whose products define *cps* clusters. The Wza and Wzc proteins play roles in the surface expression of the capsular polysaccharide. The first gene in the cluster, *orfX*, currently has no defined role. Mutations in *orfX* do not abrogate capsule formation but do influence the amount made (Drummelsmith and Whitfield, 1999). The nucleotide sequence of the *orfX–wzc* region is essentially identical in different serotypes (Rahn *et al.*, 1999). Interestingly, the colanic acid biosynthesis operon resembles those for group 1 capsules in many respects. It contains *wza–wzb–wzc* homologues but lacks *orfX*. The predicted *orfX* gene product does not resemble any protein of known function in the databases, nor is it detected in the *E. coli* K-12 genome sequence. The *orfX* gene product is therefore not required for colanic acid synthesis.

The similarities in the capsule gene clusters of *E. coli* and *K. pneumoniae* are explained by the proposal that the group 1 capsule locus in *E. coli* strains was acquired in a lateral transfer event from *K. pneumoniae* (Rahn *et al.*, 1999). This is based on several lines of evidence: (1) there is a very high degree of conservation shared in their *orfX–wza–wzb–wzc* homologues; (2) the *cps* genes are apparently located in a similar position in these bacteria (Fig. 2); and (3) IS elements flank the *E. coli* group 1 *cps* locus, providing a vehicle for transfer. The observed serotype diversification of the locus could be achieved by further lateral transfer events and recombination within the serotype-specific domain of *cps*. However, this may be an oversimplification because analysis of multiple strains provides evidence for multiple independent transfer events (Rahn *et al.*, 1999).

The region of DNA replaced in such a lateral transfer event may extend beyond *cps*. The *rfb* genes for the O8 and O9 subgroups predominate in *E. coli* strains with group 1 capsules and identical O-antigen structures are found in *K. pneumoniae* isolates. These genes are located in a position atypical for *rfb* clusters in *E. coli* (see Amor and Whitfield, 1997; Drummelsmith *et al.*, 1997) (Fig. 2), and others have already suggested that lateral transfer events between *K. pneumoniae* and *E. coli* are responsible for the evolution of the O8/O9 family of O antigens (Sugiyama *et al.*, 1998). The observed diversification of the adjacent *gnd* sequences in *E. coli* also could be explained by such events (Nelson and Selander, 1994).

4. ASSEMBLY OF GROUP 1 CAPSULES

Capsule synthesis is a complex process involving a series of reactions occurring in different cellular compartments. In overview, synthesis of group 1 capsules begins with nucleotide diphosphosugar precursors that are formed in the cytoplasm. The individual repeat units are assembled on a carrier lipid [undecaprenyl phosphate (und-P)] by the sequential activities of glycosyltransferase enzymes located at the cytoplasmic face of the plasma membrane. According to the current biosynthetic model, lipid-linked repeat units are then transferred across the plasma membrane and polymerized at the periplasmic face. The ultimate fate of lipid-linked polymerized material depends on whether it is destined for the capsular or K_{LPS} form of K-antigen, as there are distinct surface expression (translocation) pathways for each form.

4.1. Polymerization Reactions

Group 1 capsules are polymerized by a system known as "Wzy-dependent." This derives its name from the putative polymerase, Wzy. A wide variety of bacterial surface polymers are now known to be formed by Wzy-dependent processes.

These include capsular and extracellular polysaccharides in both gram-negative and gram-positive bacteria, as well as many LPS O-antigens. The role of Wzy was actually first described in the assembly of the *Salmonella* serogroups B, D, and E O-antigens (reviewed in Whitfield, 1995). A good example of a Wzy-dependent O-antigen is the B-band from *Pseudomonas aeruginosa,* described in Chapter 5, this volume.

Synthesis of the *E. coli* K30 antigen is initiated by Wba P, an enzyme that catalyzes transfer of Gal-1-P from the UDP-Gal precursor to the undecaprenyl phosphate carrier. The Wba P enzyme is a member of a family of UDP-hexose–undecaprenyl phosphate hexose-1-P transferase enzymes involved in synthesis of a variety of glucose or galactose-containing cell surface polysaccharides (Drummelsmith and Whitfield, 1999). The sequence of Wba P indicates several predicted transmembrane segments, as might be expected for an enzyme that interacts with a membrane-bound lipid acceptor. Subsequent glycosyltransferases all appear to be peripheral membrane proteins and act sequentially to assemble the completed lipid-linked repeat unit (Fig. 3A).

The Wzy-dependent polymerization reaction was described in classic experiments by Robbins *et al.* (1967). The polymer is extended in a lipid-linked form. Lipid-linked repeat units provide the substrates for the polymerase in a process where the nascent polymer grows at the reducing terminus, by addition of one repeat unit at a time. Mutations in *wzy* prevent polymerization but do not eliminate synthesis of individual repeat units. In the case of the *E. coli* K30 group 1 capsule system, a specific *wzy* mutation eliminates polymerization of both the capsular K30 polysaccharide and the K_{LPS} form, indicating they share common polymerization machinery (Drummelsmith and Whitfield, 1999). Many studies have reported the phenotypes of mutations in *wzy* homologues and these are all consistent with the assignment of Wzy as a polymerase. However, it is important to note that there is no definitive biochemical proof that Wzy is itself the polymerase. One limitation has been the inability to identify and purify the protein, as Wzy homologues typically have 10–13 transmembrane segments and the structural genes have a number of "rare" codons. Overexpression has been reported for only one Wzy homologue (Daniels *et al.,* 1998).

Work from M. J. Osborn and colleagues provided the first evidence that the Wzy-dependent polymerase reaction occurs at the periplasmic face of the plasma membrane (McGrath and Osborn, 1991). This necessitates a component in the assembly pathway that transfers lipid-linked repeat units across the plasma membrane (Fig. 3B). Preliminary data obtained with an experimental hybrid O-antigen system suggest that the *wzx* gene product may play a role in this export process (Liu *et al.,* 1996). Like Wzy, Wzx homologues also have multiple predicted transmembrane segments, and sequence similarities place them in a family of putative "polysaccharide exporter proteins" recently designated PST(1) for the capsule assembly homologues and PST(2) for those involved in O-antigen synthesis

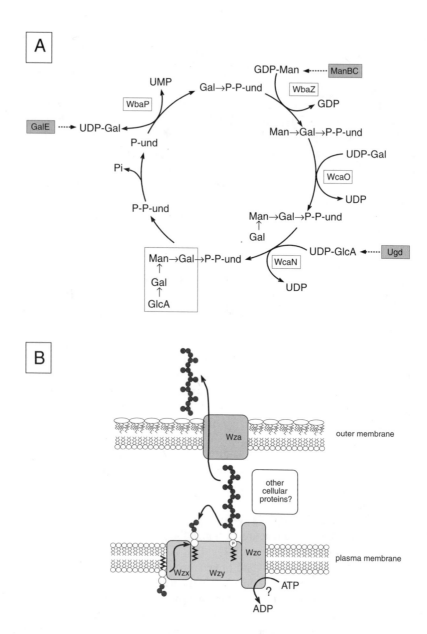

(Paulsen *et al.*, 1997). As with Wzy, unequivocal biochemical proof is lacking for the precise catalytic activity of the Wzx enzyme. It also remains to be established whether Wzx is the only protein involved in the transfer process. Indeed, there are preliminary data suggesting that the N-terminal domain of Wba P plays some role in export (Wang *et al.*, 1996).

Studies on *in vitro* biosynthesis of *Klebsiella* capsules (Sutherland and Norval, 1970; Troy *et al.*, 1971) and colanic acid (Johnson and Wilson, 1977) indicate a similar assembly system for their lipid-linked intermediates. Further examination of the sequence of the *K. pneumoniae* K2 and colanic acid *cps* clusters identifies homologues of Wba P, Wzx, and Wzc (Table I), consistent with their use of the same biosynthetic process.

For Wzy-dependent LPS O-antigens, polymerization is terminated by transfer of the polymer (or oligosaccharide) from the lipid intermediate to lipid A core acceptor by the ligase enzyme, Waa L (Heinrichs *et al.*, 1998; Whitfield *et al.*, 1997). An additional component, Wzz (O-antigen chain length determinant; Cld or Rol), controls the extent of polymerization in a process that requires Wzy and Waa L (reviewed in Whitfield *et al.*, 1997). *Escherichia coli* strains with group 1 capsules lack chromosomal *wzz* genes (Fig. 2), and the polymerization of K_{LPS} is therefore unregulated unless a plasmid containing cloned *wzz* is introduced (Dodgson *et al.*, 1996). It is this feature that results in only K-antigenic oligosaccharides being present in K_{LPS}. It is important to note that group 1 capsules are coexpressed with a limited range of LPS O-antigens, predominantly serogroups O8 and O9, and these are synthesized by ABC-transporter-dependent processes. These assembly mechanisms operate without the involvement of Wzx, Wzy, or Wzz homologues (Whitfield, 1995; Whitfield *et al.*, 1997). An example of this assembly system can be found in the A-band LPS of *P. aeruginosa* described in Chapter 5, this volume.

The length of group 1 capsular polysaccharide is not regulated by Wzz (Dodgson *et al.*, 1996; Drummelsmith and Whitfield, 1999), presumably because

---

Figure 3.   Biosynthesis of the group 1 capsule in *E. coli* K30. (A) The proposed pathway for synthesis of lipid-linked K30 repeat units. The assignment of glycosyltransferases catalyzing individual steps is the product of sequence data and preliminary biochemical investigations. Enzymes involved in the synthesis of sugar nucleotide precursors are identified in the shaded boxes. GalE is UDP-galactose-4-epimerase; ManBC are phosphomannomutase and GDP-mannose pyrophosphorylase, respectively; and Ugd is UDP-glucose pyrophosphorylase. The carrier lipid is presumed to be undecaprenyl phosphate, consistent with the many other systems that have been investigated. (B) A cartoon depicting a working model for the later steps in group 1 capsule assembly. The reactions given in panel A synthesize lipid-linked repeat units at the cytoplasmic face of the plasma membrane. The Wzx protein transfers individual lipid linked repeat units across the membrane, perhaps with the cooperation of other proteins. Wzy is thought to be the polymerase for the capsular polysaccharide and the polymerization reaction occurs at the periplasmic face of the membrane. Wzc and Wza form part of a pathway for surface assembly of the capsule, again potentially in cooperation with other cellular proteins. The limits of experimental evidence to support the model are described in the text. The Wzb protein is not shown, as its precise role in CPS assembly has not been established.

Table I

Components of the Biosynthetic Systems for Group 1 and Group 1-Related Polysaccharides in Different Bacteria.

Polysaccharide	Homologues of						Genbank accession	References
	WbaP	Wzx	Wzy	Wza	Wzb	Wzc		
E. coli K30 capsule	WbaP[a]	Wzx[c]	Wzy[b]	Wza[a]	Wzb[c]	Wzc[a]	AF104912	Drummelsmith and Whitfield (1999)
E. coli colanic acid	WcaJ[c]	Wzx[c]	WcaD[c]	Wza[c]	Wzb[a]	Wzc[a]	U38473	Stevenson et al. (1996)
K. pneumoniae K2 capsule	Orf14[c]	Orf11[c]	Orf10[c]	Orf4[c]	Orf5[c]	Orf6[c]	D21242	Arakawa et al. (1995)
E. amylovora amylovoran	AmsG[b]	AmsL[c]	AmsC[c]	AmsH[b]	AmsI[a]	AmsA[c]	X77921	Bugert and Geider (1995, 1997)
R. solanacearum EPS	—[d]	EpsE[c]	EpsF[c]	EpsA[b]	EpsP[c]	EpsB[b]	U17898	Huang and Schell (1995)
X. campestris xanthan gum	GumD[a]	GumJ[b]	GumE[b]	GumB[b]	—[e]	GumC[b]	U22511	Katzen et al. (1998); Vojnov et al. (1998)
S. meliloti succinoglycan	ExoY[b]	ExoT[b]	ExoQ[b]	ExoF[c]	—[e]	ExoP[a]	L20758, L05588	Becker et al. (1995); Becker and Pühler (1998); Glucksmann et al. (1993a,b); Gonzales et al. (1998); Müller et al. (1993); Reuber and Walker (1993)
Sphingomonas S-88 sphingan	SpsB[a]	SpsS[c]	SpsG[c]	SpsD[c]	—[e]	SpsE/SpsC[c]	U51197	Pollock et al. (1994, 1998); Yamazaki et al. (1996)

[a] Function confirmed by biochemical analysis.
[b] Function consistent with mutant phenotype.
[c] Function predicted by putative amino acid sequence homology and/or predicted membrane topology.
[d] No homologue detected but complete cluster unavailable.
[e] No homologue detected in the cluster.

ligation to lipid A core is not required for its expression on the cell surface (Dodgson *et al.*, 1996; MacLachlan *et al.*, 1993). There remains an open question concerning whether the polymerization of *E. coli* group 1 capsular polysaccharides is regulated by an alternative mechanism involving other proteins. One possible candidate is Wzc (Drummelsmith and Whitfield, 1999; Rahn *et al.*, 1999), a member of the MPA1 (cytoplasmic *m*embrane *p*eriplasmic *a*uxilliary protein) protein family (Paulsen *et al.*, 1997). Homologues of Wzc appear to be located in the plasma membrane with two transmembrane domains flanking a predicted periplasmic segment in the N-terminal region. The second transmembrane segment is followed by a C-terminal cytoplasmic domain containing a putative ATP-binding site. In terms of topology, the N-terminal region of Wzc resembles Wzz (Whitfield *et al.*, 1997). It has been shown recently that the Wzc protein encoded by the *E. coli* K-12 colanic acid gene cluster is a tyrosine autokinase (Vincent *et al.*, 1999). Recent work in our laboratory indicates that phosphorylation of multiple tyrosine residues at the carboxy-terminus of Wzc is essential for synthesis of capsular K30 antigen (Wugeditsch *et al.*, in preparation). An insertion mutation in *wzc* in *E. coli* K30 reduces the formation of the K30 capsular antigen (Drummelsmith and Whitfield, 1999) and this partial phenotype is due to the expression of an additional functional Wzc-homologue. The additional *wzc* gene is not linked to the *cps* locus and has been designated *etk* (Ilan *et al.*, 1999). A double mutant defective in both *wzc* and *etk* lacks the K30 capsule but shows increased polymerization of K_{LPS} (Wugeditsch and Whitfield, in preparation). Wzc therefore is not essential for the synthesis and polymerization of K30 repeat units in general but its role is confined to an assembly pathway for the K30 capsular polysaccharide. The *wzc* mutation could not be complemented by *wzz*. This data and the autokinase activity of the Wzc protein are consistent with a more complex role in capsule assembly.

Systems that contain Wza and Wzc often have another protein: Wzb. Wzb proteins are acid phosphatases (Bugert and Geider, 1997) and recent studies indicate that their substrates are phosphotyrosine residues in Wzc (Vincent *et al.*, 1999). The precise role played by different phosphorylation levels in capsule expression remains to be determined.

4.2. Translocation and Cell Surface Assembly of Group 1 Capsules

While many aspects of polymerization pathways are reasonably well documented, the later steps in capsule assembly are largely open questions. These include release of nascent polymer from the lipid intermediate and translocation of the polymer through the periplasm and across the outer membrane (Fig. 3B). Early studies with a group 1 capsule in *E. coli* K29 established that capsule assembly on the cell surface occurs at specific sites ("Bayer junctions" or "zones of adhesion"), where the plasma and outer membranes appear to come into close apposition when examined by electron microscopy (Bayer and Thurow, 1977). The interpretation of these regions has been the subject of some debate (see Whitfield and Valvano, 1993). However, one attractive interpretation is that these sites represent a multienzyme assembly complex that provides a physical and functional connection between the cell surface and the polymerization machinery in the plasma membrane. Such a system would overcome the practical problem of transferring high-molecular-weight capsular polymers ($M_r > 100,000$) to the surface.

One candidate component of the capsule translocation–surface assembly machinery is Wza. If *wza* is mutated, *E. coli* K30 strains are able to polymerize K30 polymer but unable to assemble a capsular layer on the cell surface (Drummelsmith and Whitfield, 1999, 2000). Consistent with a role in surface assembly, the predicted *wza* gene product is an outer membrane lipoprotein with a β-barrel structure. Structural features of Wza place it in the OMA (*o*uter *m*embrane *a*uxiliary) family of proteins (Paulsen *et al.,* 1997).

Homologues of Wza are always found in association with a member of the MPA1 family, that is, Wzc. The high degree of conservation in Wza–Wzc proteins from different group 1 capsular serotypes of *E. coli* and *K. pneumoniae* indicates that these proteins play a generic role and are not influenced by specific features in individual polymers (Drummelsmith and Whitfield, 1999; Rahn *et al.,* 1999). That both Wza and Wzc affect capsule assembly in *E. coli* K30 suggests they participate in a translocation–surface assembly pathway. We consider it likely that the pathway also involves other cellular proteins, whose roles may not be confined to capsule assembly. We have recently shown that Wza forms homomultimers and that the multimeric form creates ring-like structures in the outer membrane (Drummelsmith and Whitfield, 2000). These structures resemble "secretins" associated with filamentous phage assembly and protein translocation through type II, and type III contact-dependent secretion systems, in a range of gram-negative bacteria (reviewed in Russel, 1998). Secretins exist as large channels formed by multimeric complexes of 10 or more monomers. The capsule secretin has an apparent internal diameter of 3 nm (Drummelsmith and Whitfield, 2000). Given their size and potential impact on outer membrane integrity, the channels are presumed to be "gated." ATP-binding proteins have been implicated in gating (Russel, 1998), and it is conceivable that this is the role played by Wzc in capsule assembly. Efforts are underway in our laboratory to address this hypothesis.

It is notable that cell surface expression of K_{LPS} is not dependent on either Wza or Wzc (Drummelsmith and Whitfield, 1999). These observations indicate that the "LPS translocation" pathway operates independently of the equivalent capsule pathway. However, our current data do not preclude the two pathways sharing some common cellular proteins. The identification of components required for cell surface expression of capsules represents the first step in dissection of the relevant pathway. This information now can be exploited to seek additional interacting proteins in a putative multicomponent system.

5. REGULATION OF EXPRESSION OF GROUP 1 CAPSULES

There are a number of reasons why the expression of group 1 capsular polysaccharides would be subject to complex regulation. At the simplest level, production of these polysaccharides is energetically expensive. At a more sophisticated level, these polymers play crucial roles in the interactions between a pathogenic bacterium and its host. Modulation of the amount of polymer synthesized in response to environmental cues would facilitate differential expression as infection and colonization progress. At least some of the regulatory systems for group 1 capsules have been identified.

5.1. Antitermination

Like many other capsule gene clusters, the *E. coli* group 1 *cps* genes are preceded by a JUMPstart (*j*ust *u*pstream of *m*any *p*olysaccharide gene *start*s) sequence (Hobbs and Reeves, 1994). This sequence contains dual *ops* (*o*peron *p*olarity *s*uppressor) elements (Rahn *et al.,* 1999), a feature also seen in the *E. coli* O7 *rbf* gene cluster (Marolda and Valvano, 1998). The *ops* elements are involved in an antitermination process required to ensure that transcription proceeds to completion in long operons. The paradigm for this regulatory strategy is hemolysin synthesis (the *hyl* operon) in *E. coli* (reviewed in Bailey *et al.,* 1997). In order to function, the system requires the Rfa H protein (a Nus G homologue), which is postulated to act as a transcription elongation factor (Bailey *et al.,* 1997). To be effective, the *ops* elements must be located on the nascent transcript, where they recruit Rfa H and possibly other proteins, to permit transcription to proceed over long distances. This process may require the formation of specific stem-loop structures generated by the JUMPstart sequence (Marolda and Valvano, 1998).

In the group 1 capsule operons of *E. coli* and *K. pneumoniae,* as well as the related colanic acid biosynthesis operon, antitermination takes on a particularly important role. Available sequence data predict a single promoter upstream of the first structural gene. The transcriptional start has been identified in the colanic acid system (Stout, 1996) but not in the group 1 capsule operons. Additional promoters are not evident within the operons, although this has not been examined beyond sequence data. A large transcript has been identified for the *K. pneumoniae* K2 *cps* cluster (Wacharotayankun *et al.,* 1992), consistent with the operon yielding a single major transcriptional unit. However, located in the region immediately downstream of *wzc* is a sequence that potentially forms a strong Rho-independent terminator (Rahn *et al.,* 1999). In the absence of any additional downstream promoters, transcription through the terminator into the polymer synthesis genes is essential and this may be dependent on *ops*/Rfa H-mediated antitermination.

The terminator would potentially facilitate synthesis of higher copies of the structural (translocation) components for CPS assembly (i.e., Wza and Wzc) and lesser amounts of the highly active enzymes that synthesize the polymer itself (Rahn *et al.,* 1999).

5.2. The Role of the Rcs System in Expression of Group 1 Capsules and Colanic Acid

Putative regulatory regions upstream of *cps* are highly conserved in *E. coli* and *K. pneumoniae* strains with group 1 capsules (Rahn *et al.,* 1999), and this is reflective of common elements in their regulation. In contrast, sequences upstream of the colanic acid gene cluster share only the JUMPstart–*ops* elements with the group 1 clusters and even the position of these elements differs in the colanic acid system. The remaining upstream sequence is highly divergent when the colanic acid locus is compared to the conserved group 1 capsule operons, and this may be one factor contributing to the absence of colanic acid synthesis at 37°C (Rahn *et al.,* 1999). This is a crucial issue since the group 1 capsules are recognized virulence determinants and must be produced in significant amounts at 37°C.

One of the distinguishing features of group 1 capsule synthesis is that it is influenced by an environmental signal-responsive two-component regulatory system. This system involves Rcs (*r*egulation of *c*apsule *s*ynthesis) proteins, together with several ancillary components (Fig. 4). The Rcs regulatory system was initially identified in transcriptional control of the colanic acid gene cluster in *E. coli* K-12 (reviewed in Gottesman, 1995). At the heart of the system are Rcs C and Rcs B, the sensor and response regulator, respectively. Rcs C appears to act on RcsB both positively and negatively, presumably by phosphorylation–dephosphorylation. An additional protein, Rcs F (Gervais and Drapeau, 1992) may provide an alternative route to Rcs B activation (Gottesman, 1995). In order to activate *cps* transcription, Rcs B must interact with an additional positive regulator, Rcs A. Rcs A and Rcs B are both members of the Lux R family of transcriptional activators and the binding site for Rcs AB has been localized upstream of the *cps* operon (Stout, 1996). In *E. coli* K-12, transcription of *cps* and colanic acid synthesis is negligible at 37°C, due in part to limiting amounts of Rcs A. Rcs A has a short half-life in the cell as it provides a substrate for the Lon protease (Stout *et al.,* 1991). In addition, synthesis of Rcs A is regulated by H-NS acting as a transcriptional silencer (Sledjeski and Gottesman, 1995). A small RNA molecule, Dsr A, acts as an antisilencer.

The Rcs system is complicated by interactions with a variety of different cellular systems. Mutations that influence the balance of the system (e.g., a *lon* defect) produce a readily identifiable mucoid phenotype at 37°C. Activation of colanic acid also occurs in situations of HU imbalance (Painbéni *et al.,* 1993). Such interactions make it difficult to resolve whether some mucoid phenotypes reflect

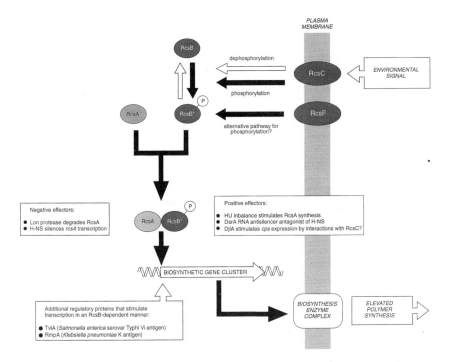

Figure 4.   Cartoon of the Rcs regulatory system. Much of the model is derived from studies of colanic acid regulation in *E. coli* K-12 . The central feature is a two-component regulatory system in which
Rcs C is the sensor and Rcs B is the response regulator. Rcs B activity appears to be modulated by
phosphorylation–dephosphorylation. In order to enhance transcription of *cps* genes, Rcs B must interact with the unstable protein, Rcs A. Additional factors that influence the system are shown in the boxes.

direct or indirect effects on colanic acid synthesis. It also should be noted that the
activity of the Rcs system is not confined to regulating expression of colanic acid
as *ftsZ* (Gervais *et al.,* 1992) and *tolQRA* (Clavel *et al.,* 1996) are both influenced
by Rcs BC.

Expression of the group 1 capsules of *E. coli* (Jayaratne *et al.,* 1993; Keenleyside *et al.,* 1992) and *K. pneumoniae* (Allen *et al.,* 1987; McCallum and Whitfield, 1991; Wacharotayankun *et al.,* 1992) is elevated by the addition of multicopy
Rcs A or Rcs B. These data are consistent with the operation of the Rcs systems as
established in the colanic acid paradigm but do not explain why these capsules are
produced at 37°C when colanic acid is not. It is clear that the nucleotide sequence
for the site of Rcs AB interaction in *E. coli* K-12 bears no similarity to the corresponding region in strains with group 1 capsules (Rahn *et al.,* 1999). Thus, the precise function of regulatory DNA sequences and their interactions with regulatory
proteins may be markedly different.

E. coli K30 *rcsA* and *rcsB* mutants still produce capsule (Jayaratne *et al.,*

1993), a result that would not be expected if the system were strictly identical to colanic acid synthesis. The basal levels of group 1 *cps* expression evident at 37°C could be achieved in the absence of positive regulators or might reflect an additional unknown component(s) missing in the colanic acid system. Such a component might alter the context of the regulatory system, allowing a different operation in strains with group 1 capsules. In highly mucoid *K. pneumoniae* K1 and K2 serotypes, a virulence plasmid-encoded ancillary regulator, Rmp A, provides an additional positive regulator (for references, see Wacharotayankun *et al.,* 1993). To date, an equivalent protein has not been identified in the *E. coli* group 1 capsule-producing strains. In another scenario, the Vi capsule system in *S. enterica* serovar Typhi is regulated by Tvi A, the product of the first gene in the Vi biosynthesis cluster (Virlogeux *et al.,* 1996). Tvi A operates in conjunction with Rcs B and Rcs C, but interestingly Rcs A is expendable in this system. There is no primary sequence similarity between Orf X (the unique first gene in group 1 capsule clusters) and Tvi A, but it is conceivable that Orf X performs a similar regulatory role.

One key question is the nature of the environmental signal(s) to which Rcs-BC sensory pairs respond. It is well established that induction of colanic acid synthesis in *E. coli* K-12 occurs in response to membrane perturbations, including defects in the inner core of LPS (Parker *et al.,* 1992) and mutations that result in misfolding of proteins (Missiakas *et al.,* 1996). Overproduction of Djl A, a member of the Hsp40 chaperone protein family, also activates colanic acid synthesis in an Rcs C-dependent manner (Clarke *et al.,* 1997; Kelley and Georgopoulos, 1997). The *djlA* message is stabilized by cold shock, and therefore may form part of an adaptive response to such stresses (Kelley and Georgopoulos, 1997). One signal that clearly activates colanic acid synthesis is an imbalance in cellular osmoregulation. Mutations that affect synthesis of membrane-derived oligosaccharides (Ebel *et al.,* 1997) or transient exposure to osmotic shock (Sledjeski and Gottesman, 1996) increase *cps* transcription. Collectively, these observations suggest that the role of the Rcs-regulated colanic acid expression is important for survival of *E. coli* outside the host. However, osmoregulation is also important in Rcs regulation of the Vi antigen in *S. enterica* serovar Typhi (Arricau *et al.,* 1998), and this polymer is certainly a virulence factor, produced at 37°C. Further work is required to determine whether the Rcs systems in *E. coli* and *K. pneumoniae* strains with group 1 capsules also respond to alterations in osmolarity, or whether additional/alternate signals are important.

6. SYNTHESIS AND REGULATION OF GROUP 1-LIKE EXTRACELLULAR POLYSACCHARIDES IN OTHER BACTERIA

Structural relationships are evident among group 1 capsules, colanic acid, and several extracellular polysaccharides from a variety of plant-associated bacteria.

It therefore is not surprising to find that patterns of synthesis and regulation are conserved. Table 1 shows that other biosynthetic systems use homologues of the central protein classes. Most can be readily identified by sequence comparison, and in some cases there are direct biochemical data to support the assignments.

The relationships to group 1 capsules are extended to the processes involved in regulating the expression of these group 1-like extracellular polysaccharides. In several examples, two-component regulatory systems play central roles, but these are often augmented by additional regulatory elements. In the case of *Erwinia* spp., regulation is mediated by an Rcs system where the components and mode of action closely resemble the colanic acid paradigm (Bereswill and Geider, 1997; Bernhard *et al.*, 1990; Coleman *et al.*, 1990; Kelm *et al.*, 1997; Poetter and Coplin, 1991; Torres-Cabassa *et al.*, 1987). A two-component system composed of Exo S–Chv I also regulates succinoglycan biosynthesis in *Sinorhizobium meliloti* (Cheng and Walker, 1998). The precise function of an additional regulator in this system, ExoR (Reed *et al.*, 1991), has not yet been clearly established, nor is its relationship to the Exo S–Chv I system currently understood. In *Ralstonia solanacearum*, there is a complex network of positive and negative regulators, with at least three different two-component systems (for references, see Clough *et al.*, 1997; Huang *et al.*, 1998). The Xps R protein is proposed to play a central role in establishing the hierarchy by acting as a signal integrator (Huang *et al.*, 1998), but an additional regulator, Eps R, is also involved (Chapman and Kao, 1998). The group 1-like extracellular polysaccharides in these plant-associated bacteria are involved in their interactions with the plant host. The complexity in the regulatory systems may reflect the intricacy and diversity of the environmental cues that must be integrated to develop an appropriate response.

7. CONCLUSIONS

Significant inroads have been made into understanding the synthesis and genetics of bacterial capsules at a descriptive level. In particular, rapid sequencing techniques and more recently genome projects have had (and will continue to have) a dramatic impact. Many of the steps in capsule synthesis are not known, primarily from mutant phenotypes. Biochemical analyses have been initiated for some enzymes, but an understanding of the enzyme mechanisms underlying most of these reactions is still elusive. As should be evident from the discussion here, there are common strategies used for assembly of capsules in bacteria with different physiologies and ecological niches. Resolution of the details of biosynthesis for one capsule synthesis system should rapidly lead to an understanding of related systems. At a regulatory level, we again have an understanding of the components involved. The next challenge is to establish the operation of these regulatory systems during growth *in vivo*.

ACKNOWLEDGMENTS

Work on the group 1 capsules in the authors' laboratory is supported by funding from the Medical Research Council of Canada (to CW). J. D. and A. R. gratefully acknowledge receipt of Postgraduate Scholarships from the Natural Sciences and Engineering Research Council. T. W. acknowledges a postdoctoral fellowship (Erwin Schrödinger Auslandsstipendium) from the Austrian Science Foundation.

REFERENCES

Allen, P., Hart, C. A., and Saunders, J. R., 1987, Isolation from *Klebsiella* and characterization of two *rcs* genes that activate colanic acid capsular biosynthesis in *Escherichia coli, J. Gen. Microbiol.* **133**:331–340.

Amor, P. A., and Whitfield, C., 1997, Molecular and functional analysis of genes required for expression of group IB K antigens in *Escherichia coli:* Characterization of the *his*-region containing gene clusters for multiple cell-surface polysaccharides, *Mol. Microbiol.* **26**:145–161.

Arakawa, Y., Wacharotayankun, R., Nagatsuka, T., Ito, H., Kato, N., and Ohta, M., 1995, Genomic organization of the *Klebsiella pneumoniae cps* region responsible for serotype K2 capsular polysaccharide synthesis in the virulent strain Chedid, *J. Bacteriol.* **177**:1788–1796.

Arricau, N., Hermant, D., Waxin, H., Ecobichon, C., Duffey, P. S., and Popoff, M. Y., 1998, The RcsB-RcsC regulatory system of *Salmonella typhi* differentially modulates the expression of invasion proteins, flagellin and Vi antigen in response to osmolarity, *Mol. Microbiol.* **29**:835–850.

Bailey, M. J., Hughes, C., and Koronakis, V., 1997, RfaH and the *ops* element, components of a novel system controlling bacterial transcription elongation, *Mol. Microbiol.* **26**:845–851.

Bayer, M. E., and Thurow, H., 1977, Polysaccharide capsule of *Escherichia coli:* Microscope study of its size, structure, and sites of synthesis, *J. Bacteriol.* **130**:911–936.

Becker, A., and Pühler, A., 1998, Specific amino acid substitutions in the proline-rich motif of the *Rhizobium meliloti* ExoP protein result in enhanced production of low-molecular-weight succinoglycan at the expense of high-molecular-weight succinoglycan, *J. Bacteriol.* **180**:395–399.

Becker, A., Niehaus, K., and Pühler, A., 1995, Low-molecular-weight succinoglycan is predominantly produced by *Rhizobium meliloti* strains carrying a mutated ExoP protein characterized by a periplasmic N-terminal domain and missing C-terminal domain, *Mol. Microbiol.* **16**:191–203.

Bereswill, S., and Geider, K., 1997, Characterization of the *rcsB* gene from *Erwinia amylovora* and its influence on exopolysaccharide synthesis and virulence of the fire blight pathogen, *J. Bacteriol.* **179**:1354–1361.

Bernhard, F., Poetter, K., Geider, K., and Coplin, D. L., 1990, The *rcsA* gene from *Erwinia amylovora:* Identification, nucleotide sequence, and regulation of exopolysaccharide biosynthesis, *Mol. Plant-Microbe Interact.* **3**:429–437.

Bugert, P., and Geider, K., 1995, Molecular analysis of the *ams* operon required for exopolysaccharide synthesis of *Erwinia amylovora, Mol. Microbiol.* **15**:917–933.

Bugert, P., and Geider, K., 1997, Characterization of the *amsI* gene product as a low molecular weight acid phosphatase controlling exopolysaccharide synthesis of *Erwinia amylovora, FEBS Lett.* **400**:252–256.

Chapman, M. R., and Kao, C. C., 1998, EpsR modulates production of extracellular polysaccharides in the bacterial wilt pathogen *Ralstonia (Pseudomonas) solanacearum, J. Bacteriol.* **180**:27–34.

Cheng, H.-P., and Walker, G. C., 1998, Succinoglycan production by *Rhizobium meliloti* is regulated through the ExoS-ChvI two-component regulatory system, *J. Bacteriol.* **180**:20–26.

Clarke, D. J., Holland, I. B., and Jacq, A., 1997, Point mutations in the transmembrane domain of DjlA, a membrane-linked DnaJ-like protein, abolish its function in promoting colanic acid production via the Rcs signal transduction pathway, *Mol. Microbiol.* **25**:933–944.

Clavel, T., Lazzaroni, J. C., Vianny, A., and Portallier, R., 1996, Expression of the *tolQRA* genes of *Escherichia coli* K-12 is controlled by the RcsC sensor protein involved in capsule synthesis, *Mol. Microbiol.* **19**:19–25.

Clough, S. J., Lee, K. E., Schell, M. A., and Denny, T. P., 1997, A two-component system in *Ralstonia (Pseudomonas) solanacearum* modulates production of PhcA-regulated virulence factors in response to 3-hydroxypalmitic acid methyl ester, *J. Bacteriol.* **179**:3639–3648.

Coleman, M., Pearce, R., Hitchin, E., Busfield, F., Mansfield, J. W., and Roberts, I. S., 1990, Molecular cloning, expression and nucleotide sequence of the *rcsA* gene of *Erwinia amylovora*, encoding a positive regulator of capsule expression: Evidence for a family of related capsule activator proteins, *J. Gen. Microbiol.* **136**:1799–1806.

Daniels, C., Vindurampulle, C., and Morona, R., 1998, Overexpression and topology of the *Shigella flexneri* O-antigen polymerase (Rfc/Wzy), *Mol. Microbiol.* **28**:1211–1222.

Dodgson, C., Amor, P., and Whitfield, C., 1996, Distribution of the *rol* gene encoding the regulator of lipopolysaccharide O-chain length in *Escherichia coli* and its influence on the expression of group I capsular antigens, *J. Bacteriol.* **178**:1895–1902.

Drummelsmith, J., and Whitfield, C., 1999, Gene products required for surface expression of the capsular form of the group 1 K antigen in *Escherichia coli* (O9a:K30), *Mol. Microbiol.* in press. 1321–1332.

Drummelsmith, J., and Whitfield, C., 2000, Translocation of group 1 capsular polysaccharide to the surface of *Escherichia coli* (O9a:K30) requires a multimeric complex in the outer membrane. *EMBO J.* **19**:57–66.

Drummelsmith, J., Amor, P. A., and Whitfield, C., 1997, Polymorphism, duplication and IS-*1*-mediated rearrangement in the *his-rfb-gnd* region of *Escherichia coli* strains with group IA capsular-K-antigens, *J. Bacteriol.* **179**:3232–3238.

Ebel, W., Vaughn, G. J., Peters, H. K., 3rd, and Trempy, J. E., 1997, Inactivation of *mdoH* leads to increased expression of colanic acid capsular polysaccharide in *Escherichia coli, J. Bacteriol.* **179**:6858–6861.

Gervais, F. G., and Drapeau, G. R., 1992, Identification, cloning, and characterization of *rcsF*, a new regulator gene for exopolysaccharide synthesis that suppresses the division mutation *ftsZ84* in *Escherichia coli* K-12, *J. Bacteriol.* **174**:8016–8022.

Gervais, F. G., Phoenix, P., and Drapeau, G. R., 1992, The *rcsB* gene, a positive regulator of colanic acid biosynthesis in *Escherichia coli,* is also an activator of *ftsZ* expression, *J. Bacteriol.* **174:**3964–3971.

Glucksmann, M. A., Reuber, T. L., and Walker, G. C., 1993a, Family of glycosyl transferases needed for the synthesis of succinoglycan by *Rhizobium meliloti, J. Bacteriol.* **175:**7033–7044.

Glucksmann, M. A., Reuber, T. L., and Walker, G. C., 1993b, Genes needed for the modification, polymerization, export, and processing of succinoglycan by *Rhizobium meliloti:* A model for succinoglycan biosynthesis, *J. Bacteriol.* **175:**7045–7055.

Gonzalez, J. E., Semino, C. E., Wang, L. X., Castellano-Torres, L. E., and Walker, G. C., 1998, Biosynthetic control of molecular weight in the polymerization of the octasaccharide subunits of succinoglycan, a symbiotically important exopolysaccharide of *Rhizobium meliloti, Proc. Natl. Acad. Sci. USA* **95:**13477–13482.

Gottesman, S., 1995, Regulation of capsule synthesis: modification of the two-component paradigm by an accessory unstable regulator, in: *Two-Component Signal Transduction* (J. A. Hoch and T. J. Silhavy, eds.), ASM Press, Washington, DC, pp. 253–262.

Grangeasse, C., Doublet, P., Vincent, C., Vaganay, E., Riberty, M., Duclos, B., and Cazzone, A. J., 1998, Functional characterization of the low-molecular-mass phosphotyrosine-protein phosphatase of *Acinetobacter johnsonii. J. Mol. Biol.* **278:**339–347.

Heinrichs, D. E., Yethon, J. A., and Whitfield, C., 1998, Molecular basis for structural diversity in the core regions of the lipopolysaccharides of *Escherichia coli* and *Salmonella enterica, Mol. Microbiol.* **30:**221–232.

Hobbs, M., and Reeves, P. R., 1994, The JUMPstart sequence: A 39 bp element common to several polysaccharide gene clusters (MicroCorrespondence), *Mol. Microbiol.* **12:**855–856.

Huang, J., and Schell, M., 1995, Molecular characterization of the *eps* gene cluster of *Pseudomonas solanacearum* and its transcriptional regulation at a single promoter, *Mol. Microbiol.* **16:**977–989.

Huang, J., Yindeeyoungyeon, W., Garg, R. P., Denny, T. P., and Schell, M. A., 1998, Joint transcriptional control of *xpsR,* the unusual signal integrator of the *Ralstonia solanacearum* virulence gene regulatory network, by a response regulator and a LysR-type transcriptional activator, *J. Bacteriol.* **180:**2736–2743.

Hull, S., 1997, *Escherichia coli* lipopolysaccharide in pathogenesis and virulence, in: *Escherichia coli: Mechanisms of Virulence* (M. Sussman, ed.), Cambridge University Press, Cambridge, England, pp. 145–167.

Ilan, O., Bloch, Y., Frankel, G., Ullrich, H., Geider, K., and Rosenshine, I., 1999, Protein tyrosine kinases in bacterial pathogens are associated with virulence and production of exopolysaccharide. *EMBO J.* **18:**3241–3248.

Jann, K., and Jann, B., 1997, Capsules of *Escherichia coli,* in: *Escherichia coli: Mechanisms of virulence* (M. Sussman, ed.), Cambridge University Press, Cambridge, England, pp. 113–143.

Jayaratne, P., Keenleyside, W. J., MacLachlan, P. R., Dodgson, C., and Whitfield, C., 1993, Characterization of *rcsB* and *rcsC* from *Escherichia coli* O9:K30:H12 and examination of the role of the *rcs* regulatory system in expression of group I capsular polysaccharides, *J. Bacteriol.* **175:**5384–5394.

Johnson, J. G., and Wilson, D. B., 1977, Role of a sugar-lipid intermediate in colanic acid synthesis by *Escherichia coli, J. Bacteriol.* **129:**225–236.

Katzen, F., Ferreiro, D. U., Oddo, C. G., Ielmini, M. V., Becker, A., Pühler, A., and Ielpi, L., 1998, *Xanthomonas campestris* pv. campestris *gum* mutants: Effects on xanthan biosynthesis and plant virulence, *J. Bacteriol.* **180:**1606–1617.

Keenleyside, W. J., Jayaratne, P., MacLachlan, P. R., and Whitfield, C., 1992, The *rcsA* gene of *Escherichia coli* O9:K30:H12 is involved in the expression of the serotype-specific group I K (capsular) antigen, *J. Bacteriol.* **174:**8–16.

Kelley, W. L., and Georgopoulos, C., 1997, Positive control of the two-component RcsC/B signal transduction network by DjlA: A member of the DnaJ family of molecular chaperones in *Escherichia coli, Mol. Microbiol.* **25:**913–931.

Kelm, O., Kiecker, C., Geider, K., and Bernhard, F., 1997, Interaction of the regulator proteins RcsA and RcsB with the promoter of the operon for amylovoran biosynthesis in *Erwinia amylovora, Mol. Gen. Genet.* **256:**72–83.

Liu, D., Cole, R., and Reeves, P. R., 1996, An O-antigen processing function for Wzx (RfbX): A promising candidate for O-unit flippase, *J. Bacteriol.* **178:**2102–2107.

MacLachlan, P. R., Keenleyside, W. J., Dodgson, C., and Whitfield, C., 1993, Formation of the K30 (group I) capsule in *Escherichia coli* O9:K30 does not require attachment to lipopolysaccharide lipid A-core. *J. Bacteriol.* **175:**7515–7522.

Marolda, C. L., and Valvano, M. A., 1998, Promoter region of the *Escherichia coli* O7-specific lipopolysaccharide gene cluster: structural and functional characterization of an upstream untranslated mRNA sequence, *J. Bacteriol.* **180:**3070–3079.

McCallum, K. L., and Whitfield, C., 1991, The *rcsA* gene of *Klebsiella pneumoniae* O1:K20 is involved in expression of the serotype-specific K (capsular) antigen, *Infect. Immun.* **59:**494–502.

McGrath, B. C., and Osborn, M. J., 1991, Localization of terminal steps of O-antigen synthesis in *Salmonella typhimurium, J. Bacteriol.* **173:**649–654.

Missiakas, D., Betton, J. M., and Raina, S., 1996, New components of protein folding in extracytoplasmic compartments of *Escherichia coli* SurA, FkpA and Skp/OmpH, *Mol. Microbiol.* **21:**871–884.

Müller, P., Keller, M., Weng, W. M., Quandt, J., Arnold, W., and Pühler, A., 1993, Genetic analysis of the *Rhizobium meliloti exoYFQ* operon: ExoY is homologous to sugar transferases and ExoQ represents a transmembrane protein, *Molc. Plant-Microbe Interact.* **6:**55–65.

Nelson, K., and Selander, R. K., 1994, Intergeneric transfer and recombination of the 6-phosphogluconate dehydrogenase gene (*gnd*) in enteric bacteria, *Proc. Natl. Acad. Sci. USA* **91:**10227–10231.

Painbéni, E., Mouray, E., Gottesman, S., and Rouviere-Yaniv, J., 1993, An imbalance of HU synthesis induces mucoidy in *Escherichia coli, J. Mol. Biol.* **234:**1021–1037.

Parker, C. T., Kloser, A. W., Schnaitman, C. A., Stein, M. A., Gottesman, S., and Gibson, B. W., 1992, Role of the *rfaG* and *rfaP* genes in determining the lipopolysaccharide core structure and cell surface properties of *Escherichia coli* K-12, *J. Bacteriol.* **174:**2525–2538.

Paulsen, I. T., Beness, A. M., and Saier, M. H. J., 1997, Computer-based analyses of the protein constituents of transport systems catalysing export of complex carbohydrates in bacteria, *Microbiology* **143:**2685–2699.

Podschun, R., and Ullmann, U., 1998, *Klebsiella* spp. as nosocomial pathogens: Epidemiology, taxonomy, typing methods, and pathogenicity factors, *Clin. Microbiol. Rev.* **11:**589–603.

Poetter, K., and Coplin, D. L., 1991, Structural and functional analysis of the *rcsA* gene from *Erwinia stewartii, Mol. Gen. Genet.* **229**:155–160.

Pollock, T. J., Thorne, L., Yamazaki, M., Mikolajczak, M. J., and Armentrout, R. W., 1994, Mechanism of bacitracin resistance in gram-negative bacteria that synthesize exopolysaccharides, *J. Bacteriol.* **176**:6229–6237.

Pollock, T. J., van Workum, W. A., Thorne, L., Mikolajczak, M. J., Yamazaki, M., Kijne, J. W., and Armentrout, R. W., 1998, Assignment of biochemical functions to glycosyl transferase genes which are essential for biosynthesis of exopolysaccharides in *Sphingomonas* strain S88 and *Rhizobium leguminosarum, J. Bacteriol.* **180**:586–593.

Rahn, A., Drummelsmith, J., and Whitfield, C., 1999, Conserved organization in the gene clusters for expression of *Escherichia coli* group 1 K antigens: Relationship to the colanic acid biosynthesis locus and the genes from *Klebsiella pneumoniae, J. Bacteriol.* **181**:2307–2313.

Reed, J. W., Glazebrook, J., and Walker, G. C., 1991, The *exoR* gene of *Rhizobium meliloti* affects RNA levels of other *exo* genes but lacks homology to known transcriptional regulators, *J. Bacteriol.* **173**:3789–3794.

Reuber, T. L., and Walker, G. C., 1993, Biosynthesis of succinoglycan, a symbiotically important exopolysaccharide of *Rhizobium meliloti, Cell* **74**:269–280.

Rick, P. D., and Silver, R. P., 1996, Enterobacterial common antigen and capsular polysaccharides, in: *Escherichia coli and Salmonella: Cellular and Molecular Biology* (F. C. Neidhardt, ed.), ASM Press, Washington, DC, pp. 104–122.

Robbins, P. W., Bray, D., Dankert, M., and Wright, A., 1967, Direction of chain growth in polysaccharide synthesis, *Science* **158**:1536–1542.

Russel, M., 1998, Macromolecular assembly and secretion across the bacterial cell envelope: Type II protein secretion systems, *J. Mol. Biol.* **279**:485–499.

Sledjeski, D., and Gottesman, S., 1995, A small RNA acts as an antisilencer of the H-NS-silenced *rcsA* gene of *Escherichia coli, Proc. Natl. Acad. Sci. USA* **92**:2003–2007.

Sledjeski, D. D., and Gottesman, S., 1996, Osmotic shock induction of capsule synthesis in *Escherichia coli* K-12, *J. Bacteriol.* **178**:1204–1206.

Stevenson, G., Andrianopoulos, K., Hobbs, M., and Reeves, P. R., 1996, Organization of the *Escherichia coli* K-12 gene cluster responsible for production of the extracellular polysaccharide colanic acid, *J. Bacteriol.* **178**:4885–4893.

Stout, V., 1996, Identification of the promoter region for the colanic acid polysaccharide biosynthetic genes in *Escherichia coli* K-12, *J. Bacteriol.* **178**:4273–4280.

Stout, V., Torres-Cabassa, A., Maurizi, M. R., Gutnick, D., and Gottesman, S., 1991, RcsA, an unstable positive regulator of capsular polysaccharide synthesis., *J. Bacteriol.* **173**:1738–1747.

Sugiyama, T., Kido, N., Kato, Y., Koide, N., Yoshida, T., and Yokochi, T., 1998, Generation of *Escherichia coli* O9a serotype, a subtype of *E. coli* O9, by transfer of the *wb** gene cluster of *Klebsiella* O3 into *E. coli* via recombination, *J. Bacteriol.* **180**:2775–2778.

Sutherland, I. W., and Norval, M., 1970, The synthesis of exopolysaccharide by *Klebsiella aerogenes* membrane preparations and the involvement of lipid intermediates, *Biochem. J.* **120**:567–576.

Torres-Cabassa, A., Gottesman, S., Frederick, R. D., Dolph, P. J., and Coplin, D. L., 1987, Control of extracellular polysaccharide synthesis in *Erwinia stewartii* and *Escherichia coli* K-12: A common regulatory function, *J. Bacteriol.* **169**:4525–4531.

Troy, F. A., Frerman, F. E., and Heath, E. C., 1971, The biosynthesis of capsular polysaccharide in *Klebsiella aerogenes, J. Biol. Chem.* **246**:118–133.

Vincent, C., Doublet, P., Grangeasse, C., Vaganay, E., Cozzone, A. J., and Duclos, B., 1999, Cells of *Escherichia coli* contain a protein-tyrosine kinase, Wzc, and a phosphotyrosine-protein phosphatase, Wzb. *J. Bacteriol.* **181**:3472–3477.

Virlogeux, I., Waxin, H., Ecobichon, C., Lee, J. O., and Popoff, M. Y., 1996, Characterization of the *rcsA* and *rcsB* genes from *Salmonella typhi: rcsB* through *tviA* is involved in regulation of Vi antigen synthesis, *J. Bacteriol.* **178**:1691–1698.

Vojnov, A. A., Zorreguieta, A., Dow, J. M., Daniels, M. J., and Dankert, M. A., 1998, Evidence for a role for the *gumB* and *gumC* gene products in the formation of xanthan from its pentasaccharide repeating unit by *Xanthomonas campestris, Microbiology* **144**:1487–1493.

Wacharotayankun, R., Arakawa, Y., Ohta, M., Hasegawa, T., Mori, M., Horii, T., and Kato, N., 1992, Involvement of *rcsB* in *Klebsiella* K2 capsule synthesis in *Escherichia coli* K-12, *J. Bacteriol.* **174**:1063–1067.

Wacharotayankun, R., Arakawa, Y., Ohta, M., Tanaka, K., Akashi, T., Mori, M., and Kato, N., 1993, Enhancement of extracapsular polysaccharide synthesis in *Klebsiella pneumoniae* by RmpA2, which shows homology to NtrC and FixJ, *Infect. Immun.* **61**:3164–3174.

Wang, L., Liu, D., and Reeves, P. R., 1996, C-terminal half of *Salmonella enterica* WbaP (RfbP) is a galactosyl-1-phosphate transferase domain catalyzing the first step of O-antigen synthesis, *J. Bacteriol.* **178**:2598–2604.

Whitfield, C., 1995, Biosynthesis of lipopolysaccharide O-antigens, *Trends Microbiol.* **3**:178–185.

Whitfield, C., and Roberts, I. S., 1999, Structure, assembly, and regulation of expression of capsules in *Escherichia coli, Mol. Microbiol.* **31**:1307–1319.

Whitfield, C., and Valvano, M. A., 1993, Biosynthesis and expression of cell surface polysaccharides in gram-negative bacteria, *Adv. Micro. Physiol.* **35**:135–246.

Whitfield, C., Keenleyside, W. J., and Clarke, B. R., 1994, Structure, function and synthesis of cell surface polysaccharides in *Escherichia coli*, in: *Escherichia coli in Domestic Animals and Man* (C. L. Gyles, ed.), CAB International, Wallingford, England, pp. 437–494.

Whitfield, C., Amor, P. A., and Köplin, R., 1997, Modulation of surface architecture of gram-negative bacteria by the action of surface polymer: Lipid A-core ligase and by determinants of polymer chain length, *Mol. Microbiol.* **23**:629–638.

Wugeditsch, T., Hocking, J., Drummelsmith, J., and Whitfield, C., in preparation, Phosphorylation of Wzc, a tyrosine autokinase, is essential for assembly of group 1 capsular polysaccharides in *Escherichia coli*.

Yamazaki, M., Thorne, L., Mikolajczak, M., Armentrout, R. W., and Pollock, T. J., 1996, Linkage of genes essential for synthesis of a polysaccharide capsule in *Sphingomonas* strain S88, *J. Bacteriol.* **178**:2676–2687.

10

Mutans Streptococci Glucosyltransferases

Gregory Mooser and Kumari S. Devulapalle

1. INTRODUCTION

There have been a number of important reviews on streptococcal glucosyltransferases (GTFs). The historical value of these reviews provides a significant contribution in the study of glucosyltransferases (Loesche, 1986; Hamada and Slade, 1980; Gibbons, 1975; Gibbons and Van Houte, 1975). During World War II, even greater attention was focused on the enzyme due to the widespread application of dextran, a by-product of glucosyltransferase, which served as a plasma support (Jeanes, 1977). Subsequently, Aoki *et al.* (1986) and Ferretti *et al.* (1987) cloned glucosyltransferase genes, which opened up the possibility to explore the structure and function of glucosyltransferases in greater detail.

2. CATALYTIC MECHANISM OF GLUCOSYLTRANSFERASE

The characterization of any enzyme should begin with thorough kinetic analyses. This analysis is essential because it provides clues to the kinetic mechanism of the enzyme. The earliest kinetic description of glucosyltransferase was reported by Mooser *et al.* (1985). The mechanism was established as a modified

Gregory Mooser and Kumari S. Devulapalle • School of Dentistry, Department of Basic Sciences, University of Southern California, Los Angeles, California 90089-0641.

Glycomicrobiology, edited by Doyle.
Kluwer Academic/Plenum Publishers, New York, 2000.

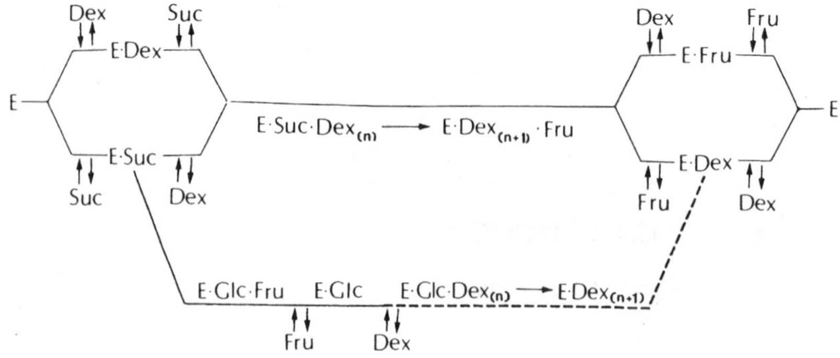

Figure 1. The enzymatic catalytic mechanism for glucosyltransferase. Dex, Suc, Fru, and Glc are dextran, sucrose, fructose, and glucose, respectively (Mooser *et al.,* 1985).

ping-pong mechanism. It required steady-state kinetics, fructose isotope exchange at equilibrium, and sucrose hydrolysis. Each of these reactions contributed to the kinetic mechanism that ultimately established a modified ping-pong mechanism (Fig.1). Identifying a covalent intermediate of glucosyltransferase provided insight into the mechanism and ultimately led to identifying a catalytic residue of the enzyme (Mooser and Iwaoka, 1989). In a report by Mooser and Iwaoka (1989), a catalytic aspartic acid was identified. The aspartic acid identified in this report is shown as the residue that stabilizes the developing carbonium ion in equilibrium with the covalent glucosyl–enzyme complex. A general acid donates a proton to the glycoside oxygen to facilitate fructose departure and in deglucosylation serves as a general base to activate water or another glucosyl acceptor (Fig. 2) (Mooser *et al.,* 1991).

2.1. Subsites of Glucosyltransferase

During the kinetic analysis, an interesting kinetic phenomenon was observed with regard to probing the subsites of the active site with the respective subsite binding ligands. The binding specificity of the glucosyl and fructosyl subsites of the sucrose-binding site was examined to identify ligands that bind exclusively to each subsite. Fructose was found to be a moderate glucosyltransferase (GTF) inhibitor, but free glucose, α-methylglucoside, and glucose epimers were very weak inhibitors. In contrast, glucose transition-state analogues, D-glucano-1,5-lactone, 1-deoxynojirimycin (dNJ), and most N-alkyl derivatives of dNJ were moderate to strong inhibitors; in particular N-methyl-dNJ was found to be the strongest GTF inhibitor identified to date. Multiple inhibitor kinetic analysis established nonex-

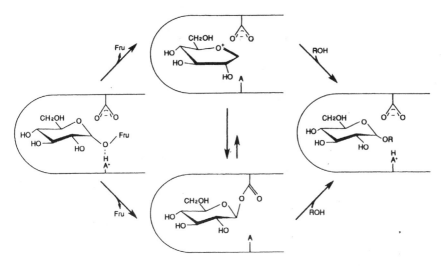

Figure 2. Proposed mechanism of Gtase catalysis. The aspartic acid identified in this report is shown as the residue that stabilizes the developing carbonium ion in equilibrium with the covalent glucosyl–enzyme complex. A general acid donates a proton to the glycoside oxygen to facilitate fructose departure and in deglucosylation (not shown) serves as a general base to activate water or another glucosyl acceptor (Mooser *et al.*, 1991).

clusive binding of fructose and dNJ at the respective subsites. Binding of fructose and *N*-alkyl-dNJ derivatives to a small degree was partially exclusive. Fructose and dNJ were used as reporter ligands to localize the subsite specificity of two test inhibitors: a reversible inhibitor, Zn^{2+}, and an irreversible inhibitor, diethyl pyrocarbonate (DEP). Zn^{2+} paired with dNJ in multiple inhibitor kinetic analysis showed no competition between the inhibitors, while Zn^{2+} paired with fructose decreased ligand affinity sevenfold, establishing Zn^{2+} binding exclusively at the fructose subsite. Analogous experiments adapted to the irreversible inhibitor DEP indicated that it reacts at both subsites or induces a protein conformational change at one subsite that alters ligand binding at the adjacent subsite (Devulapalle and Mooser, 1994).

2.2. Inhibitors of Glucosyltransferase

The earlier studies on glucosyltransferase inhibition were reported by Newbrun *et al.* (1983). They reported that acarbose and 1-deoxynojirimycin were strong inhibitors of GTF (Newbrun *et al.*, 1983). The most important inhibitors of glucosyltransferase (Fig. 3) are listed in Table I with their dissociation constants.

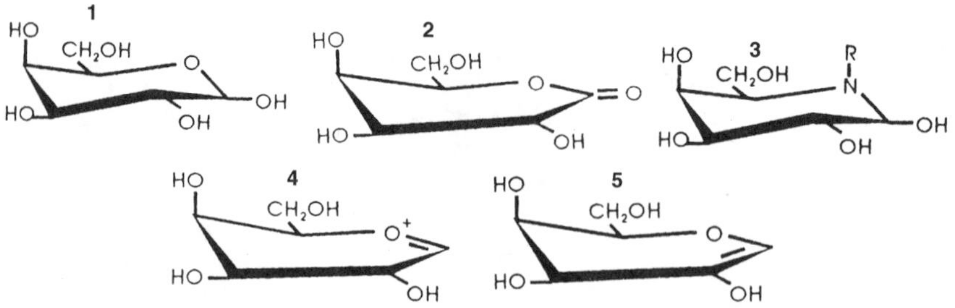

Figure 3. Structures of glucosyltransferase inhibitors (1–5). (1) D-glucose; (2) D-glucano-1, 5-lactone; (3) 1-deoxynojirimycin (5-imino-1, 5-dideoxy D-glucose (dNJ); (4) *N*-alkyl-dNJ derivative; (5) D-glucal (1, 2-dideoxy-1-enopyranose) (Devulapalle and Mooser, 1994).

Kobayashi *et al.* (1995) recently identified novel cyclic isomalto-oligosaccharides, cyclodextran, that strongly inhibited the dextransucrase reaction. Their results showed that cyclodextran inhibited both *Leuconostoc* dextranase and glucosyltransferase from *S. mutans*.

2.3. Peptide Combinatorial Libraries

A peptide combinatorial library is a relatively new approach and powerful tool to the systematic synthesis and screening of peptide inhibitors (Houghten *et al.*, 1991). Appropriate scanning strategies can easily identify the stronger peptide inhibitors in the library. Synthetic peptide combinatorial libraries also allow for further identification of peptides that have strong inhibition on glucosyltransferase activity (Devulapalle and Mooser, 1998).

Table I
Dissociation Constants of Glucose and Fructose Subsite Inhibitors

Inhibitor	K_i		
	GTF-ImM	GTF-SmM	GTF-S/GTF-I
Fructose	6.0 ± 0.9	12.3 ± 1.0	2
dNJ[a]	0.56 ± 0.07	8.8 ± 1.1	16
N-methyl-dNJ	0.03 ±0.008	1.1 ± 0.1	37
N-ethyl-dNJ	0.24 ± 0.05	28.4 ± 3.8	118
N-butyl-dNJ	1.73 ± 0.51	102.0 ± 17	59

[a]dNJ; 1-Deoxynojirimycin.

3. STUDIES ON THE STRUCTURE OF GLUCOSYLTRANSFERASES

Glucosyltransferases are very large proteins that suggest a multidomain structure.This was established in part when the C-terminal domain was purified and cloned. It was shown to have almost identical binding properties as that of the native enzyme (Mooser and Wong, 1988). Additionally, Ma *et al.* (1996) observed that *S. sobrinus* produced glucan-binding lectin (GBL), which is capable of binding glucans. They also isolated a 60-kDa GBL and identified that this GBL is responsible for glucan-induced aggregation (Ma *et al.*, 1996). The C-terminal domain was analyzed by Giffard and Jacques (1994). They established that glucosyltransferases are characterized by "YG" repeating units (Giffard and Jacques, 1994).

3.1. Knowledge-Based Model of Glucosyltransferase

Since the structure of glucosyltransferase has not been defined, knowledge-based modeling can be a powerful tool in identifying structural similarities among protein families. The crystal structures of several α-amylases have shown significant similarity with the catalytic domain of glucosyltransferase. This allows for the modeling of large sections of the glucosyltransferase catalytic domain (Devulapalle *et al.*, 1997). An aspartyl residue was found to contribute to the stabilization of the glucosyl transition state. It was the first of three catalytic residues that comprise the glucosyltransferase active site (Mooser *et al.*, 1991). The sequence surrounding the aspartate was found to have substantial sequence similarity with members of the α-amylase family. Because little is known of the protein structure beyond the amino acid sequence, a knowledge-based interactive algorithm was used to identify significant levels of homology with α-amylases and glucosyltransferase from *Streptococcus downei* GTF (Devulapalle *et al.*, 1997). The significance of GTF similarity is underlined by GTF–α-amylase residue conserved in all but one α-amylase invariant residue. Seven solved crystal structures were aligned with the catalytic domain of GTF from *S. downei*. The algorithm produced eight blocks of significant amino acid sequences (Fig. 4A–H).

Block **A** begins near the N-terminus of the GTF and α-amylase sequences. The block traverses α_1, β_2, and the associated loop. Within the block is a conserved pattern identified by Janeček (1996) that is common to $(\alpha/\beta)_8$-barrels. The pattern begins with a conserved Gly (or less commonly Ala or Ser) and terminates with a Pro some 6 to 14 residues further in the sequence (Janeček, 1996). The pattern is found in all of the represented α-amylases and GTF; both the Gly and the Pro are invariant residues Gly or Ala (78/70) and Pro(85/79) (Janeček, 1996).

Block **B** is a highly conserved segment that encompasses secondary elements

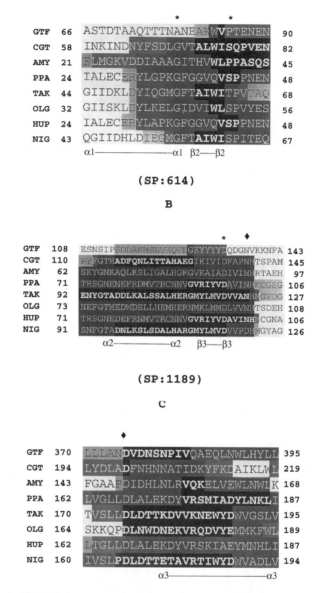

Figure 4. (A–H). Multiple sequence alignment of glucosyltransferase with α-amylases (Devulapalle *et al.*, 1997). GTF, Glucosyltransferase; CGT, cyclodextrin glycosyltransferase; AMY, barley α-amylase 2; PPA, porcine pancreatic α-amylase; TAK, *A. oryzae* taka-amylase A; OLG, *B. cereus* oligo-1,6-glucosidase; HUP, human pancreatic α-amylase; NIG, *A. niger* α-amylase. Single-letter code is used to identify amino acids and residue numbering is given for each block separately. The similarity significance is indicated by the level of shading from light to dark, with dark shades denoting the most significant similarity. Numbers in parentheses indicate the statistical significance of sum of the pairs. Invariant residues (∗) and catalytic residues (●) are indicated. The calcium (◆) and chloride (◇) sites are also indicated (Brayer *et al.*, 1995).

D

```
                     ◇   •
                     *   *
GTF  406  DANFDSIRVDAVDNVDADLLQISSDYL  432
CGT  220  DMGVDGIRVDAVKHMPLGWQKSWMSSI  246
AMY  170  DIGFDGWRFDFAKGYSADVAKIYIDRS  196
PPA  188  DIGVAGFRLDASKHMWPGDIKAVLDKL  214
TAK  197  NYSIDGLRIDTVKHVQKDFWPGYNKAA  223
OLG  190  EKGIDGFRMDVINFISKEEGLPTVETE  216
HUP  188  DIGVAGFRLDASKHMWPGDIKAILDKL  214
NIG  196  NYSVDGLRIDSVLEVQPDFFPGYNKAS  222
             β4——β4        α4————————α4
```

(SP:853)

E

```
              •
              *
GTF  449  VSIVEA  454
CGT  253  FTFGEW  258
AMY  200  FAVAEI  205
PPA  229  FIFQEV  234
TAK  226  YCIGEV  231
OLG  251  MTVGEM  256
HUP  229  FIYQEV  234
NIG  225  YCVGEI  230
          β5————β5
```

(SP:260)

F

```
                                       •
                                     * **
GTF  500  HNSLVDREVDDREVETVPSYSFARAHDSEV  529
CGT  302  ALDSMINSTATDYNQVNDQVTFIDNHDMDR  331
AMY  263  LRGTDGKAPGMIGWWPAKAVTFVDNHDTGS  292
PPA  274  MSYLKNWGEGWGFMPSDRALVFVDNHDNQR  303
TAK  271  DLYNMINTVKSDCPDSTLLGTFVENHDNPR  300
OLG  303  TLKENLTKWQKALEHTGWNSIYWNNHDQPR  332
HUP  274  MSYLKNWGEGWGFVPSDRALVFVDNHDNQR  303
NIG  270  NLYNMIKSVASDCSDPTLLGNFIENHDNPR  299
          α6————————————α6       β7β7
```

(SP:461)

Figure 4. (*continued*)

G

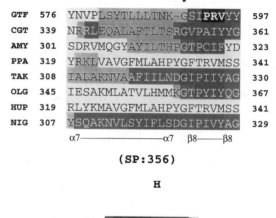

(SP:356)

H

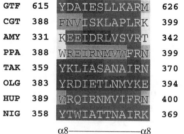

(SP:308)

Figure 4. (*continued*)

α2, β3, and the connecting loop. The block includes two α-amylase invariant residues, an Asp (133/135), and a His. The Asp is conserved in GTF. His is not conserved and instead a Val is present. Of the ten invariant α-amylase residues, His is the only α-amylase invariant residue that is not conserved in GTF (Janeček, 1994; Svensson, 1994). β3 begins the B domain in α-amylases, which is manifested as a long loop terminating at α3. The residue Asn (137/139) also participates in Ca^{2+} binding.

Block **C** includes the terminal segment of the B domain and α3. The conserved residue Asp (375/199) participates in Ca^{2+} binding.

Block **D** is a highly conserved segment beginning with β4 and continues through α4. The block includes two invariant α-amylase residues, Arg (413/227) and Asp (415/229); both are conserved in GTF. The Asp is one of the three residues that form the catalytic site. The invariant Arg (413/227) contributes a ligand to a Cl^{R}MI binding site. The presence of chloride is not confirmed in all of the α-amylases.

Block **E** contains the secondary element β5 and an invariant Glu (453/257). The residue is the second of the three α-amylase catalytic residues.

Block **F** includes secondary elements, α6 and β7, and two α-amylase invariant residues, a His (525/327) and an Asp (526/328); both are conserved in GTF. The Asp completes the α-amylase catalytic triad.

Block **G** encompasses α7, β8, and the intervening loop. The block also includes an α-amylase invariant Gly (590/354).

Block **H** concludes the barrel with α8.

Similarity blocks D, E and F, respectively, include β4, β5, and β7, which are the locations of the three α-amylase catalytic sites. The invariant catalytic residues are positioned at the C-terminus of the β-structure as is common in (β/α)$_8$-barrels (Farber and Petsko, 1990). The relevant residue in each of the three blocks is conserved in GTF.

The three GTF residues homologous with the α-amylase catalytic triad were subjected to conservative site-directed mutagenesis. The mutated enzymes lost almost all the activity. These data provide evidence for a catalytic role in each of the three GTF residues and further support evidence of similarity of GTF and α-amylase structure and function. In GTF, most of the secondary structure elements were identified, with the exception of α5, β1, and β6, and these elements are consistently weak and conserved in α-amylase structures. In addition, 9 of the 10 α-amylase invariant residues are conserved in GTF. The results of GTF site-directed mutagenesis establish that the GTF catalytic mechanism is comparable to α-amylases. The consistency of these data strongly support the (β/α)$_8$-barrel structure for the catalytic domain of mutans streptococci glucosyltransferases.

4. EVOLUTION OF GLUCOSYLTRANSFERASE

Giffard *et al.* (1993) established an evolutionary tree of the eight glucosyltransferases based on constructing a phylogenetic tree from a highly homologous region in the catalytic domain of these enzymes (Fig. 5). The glycohydrolases (EC3.2.1.-) are classified based on sequence and structural homology. There are currently 81 families that are updated continually through the worldwide web server (Henrissat and Bairoch, 1996). The mutans streptococci glucosyltransferases are currently classified under family 70 and the family is termed dextransucrase (Henrissat and Bairoch, 1996). Family 13 consists of α-amylases and cyclodextrin glucanotransferases. Families 13 and 17 belong to GH-H 'clan' (Henrissat and

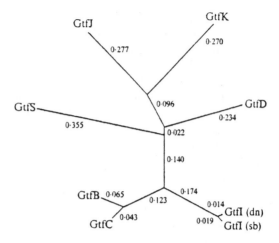

Figure 5. Glucosyltransferases unrooted phylogenetic tree. Reproduced with permission from Giffard *et al.* 1993).

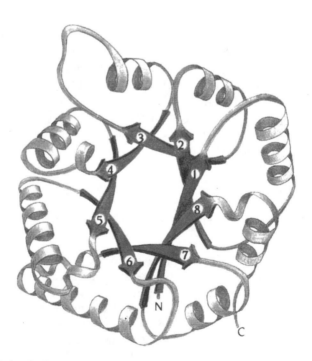

Figure 6. A closed α/β barrel diagram of the enzyme triosephosphate isomerase. Reproduced with permission from Branden and Tooze (1991). Glucosyltransferase catalytic domain model is similar to triosephosphate isomerase (Devulapalle *et al.*, 1997).

Bairoch, 1996). The members of the families and 13 and 70 belong to the 4/7 superfamily, An important characteristic of the superfamily is the presence of aspartates/glutamates at the C-terminal of the fourth and seventh strands of the $(\beta/\alpha)_8$-barrel (Jenkins *et al.*, 1995; Wiesmann *et al.*, 1995). There has been an alternative alignment based on circular permutation by MacGregor *et al.* (1996).

The relevant residues in GTF Asp 415 and Asp 526 are located close to the C-terminus of the β-strand four and β-strand seven. The catalytic glutamate residue of glucosyltransferase (Glu 453) is located in the fifth β-strand, as is the case with α-amylases and cyclodextrin glucanotransferases. The consistency of these data strongly support the $(\beta/\alpha)_8$-barrel structure (Fig. 6) for the catalytic domain of mutans streptococci glucosyltransferases. This establishes that glucosyltransferase belongs to 4/7-superfamily (Devulapalle *et al.*, 1997).

REFERENCES

Aoki, H., Shiroza, T., Hayakawa, M., Sato, S., and Kuramitsu, H. K., 1986, Cloning of a *Streptococcus mutans* glucosyltransferase gene coding for insoluble glucan synthesis, *Infect. Immun.* **53**:587–594.

Branden, C., and Tooze, J., 1991, *Introduction to Protein Structure,* Garland Publishing, New York.

Brayer, G. D., Luo, Y., and Withers, S. G., 1995, The structure of human pancreatic α-amylase at 1.8 Å resolution and comparisons with related enzymes, *Protein Sci.* **4**:1730–1742.

Devulapalle, K. S., and Mooser, G., 1994, Subsite specificity of the active site of glucosyltransferases from *Streptococcus sobrinus, J. Biol. Chem.* **269**:11967–11971.

Devulapalle, K. S., and Mooser, G., 1998, Preliminary screening of a hexapeptide combinatorial library for glucosyltransferase (GTF-I) inhibition, *Protein Peptide Lett.* **5**:159–162.

Devulapalle, K. S., Goodman, S. D., Gao, Q., Hemsley, A., and Mooser, G., 1997, Knowledge-based model of a glucosyltransferase from the oral bacterial group of mutans streptococci, *Protein Sci.* **6**:2489–2493.

Farber, G. K., and Petsko, G. A., 1990, The evolution of α/β barrel enzymes, *Trends. Biochem. Sci.* **15**:228–234.

Ferretti, J. J., Gilpin, M. L., and Russell, R. R. B., 1987, Nucleotide sequence of a glucosyltransferase gene from *Streptococcus sobrinus* MFe28, *J. Bacteriol.* **169**:4271–4278.

Gibbons, R. J., 1975, Dental caries, *Annu. Rev. Med.* **26**:121–136.

Gibbons, R. J., and Van Houte, J., 1975, Bacterial adherence in oral microbial ecology, *Annu. Rev. Microbiol.* **29**:19–44.

Giffard, P. M., and Jacques, N. A., 1994, Definition of a fundamental repeating unit in streptococcal glucosyltransferase glucan-binding regions and related sequence, *J. Dent. Res.* **73**:1133–1141.

Giffard, P. M., Allen, D. M., Milward, C. P., Simpson, C. L., and Jacques, N. A., 1993, Sequence of the *gtf*K gene of *Streptococcus salivarius* ATCC 25975 and evolution of the *gtf* genes of oral streptococci, *J. Gen. Microbiol.* **139**:1511–1522.

Hamada, S., and Slade, H. D., 1980, Biology, immunology, and cariogenicity of *Streptococcus mutans*, *Microbiol. Rev.* **44**:331–384.

Henrissat, B., and Bairoch, A., 1996, Updating the sequence-based classification of glycosyl hydrolases, *Biochem. J.* **316**:695–696.

Houghten, R. A., Pinilla, C., Blondelle, S. E., Appel, J. R., Dooley, C. T., and Cuervo, J. H., 1991, Generation and use of synthetic peptide combinatorial libraries for basic research and drug discovery, *Nature* **354**:84–86.

Janeček, Š., 1994, Parallel β/α-barrels of α-amylase, cyclodextrin glycosyltransferase and oligo-1,6-glucosidase versus the barrel of β-amylase: Evolutionary distance is a reflection of unrelated sequences, *FEBS. Lett.* **353**:119–123.

Janeček, Š., 1996, Invariant glycines and prolines flanking in loops the strand β2 of various (α/β)$_8$-barrel enzymes: A hidden homology? *Prot. Sci.* **5**:1136–1143.

Jeanes, A., 1977, Dextrans and pullulans: Industrially significant α-D-glucans. in: *Extracellular Microbial Polysaccharides*, Vol. 45 (P. A. Sandford and A. Laskin, eds.), American Chemical Society, Washington, DC, pp. 284–298.

Jenkins, J., Leggio, L. L., Harris, G., and Pickersgill, R., 1995, β-Glucosidase, β-galactosidase, family A cellulases, family F xylanases and two barley glycanases form a superfamily of enzymes with 8-fold β/α architecture and with two conserved glutamates near the carboxy-terminal ends of β-strands four and seven, *FEBS. Lett.* **362**:281–285.

Kobayashi, M., Funane, K., and Oguma, T., 1995, Inhibition of dextran and mutan synthesis by cycloisomaltooligosaccharides, *Biosci. Biotech. Biochem.* **59**:1861–1865.

Loesche, W. J., 1986, Role of *Streptococcus mutans* in human dental decay, *Microbiol. Rev.* **50**:353–380.

Ma, Y., Lassiter, M. O., Banas, J. A., Galperin, M. Y., Taylor, K. G., and Doyle, R. J., 1996, Multiple glucan-binding proteins of *Streptococcus sobrinus*, *J. Bacteriol.* **178**:1572–1577.

MacGregor, E. A., Jespersen, H. M., and Svensson, B., 1996, A circularly permuted α-amylase-type α/β-barrel structure in glucan-synthesizing glucosyltransferases. *FEBS Lett.* **378**:263–266.

Mooser, G., and Iwaoka, K. R., 1989, Sucrose 6-α-D-glucosyltransferase from *Streptococcus sobrinus:* Characterization of a glucosyl-enzyme complex, *Biochemistry* **28**:443–449.

Mooser, G., and Wong, C., 1988, Isolation of a glucan-binding domain of glucosyltransferase (1,6-α-glucan synthase) from *Streptococcus sobrinus*, *Infect. Immun.* **56**:880–884.

Mooser, G., Shur, D., Lyou, M., and Watanabe, C., 1985, Kinetic studies on dextransucrase from the cariogenic oral bacterium *Streptococcus mutans*, *J. Biol. Chem.* **260**:6907–6915.

Mooser, G., Hefta, S. A., Paxton, R. J., Shively, J. E., and Lee, T. D., 1991, Isolation and sequence of an active-site peptide containing a catalytic aspartic acid from two *Streptococcus sobrinus* α-glucosyltransferases, *J. Biol. Chem.* **266**:8916–8922.

Newbrun, E., Hoover, C. I., and Walker, G. J., 1983, Inhibition by acarbose, nojirimycin and 1-deoxynojirimycin of glucosyltransferase produced by oral streptococci, *Arch. Oral. Biol.* **28**:531–536.

Svensson, B., 1994, Protein engineering in the α-amylase family: Catalytic mechanism, substrate specificity, and stability, *Plant Mol. Biol.* **25:**141–157.

Wiesmann, C., Beste, G., Hengstenberg, W., and Schulz, G. E., 1995, The three-dimensional structure of 6-phospho-β-galactosidase from *Lactococcus lactis, Structure* **3:**961–968.

11

Glycosyl Hydrolases from Extremophiles

Constantinos E. Vorgias
and Garabed Antranikian

1. INTRODUCTION

A feature of extremophilic organisms is their ability to survive and grow in an environment that can be considered extreme from the anthropocentric point of view. The survival mechanisms of these organisms are partially due to the proper adaptation of the individual components by the organisms. Extremophilic microorganisms are adapted to live at high temperatures (as in volcanic springs), at low temperatures (as in the cold polar seas), at high pressure in the deep sea, at extreme pH values (pH 0–1 or pH 10–11), or at very high salt concentrations. Presently, more than 60 species of hyperthermophilic bacteria and archaea have been isolated and characterized. They consist of anaerobic and aerobic chemolithototrophs and heterotrophs. Various heterotrophs are able to utilize various biopolymers such as starch, hemicellulose, chitin, proteins, and peptides.

Several hyperthermophilic archaea have been isolated with growth temperatures (103–110°C) and have been classified as members of the genera *Pyrobaculum, Pyrodictium, Pyrococcus,* and *Methanopyrus* (Stetter, 1996). Within the bacteria, *Aquifex pyrophilus* and *Thermotoga maritima* exhibit the highest growth temperatures.

Metabolic processes and specific biological functions of these microorgan-

Constantinos E. Vorgias • National and Kapodistrian University of Athens, Faculty of Biology, Department of Biochemistry-Molecular Biology, Panepistimiopolis-Kouponia, 157 01 Athens, Greece. *Garabed Antranikian* • Technical University Hamburg-Harburg, Department of Biotechnology, Institute for Technical Microbiology, 21071 Hamburg, Germany.

Glycomicrobiology, edited by Doyle.
Kluwer Academic/Plenum Publishers, New York, 2000.

isms are facilitated by enzymes and proteins that function under extreme conditions. The enzymes that have been recently isolated from these exotic microorganisms show unique features. Generally, the enzymes are: (1) extremely thermostable; and (2) resist chemical denaturants such as detergents, chaotropic agents, organic solvents, and pH extremes (Friedrich and Antranikian 1996; Jørgensen *et al.*, 1997; Leuschner and Antranikian 1995; Rüdiger *et al.*, 1995). Therefore, these enzymes offer an exceptional opportunity to be used as models, where it is possible to study their features in terms of stability, specificity, and enzymatic mechanisms, in order to learn how to design and construct proteins with properties that are of particular interest for industrial applications.

Biotechnological processes at elevated temperatures have many advantages. The increase of temperature has a significant influence on the bioavailability and solubility of organic compounds. The elevation of temperature is accompanied by a decrease in viscosity and an increase in the diffusion coefficient of organic compounds. Consequently, higher reaction rates, due to smaller boundary layers, are expected (Becker *et al.*, 1997; Krahe *et al.*, 1996). Of special interest are reactions involving less soluble hydrophobic substrates such as polyaromatic, aliphatic hydrocarbons, and fats and polymeric compounds such as starch, cellulose, hemicellulose, chitin, and proteins. The bioavailability of hardy biodegradable and insoluble environmental pollutants also can be improved dramatically at elevated temperatures allowing efficient bioremediation. Furthermore, by performing biological processes at temperatures above 60°C, the risk of contamination is reduced and controlled processes under strict conditions can be carried out.

The determination of the mechanism of enzyme adaptation to extreme conditions is strategic, since extremophiles are unique models for investigations on how biomolecules are stabilized and constitute a valuable resource for the exploitation of novel biotechnological processes. The number of genes from thermophiles that were cloned and expressed in mesophiles is sharply increasing (Ciaramella *et al.*, 1995). The proteins produced in mesophilic hosts are able to maintain their thermostability, are correctly folded at low temperature, are not hydrolyzed by host proteases, and can be purified by using thermal denaturation of the mesophilic host proteins. The obtained degree of enzyme purity is generally adequate for most industrial applications.

In this chapter we will briefly discuss the enzymatic action and properties of starch, pullulan, cellulose, xylan, pectin, and chitin hydrolases and focus only on those enzymes that have been isolated and characterized from extreme thermophilic (optimal growth 70–80°C) and hyperthermophilic (optimal growth 85–100°C) archaea and bacteria. We also are going to discuss their biotechnological significance. Some of these aspects already have been presented in recent reviews (Antranikian, 1992; Ladenstein and Antranikian, 1998; Moracci *et al.*, 1998; Müller *et al.*, 1998; Rüdiger *et al.*, 1994; Sunna and Antranikian, 1997).

2. STARCH-DEGRADING ENZYMES FROM EXTREMOPHILIC MICROORGANISMS

Starch is a widespread natural nutrient storage polysaccharide consisting of glucose residues. In plant cells or seeds, starch is usually deposited in the form of large granules in the cytoplasm. Starch occurs in two forms: (1) α-amylose (15–25% of starch), which is a linear polymer of α-1,4-linked glucopyranose residues, and (2) amylopectin (75–85% of starch), which is highly branched containing α-1,6-glycosidic linkages at branching points (Fig. 1). α-Amylose chains are polydisperse and vary in molecular weights from a few to thousands. They are not soluble in water, but form hydrated micelles. In amylopectin the average length of the branches is from 24 to 30 glucose residues, depending on the species, and in solution, amylopectin yields colloidal or micellar forms. The molecular weight of amylopectin may be as high as 100 million.

2.1. Starch-Degrading Enzymes

In order to be able to utilize starch cells must employ a number of enzymes for its degradation and bioconversion to smaller sugars and oligosaccharides, such as glucose and maltose (Antranikian, 1992, Rüdiger et al., 1994). Starch-hydrolyzing enzymes can be distinguished as endo-acting or endo-hydrolases and exo-acting enzymes or exo-hydrolases, as summarized in Fig. 2.

Endo-acting starch-degrading enzyme is α-amylase (E.C.3.2.1.1) or α-1,4 glucan glucanohydrolase, which hydrolyzes α-1,4 glucosidic linkages in the interior of the starch polymer or oligosaccharides in a random manner. The action of this enzyme leads to the formation of linear and branched oligosaccharides and the sugar-reducing groups are liberated in the α-anomeric configuration.

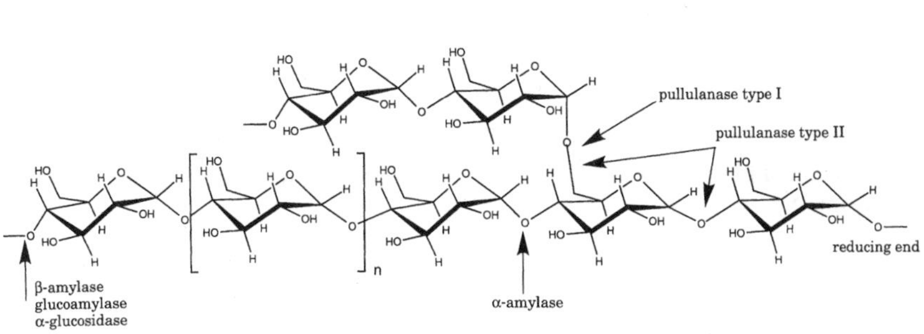

Figure 1.   Chemical structure of starch and the enzymes involved in its degradation.

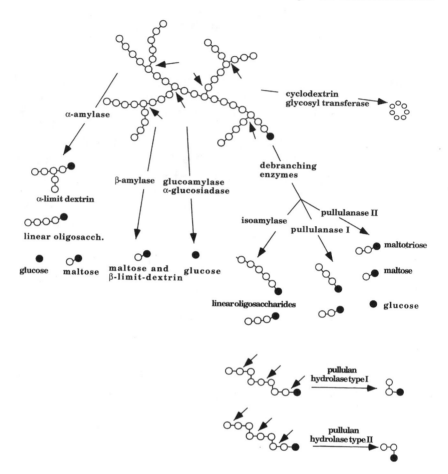

Figure 2.   Schematic presentation of the action of starch hydrolyzing enzymes and their degradation products.

α-Amylases belong to the families, numbers 13 and 57 of the glycosyl hydrolases superfamily (Henrissat, 1991). Family 13 has around 150 members from eukaryotae and prokaryotae, unlike family 57, which has only three members from prokaryotae and archaea. In the structural level there are several X-ray structures determined and the mode of the enzymatic action is well studied.

Exo-acting starch hydrolases are β-amylases, glucoamylases, α-glucosidases, and isoamylase. These enzymes attack the substrate from the nonreducing end, producing small and well-defined oligosaccharides.

β-Amylase (E.C.3.2.1.2) or α-1,4-D-glucan maltohydrolase or saccharogen amylase hydrolyzes α-1,4 glucosidic linkages to remove successive maltose units from the nonreducing ends of the starch chains producing β-maltose by an inversion of the anomeric configuration of the maltose (Fig.1). β-Amylase belongs to family 14 of the glycosyl hydrolases, having 11 members from eukaryotae and prokaryotae.

Glucoamylases (E.C.3.2.1.3) hydrolyze terminal α-1,4-linked-D-glucose residues successively from nonreducing ends of the chains with release of β-D-glucose (Fig. 2). The enzyme has several names: α-1,4-D-glucan glucohydrolase, amyloglucosidase, and γ-amylase and it is a typical fungal enzyme. Most forms of the enzyme can hydrolyze α-1,6-D-glucosidic bonds when the next bond in sequence is α-1,4. However, some preparations of this enzyme hydrolyze α-1,6- and α-1,3-D-glucosidic bonds in other polysaccharides. In contrast to α-glycosidase, glucoamylases preferentially degrade polysaccharides with high molecular weights.

α-Glucosidases (E.C.3.2.1.20) or α-D-glucoside glucohydrolase attacks the α-1,4 or α-1,6 linkages from the nonreducing end in short saccharides that are produced by the action of other amylolytic enzymes (Fig. 2). Unlike glucoamylase, α-glucosidase liberates glucose with an α-anomeric configuration. α-Glucosidases are members of the family 15 and the very diverse family 31 of the glycosyl hydrolases (Henrissat, 1991). Isoamylase (E.C.3.2.1.68), or glycogen-6-glucanohydrolase, is a debranching enzyme specific for α-1,6 linkages in polysaccharides such as amylopectin, glycogen, and β-limit dextrins, but it is unable to hydrolyze the α-1,6 linkages in pullulan; therefore, it has limited action on α-limit dextrins.

2.2. Pullulan-Degrading Enzymes

Pullulan is a linear α-glucan consisting of maltotriose units joined by α-1,6 glycosidic linkages and it is produced by *Aureobasidium pullulan* with a chain length of 480 maltotriose units (Fig. 3). Enzymes capable of hydrolyzing α-1,6 glycosidic bonds in pullulan and amylopectin are defined as debranching enzymes or pullulanases. On the basis of substrate specificity, pullulanases have been classified into pullulanases types I and type II. Pullulanase type I (E.C.3.2.1.41) specifically hydrolyzes the α-1,6-linkages in pullulan and in the branched oligosaccharides, and its degradation product is α-limit dextrin (Fig. 2). These type I enzymes are unable to attack α-1,4-linkages in α-glucans and belong to family 13 of the glycosyl hydrolases. Pullulanase type II or amylopullulanases attack α-1,6-glycosidic linkages in pullulan and α-1,4-linkages in other oligosaccharides (Fig. 2). This enzyme has a relatively multiple specificity and is able to fully convert polysaccharides to small sugars in the absence of other enzymes such as α-amylases or β-amylases.

In contrast to the previously described pullulanases, pullulan hydrolases type I and type II are unable to hydrolyze α-1,6-glycosidic linkages in branched substrates or in pullulan. Because they can hydrolyze the α-1,4-linkages in pullulan,

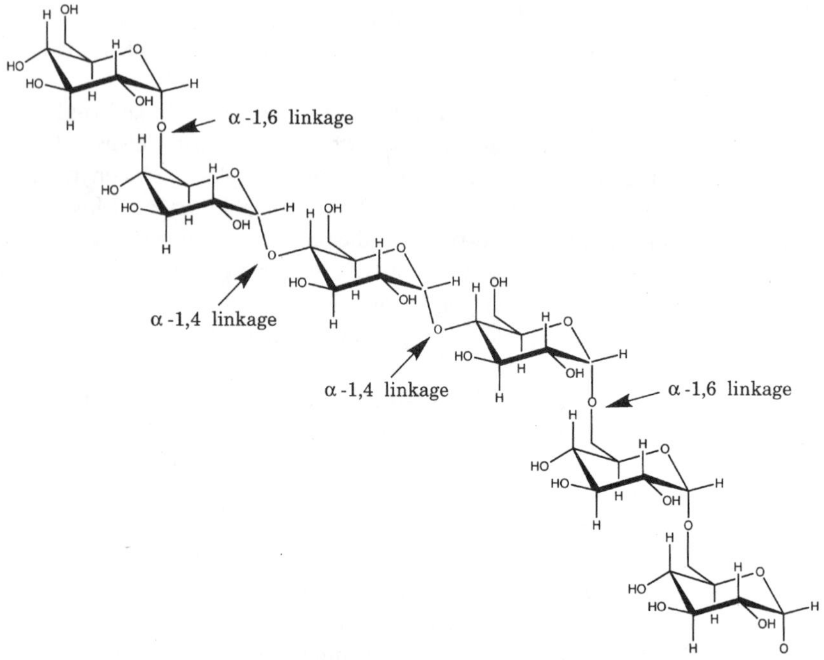

Figure 3.   Chemical structure of pullulan.

they were incorrectly named pullulanases. They can attack α-1,4-glycosidic linkages but not α-1,6-linkages in pullulan. Pullan hydrolase type I or neopullulanase
(E.C.3.2.1.135) hydrolyzes pullulan to panose (α-6-D-glucosylmaltose). Pullulan
hydrolase type II or isopullulanase (E.C.3.2.1.57) hydrolyzes pullulan to isopanose (α-6-maltosylglucose) (Fig. 2).

Finally, cyclodextrin glycotransferase (E.C.2.4.1.19) or α-1,4-D-glucan α-4-
D-(α-1,4-D-glucano)-transferase is an enzyme that has been found only in bacteria. This enzyme produces a series of nonreducing cyclic dextrins from starch,
amylose, and other polysaccharides (Fig. 2). α-, β-, and γ-Cyclodextrins are rings
formed by 6, 7, and 8 glucose units that are linked by α-1,4-bonds, respectively.

2.3. Ability of Hyperthermophilic Microorganisms
to Produce Starch-Degrading Enzymes

Many hyperthermophilic organisms are known to utilize natural carbohydrates polymers for nutritional purposes. Therefore, it is apparent that these

organisms can facilitate the enzymatic degradation of carbohydrates, producing enzymes that are capable of hydrolyzing them very effectively at elevated temperatures.

Hyperhermophilic bacteria and archaea able to grow on starch at temperatures over 70°C have been identified and the corresponding starch-degrading enzymes have been isolated and characterized. In several cases genes encoding these enymes have been isolated, cloned, and overexpressed in heterologous hosts.

2.3.1. α-AMYLASES

Extremely thermostable α-amylases have been characterized from *Pyrococcus woesei, Pyrococcus furiosus* (Koch *et al.,* 1991), and *Thermococcus profundus* (Chung *et al.,* 1995; Lee *et al.,* 1996), with optimum activities at temperatures of 100°C, 100°C, and 80°C, respectively (Table I). In the hyperthermophilic archaea of the genera *Sulfolobus, Thermophilum, Desulfurococcus, Thermococcus,* and *Staphylothermus* (Bragger *et al.,* 1989; Canganella *et al.,* 1994) amylolytic activities have been detected. Due to the high activity of these hydrolases, the cells do not produce high amounts of protein, and therefore it is essential to clone the corresponding genes and express them in heterologous hosts.

The gene encoding an extracellular α-amylase from *P. furiousus* recently has been cloned and the recombinant enzyme expressed in *Escherichia coli* and *Bacillus subtilis* (Jørgensen *et al.,* 1997). This enzyme is an interesting one be-

Table I

α-Amylases Produced by Extremely Thermophilic and Hyperthermophilic Archaea and Bacteria

Organism	Growth Temp. (°C)	Enzyme Optimal Temp. (°C)	Enzyme Optimal pH	Enzyme MW (kDa)
Desulfurococcus mucosus	85	100	5.5	
Pyrococcus furiosus	100	100	6.5–7.5	129
		100	6.5–7.5	129
Pyrococcus sp. KOD1		90	6.5	68
Pyrococcus woesei	100	100	5.5	—
Pyrodictium abyssi	98	100	5.0	—
Staphylothermus marinus	90	100	5.0	240
Sulfolobus solfatatrcus	88			
Thermococcus celef	85	90	5.5	
Thermococcus profundus DT5432	80	80	5.5	42
Thermococcus profundus	80	80	4.0–5.0	42
Thermococcus aggregans	85	100	5.5	
Dictyoglomus thermophilum Rt46B.1		90	5.5	75
Thermotoga maritima MSB8	90	85–90	7.0	61

cause besides its high thermostability (thermal activity even at 130°C) it does not require metal ions either for stability or for optimal activity. Pyrococcal α-amylase also has a unique product pattern and substrate specificity that makes it a unique enzyme and an interesting candidate for industrial applications. The intracellular α-amylase gene from *P. furiousus* also has been cloned and sequenced (Laderman *et al.*, 1993).

α-Amylases with lower thermostability and thermoactivity have been isolated and characterized from the archeon *Pyrococcus* sp. KOD1, and the bacterium *Thermotoga maritima*. The genes encoding these enzymes have been well-expressed in *E. coli* (Liebl *et al.*, 1997; Tachibana *et al.*, 1996). *T. maritima* amylase requires the presence of Ca^{2+} for its enzymatic activity (Liebl *et al.*, 1997), similar to amylase from *Bacillus licheniformis*.

Recent investigations have shown that the hyperthermophilic archaeon *Pyrodictium abyssi* can grow anaerobically on various polymeric substrates and it secretes a heat-stable amylase that is active even above 100°C and in a wide pH range (Table 1).

2.4. Pullulan-Degrading Enzymes from Hyperthermophilic Organisms

Thermostable and thermoactive pullulanases from extremophilic microorganisms have been detected in *Thermococcus celer, Desulfurococcus mucosus, Staphylothermus marinus,* and in the novel archaeal strain *Thermococcus aggregans* (Table II). These pullulanases show temperature optima between 90°C and 105°C and high thermostability in the absence of substrate and calcium ions (Canganella *et al.*, 1994). Most of the thermophilic pullulanases studied to date belong to type II and have been purified from *P. woesei, P. furiosus,* and *Thermococcus litoralis* (Brown and Kelly, 1993) and ES4 (Schuliger *et al.*, 1993). The extreme thermostability of these enzymes, coupled with their ability to attack both α-1,6 and α-1,4 glycosidic linkages, may improve the industrial starch hydrolysis process.

The enzyme from *P. woesei* has been overexpressed in *E. coli*. The recombinant purified enzyme has a temperature optimum at 100°C and it is extremely thermostable, with a half life of 7 min at 110°C (Rüdiger *et al.*, 1995). The aerobic thermophilic bacterium *Thermus caldophilus* GK-24 produces a thermostable pullulanase of type I when grown on starch. This pullulanase is optimally active at 75°C and pH 5.5, and is thermostable up to 90°C and does not require Ca^{2+} ions either for activity or stability.

The first starch-debranching enzyme from an anaerobe was identified in the thermophilic bacterium *Fervidobacterium pennavorans* Ven5 (Koch *et al.*, 1997), and the corresponding gene was cloned and expressed in *E. coli* (unpublished results). In contrast to the pullulanase from *P. woesei,* the enzyme from *F. penavo-*

Table II

Pullulanases (Type I and II) from Extremely Thermophilic and Hyperthermophilic Archaea and Bacteria

Enzymes	Organism	Growth Temp. (°C)	Enzyme Optimal Temp. (°C)	Enzyme Optimal pH	Enzyme MW (kDa)
Pullulanase type I	*Fervidobacterium pennavorans Ven5*	75	80	6.0	190 (93 sub)
	Thermotoga maritima MSB8	90	90	6.0	93
	Thermus caldophilus GK24	75	75	5.5	65
Pullulanase type II	*Desulfurococcus*	—	100	5.0	—
	Pyrococcus woesei	100	100	6.0	90
	Pyrodictium abyssi	98	100	9.0	—
	Thermococcus celer	85	90	5.5	—
	Thermococcus litoralis	90	98	5.5	119
	Thermococcus aggregans	85	100	6.5	—

rans Ven5 attacks exclusively the α-1,6-glycosidic linkages in polysaccharides. This is the only thermostable debranching enzyme known to date that hydrolyzes amylopectin leading to the formation of long linear chain polysaccharides, which are the ideal substrates for the enzymatic action of glucoamylase.

2.5. Cyclodextrin Glycosyl Transferases

Cyclodextrin glycosyl transferase (CGTase) attacks α-1,4-linkages in polysaccharides in a random fashion and acts on starch by an intramolecular transglycosylation reaction. The nonreducing cyclization products of this reaction are α-, β-, or γ-cyclodextrin consisting of 6, 7, or 8 glucose molecules, respectively. Thermostable CGTases already have been found in *Thermoanaerobacter* and *Thermoanaerobacterium thermosulfurogenes* (Petersen *et al.*, 1995; Wind *et al.*, 1995). Recently a heat- and alkali-stable CGTase (65°C, pH 4–10) was purified from a newly identified strain that was isolated from Lake Bogoria, Kenya (Prowe *et al.*, 1996).

2.6. Biotechnological Relevance

Industrial production of fructose from starch, consists of three steps: liquefaction, saccharification, and isomerization. This multistage process (step 1: pH 6.5, 98°C; step 2: pH 4.5, 60°C: step 3; pH 8.0, 65°C) leads to the conversion of starch to fructose with concurrent formation of high concentrations of salts that have to be removed by ion exchangers. The application of thermostable enzymes such as amylases, glucoamylases, pullulanases, and glucose isomerases that are active and stable above 100°C and at acidic pH values can simplify this complicated process. Therefore, strong efforts have been invested in the isolation of thermostable and thermoactive amylolytic enzymes from hyperthermophiles, since they could improve the starch conversion process and lower the cost of sugar syrup production.

The predominant biotechnological application of CGTase occurs in the industrial production of cyclodextrins. Due to the ability of cyclodextrins to form inclusion complexes with a variety of organic molecules, they improve the solubility of hydrophobic compounds in aqueous solutions. Cyclodextrin production occurs in a multistage process in which in the first step starch is liquefied by a heat-stable amylase and in the second step the cyclization reaction with the CGTase from *Bacillus* sp. takes places. Due to the low stability of this enzyme, the process must run at lower temperatures. The finding of heat-stable and more specific

CGTases from extremophiles will solve this problem. The application of heat-stable CGTase in jet cooking, where temperatures up to 105°C are used, will allow the liquefaction and cyclization to take place in one step.

3. CELLULOSE-DEGRADING ENZYMES FROM EXTREMOPHILIC MICROORGANISMS

Cellulose is the most abundant organic biopolymer in nature since it is the structural polysaccharide of the cell wall in the plant kingdom. It consists of glucose units linked by β-1,4-glycosidic bonds in a polymerization grade up to 15,000 glucose units in an absolutely linear mode. The minimal molecular weight of cellulose from different sources has been estimated to vary from about 50,000 to 2,500,000 in different species, which is equivalent to 300 to 15,000 glucose residues (Fig. 4). Although cellulose has a high affinity for water, it is completely insoluble in it.

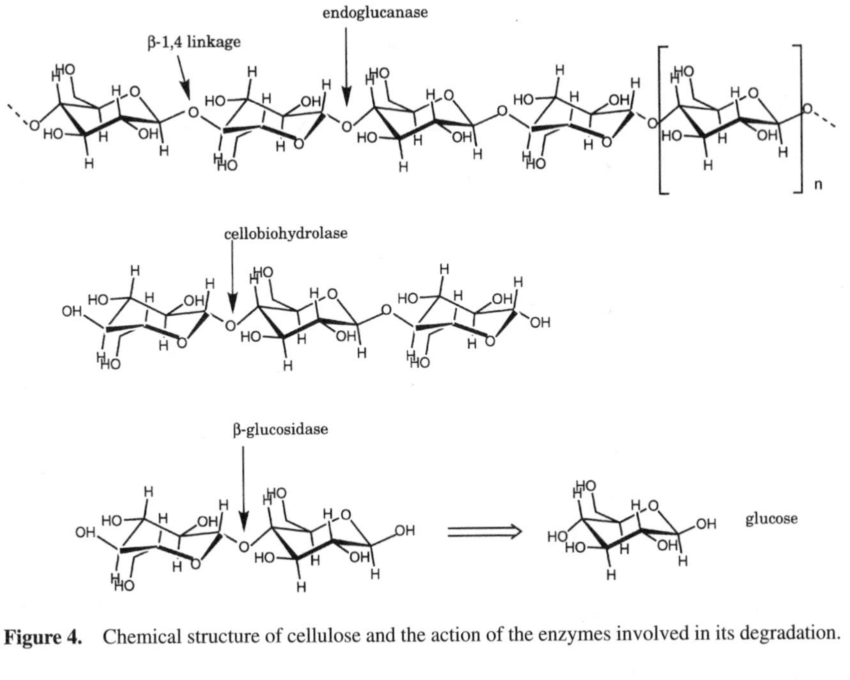

Figure 4.   Chemical structure of cellulose and the action of the enzymes involved in its degradation.

3.1. Cellulose-Degrading Enzymes

Cellulose can be hydrolyzed to its monomeric glucose units by the synergistic action of at least three different enzymes: endoglucanases, exoglucanase (cellobiohydrolase), and β-glucosidase (Fig. 4). Cellulose hydrolyzing enzymes are widespread in fungi and bacteria, but until now have not been found in hyperthermophilic archaea.

3.1.1. CELLULASES

Cellulase (EC 3.2.1.4) or β-1,4-D-glucan glucanohydrolase or endo-β-1,4-glucanase or carboxymethyl cellulase, is an endoglucanase that hydrolyzes cellulose in a random manner, producing oligosaccharides, cellobiose, and glucose. The enzyme catalyzes the endohydrolysis of β-1,4-D-glucosidic linkages in cellulose, but it also will hydrolyze 1,4-linkages in β-D-glucans containing some 1,3-linkages. Cellulases belong to the family 5 of the glycosyl hydrolases (Henrissat, 1991).

3.1.2. CELLOBIOHYDROLASE

Exoglucanase (or β-1,4-cellobiosidase) or exocellobiohydrolase (or β-1,4-cellobiohydrolase) (EC 3.2.1.91) hydrolyses β-1,4 D-glucosidic linkages in cellulose and cellotetraose, releasing cellobiose from the nonreducing end of the chain; they belong to family 6 of the glycosyl hydrolases.

3.1.3. β-GLUCOSIDASE

β-Glucosidase (EC 3.2.1.21) or gentobiase, cellobiase, or amygdalase catalyze the hydrolysis of terminal, nonreducing β-D-glucose residues with release of β-D-glucose. The enzymes belong to family 3 of the glycosyl hydrolases and have a broad specificity for β-D-glucosides. They are able to hydrolyze β-D-galactosides, α-L-arabinosides, β-D-xylosides, and β-D-fucosides.

3.2. Distribution of Cellulose-Degrading Enzymes in Hyperthermophilic Microorganisms

Several cellulose-degrading enzymes from various thermophilic organisms have been detected, purified, characterized, and cloned. Table III summarizes the properties of these enzymes. A thermostable cellulase from *Thermotoga maritima*

Table III

Cellulose-Hydrolyzing Activities from Various Thermophic
and Thermoacidophilic Microorganisms

Organism	Growth Temp. (°C)	Enzyme Optimal Temp. (°C)	Enzyme Optimal pH	Enzyme MW (kDa)
Thermotoga maritima MSB8	80	95	6.0–7.0	29
Thermotoga neapolitana	80	110	—	
Thermotoga sp FjSS3-B1	80	115	6.8–7.8	36
Pyrococcus furiosus	100		6.8–7.8	224(56 sub)
Sulfolobus sulfataricus MT4	88		6.8–7.8	224(56 sub)
Sulfolobus shibatae	95		6.8–7.8	224(56 sub)
Sulfolobus acidocaldaricus	88		6.8–7.8	224(56 sub)
Caldocellum saccharolyticum	—		6.8–7.8	200 (1751 aa)
Anaerocellum thermophilum	85	85–95	5.0–6.0	230

MSB8 has been characterized (Bronnenmeier *et al.,* 1995). The enzyme is rather small, with a molecular weight of 27 kDa, and it is optimally active at 95°C and pH between 6.0 and 7.0. Another marine eubacterium, *Thermotoga neapolitana,* on cultivation in the presence of cellobiose produces two endoglucanases, the endoglucanases A and B (Bok *et al.,* 1994). Purified endoglucanase B shows a remarkable thermostability between 100 and 106°C, and both enzymes show high specificity for carboxy-methyl (CM)-cellulose (CMC).

Cellulase and hemicellulase genes have been found clustered together on the genome of the extremely thermophilic anaerobic bacterium *Caldocellum saccharolyticum,* which is capable of growing on cellulose and hemicellulose as sole carbon sources (Teo *et al.,* 1995). The gene for one of the cellulases (*Cel*A) was isolated and consists of 1751 amino acids. This is the largest known cellulase gene (Teo *et al.,* 1995).

A large cellulolytic enzyme (*Cel* A) with the ability to hydrolyze microcrystalline cellulose was isolated from the extremely thermophilic, cellulolytic bacterium *Anaerocellum thermophilum* (Zverlov *et al.,* 1998). The enzyme has an apparent molecular mass of 230 kDa. It exhibits significant activity toward Avicel and is most active toward soluble substrates such as CMC and β-glucan. Maximal activity was observed between pH values of 5 and 6 and temperatures of 95°C (CMC) and 85°C (Avicelase).

A thermostable cellobiohydrolase was also reported from *Thermotoga maritima* MSB8 (Bronnenmeier *et al.,* 1995). The enzyme has a molecular weight of 29 kDa, an optimal activity at 95°C at pH 6.0–7.5, and the half-live of 2 hr at 95°C in the absence of substrate. Cellobiohydrolase hydrolyzes Avicel with main product cellobiose and cellotriose, as well as CMC and β-glucan.

Thermotoga sp. FjSS3-B1 (Ruttersmith and Daniel, 1991) produces cellobiase. The enzyme is highly thermostable and shows maximal activity at 115°C, pH:

6.8–7.8. The thermostability of this enzyme is salt dependent; at 0.5 M NaCl the life time is almost doubled, from 70 min to 130 min at 108°C and from 7 min to 15 min at 113°C. This cellobiase is active against amorphous cellulose and CMC.

β-Glucosidases have been detected in *Sulfolobus sulfataricus* MT4, *Sulfolobus acidocaltaricus,* and *Sulfolobus shibatae* (Grogan, 1991), as well as in *Pyrococcus furiosus* (Kengen *et al.,* 1993). Among these enzymes the β-glucosidase from *Sulfolobus sulfataricus* MT4 has been purified and characterized (Pisani *et al.,* 1990; Nucci *et al.,* 1993). This β-glucosidase is a homotetramer with 56 kDa for each subunit; it very resistant to various denaturants with activity up to 85°C (Pisani *et al.,* 1990). The gene for this β-glucosidase has been cloned and overexpressed in *E. coli* (Cubellis *et al.,* 1990; Moracci *et al.* 1992; Prisco *et al.,* 1994).

Pyrococcus furiosus β-glucosidase is expressed in high levels in cells grown on cellobiose (Kengen *et al.,* 1993). β-Glucosidase is a homotetramer with molecular weight of 58 kDa for each subunit. It is very stable and shows optimal activity at 102°C to 105°C, while the half life is 3.5 days at 100°C and 13 hr at 110°C. Previously, the gene of *P. furiosus* β-glucosidase had been cloned and expressed in *E. coli* (Voorhorst *et al.,* 1995)

3.3. Biotechnological Relevance

Cellulases have found various applications in several biotechnological applications. The most effective commercial cellulase is the one produced by *Trichoderma* sp. Other cellulases of commercial interest are obtained from strains of *Aspergillus, Penicillium,* and *Basidiomycetes.* Fungal cellulases are used in alcohol production. Cellulolytic enzymes also can be used to improve juice yields and effective color extractions of juices. The presence of cellulases in detergents cause color brightening and softening and improve particulate soil removal. A novel application of cellulases in textile industry is the use of Denimax (Novo Nordisk) for the "biostoning" of jeans instead of the use of abrasive stones in stonewashed jeans. Also, other significant applications of cellulases include the pretreatment of cellulosic biomass and forage crops to improve nutritional quality and digestibility, enzymatic saccharification of agricultural and industrial wastes, and production of fine chemicals.

4. XYLAN-DEGRADING ENZYMES FROM EXTREMOPHILES

Xylan is a heterogeneous molecule that comprises the major polymeric compound of hemicellulose. Hemicellulose is a fraction of plant cell walls; is associated with cellulose, lignin, and other polysaccharides; and functions as the major

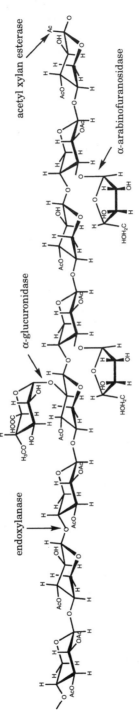

Figure 5. Chemical structure of substituted xylan and the enzymes involved in its degradation.

reservoir of fixed carbon in nature. The main chain of the xylan heteropolymer is composed of xylose residues linked by β-1,4-glycosidic bonds. Approximately half the xylose residues have substitutions at O-2 or O-3 positions of acetyl, arabinosyl and glucuronosyl groups, as shown in Fig. 5. Due to the heterogeneity of xylan, its degradation requires the synergistic action of a xylanolytic enzyme system, which consists of five distinct activities with endoxylanase as the major enzyme. Xylanolytic enzymes are widespread in the terrestrial as well as the marine environment. The concerted action of these enzymes degrades xylan to its constituent sugars; for a detailed description, see the reviews of Sunna *et al.* (1996b) and Sunna and Antranikian (1997).

4.1. Xylan-Degrading Enzymes

The endo-β-1,4-xylanase (E.C.3.2.1.8) or β-1,4-xylan xylanohydrolase hydrolyzes β-1,4-xylosidic linkages in xylans, whereas β-1,4-xylosidase, β-xylosidase, β-1,4-xylan xylohydrolase, xylobiase, or exo-β-1,4-xylosidase (E.C.3.2.1.37) hydrolyzes β-1,4-xylans and xylobiose by removing the successive xylose residues from the nonreducing termini. α-Arabinofuranosidase or arabinosidase (E.C.3.2.1.55) hydrolyzes the terminal nonreducing α-L-arabinofuranoside residues in α-L-arabinosides. The enzyme also acts on α-L-arabinofuranosides, α-L-arabinans containing either (1,3) or (1,5)-linkages, arabinoxylans, and arabinogalactans. Glucuronoarabinoxylan endo-β-1,4-xylanase, feraxan endoxylanase, glucuronoarabinoxylan β-1,4-xylanohydrolase (E.C.3.2.1.136) attacks β-1,4-xylosyl links in some glucuronoarabinoxylans. This enzyme also shows high activity toward feruloylated arabinoxylans from cereal plant cell walls. Acetyl xylan esterase (E.C.3.1.1.6) removes acetyl groups from xylan. Xylanases from prokaryotae and eukaryotae comprise families 10 and 11 of the glycosyl hydrolases.

4.2. Formation of Xylan-Degrading Enzymes by Hyperthermophilic Microorganisms

So far, only a few extreme thermophilic microorganisms are able to grow on xylan and secrete thermoactive xylanolytic enzymes (Table IV). The members of the order of *Thermotogales* and the species *Dictyoglomus thermophilum* have been described to produce xylanases that are active and stable at high temperature (Sunna *et al.,* 1996a; Sunna and Antranikian, 1996; Gibbs *et al.,* 1995).

The most thermostable endoxylanases described so far are those derived from *Thermotoga* sp. strain FjSS3-B.1 (Simpson *et al.,* 1991), *Thermotoga maritima* (Winterhalter and Liebl, 1995), *Thermotoga neapolitana* (Bok *et al.,* 1994), and

Table IV
Production of Heat-Stable Xylanases by Extreme Thermophilic and Hyperthermophilic
Archaea and Bacteria

Organism	Growth Temp. (°C)	Enzyme Optimal Temp. (°C)	Enzyme Optimal pH	Enzyme MW (kDa)
Pyrodictium abyssi	98	110	5.5	
Dictyoglomus thermophilum Rt46B.1	73	85	6.5	31 (recombinant)
Thermotoga maritima MSB8	80	92	6.2	120 Xyn A
		105	5.4	40 XynB
Thermotoga sp strain FjSS3-B.1	80	105	5.3	31 (wild type)
		85	6.3	40 (recombinant)
Thermotoga neapolitana	80	85	5.5	37 (wild type)
		102	5.5–6.0	119 (recombinant)
Thermotoga thermarum	77	80	6.0	105/150 Endoxyl 1
		90–100	7.0	35 Endoxyl 2
Caldocellum saccharolyticum	—	70–75	5.5–6.0	40 XynA

Thermotoga thermarum (Sunna *et al.,* 1996b). These enzymes, active between 80 and 105°C, are mainly cell associated and most probably localized within the toga (Ruttersmith *et al.,* 1992; Schumann *et al.,* 1991; Sunna *et al.,* 1996a; Winterhalter and Liebl, 1995).

Several genes encoding xylanases have been cloned and sequenced. The gene from *T. maritima,* encoding a thermostable xylanase, has been cloned and expressed in *E. coli.* Comparison between the *T. maritima* recombinant xylanase and the commerially available xylanase, Pulpenzyme TM (Yang and Eriksson, 1992), indicates that the thermostable enzyme has properties that make it an attractive candidate for application in pulp and paper industry (Chen *et al.,* 1997) .

Archaea growing at temperatures above 90°C seem to be unable to utilize xylan as a carbon source. Recently, however, it was shown that the hyperthermophilic archaeon *Pyrodictium abyssi* is able to produce a unique thermostable endoxylanase upon growth in the presence of xylan, xylose, or arabinose (Andrade *et al.,* 1996). This extracellular enzyme, which is inducible, displays optimal activity at 110°C and pH 5.5. Several years ago, Luthi *et al.* (1990) reported the isolation of a clone from the extremely thermophilic anaerobe *Caldocellum saccharolyticum.* Five open reading frames were found in this clone that appear to code for a xylanase (*Xyn* A; 40.4 kDa) and β-xylosidase (*Xyn* B; 56.3 kDa). The *xyn*A gene product shows significant homology to the xylanases from the alkalophilic *Bacillus* sp. strain C125 and *Clostridium thermocellum.* The enzymes of the *Caldocellum saccharolyticum,* however, have not yet been biochemically characterized.

4.3. Biotechnological Relevance

Xylanases have a wide range of potential biotechnological applications. They already are produced on industrial scale and are used as food additives in poultry, for increasing feed efficiency diets (Annison, 1992; Classen, 1996), and in wheat flour for improving dough handling and the quality of baked products (Maat et al., 1992).

In recent years the major interest in thermostable xylanases is in enzyme-aided bleaching of paper (Viikari et al., 1994). More than 2 million tons of chlorine and chlorine derivatives are used annually in the United States for pulp bleaching. The chlorinated lignin derivatives generated by this process constitute a major enviromental problem caused by the pulp and paper industry (McDonough, 1992). Recent investigations have demonstrated the feasibility of enzymatic treatments as alternatives to chlorine bleaching for removal of residual lignin from pulp (Viikari et al., 1994). Treatment of kraft pulp with xylanase leads to a release of xylan and residual lignin without undue loss of other pulp components. Xylanase treatment opens up the cell wall structure, thereby facilitating lignin removal in subsequent bleaching stages. In addition, fragmentation of the xylan polymer allows free diffusion of those portions of the residual lignin that are covalently attached to xylan.

Candidate xylanases for this important, potential market would have to satisfy several criteria: (1) they must lack cellulolytic activity, to avoid hydrolysis of the cellulose fibers; (2) their molecular weight should be low enough to facilitate their diffusion in the pulp fibers; (3) they must be stable and active at high temperature and at alkaline pH; and (4) one must be able to obtain high yields of enzyme at very low cost. All the xylanases currently available from commercial suppliers can only partially fulfill the criteria. Xylanases from moderate thermophilic microorganisms are rapidly denatured at temperatures above 70°C. Several nonchlorine bleaching stages used in commercial operations are performed well above this temperature; consequently, pulp must be cooled before treatment with the available enzymes and reheated for subsequent processing steps. This adds substantially to the cost of bleaching processes. Thus, xylanases active and stable at temperatures of 90°C or more have enormous potential for use in the pulp and paper industry (Chen et al., 1997).

5. PECTIN-DEGRADING ENZYMES FROM EXTREMOPHILIC ORGANISMS

Pectin is a branched heteropolysaccharide consisting of a main chain of α-1,4-D-polygalacturonate, which is partially methyl esterified (Fig. 6). Along the

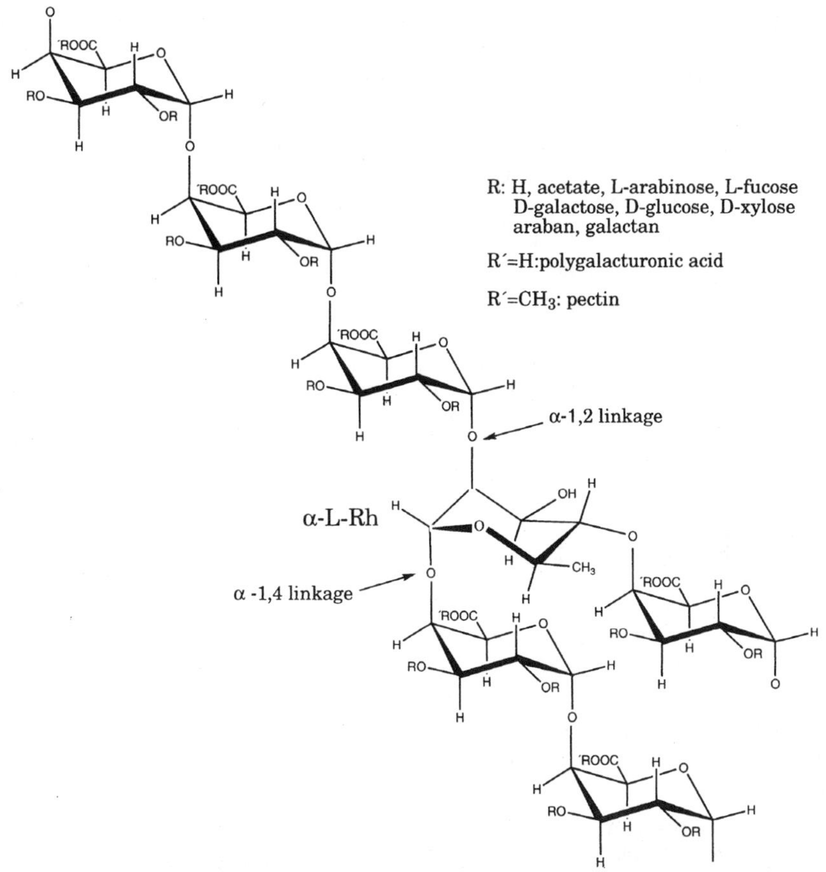
R: H, acetate, L-arabinose, L-fucose
D-galactose, D-glucose, D-xylose
araban, galactan
R′=H:polygalacturonic acid
R′=CH3: pectin
α-1,2 linkage
α-L-Rh
α -1,4 linkage
ROOC
RO
OR
OH
CH3

Figure 6. Chemical structure of pectin.

chain, L-rhamnopyranose residues are present that are the binding sites for side chains composed of neutral sugars. Pectin is an important plant material that is present in the middle lamellae as well as in the primary cell walls.

5.1. Pectin-Degrading Enzymes

Pectin is degraded by pectinolytic enzymes that can be classified into two major groups. The first group comprises methylesterases, whose function is to remove the methoxy groups from pectin. The second group comprises the depolymerases (hydrolases and lyases), which attack both pectin and pectate (polygalacturonic acid).

Pectinase, pectin depolymerase, or polygalacturonase (E.C.3.2.1.15) randomly hydrolyzes α-1,4-galactosiduronic linkages in pectate and other galacturonans. Exopolygalacturonase, Galacturan α-1,4-galacturonidase or poly-galacturonate hydrolase (E.C.3.2.1.67) removes one molecule galacturonate from polygalacturonate. Exo-poly-α-galacturonosidase (E.C.3.2.1.82) hydrolyzes pectic acid from the nonreducing end of pectin, releasing digalacturonate. Pectinesterase, pectin methylesterase, pectin demethoxylase, or pectin methoxylase (E.C.3.1.1.11) removes methyl groups from pectin, producing methanol and pectate. Oligogalacturonide lyase (E.C.4.2.2.6) catalyzes the eliminative removal of unsaturated terminal residues from oligosaccharides of D-galacturonate. Pectin lyase (E.C.4.2.2.10) catalyzes the eliminative cleavage of pectin to give oligosaccharides with terminal 4-deoxy-6-methyl-α-D-galact-4-enuronosyl groups. This enzyme does not act on deesterified pectin.

5.2. Production of Pectin-Degrading Enzymes by Hyperthermophiles

A great variety of pectinolytic bacteria have been isolated from various habitats such as trees, lakes, soil, tumen, mullet gut, and the human intestinal track. Pectin hydrolases are predominantly synthesized by fungi (Albersheim et al., 1960; Itoh et al., 1982; Kamamiya et al., 1974; Schlemmer et al., 1987; Sone et al., 1988), whereas pectate lyases are mostly produced by bacteria and usually act at alkaline pH and are Ca^{2+} dependent (Whitaker, 1991). Pectin degradation from bacteria has been reported for *Thermoanaerobacter thermohydrosulfuricus* (Wiegel et al., 1979), *Thermoanaerobacter thermosulfurigenes* (Schink and Zeikus, 1983), *Clostridium thermocellum* (Spinnler et al., 1986), *Desulfurococcus amylolyticus* (Bonch-Osmolovsskaya et al., 1988), *Clostridium thermosaccharolyticum* (van Rijssel and Hansen, 1989), and *Bacillus stearothermophilus* (Karbassi and Vaughn, 1980).

Although many microorganisms have been screened for pectinolytic activity, little attention has be given to pectinolytic enzymes from thermophilic and hyperthermophilic bacteria (Kozianowski et al., 1997). Previously a novel anaerobic strain from a thermal spa in Italy was isolated that produces two thermoactive lyases that have a very high affinity for polygalacturonate. This is a spore-forming anaerobic microorganism able to grow on citrus pectin and pectate optimally at 70°C, which has been identified as *Thermoanaerobacter italicus*. After growth on citrus pectin, two pectate lyases (α and β) were induced, purified, and biochemically characterized. Pectate lyase α is a single polypeptide of 135 kDa, whereas pectate lyase β is a heterodimer with 93 and 158 kDa molecular weight for the two subunits. Both enzymes display similar catalytic properties and can function at temperatures up to 80°C. An increase in the enzymatic activity of both pectate lyases was observed upon addition of Ca^{2+} at 1 mM concentration.

Another anaerobic, extremely thermophilic, non–spore-forming bacterium was isolated from a sediment sample taken from Owens Lake, California, and designated strain OLT. It grows between 50 and 80°C, with a temperature optimum at 75°C and at pH range 5.5 to 9.0, with a pH optimum at about 7.5. The isolate utilized pectin, sucrose, xylose, fructose, ribose, xylan, starch, and cellulose. It has been proposed that OLT be designated *Caldicellulosiruptor owensensis* sp. nov., based on 16S rDNA sequence analysis.

Unfortunately, pectin-hydolyzing enzymes from archaea have not yet been identified and characterized (Huang *et al.*, 1998). However, due to the lack of biochemical data on various pectin hydrolases, it is impossible to make proper comparisons from various organisms.

5.3. Biotechnological Relevance

Enzymatic pectin degradation is widely applied in food technology processes, as in fruit juice extraction, in order to increase the juice yield, reduce its viscosity, improve color extraction from the skins, as well as to macerate fruit and vegetable tissues.

6. CHITIN-DEGRADING ENZYMES FROM EXTREMOPHILIC ORGANISMS

Chitin is a linear β-1,4 homopolymer of *N*-acetyl-glucosamine residues and it is the second-most abundant natural biopolymer after cellulose on earth (Fig. 7). Chitin is produced in enormous amounts, particularly in the marine environment (Gooday 1990, 1994), and its turnover is due to the action of chitinolytic enzymes. Chitin is the major structural component of most fungi and invertebrates (Gooday, 1990, 1994), whereas chitin serves as a nutrient for soil or marine bacteria.

6.1. Chitin-Degrading Enzymes

Chitin degradation is known to proceed with the endo-acting chitin hydrolase chitinase A (E.C.3.2.1.14) and the chitin oligomer exo-acting hydrolases chitinase B and *N*-acetyl-D-glucosaminidase (trivial name: chitobiase) (E.C.3.2.1.52). The chemical structure and the site of action of chitinolytic enzymes is shown in Fig. 7.

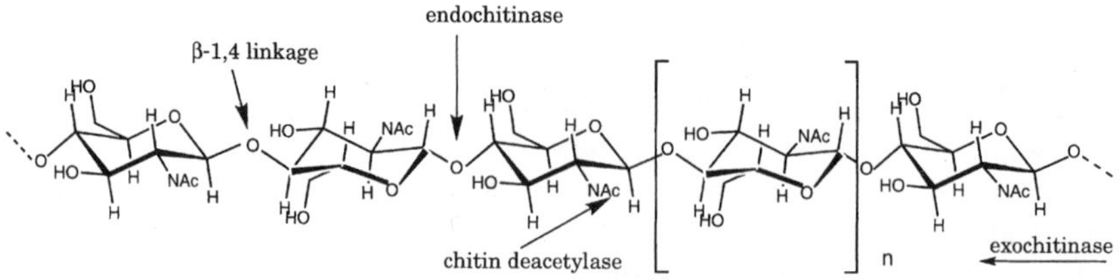

Figure 7.   Chemical structure of chitin and the action of the enzymes involved in its degradation.

Endo- and exochitinases comprise three glycosyl hydrolase families: families 18, 19, and 20. Family 18 consists of chitinases, endo-β-N-acetyl-D-glucosaminidases (EC 3.2.1.96) and di-N-acetylchitobiases from eukaryotae, prokaryotae, and viridae. The N-acetyl-D-glucosamine oligomeric product retains its C1 anomeric configuration. Family 19 contains only chitinases from eukaryotae and prokaryotae, and in contrast to the family 18, the product has inverted anomeric configuration. Finally, family 20 contains β-hexosaminidases and chitobiases. Chitobiases degrade only small N-acetyl-D-glucosamine oligomers (up to pentamers) and the released N-acetyl-D-glucosamine monomers retain their C1 anomeric configuration. The X-ray structure of these enzymes previously has been determined and the enzymatic mechanism of hydrolyis of chitin by chitinase A and chitobiase from *Serratia marcescens* has been resolved at the atomic level (Perrakis *et al.*, 1994; Tews *et al.*, 1996).

6.2. Chitin-Degrading Enzymes from Hyperthermophilic Organisms

Although a large number of chitin-hydrolyzing enzymes have been isolated and their corresponding genes have been cloned and characterized, only a few chitin-hydrolyzing enzymes that are thermostable are known. These enzymes have been isolated from the thermophilic bacterium *Bacillus licheniformis* X-7u (Takayanagi *et al.*, 1991), *Bacillus sp.* BG-11 (Bharat and Hoondal, 1998), and *Streptomyces thermoviolaceus* OPC-520 (Tsujibo *et al.*, 1995)

The extremophilic anaerobic archeon *Thermococcus chitonophagus* has been reported to hydrolyze chitin (Huber *et al.*, 1995). This is the first extremophilic archeon that produces chitinase(s) and N-acetylglucosaminidase(s) in order to degrade chitin for nutritional purposes. *Thermococcus chitonophagus* can grow up to 93°C under nitrogen and the chitinolytic enzyme system is cell associated and inducible by chitin. The chitin-degrading enzymes have been identified and their biochemical characterization as well as their molecular cloning is underway (C. Vorgias, unpublished results).

6.4. Biotechnological Relevance

Although chitin and its partially deacetylated derivative chitosan are not well
established as products with a particular biotechnological interest, there are a number of scientific works reporting that the natural polymer chitin exhibits interesting properties that make it a valuable raw material for several applications
(Cohen-Kupiec and Chet, 1998; Kramer and Muthukrishnan, 1997; Spindler *et al.,*
1990; Muzzarelli, 1997; Shigemasa and Minami, 1996; Georgopapadakou and
Tkacz, 1995; Benhamou, 1995; Chandy and Sharma, 1990; Kas, 1997).

ACKNOWLEDGMENTS

Thanks are due to European Union in the framework of the BIOTECH project
"Extremophilies as Cell Factories" and the Deutsches Bundesamt für Umwelt for
financial support.

REFERENCES

Albersheim, P., Neukom, H., and Deuel, H., 1960, Über die Bindung von ungesättingen Ab-
bauprodukten durch ein pektinabbauendes Enzym, *Helv. Chim. Acta* **43**:1422–1426.
Andrade, C. M., Morana, A., De Rosa, M., and Antranikian, G., 1996, Production and characterization of amylolytic and xylanolytic enzymes from the hyperthermophilic ar-
chaeon *Pyrodictium abyssi.* First International Congress on Extremophiles, Estoril
Portugal, 2–6 June 1996.
Annison, G., 1992, Commercial enzyme supplementation of wheat-based diets raises ileal
glycanase activitites and improves apparent metabolisable energy starch and pentosan
digestibilities in broiler chickens, *Anim. Feed Sci. Technol.* **38**:105–121.
Antranikian, G., and Koch, R., 1990, Thermostable α-amylase from the hyperthermophilic
Pyrococcus sp. International Patent WO 95/11352
Antranikian, G., 199, Microbial degradation of starch. In: Winkelmann, G. (ed.) *Microbial
Degradation of Natural Products* VCH Weinheim; pp. 27–56.
Becker, P., Abu-Reesh, I., Markossian, S., Antranikian, G., and Märkl, H., 1997, Determi-
nation of the kinetic parameters during continuous cultivation of the lipase producing
thermophile *Bacillus* sp. IHI-91 on olive oil, *Appl. Microbiol. Biotechnol.* **48**:184–
190.
Benhamou, N., 1995, Immunocytochemistry of plant defense mechanisms induced upon
microbial attack, *Microsc. Res. Tech.* **31**:63–78.
Bharat, B., and Hoondal, G. S., 1998, Isolation, purification and properties of a thermostable
chitinase from an alkalophilic *Bacillus* sp. BG-11, *Biotechnol. Lett.* **20**:157–159.
Bok, J. D., Goers, S. K., and Eveleigh, D. E., 1994, Cellulase and xylanase systems of *Thermotoga neapolitana, ACS Symp. Ser.* **566**:54–65.
Bonch-Osmolovskaya, E. A., Sleraseev, A. I., Miroshnichenko, M. L., Sveltichnaya, T. P.,

and Alekseev, V. A., 1988, Characteristics of *Desulfurococcus amyloliticus* sp. nov., a new extremely thermophilic archaebacterium isolated from thermal springs of Kamchatka and Kunashir Island, *Mikrobiologyia* **57**:94–101.

Bragger, J. M., Daniel, R. M., Coolbear, T., and Morgan, H. W, 1989, Very stable enzyme from extremely thermophilic archaebacteria and eubacteria, *Appl. Microbiol. Biotechnol.* **31**:556–561.

Bronnenmeier, K., Kern, A., Leible, W., and Staudenbauer, W. L., 1995, Purification of *Thermotoga maritima* enzymes for the degradation of cellulose materials, *Appl. Env. Microbiol.* **61**:1399–1407.

Brown, S. H., and Kelly, R. M., 1993, Characterization of amylolytic enzymes, having both α-1,4 and α-1,6 hydrolytic activity, from the thermophilic archaea *Pyrococcus furiosus* and *Thermococcus litoralis, Appl. Env. Microbiol.* **59**:2614–2621.

Canganella, F., Andrade, C., and Antranikian, G., 1994, Characterization of amylolytic and pullulytic enzymes from thermophilic archaea and from a new *Fervidobacterium* sp., *Appl. Microbiol. Biotechnol.* **42**:239–245.

Chandy, T., and Sharma, C. P., 1990, Chitosan as a biomaterial *Biomater, Artif. Cells Artif. Organs* **31**:1–24.

Chen, C. C., Adolphson, R., Dean, F. D. J., Eriksson, K. E. L., Adamas, M. W. W., and Westpheling, J., 1997, Release of lignin from kraft pulp by a hyperthermophilic xylanase from *Thermotoga maritima, Enzyme Microbiol. Technol.* **20**:39–45.

Chung, Y. C., Kobayashi, T., Kanai, H., Akiba, T., and Kudo, T., 1995, Purification and properties of extracellular amylase from the hyperthermophilic archeon *Thermococccus profundus* DT5432, *Appl. Env. Microbiol.* **61**:1502–1506.

Ciaramella, M., Cannio, R., Moracci, M., Pisani, F. M., and Rossi, M., 1995, Molecular biology of extremophiles, *World J. Microbiol. Biotechnol.* **11**:71–84.

Classen, H. L., 1996, Cereal grain starch and exogenous enzymes in poultry diets, *Anim. Feed Sci. Technol.* **33**:791–794.

Cohen-Kupiec, R., and Chet, I., 1998, The molecular biology of chitin digestion, *Curr. Opin. Biotechnol.* **9**:270–277.

Cubellis, M. V., Rozzo, C., Montecucchi, P., and Rossi, M., 1990, Isolation and sequencing of a new β-galactosidase-encoding archaebacterial gene, *Gene* **94**:89–94.

Friederich, A., and Antranikian, G., 1996, Keratin degradation by *Fervidobacterium pennavorans,* a novel thermophilic anaerobic species of the order *Thermotogales, Appl. Env. Microbiol.* **62**:2875–2882.

Georgopapadakou, N. H., and Tkacz, J. S., 1995, The fungal cell wall as a drug target, *Trends Microbiol.* **3**:98–104.

Gibbs, M. D., Reeves, R. A., and Bergquist, P. L., 1995, Cloning, sequencing and expression of a xylanase gene from the extreme thermophile *Dictyoglomus thermophilum* Rt46B.1 and activity of the enzyme on fiber-bound substrate, *Appl. Env. Microbiol.* **61**:4403–4408.

Gooday, G. W., 1990, Physiology of microbial degradation of chitin and chitosan, *Biodegradation* **1**:177–190.

Gooday, G. W., 1994, Physiology of microbial degradation of chitin and chitosan, in: Biochemistry of Microbial Degradation (C. Ratledge ed.), Kluwer, Dordrecht, pp. 279–312.

Grogan, D. W., 1991, Evidence that β-galactosidase of *Sulfolobus solfataricus* is only one of several activities of a thermostable β-D-glycosidase, *Appl. Env. Microbiol.* **57**:1644–1649.

Henrissat, B., 1991, A classification of glycosyl hydrolases based on amino acid sequence similarity, *Biochem. J.* **280**:309–316.

Huang, C. Y., Patel, B. K., Mah, R. A., and Baresi, L., 1998, *Caldicellulosiruptor owensensis* sp. nov., an anaerobic, extremely thermophilic, xylanolytic bacterium, *Int. J. Syst. Bacteriol.* **48**:91–97.

Huber, R., Stöhr J., Hohenhaus, S., Rachel, R., Burggraf S., Jannasch, H. W., and Stetter, K. O., 1995, *Thermococcus chitonophagus* sp. nov., a novel chitin degrading, hyperthermophilic archeum from the deep-sea hydrodermal vent enviroment, *Arch. Microbiol.* **164**:255–264.

Itoh, Y., Suguira, J., Izaki, K., and Takahashi, H., 1982, Enzymological and immunological properties of pectin lyases from bacteriocinogenic strains of *Erwinia carotovora,* *Agric. Biol. Chem.* **46**:199–205.

Jørgensen, S., Vorgias, C. E., and Antranikian, G., 1997, Cloning, sequencing and expression of an extracellular α-amylase from the hyperthermophilic *archeon Pyrococcus fusiosus* in *Escherichia coli* and *Bacillus subtilis, J. Biol. Chem.* **272**:16335–16342.

Kamamiya, S., Nishiya, T., Izaki, K., and Takahashi, H., 1974, Purification and properties of a pectin *trans*-eliminase in *Erwinia aroidae* formed in the presence of nalidixic acid, *Agric. Biol. Chem.* **38**:1071–1078.

Karbassi, A., and Vaughn, R. H., 1980, Purification and properties of polygalacturonic *trans*-eliminase from *Bacillus stearothermophilus, Can J. Microbiol.* **26**:377–384.

Kas, H. S., 1997, Chitosan: Properties, preparations and application to microparticulate systems, *J. Microencapsul.* **31**:689–711.

Kengen, S. W. M., Luesink, E. J., Stams, A, J. M., and Zehnder, A. J. B., 1993, Purification and characterization of an extremely β-glucosidase from the hyperthermophilic archaeon *Pyrococcus furiosus, Eur. J. Biochem.* **213**:305–312.

Koch, R., Spreinat, K., Lemke, K., and Antranikian, G., 1991, Purification and properties of a hyperthermoactive α-amylase from the archaeobacterium *Pyrococcus woesei, Arch. Microbiol.* **155**:572–578.

Koch, R., Canganella, F., Hippe, H., Jahnke, K. D., and Antranikian, G., 1997, Purification and propeties of a thermostable pullanase from a newly isolated thermophilic anaerobic bacterium *Fervidobacterium pennavorans* Ven5, *Appl. Env. Microbiol.* **63**:1088–1094.

Kozianowski, G., Canganella, F., Rainey, F. A., Hippe, H., and Antranikian, G., 1997, Purification and characterization of thermostable pectate-lyase from a newly isolated thermophilic bacterium, *Thermoanaerobacter italicus* sp. nov., *Extremophiles* **1**:171–182.

Krahe, M., Antranikian, G., and Märkl, H., 1996, Fermentation of extremophilic microorganisms, *FEMS Microbiol. Rev.* **18**:271–285.

Kramer, K. J., and Muthukrishnan, S., 1997, Insect chitinases: Molecular biology and potential use as biopesticides, *Insect Biochem. Mol. Biol.* **27**:887–900.

Ladenstein, R., and Antranikian, G., 1998, Proteins from hyperthermophiles: stability and enzymatic catalysis close to the boiling point of water, *Adv. Biochem. Eng. Biotechnol.* **61**:37–85.

Laderman, K. A., Asada, K., Uemori, T., Mukai, H., Taguchi, Y., Kato, I., and Anfinsen,

C. B., 1993, Alpha-amylase from the hyperthermophilic archaebacterium *Pyrococcus furiosus.* Cloning and sequencing of the gene and expression in *Escherichia coli. J. Biol. Chem.* **268**:24402–24407.

Lee, J. T., Kanai, H., Kobayashi, T., Akiba, T., and Kudo, T., 1996, Cloning, nucleotide sequence and hyperexpression of α-amylase gene from an archaeon, *Thermococcus profundus, J. Ferment. Bioeng.* **82**:432–438.

Liebl, W., Stemplinger, I., and Ruile, P., 1997, Properties and gene structure of the *Thermotoga maritima* α-amylase Amy A, a putative lipoprotein of a hyperthermophilic bacterium, *J. Bacteriol.* **179**:941–948.

Luthi, E., Jasmat, N. B., and Bergquist, P. L., 1990a, Xylanases from the thermophilic bacterium *Caldocellum saccharolyticum:* Overexpression of the gene in *Escherichia coli* and characterization of the gene product, *Appl. Env. Microbiol.* **1990**:2677–2683.

Luthi, E., Love, D. R., McAnulty, J., Wallace, C., Caughey, P. A., Saul, D., and Bergquist, P. L., 1990b, Cloning, sequence analysis, and expression of genes encoding xylandegrading enzymes from the thermophile *Caldocellum saccharolyticum, Appl. Env. Microbiol.* **56**:1017–1024.

Maat, J., Roza, M., Verbakel, J., Stam, H., Santos, H., da Silva, M. J., Bosse, M., Egmond, M. R., Hagemans, M. L. D., Gorcom, R. F. M., Hessing, J. G. M., Hondel, C. A. M. J. J., and Rotterdam, C., 1992, Xylanases and their application in bakery. *Prog. Biotechnol.* **7**:349–360.

McDonough, T. J., 1992, A survey of mechanical pulp bleaching in Canada: Single-stage hydrosulfite lines are the rule, *Pulp. Pap. Can.* **93**:57–63.

Moracci, M., Ciaramella, M., Nucci, R., Pearl, L. H., Sanderson, I., Trincone, A., and Rossi, M., 1993, Thermostable β-glycosidase from *Sulfolobus solfataricus, Biocatalysis* **11**:89–103.

Moracci, M., Di Lernia, I., Antranikian, G., De Rosa, M., and Rossi, M., 1998, Thermozymes in carbohydrate processing. VTT Symposium 177, VTT Biotechnology and Food Research UDC 664: 663.1: 579.67: 56–65.

Müller, R., Antranikian, G., Maloney, S., and Sharp, R., 1998, Thermophilic degradation of environmental pollutants. *Adv. Biochem. Eng. Biotechnol.* **61**:155–169.

Muzzarelli, R. A., 1997, Human enzymatic activities related to the therapeutic administration of chitin derivatives, *Cell. Mol. Life Sci.* **53**:131–140.

Nucci, R., Moracci, M., Vaccaro, C., Vespa, N., and Rossi, M., 1993, Exoglucosidase activity and substrate specificity of the β-glycosidase isolated from the extreme thermophilic *Sulfolobus solfataricus, Biotechnol. Appl. Biochem.* **17**:239–250.

Perrakis, A., Tews, I., Dauter, Z., Oppenheim, A., Chet, I., Wilson, K. S., and Vorgias, C. E., 1994, Crystal structure of a bacterial chitinase at 2.3 Å resolution, *Structure* **2**:1169–1180.

Petersen, S., Jensen, B., Dijkhuzen, L., Jørgensen, S., and Dijkstra, B., 1995, A better enzyme for cyclodextrins, *Chemtech* **12**:19–25.

Pisani, F. M., Rella, R., Raia, C. A., Rozzo, C., Nucci, R., Gambacorta, A., De Rosa, M., and Rossi, M., 1990, Thermostable β-galactosidase from the archaebacterium *Sulfolobus solfataricus*—Purification and properties, *Eur. J. Biochem.* **187**:321–328.

Prisco, A., Moracci, M., Rossi, M., and Ciaramella, M., 1994, A gene encoding a putative membrane protein homologous to the major facilitator superfamily of transporters of maps upstream of the β-glycosidase gene in the archaeon *Sulfolobus solfataricus, J. Bacteriol.* **177**:1616–1619.

Prowe, S., Van de Vossenberg, J., Driessen, A., Antranikian, G., and Konings, W., 1996, Sodium-coupled energy transduction in the newly isolated thermoalkaliphic strain LBS3, *J. Bacteriol.* **178**:4099–4104.

Rüdiger, A., Sunna, A., and Antranikian, G., 1994, Enzymes from extreme thermophilic and hyperthermophilic archea and bacteria, in: *Carbohydrases. Handbook of Enzyme Catalysis in Organic Synthesis* (K. Drauz and H. Waldmann, eds.), VCH Weinheim, pp. 946–961.

Rüdiger, A., Jørgensen, P. L., and Antranikian, G., 1995, Isolation and characterization of a heat stable pullulanase from the hyperthermophilic archeon *Pyrococcus woesei* after cloning and expression of its gene in *Escherichia coli, Appl. Env. Microbiol.* **61**:567–575.

Ruttersmith, L. D., and Daniel, R. M., 1991, Thermostable cellobiohydrolase from the thermophilic cubacterium *Thermotoga* sp. strain Fj533-B1 *J Biochem* **277**:887–890.

Ruttersmith, L. D., Daniel, R. M., and Simpson, H. D., 1992, Cellulolytic and hemicellulolytic enzymes functional above 100°C, *Ann. NY Acaa. Sci.* **672**:137–141

Schink, B., and Zeikus, J. G., 1983, Characterization of pectinolytic enzymes of *Clostridium thermosulfurogenes, FEMS Microbiol. Lett.* **17**:295–298.

Schlemmer, A, F., Ware, C. F., and Keen, N., 1987, Purification and characterization of a pectin lyase produced by *Pseudomonas fluorescens* W51, *J. Bacteriol.* **169**:4493–4498.

Schuliger, J. W., Brown, S. H., Baross, J. A., and Kelly, R. M., 1993, Purification and characterisation of a novel amylolytic enzyme from ES4, a marine hyperthermophilic archaeum, *Mol. Mar. Biol. Biotechnol.* **2**:76–87.

Schumann, J., Wrba, A., Jaenicke, R., and Stetter, K. O., 1991, Topographical and enzymatic characterization of amylase from the extreme thermophilic eubacterium *Thermotoga maritima, FEBS Lett* **282**:122–126.

Shigemasa, Y., and Minami, S., 1996, Applications of chitin and chitosan for biomaterials, *Biotechnol. Genet. Eng. Rev.* **1**:383–420.

Simpson, H. D., Haufler, U. R., and Daniel, R. M., 1991, An extremely thermostable xylanase from the thermophilic eubacterium *Thermotoga, Biochem. J.* **277**:177–185.

Sone, H., Sugiura, J., Itoh, Y., Izaki, K., and Takahashi, H., 1988, Production and properties of pectin lyase in *Pseudomonas marginalis* induced with mitomycin C, *Agric. Biol. Chem.* **52**:3205–3207.

Spindler, K. D., Spindler-Barth, M., and Londershausen, M., 1990, Chitin metabolism: A target for drugs against parasites, *Parasitol. Res.* **76**:283–288.

Spinnler, H. E., Lavigne, B., and Blachere, H., 1986, Pectinolytic activity of *Clostridium thermocellum:* Its use for anaerobic fermentation of sugar beet purple, *Appl. Microbiol. Biotechnol.* **23**:434–437.

Stetter, K. O., 1996, Hyperthermophiles in the history of life, *CIBA Found. Symp.* **202**:1–18.

Sunna, A., and Antranikian, G., 1996, Growth and production of xylanolytic enzymes by the extreme thermophilic anaerobic bacterium *Thermotoga maritima, Appl. Microbiol. Biotechnol.* **45**:671–676.

Sunna, A., and Antranikian, G., 1997, Xylanolytic enzymes from fungi and bacteria, *Crit. Rev. Biotechnol.* **17**:39–67.

Sunna, A., Moracci, M., Rossi, M., and Antranikian, G., 1996a, Glycosyl hydrolases from hyperthermophiles, *Extremophiles* **1**:2–13.

Sunna, A., Puls, J., and Antranikian, G., 1996b, Purification and characterization of two thermostable endo-1,4-β-D-xylanases from *Thermotoga thermarum, Biotechnol. Appl. Biochem.* **24**:177–185.

Tachibana, Y., Mendez, L. M., Fujiwara, S., Takagi, M., and Imanaka, T., 1996, Cloning and expression of the α-amylase gene from the hyperthermophilic archeon *Pyrococcus* sp. KOD1 and characterization of the enzyme, *J. Ferment. Bioeng.* **82**:224–232.

Takayanagi, T., Ajisaka, K., Takiguchi, Y., and Shimahara, K., 1991, Isolation and characterization of thermostable chitinases from *Bacillus licheniformis* X-7u, *Biochim. Biophys. Acta* **1078**:404–410.

Teo, V. S., Saul, D. J., and Bergquist, P. L., 1995, *Cel*A, another gene coding for a multidomain cellulase from the extreme *thermophile Caldocellum saccharolyticum, Appl. Microbiol. Biotechnol.* **43**:291–296.

Tews, I., Perrakis, A., Dauter, Z., Oppenheim, A., Wilson, K. S., and Vorgias, K. S., 1996, Bacterial chitobiase structure provides insight into catalytic mechanism and the basis of Tay-Sachs disease, *Nature Struct. Biol.* **3**:638–648.

Tsujibo, H., Endo, H., Miyamoto, K., and Inamori, Y., 1995, Expression in *Escherichia coli* of a gene encoding a thermostable chitinase from *Streptomyces thermoviolaceus* OPC-520. *Biosci. Biotechnol. Biochem.* **59**:145–146.

van Rijssel, M., and Hansen, T. A., 1989, Fermentation of pectin by a newly isolated *Clostridium thermosaccharolyticum* strain, *FEMS Microbiol. Lett.* **61**:41–46.

Viikari, L., Kantelinen, A., Sundquist, J., and Linko, M., 1994, Xylanases in bleaching. From an idea to industry, *FEMS Microbiol. Lett.* **13**:335–350.

Voorhorst, W. G. B., Eggen, R. I. L., Luesink, E. J., and de Vos, W. M., 1995, Characterization of the *cel*B gene coding for β-glucosidase from the hyperthermophilic archaeon *Pyrococcus furiosus* and its expression and site-directed mutation in *Escherichia coli, J. Bacteriol.* **177**:7105–7111.

Whitaker, J. R., 1991, Microbial pectolytic enzymes. in: *Microbial Enzymes and Biotechnology* (W. M. Fogarty and C. T. Kelly, eds.), Applied Science, London, New York, pp. 133–176.

Wiegel, J., Ljungdahl, L. G., and Rawson, J. R., 1979, Isolation from soil and properties of the extreme thermophilic *Clostridium thermohydrosulfuricum, J. Bacteriol.* **139**:800–810.

Wind, R., Liebl, W., Buitlaar, R., Penninga, D., Spreinat, A., Dijkhuzen, L., and Bahl, H., 1995, Cyclodextrin formation by the thermostable α-amylase of *Thermoanaerobacterium thermosulfurigenes* EM1 and reclassification of the enzyme as a cyclodextrin glycosyltransferase, *Appl. Env. Microbiol.* **61**:1257–1265.

Winterhalter, C., and Liebl, W., 1995, Two extremely themostable xylanases of the hyperthermophilic bacterium *Thermotoga maritima* MSB8, *Appl. Env. Microbiol.* **61**:1810–1815.

Yang, J. L., and Eriksson, K. E. L., 1992, Use of hemicellulolytic enzymes as one stage in bleaching of kraft pulps, *Holzforschung* **46**:481–488.

Zverlov, V., Mahr, S., Riedel, K., and Bronnenmeier, K., 1998, Properties and gene structure of a bifunctional cellulolytic enzyme (*Cel* A) from the extreme thermophile *Anaerocellum thermophilum* with separate glycosyl hydrolase family 9 and 48 catalytic domains, *Microbiology* **144**:457–465.

12

Profiling and Trace Detection of Bacterial Cellular Carbohydrates

Alvin Fox

1. INTRODUCTION

Many of the unusual macromolecules found in bacteria, including peptidoglycans, teichoic acids, and lipopolysaccharides, contain carbohydrates as major
constituents. Structural characterization of these molecules, including identifying
their monomeric sugar components, has helped provide an understanding of their
physiological role in the bacterial cell. The uniqueness of many of these sugars
and their association with particular species and genera has been important in the
area of chemotaxonomy. The advent of modern analytical chemical methods,
based on chromatography and mass spectrometry, has clarified and expanded
on this previously known microbial chemistry. Additionally, trace detection of
some of these bacterial compounds in environmental and clinical samples has
opened opportunities to answer fundamental questions of the role of bacterial
constituents in disease that previously could not be addressed adequately (Fox *et
al.,* 1990a)

A novel aspect of work presented here is the use of high-resolution gas
chromatographic (GC) and high-performance liquid chromatographic (LC)
separations, both coupled with the power of online mass spectrometric (MS) or
tandem mass spectrometric (MS/MS) analysis. In studying sugars present in

---

Alvin Fox • Department of Microbiology and Immunology, University of South Carolina, School
of Medicine, Columbia, South Carolina 29208.

Glycomicrobiology, edited by Doyle.
Kluwer Academic/Plenum Publishers, New York, 2000.

complex biological matrices, contaminating compounds are commonplace, masking detection and causing false-positives. Interferences are common in colorimetric methods. A chromatographic peak at the correct retention time (using a nonselective detection method) does not constitute definitive identification.

The utility of MS analysis is that in the selected ion monitoring (SIM) mode chromatograms are free of extraneous background peaks and allow major and minor components to be reliably detected. In the total ion mode, sugars are readily identified from their mass spectrum. The selectivity of GC-MS (and LC-MS) also has allowed the analyses of whole cell hydrolyzates (rather than cellular or cell wall fractions) for their carbohydrate composition. This avoids possible differential losses of characteristic components during the isolation–fractionation procedures. Such losses might help explain some discrepancies between the results of certain studies performed in the past.

As noted above, the mass spectrometer is widely used as a selective chromatographic detector to ignore extraneous chromatographic peaks. In tandem mass spectrometry, the instrument further decreases nonspecific peaks by screening out background peaks twice (monitoring mode). Additionally, when present at the relatively high levels, it is possible to categorically identify sugar markers for bacteria in a chromatographic peak by the mass spectrum (identification mode). However, at the low levels present in human tissues or body fluids, upon infection, or in certain environmental samples, it is not possible to obtain a full mass spectrum for "absolute" identification. Categorical identification at trace levels has awaited the development of more advanced instrumentation. In the GC-MS/MS identification mode one can also obtain absolute chemical confirmation by means of a product ion spectrum (chemical fingerprint). Thus, in both monitoring and identification modes, for trace analysis GC-MS/MS is always preferred over GC-MS.

The emphasis here is on providing examples of work involving glycomicrobiology that have been performed using MS and MS/MS. Recent developments in use of GC-MS and GC-MS/MS methodology for bacterial carbohydrate analysis have been reviewed elsewhere (Fox, 1999). Other earlier reviews provided the history of developments in the preparation of alditol acetates of bacterial sugars and their analysis using GC-MS (Fox and Black, 1994; Fox et al., 1989). GC-MS/MS was readily adapted from existing GC-MS procedures (Fox et al., 1995, 1996a,b). LC-MS and LC-MS/MS for bacterial carbohydrate analysis only have been recently introduced and are developmental in status (Shahgholi et al., 1997; Wunschel et al., 1997). An advantage of LC-MS (and LC-MS/MS) over GC-MS (and LC-MS-MS) is that sugars are analyzed in native form. This simplifies sample preparation by eliminating derivatization. However, GC-MS and GC-MS/MS instruments are more amenable to routine analysis.

2. METHODOLOGY

2.1. Analysis of Neutral and Amino Sugars Using GC-MS and GC-MS/MS

Neutral and amino sugar profiles were determined using the alditol acetate method (Fox and Black, 1994; Fox *et al.,* 1989). In brief, 10 mg of bacteria were hydrolyzed at 100°C for 3 hr in 2 *N* sulfuric acid to release sugar monomers. Arabinose and methylglucamine (1-deoxy-1-methylamino glucose) were added as internal standards for neutral and amino sugars, respectively. The solution was neutralized with 50% *N,N*-dioctylmethylamine in chloroform and hydrophobic materials removed by extraction on C18 columns. The sugars were reduced with sodium borodeuteride at 4°C. Borodeuteride was removed by multiple methanol:acetic acid (200:1) evaporations under nitrogen. Samples were dried under vacuum for 3 hr at 60°C. After drying, sugars were acetylated for 15 hr at 100°C. Acidic sugars retain a free carboxylate group, and thus are not detected on GC analysis.

Hydrophilic postderivatization cleanup included acid and alkaline extractions. GC-MS analyses were carried out with a mass selective detector (model 5970; Hewlett-Packard Co., Palo Alto, CA). GC-MS/MS analyses were performed with an ion trap MS/MS (GCQ, Finnigan, Atlanta GA) and a triple quadrupole (Quattro 1, Micromass, Boston, MA). Each GC instrument was equipped with an automated sample injector and a fused-silica capillary column (currently a nonpolar DB-5MS is recommended). For GC-MS analysis, electron ionization (EI) was performed at 70 eV for both total spectrum scanning and SIM. For GC-MS/MS analysis, EI was performed followed by collision-induced dissociation (CID) of the precursor ion to generate product ions (Fox *et al.,* 1995, 1996b).

2.2. Analysis of Neutral and Acidic Sugars Using LC-MS and LC-MS/MS

Acidic and neutral sugars were analyzed as described in detail elsewhere (Wunschel *et al.,* 1997). In brief, hydrolysis, neutralization and C18 column extraction was performed as described above. Arabinose and galacturonic acid were used as internal standard for neutral and acidic sugars, respectively (Sigma Chemical Co., St. Louis, MO). After hydrophobic extraction, the samples were acidified to 0.1 *N* H_2SO_4 in preparation for removal of the cationic contaminants. The cationic species were removed using an SCX (strong cation exchange) solid phase extension column (J. T. Baker, Phillipsburg, NJ). The samples were dried under a nitrogen gas flow at 60°C. Finally, samples were dissolved in 200 μl of water.

Separation of neutral and acidic sugars was performed on a 4-mm Carbopac

PA-1 pellicular anion exchange column in a sodium hydroxide–sodium acetate gradient (Dionex Corp, Sunnyvale, CA). Sodium ions were removed online using a Dionex 2-mm online anion suppressor run in the external water mode prior to passage into a Micromass Quattro I (Danvers, MA) triple quadrupole mass spectrometer. At this stage, amino sugars are lost. The LC pump and mass spectrometer were interfaced to an AS3500 autosampler. After electrospray ionization, selected ion monitoring was performed for molecular ions [M-H]⁻. For identification of sugars, parent ions were subjected to CID.

3. CHEMOTAXONOMIC CHARACTERIZATION OF BACTERIAL SPECIES

Carbohydrate profiling, using GC-MS, has proved useful in chemotaxonomy; some of the sugars identified are listed in Table I. The major sources of sugars present in bacteria are different in gram-positive and gram-negative bacteria. For gram-positive bacteria, sugars are derived from teichoic acids, teichuronic acids, or neutral polysaccharides [usually bound to the peptidoglycan (PG) layer] and membrane lipoteichoic acid. For gram-negative bacteria, the core and O-antigen of the lipopolysaccharide (LPS) serve as a source of discriminating carbohydrates. Capsular polysaccharides can be present in both gram-positive and gram-negative bacteria and serve as another source for carbohydrate profiles.

Table I
Sugar Markers in Whole Bacterial Cell Hydrolysates Analyzed as Alditol Acetates
Using Gas Chromatography–Mass Spectrometry

Compound	Source	Organism	Reference
Muramic acid	Peptidoglycan	Bacteria but not mammalian tissues or fungi	Fox *et al.* (1980, 1995)
Heptoses	Lipopolysaccharide	Gram-negative but not gram-positive bacteria	Fox *et al.* (1984, 1993b)
Quinovose	Spore polysaccharide	*Bacillus subtilis* but not *Bacillus cereus*	Wunschel *et al.* (1994)
Galactose	Vegetative cell wall polysaccharide	*Bacillus anthracis* but not *B. cereus*	Fox *et al.* (1993a)
Fucosamine	Lipopolysaccharide	*Tatlockia* but not *Legionella*	Fox *et al.* (1991)
Quinovosamine	Lipopolysaccharide	*Brucella abortis, B. suis,* and *B. melitiensis* but not *B. canis*	Fox *et al.* (1998a)

Deoxyribose and ribose are major microbial sugars, but because they are present in DNA and RNA in all organisms, they are not used for taxonomic discrimination. Muramic acid (3-*O*-lactyl glucosamine) and glucosamine, components of the glycan backbone of PG, are present universally in bacterial chromatograms. Glucosamine also is commonly found in other bacterial and nonbacterial constituents. The presence of these two sugars thus is not generally characteristic of particular bacterial species. It is widely accepted that gram-positive bacteria contain high levels of muramic acid because of the multilayered PG present in gram-positive but not gram-negative bacteria. However, few studies have directly compared the amount of muramic acid present among bacterial species. Gram-positive bacteria generally contain more muramic acid than gram-negative bacteria as predicted based on cell envelope composition. However, certain gram-negative and gram-positive organisms have intermediate levels and can not be distinguished by muramic acid content (Eudy *et al.,* 1985).

The core region of many gram-negative bacteria contain L-glycero-D-mannoheptose or less commonly D-glycero-D-mannoheptose. These two sugars can be used to distinguish among certain gram-negative bacterial species. Based on studies of the LPS of *Escherichia coli,* it is often assumed that heptoses are ubiquitous in gram-negative bacteria. However, it is not uncommon for heptoses to be absent in gram-negative bacteria, for example, legionellae (Fox *et al.,* 1984, 1990b) and brucellae (Fox *et al.,* 1998a; Moreno *et al.,* 1979).

Many gram-positive bacterial cells contain ribitol or glycerol as part of their teichoic acids. The presence of ribitol can be useful for taxonomic discrimination. However, glycerol also is found in bacterial glycerides. Thus glycerol cannot be used for teichoic acid detection in cellular hydrolysates. Carbohydrates less widely distributed among bacterial species have greater utility for chemotaxonomic discrimination. Examples are given below from our studies of bacilli (gram-positive bacteria) legionellae and brucellae (both gram-negative organisms). In all three instances there have been numerous unresolved questions regarding the taxonomic interrelationships within and among the constituent species/genera. Carbohydrate profiling has helped resolve a number of these questions.

Legionellae are important environmental agents that often colonize hot water towers. After airborne transmission, these microbes can initiate disease in susceptible individuals. *Legionella pneumophila* is the major pathogen in the family *Legionellaceae* and the causative agent of Legionnaire's disease. It has been difficult to use conventional biochemical tests in the differentiation of members of the *Legionellaceae.* We first differentiated legionellae by analysis of their carbohydrate content using GC with flame ionization detection (*Fox et al.,* 1984). Subsequently, total ion mode GC-MS was used to detect a number of unusual sugars, including a branched octose (Fox *et al.,* 1990b), which was subsequently identified by others as yersiniose A, 3,6-dideoxy-4-hydroxyethyl-D-xylo-hexose, (Sonesson and Jantzen, 1992). Increased sensitivity and selectivity for carbohydrate detection was subsequently achieved using SIM GC-MS (Fox *et al.*, 1990b; Walla *et al.*, 1984). Two of the uncommon aminodideoxyhexoses discovered in the legionellae were later identified as quinovosamine (2-amino-2,6-dideoxygalactose) and fucosamine (2-amino-2,6-dideoxyglucose) (Sonesson *et al.*, 1989). These two sugars help discriminate among two of the major groups present among the *Legionellaceae (Legionella* and *Tatlockia,* respectively). The two genera also are readily discriminated by 16S rRNA sequencing and physiological tests. Among the strong evidence for transferring *Legionella maceachernii* into the genus *Tatlockia* (along with *Tatlockia micdadei)* was the similarity of the carbohydrate profiles including the presence of large amounts of rhamnose and fucose and the presence of yersiniose A and fucosamine, which agreed with the similarities of their 16S rRNA sequences (Fox *et al.*, 1991).

Subsequently, we focused on another group of environmental pathogens, the gram-positive bacilli. Many aspects of the taxonomic characterization and clinical identification of bacilli are only recently being resolved. As an example, differentiation of the environmental organisms, *Bacillus thuringiensis, B. anthracis* and *B. cereus,* presents a taxonomic challenge for they display few distinguishing physiological characteristics and genetic relatedness, including DNA–DNA homology (Kaneko *et al.*, 1978) and high degree of 16S and 23S ribosomal RNA and 16S-23S interspace rRNA sequence relatedness (Harrell *et al.*, 1995; Ash and Collins, 1992; Ash *et al.*, 1991). As a result of the extreme similarity within this group, these organisms have been referred to as the *B. cereus* group. Two of these species are human pathogens (*B. anthracis,* the causative agent of anthrax, and *B. cereus,* a food-poisoning organism). A distinguishing characteristic of *B. thuringiensis* is its ability to produce a class of insecticidal proteins, known as crystallins or δ-toxins (Gonzales *et al.*, 1981). In contrast, *B. subtilis* and related species including *B. atrophaeus* are not human or insect pathogens and are readily differentiated from the *B. cereus* group, providing a closely related but distinct organism for study (Seki *et al.*, 1975, 1978; Nakamura, 1989; Wunschel *et al.*, 1994; Nagpal *et al.*, 1998).

Using GC-MS, sugar profiles of vegetative cells are similar for *B. cereus* and *B. thuringiensis. B. anthracis* contains high levels of galactose, which generally distinguishes it from *B. cereus/B. thuringiensis,* whereas *B. subtilis* is distinguished from the *B. cereus* group by low mannosamine levels (Fox *et al.*, 1993a; Wunschel *et al.*, 1994). The presence of galactose in *B. anthracis* has been noted previously (Cole *et al.*, 1984; Ezzell *et al.*, 1990). Spore profiles differ from vegetative profiles in all four species. Like vegetative profiles, spore profiles are distinctive for *B. cereus/B. thuringiensis, B. anthracis,* and *B. subtilis. B. cereus* and *B. thuringiensis* spores both contain rhamnose, fucose, 2-*O*-methyl rhamnose and 3-*O*-methyl rhamnose, unlike *B. anthracis* spores that contain only rhamnose and 3-O-methyl rhamnose. *B. subtilis* strains are heterogeneous with some resembling *B. anthracis* and others *B. cereus/B. thuringiensis,* although *B. subtilis* strains typically contain the rare sugar quinovose.

Molecular and chemical characteristics often provide complementary information in differentiation of closely related organisms. The genus *Brucella* is a highly conserved group of organisms. Identification of the four species pathogenic for man (*B. melitensis, B. abortus, B. suis,* and *B. canis*) is problematic for many clinical laboratories depending primarily on serology and phenotypic characteristics. The six species of *Brucella* are sufficiently related by DNA–DNA hybridization that a monospecies genus has been suggested (Verger *et al.,* 1985). 16S rRNA sequences of *B. abortus* and the other five species are also 98.5–99.7% similar (Dorsch *et al.,* 1989; Moreno *et al.,* 1990; Romero *et al.,* 1995a,b). The 16S rRNA sequence places this genus as a member of the alpha-2 subdivision of the *Proteobacteria,* closely related to *Bartonella* and *Agrobacterium.*

Polymerase chain reaction (PCR) amplification of the less genetically conserved 16S/23S rDNA interspace region was evaluated for species-specific polymorphism. *B. abortus, B. melitensis, B. suis,* and *B. canis* produced identical PCR interspace profiles (Rijpens *et al.,* 1996; Fox *et al.,* 1998a). However, these PCR products were unique to brucellae, allowing them to be readily distinguished from other gram-negative bacteria (including *Bartonella and Agrobacterium*). Carbohydrate profiles of multiple strains differentiated *B. canis* from the other three *Brucella* species due to the absence of the rare amino sugar quinovosamine, previously noted to be a component of LPS (Bowser *et al.,* 1974; Moreno *et al.,* 1981). Many strains contain mannose and some galactose, as previously found in LPS (Kreutzer *et al.,* 1979; Moreno *et al.,* 1981), but this does not contribute to chemotaxonomic discrimination. PCR of the rRNA interspace region is useful in identification of the genus *Brucella,* whereas carbohydrate profiling is capable of differentiating *B. canis* from the other *Brucella* species (Fox *et al.,* 1998a).

4. PHYSIOLOGICAL CHARACTERIZATION OF BACTERIAL CELLS

B. subtilis and related organisms synthesize a teichoic acid in the presence of phosphate excess and a glucuronic acid containing teichuronic acid under phosphate limitation (Ellwood and Tempest, 1969; Wright and Heckels, 1975; Lang *et al.,* 1982). *B. subtilis* strain W23 produces a ribitol containing teichoic acid (Wright and Heckels, 1975), whereas the teichoic acid of strain 168 (Lang *et al.,* 1982) and *B. atrophaeus* contain glycerol (Ellwood and Tempest, 1969). Galactosamine also is a major component of the cell wall of *B. subtilis,* but its source is somewhat controversial. It has been suggested that galactosamine is derived from a so-called "secondary" teichuronic acid (Duckworth *et al.,* 1972), whereas others have noted it is a component of teichuronic acid (Wright and Heckels, 1975; Mauck and Glaser, 1972).

Both GC-MS and LC-MS produced three characteristic carbohydrate profiles for bacilli having different physiological states. GC-MS allowed demonstration of a teichoic acid to teichuronic acid shift by the disappearance of anhydroribitol (derived from ribitol) on phosphate limitation. Galactosamine was not detected in cells grown in phosphate excess but was found in cells grown in phosphate limitation. This was consistent with galactosamine being derived from teichuronic acid. LC-MS analysis also allowed determination of the appearance of glucuronic acid, known to be characteristic of teichuronic acid, on phosphate limitation. Quinovose (6-deoxyglucose), characteristic of spores, is readily detected by GC-MS or LC-MS analysis. Spores are readily differentiated from vegetative cells by the appearance of high concentrations of quinovose often associated with rhamnose and 3−O-methyl rhamnose and the disappearance of ribitol and glucuronic acid. The source of quinovose and other unique sugars observed in the spore of B. subtilis (and the closely related organism B. atrophaeus) is currently not defined but most likely represents a previously undescribed spore polysaccharide. As discussed earlier, the B. cereus group of organisms also switches carbohydrate composition on changing from vegetative form to spore. This includes production of methylpentoses including rhamnose, an isomer of quinovose (Fox et al., 1993a, 1998c; Wunschel et al., 1994). A rhamnose containing polysaccharide has been isolated from the exosporium of B. cereus (Matz et al., 1970).

Surface carbohydrates present in Staphylococcus aureus include microcapsule and teichoic acid. The teichoic acid of S. aureus consists of a polyribitol phosphate backbone with side chains consisting of N-acetyl glucosamine and D-alanine. Both N-acetyl glucosamine and D-alanine are found in other cell wall polymers including the peptidoglycan. Ribitol is a specific chemical marker for teichoic acid. The microcapsule of S. aureus is more variable in composition, but many serotypes are characterized by the presence of fucosamine (Liau and Hash, 1977; Wu and Parks, 1971). It has been suggested that production of microcapsule varies with growth conditions (including phosphate limitation) and this may affect immunogencity and pathogenicity (Robbins et al., 1995). As noted above, for certain bacilli grown under conditions of phosphate limitation, it is well documented that there is a shift from a teichoic acid to a teichuronic acid. Whether this shift occurs in S. aureus is more controversial (Dobson and Archibald, 1978).

Fucosamine served as a marker for microcapsule and anhydroribitol for teichoic acid, respectively, for S. aureus. The presence of teichoic acids and microcapsule in S. aureus is established, but as stated, production of a teichuronic acid has not been proved. As noted, the shift from a teichoic acid to teichuronic acid in B. subtilis W23 is well documented, and this served as a positive control proving that cells were grown under optimal phosphate excess or limitation conditions.

Anhydroribitol and fucosamine were present at high concentrations under all growth conditions for S. aureus. Fucosamine was produced equally well under phosphate excess and limitation. This behavior was inconsistent with fucosamine being derived from a teichuronic acid. Furthermore, there was no evidence of a shift from teichoic acid to a teichuronic acid. LC-MS analyses were performed, showing that staphylococci cultured under phosphate excess produced a large amount of ribitol. Ribitol production was essentially unchanged under phosphate-limiting conditions. Trace levels of a peak at the retention time for glucuronic acid were observed under phosphate excess conditions. There was only a slight increase in the amount of glucuronic acid under the phosphate-limiting conditions (total levels of 0.06%). Our results agree with Dobson and Archibald (1978), and we can provide no evidence for a teichoic acid to teichuronic acid switch in *S. aureus* (Fox *et al.*, 1998b).

5. TRACE ANALYSIS IN COMPLEX MATRICES

During the past few years, GC-MS/MS was introduced for trace detection of carbohydrate markers for bacteria and their constituents in complex clinical and environmental matrices (Fox *et al.*, 1995, 1996a,b; Saraf and Larsson, 1996). GC-MS analysis was developed much earlier for this purpose (Fazio *et al.*, 1979; Fox *et al.*, 1980). High-resolution chromatographic separations coupled with selective cleanup steps are important in improving the specificity of detection of marker compounds (e.g., muramic acid) in complex matrices. However, chromatographic separation is not sufficient to eliminate extraneous peaks when nonselective detectors are employed. The use of the mass spectrometer, as a selective GC detector (e.g., GC-MS analysis in the SIM mode), helps greatly in diminishing background noise. However, even using SIM, it is not uncommon to find extraneous background peaks. The specificity of the tandem mass spectrometer in multiple reaction monitoring (MRM) mode as a GC detector provides even further specificity in detection at trace levels in complex matrices. Absolute identification is achieved by GC-MS/MS by the product ion spectrum. A particular focus of these studies has been muramic acid, an unusual sugar, found in bacterial PG but not in mammals or fungi.

5.1. Levels of Muramic Acid as a Measure of Biopollution in Environmental Samples

Exposure to high levels of airborne organic dust in indoor environment, such as cotton factories, poultry houses, and swine confinement buildings, can lead to pulmonary diseases including byssinosis and allergic hypersensitivity. The respiratory problems associated with these situations are often related to bacterial and fungal contamination (Castellan *et al.*, 1984; Sandstrom *et al.*, 1994). In many office and public buildings in the United States, interchange between outside and inside air is limited, and a primary source of interchange is through the filtration system of the air conditioner. The limited introduction of fresh air, constant recirculation, and portions of heating, ventilation, and air conditioning (HVAC) systems with high humidity or standing water create a fertile breeding ground for bacteria and other microbes. The consequent microbial contamination of indoor air can create health problems.

Currently, culture remains the most widely used procedure of assessing the microbial content of indoor air. It provides quantitative measures of viable bacteria as well as population diversity. Unfortunately, microbial culture techniqu ‎s may fail to detect the majority of the original microbial population. In fact, mɛ ꞁy airborne microbes are likely to be nonviable due to the hostile indoor air envirᴐn-ment Alternatively, non–culture-based methods measure components of the bacteria that can be derived from viable or nonviable bacteria or partially degraded cells (Dillon *et al.,* 1996).

Muramic acid, as a marker for PG levels, has been assayed in surface dust and airborne levels monitored (Fox *et al.,* 1993b, 1995; Krahmer *et al.,* 1998). While PG is present in both gram-negative and gram-positive bacteria, LPS is only present in gram-negative organisms. PG and LPS are both highly inflammatory substances and capable of activating the humoral and cellular arms of the immune system. PG displays toxicity for the hamster tracheal epithelium *in vitro* (Cookson *et al.,* 1989). In one recent study in swine houses, there was a correlation between airborne PG but not LPS levels with blood granulocyte levels and body temperature (Zhipping *et al.,* 1996). In a later study, upon exposure to swine dust, granulocyte levels increased dramatically in bronchoalveolar lavage and nasal lavage fluid. Interleukin-6 (IL-6) and tumor necrosis factor (TNF) levels also increased significantly. LPS but not PG levels correlated with IL-6 levels. Otherwise, there were no significant correlations (Wang *et al.,* 1997). Thus the relative contribution of PG and LPS to health effects during environmental exposure is not clearly established.

5.2. Muramic Acid Detection in Mammalian Tissues and Body Fluids

After systemic administration, persisting bacterial peptidoglycan–polysaccharide complexes cause chronic inflammation in animal models (Fox *et al.,* 1980; Gilbart and Fox, 1987; Gilbart *et al.,* 1986). In contrast, small subunits of PG are rapidly eliminated *in vivo* and do not cause chronic inflammation (Parant *et al.,* 1979; Ambler and Hudson, 1984; Tomasic *et al.,*1980; Fox and Fox, 1991). By extrapolation, it can be hypothesized that there are human diseases in which bacterial remnants play a role. However, direct detection of these bacterial components *in vivo* by conventional means has been difficult. It also has been entirely possible that in healthy mammals bacteria degraded in the respiratory tract or gut might pass into the bloodstream and localize in tissues. There is a real need to categorically prove whether bacterial remnants are indeed present and the precise levels of material not only during the infectious process but also in "normal" tissues and body fluids.

Unfortunately, the levels of muramic acid present (even in grossly infected tissues and body fluids) are so low that it has been extremely difficult to assay. This has resulted in much of the information available often being confusing and contradictory. In trace analysis of muramic acid in complex biological matrices contaminating compounds are commonplace, masking detection and causing false-positives. Therefore, observing a chromatographic peak at the correct retention time (using a nonselective detector) does not constitute definitive identification. One may merely be detecting a coeluting contaminant. An attempt was made to isolate muramic acid from normal tissues using thin-layer chromatography. The fluorescamine derivative of muramic acid was identified by periodate oxidation (indicating the presence of a sugar or other diol) and alkaline release of lactic acid (which is characteristic but not specific for muramic acid). It is entirely possible that a substance or mixture of substances other than muramic acid were detected (Sen and Karnovsky, 1984). A higher-resolution chromatographic technique, high-performance liquid chromatography, was used for analysis of human spleen. Muramic acid was analyzed as a dansyl derivative, which detects amino sugars, amino acids, and other compounds containing amino groups. The peak isolated at the retention time for muramic acid contained numerous components on rechromatography (Hoijer et al., 1995).

The detection of muramic acid in peripheral blood leukocytes of healthy human subjects has been reported using SIM GC-MS, showing that 21% of samples were positive. It is encouraging that cells from umbilical vein blood of healthy newborns, used as negative controls, were found not to contain muramic acid (Lehtonen et al., 1995). Subsequent work from the same group, however, noted the presence of muramic acid in fewer than 5% of people, regardless of disease state, in all people over 40 years of age (Lehntonen et al., 1997). It needs to be confirmed that 1 in 20 adult human samples do indeed contain muramic acid and that these peaks do not represent extraneous background peaks randomly showing up in samples. Such confirmation will include the use of GC-MS/MS in these analyses.

Muramic acid was first categorically detected in the tissues of polyarthritic rats previously injected with streptocococcal cell wall components (Fox et al., 1980). Tissues from nonarthritic rats served as negative controls in these experiments. When present at relatively high levels, it is possible to identify muramic acid in a chromatographic peak by the total ion mass spectrum (GC-MS analysis). For example, in a 1980 report using GC-MS (after systemic administration of streptococcal cell wall components), a peak at the retention time for muramic acid found in rat spleen (70 μg/g wet weight of tissue) had an identical "mass spectrum" to that of standard muramic acid. In joints of cell-wall-injected rats, a peak was observed (at 1 μg/g levels using SIM) at the retention time for muramic acid, but a peak was not present in normal joints used as negative controls (Fox et al., 1980). Muramic acid was subsequently detected in human septic fluids also using SIM GC-MS but not in control synovial fluids (Christensson et al., 1989; Lehtonen et al., 1994). However, at the low levels present in these biological samples it proved impossible to obtain a full mass spectrum for "absolute" identification.

Absolute identification at trace levels has awaited the development of more advanced GC-MS/MS instrumentation. Ion trap GC-MS/MS has been used for "absolute" identification at trace levels of muramic acid in human body fluids (Fox et al., 1986). This is the only report to date using GC-MS/MS to detect muramic acid or indeed any other bacterial constituent in a human–animal body fluid or tissue. Product ion spectra of muramic acid peaks ($\geq$30 ng/ml) in infected human body fluids were identical to those of pure muramic acid. Muramic acid was positively identified in synovial fluids during infection and was eliminated over time, during antibiotic therapy, but was absent from aseptic fluids (Fox et al., 1996a).

6. CONCLUDING REMARKS AND PERSPECTIVE

It has been demonstrated that analysis of the carbohydrate composition of bacteria provides important information in taxonomic discrimination and determination of physiological characteristics. This is readily obtained with GC-MS analysis. Extrapolation of such analyses to more complex environmental and clinical samples is best achieved with more sophisticated GC-MS/MS instrumentation. The primary difficulty has been in converting carbohydrate monomers into a suitable form for GC-MS (or GC-MS/MS). Due to the time-consuming and complex nature of derivatization, such analytical microbiology techniques are not routinely used outside specialist laboratories. Automation of derivatization reactions would help enormously in popularizing these techniques. An automated derivatization instrument (patent pending) that eliminates the entire manual derivatization process has been developed (Steinberg and Fox, 1999). Alternatively, sugars can be analyzed in their native form (without derivatization) by LC-MS or LC-MS/MS. At the current time, LC-based methods are more difficult to perform routinely. Furthermore, sensitivity for LC-MS and LC-MS/MS, although vastly improved, is still limited compared to GC-MS and GC-MS/MS. Mass spectrometry also has not been in the mainstream of instrumentation generally used by microbiologists. However, modern instruments are simple to use, being run by Windows-based personal computers. It is with great expectation that we look forward to the future and foresee an expansion of the role of mass spectrometry for the identification and trace detection of bacterial carbohydrates.

ACKNOWLEDGMENTS

I would like to thank the Army Research Office and Center for Indoor Air Research; the primary source of funds while many of these studies were conducted. Recent work has been supported by the Office of Naval Research. So many people have contributed to this work that it is impossible to thank them all. Hopefully the references at least mention their contributions. Molecular biology and microbial physiology have contributed immensely to much of the work presented. I would like to thank my wife, Karen Fox, in particular. Without her knowledge and insight, many of these studies would have been impossible.

REFERENCES

Ambler, L., and Hudson, A., 1984, Pharmacokinetics and metabolism of muramyl dipeptide and nor-muramyl dipeptide [^{3}H-labelled] in the mouse, *Int. J. Immunopharmacol.* **6:**133–139.

Ash, C., and Collins, M. D., 1992, Comparative analysis of 23S ribosomal RNA gene sequences of *Bacillus anthracis* and emetic *Bacillus cereus* determined by PCR-direct sequencing, *FEMS Microbiol. Lett.* **94:**75–80.

Ash, C., Farrow, J. A. E., Dorsch, M., Stackebrandt, E., and Collins, M. D, 1991, Comparative analysis of *Bacillus anthracis, Bacillus cereus* and related species on the basis of reverse transcriptase sequencing of 16S rRNA, *Int. J. Syst. Bacteriol.* **41:**343–346.

Bowser, D., Wheat, R., Foster, J., and Leong, D., 1974, Occurrence of quinovosamine in lipopolysaccharides of *Brucella* species, *Infect. Immun.* **9:**772–774.

Castellan, R., Olenchok, S. A., Hankinson, J., Millner, P., Cocke, J. B., Brag, C. K., Perking, H., and Jacobs, R. R., 1984, Acute bronchoconstriction induced by cotton dust: Dose-related responses to endotoxin and other dust factors, *Ann. Intern. Med.* **101:**157–163.

Christensson, B., Gilbart, J., Fox, A., and Morgan, S. L, 1989, Mass spectrometric quantitation of muramic acid, a bacterial cell wall component in septic synovial fluids, *Arth. Rheum.* **32:**1268–1272.

Cole, H. B., Ezzell, Jr., J. W., Keller, K. F., and Doyle, R. J, 1984, Differentiation of *Bacillus anthracis* and other *Bacillus* species by lectins, *J. Clin. Microbiol.* **19:**48–53.

Cookson, B., Cho, H., Herwaldt, L. and Goldman, W., 1989, Biological activities and chemical composition of purified tracheal cytotoxin of *Bordetella pertussis, Infect. Immun.* **57:**2223–2229.

Dillon, H., Heinsohn P. and Miller D, eds., 1996, *Field Guide for the Determination of Biological Contaminants in Environmental Samples,* American Industrial Hygiene Association, pp. 139–148.

Dobson, B. C., and Archibald, A. R., 1978, Effect of growth limitations on cell wall composition of *Staphylococcus aureus* H, *Arch. Microbiol.* **119:**295–301.

Dorsch, M., Moreno, E., and Stackebrandt, E., 1989, Nucleotide sequences of the 16S rRNA from *Brucella abortis, Nucleic Acids Res.* **17:**1765.

Duckworth, M., Archibald, A., and Baddiley, J., 1972, The location of N-acetylgalactosamine in the walls of *Bacillus subtilis* 168, *Biochem J.* **130**:691–696.

Ellwood, D. C., and Tempest, D. W., 1969, Control of teichoic acid and teichuronic acid biosynthesis in chemostat cultures of *Bacillus subtilis* var. *niger, Biochem. J.* **111**: 1–5.

Eudy, L., Walla, M., Fox, A., and Morgan, S. L, 1985, Gas chromatographic–mass spectrometric determination of muramic acid content and pyrolysis profiles for a group of gram-positive and gram-negative bacteria, *Analyst* **110**:381–385.

Ezzell, J. W., Abshire, T. G., Little, S. F., Lidgerding, B. C., and Brown, C., 1990, Identification of *Bacillus anthracis* by using monoclonal antibody to cell wall galactose-N-acetylglucosamine polysaccharide, *J. Clin. Microbiol.* **28**:223–231.

Fazio, S. D, Mayberry W. R., and White, D. C., 1979, Muramic acid in sediments, *Appl. Env. Microbiol.* **38**:348–350.

Fox, A., 1999, Carbohydrate profiling of bacteria by gas chromatography-mass spectrometry and their trace detection in complex matrices by gas chromatography-tandem mass spectrometry. *J. Chromatogr.* **843**:287–300.

Fox, A., and Black, G., 1994, Identification and detection of bacteria after derivatization and gas chromatography–mass spectrometry, in: *Mass Spectrometry for the Characterization of Microorganisms* (C. Fenselau, ed.), American Chemical Society, Washington, DC, pp. 107–131.

Fox, A. and Fox, K., 1991, Rapid elimination of synthetic adjuvant peptide from the circulation after systemic administration and absence of detectable natural muramyl dipeptides in the circulation at current analytical limits, *Infect. Immun.* **59**:1202–1205.

Fox, A., Schwab, J. H., and Cochran, T., 1980, Muramic acid detection in mammalian tissues by gas-liquid chromatography–mass spectrometry, *Infect. Immun.* **29**:526–531.

Fox, A., Lau, P., Morgan, S. L., Brown, A., Zhu, Z-T., Lema, M., and Walla, M., 1984, Capillary gas chromatographic analysis of carbohydrates of *Legionella pneumophila* and other members of the *Legionellaceae, J. Clin. Microbiol.* **19**:326–332.

Fox, A., Morgan, S. L., and Gilbart, J, 1989, Preparation of alditol acetates and their analysis by gas chromatography and mass spectrometry, in: *Analysis of Carbohydrates by GLC and MS* (J. Biermann and G. McGinnis, eds.), CRC Press, Boca Raton, FL, pp. 87–117.

Fox A., Morgan S. L., Larsson L., and Odham G., eds., 1990a, *Analytical Microbiology Methods: Chromatography and Mass Spectrometry,* Plenum Press, New York.

Fox, A. Rogers, J. C., Fox, K. F., Schnitzer, G., Morgan, S. L., Brown, A., and Aono, A., 1990b, Chemotaxonomic differentiation of legionellae by detection and characterization of aminodideoxyhexoses and other unique sugars using gas chromatography–mass spectrometry, *J. Clin. Microbiol.* **28**:546–552.

Fox, A., Black, G., Fox, K., and Rostovtseva, S., 1993a, Determination of carbohydrate profiles of *Bacillus anthracis* and *Bacillus cereus* including identification of O-methyl methylpentoses by using gas chromatography-mass spectrometry, *J. Clin Microbiol.* **31**:887–894.

Fox, A., Rosario, R., and Larsson, L., 1993b, Monitoring of bacterial sugars and hydroxy fatty acid markers in dust from air conditioners by gas chromatography–mass spectrometry, *Appl. Env. Microbiol.* **59**:4354–4360.

Fox, A., Wright, L., and Fox, K., 1995, Gas chromatography tandem mass spectrometry for detection of muramic acid, a peptidoglycan marker in organic dust, *J. Microbiol. Methods* 22:11–26.

Fox A., Fox K., Christensson B., Krahmer M., Harrelson D., and Krahmer, K., 1996a, Absolute identification of muramic acid at trace levels in human septic fluids *in vivo* and absence in aseptic fluids, *Infect. Immun.* **64:**3911–3955.

Fox, A., Krahmer, M., and Harrison, D., 1996b, Monitoring muramic acid in air (after alditol acetate derivatization) using a gas chromatograph-ion trap tandem mass spectrometer, *J. Microbiol. Methods* **27:**129–138.

Fox, K., Brown, A., Fox, A., and Schnitzer, G., 1991, *Tatlockia,* a genetically and chemically distinct group of bacteria. Proposal to transfer *Legionella maceachernii* (Brenner et al.) to the genus *Tatlockia,* as *Tatlockia maceachernii* comb nov., *Syst. Appl. Microbiol.* **14:**52–56.

Fox K., Fox A., Nagpal M., Steinberg, P. and Heroux, K, 1998a, Identification of *Brucella* by ribosomal spacer region PCR and differentiation of *B. canis* from other *Brucella* pathogenic for man by carbohydrate profiles, *J. Clin. Microbiol.* **36:**3217–3222.

Fox, K., Stewart, G., and Fox, A., 1998b, Synthesis of microcapsule by *Staphylococcus aureus* is not responsive to environmental phosphate concentrations, *Infect. Immun.* **66:**4004–4007.

Fox, K., Wunschel, D., Fox, A., and Stewart, G., 1998c, Complementarity of GC-MS and LC-MS analyses for determination of carbohydrate profiles of vegetative cells and spores of bacilli, *J. Microbiol. Methods* **33:**1–12.

Gilbart, J., and Fox, A., 1987, Elimination of group A streptococcal cell walls from mammalian tissues, *Infect. Immun.* **55:**1526–1528.

Gilbart, J., Fox, A., Whiton, R., and Morgan, S. L., 1986, Rhamnose and muramic acid: chemical markers for bacterial cell walls in mammalian tissues, *J. Microbiol. Methods* **5:**271–282.

Gonzalez, J. M., Dulmage, H. T., and Carlton, B. C., 1981, Correlation between specific plasmids and δ-endotoxin production in *Bacillus thuringiensis, Plasmid* **5:**351–365.

Harrell, L. J., Andersen, G. L., and Wilson, K. H, 1995, Genetic variability of *Bacillus anthracis* and related species, *J. Clin. Microbiol.* **33:**1847–1850.

Hoijer, M., Melief, M. J., van Helden-Meeuwsen, Eulderink, C. F., and Hazenberg, M.,1995, Detection of muramic acid in carbohydrate fraction of human spleen, *Infect. Immun.* **63:**1652–1657.

Kaneko, T., Nozaki, R., and Aizawa, K., 1978, Deoxyribonucleic acid relatedness between *Bacillus anthracis, Bacillus cereus* and *Bacillus thuringiensis, Microbiol. Immunol.* **22:**639–641.

Krahmer, M., Fox, K., Fox, A., Saraf, A., and Larsson, L., 1998, Total and viable bacterial load in two different agricultural environments using gas chromatography–tandem mass spectrometry and culture: A prototype study, *Am. Indust. Hyg. Assoc. J.* **59:**524–531.

Kreutzer, D., Buller, C., and Robertson, D., 1979, Chemical characterization and biological properties of lipopolysaccharides isolated from smooth and rough strains of *Brucella abortus, Infect. Immun.* **23:**811–818.

Lang, W. Glassey, K., and Archibald, A., 1982, Influence of phosphate supply on teichoic acid and teichuronic acid content of *Bacillus subtilis* cell walls, *J. Bacteriol.* **151**:367–375.

Lehtonen, L., Kortkangas, P., Oksman, P., Ereola, E., Aro, H., and Toivanen, T, 1994, Synovial fluid muramic acid in acute inflammatory arthritis, *Br. J. Rheumatol.* **33**:1127–1130.

Lehtonen, L., Eerola, E., Oksman, P., and Tovanen, P., 1995, Muramic acid in peripheral blood leukocytes of healthy human subjects, *J. Infect. Dis.* **171**:1060–1064.

Lehtonen, L., Eerola, E., and Toivanen, P., 1997, Muramic acid in human peripheral blood leukocytes in different age groups, *Eur. J. Clin Investig.* **27**:791–792.

Liau, D. F., and Hash, J., 1977, Structural analysis of the surface polysaccharide of *Staphylococcus aureus* M, *J. Bacteriol.* **131**:194–200.

Matz, L. L., Cabrera Beaman, T., and Gerhardt, P., 1970, Chemical composition of exosporium from spores of *Bacillus cereus, J. Bacteriol.* **101**:196–201.

Mauck, J., and Glaser, L., 1972, On the mode of *in vivo* assembly of the cell wall of *B. subtilis, J. Biol. Chem.* **25**:1180–1187.

Moreno, E., Pitt, M., Jones, L., Schuring, G., and Berman, D., 1979, Purification and characterization of smooth and rough lipopolysaccharides from *Brucella abortus, Infect. Immun.* **138**:361–369.

Moreno, E., Speth, S., Jones, L., and Berman, D., 1981, Immunochemical characterization of *Brucella* lipopolysaccharides and polysaccharides, *Infect. Immun.* **31**:214–222.

Moreno, E., Stackebrandt, E., Dorsch, M., Wolters, J., Busch, M., and Mayer, H., 1990, *Brucella abortus* 16S rRNA and lipid A reveal phylogenetic relationship with members of the alpha-2 subdivision of the class *Proteobacteria, J. Bacteriol.* **172**:3569–3576.

Nagpal, M. L., Fox, K. F., and Fox, A., 1998, Utility of 16S-23S rRNA spacer region methodology: How similar are interspace regions within a genome and between strains for closely related organisms? *J. Microbiol. Methods* **33**:211–219.

Nakamura, L., 1989, Taxonomic relationship of black-pigmented *Bacillus subtilis* strains and a proposal for *Bacillus atrophaeus* spec. nov., *Int. J. Syst. Bacteriol.* **39**:395–300.

Parant, M., Parant, F., Chedid, L., Yapo, A., Petit, J. F., and Lederer E., 1979, Fate of the synthetic immunoadjuvant, muramyl dipeptide (^{14}C-labelled) in the mouse, *Int. J. Immunopharmacol.* **1**:35–41.

Rijpens, N. P., Jannes, G., Van Asbroeck, M., Rossau, R., and Herman, L. M. F., 1996, Direct detection of *Brucella* spp in raw milk by reverse hybridization with 16S-23S rRNA spacer probes, *Appl. Environ. Microbiol.* **62**:1683–1688.

Robbins, J. B., Schneerson, R., Vann, W. F., Bryla, D., and Fattom A., 1995, Prevention of systemic infections caused by group B streptococcus and *Staphylococcus aureus* by multivalent polysaccharide-protein vaccines, *Ann. NY Acad. Sci.* **754**:68–82.

Romero, C., Gamazo, C., Pardo, M., and Lopez-Goni, I., 1995a, Specific detection of *Brucella* DNA by PCR, *J. Clin. Microbiol.* **33**:615–617.

Romero, C., Pardo, M., Grillo, M., Diaz, R., Blasko, J., and Lopez-Goni, I., 1995b, Evaluation of PCR and indirect enzyme immunoassay on milk samples for diagnosis of brucellosis in dairy cattle, *J. Clin. Microbiol.* **33**:3198–3200.

Sandstrom, T., Bjermer, L., and Rylander, R., 1994, Lipopolysaccharide (LPS) inhalation in healthy subjects causes bronchoalveolar neutrophilia, lymphocytosis and fibronectin, *Am. J. Intern. Med.* **25**:103–104.

Saraf, A., and Larsson, L., 1996, Use of a gas chromatography/ion trap tandem mass spectrometer for the determination of microorganisms in organic dust, *J. Mass Spectrom.* **31:**389–396.

Seki, T., Oshima, T., and Oshima, Y., 1975, Taxonomic study of *Bacillus* by deoxyribonucleic acid-deoxyribonucleic acid hybridization and interspecific transformation, *Int. J. Syst. Bacteriol.* **25:**258–270.

Seki, T., Chung, C.-K., Mikami, H., and Oshima, Y., 1978, Deoxyribonucleic acid homology and taxonomy of the genus *Bacillus, Int. J. Syst. Bacteriol.* **28:**182–189.

Sen, Z., and Karnovsky, M. L., 1984, Qualitative detection of muramic acid in normal mammalian tissues, *Infect. Immun.* **43:**937–941.

Shahgholi, M., Ohorodnik, S., Callahan, J., and Fox, A., 1997, Trace detection of underivatized muramic acid in environmental dust samples by microcolumn liquid chromatography–electrospray tandem mass spectrometry, *Anal. Chem.* **69:**1956–1960.

Sonesson, A., and Jantzen, E., 1992, The branched octose yersiniose A is a lipopolysaccharide constituent of *Legionella micdadei* and *Legionella maceachernii chain octose, J. Microbiol. Methods* **15:**241–248.

Sonesson, A., Jantzen, E., Bryn, K., Larsson, L., and Eng, J., 1989, Chemical composition of a lipopolysaccharide from *Legionella pneumophila, Arch. Microbiol.* **153:**72–78.

Steinberg, P., and Fox, A., 1999, Automated derivatization instrument: Preparation of alditol acetates using gas chromatography-mass spectrometry, *Analyt. Chem.* **71:**1914–1917.

Tomasic, J., Ladesic, B., Valinger, Z., and Hrsak, I., 1980, The metabolic fate of ^{14}C-labelled peptidoglycan monomer in mice. I. Identification of the monomer and the corresponding pentapeptide in urine, *Biochim. Biophys. Acta* **629:**77–82.

Verger, J.-M., Grimont F., Grimont, P., and Grayon, M., 1985, *Brucella,* a monospecific genus as shown by dexoyribonucleic acid hybridization, *Int. J. Syst. Bacteriol.* **35:**292–295.

Walla, M., Lau, P. Y., Morgan, S. L., Fox, A., and Brown, A., 1984, Capillary gas chromatography–mass spectrometry of carbohydrate components of *Legionella* and other bacteria, *J. Chromatogr.* **288:**399–413.

Wang, Z., Larsson, K., Palmberg, L., Malmberg, P., Larsson, P., and Larrson, L., 1997, Inhalation of swine dust induces cytokine release in the upper and lower airways, *Eur Respir. J.* **10:**381–387.

Wright, J., and Heckels, J., 1975, The teichuronic acid of cell walls of *Bacillus subtilis* W23 grown in chemostat under phosphate limitation, *Biochem. J.* **147:**187–189.

Wu, T., and Park, J. T., 1971, Chemical characterization of a new surface polysaccharide antigen from a mutant of *Staphylococcus aureus, J. Bacteriol.* **108:**874–884.

Wunschel, D., Fox, K., Black, G., and Fox, A., 1994, Discrimination among the *B. cereus* group, in comparison to *B. subtilis,* by structural carbohydrate profiles and ribosomal RNA spacer region PCR, *Syst. Appl. Microbiol.* **17:**625–635.

Wunschel D., Fox K., Fox A., Nagpal, M., Kim, K., and Stewart, G., 1997, Quantitative analysis of neutral and acidic sugars in whole bacterial cell hydrolysates using high-performance anion exchange liquid chromatography-electrospray ionization tandem mass spectrometry, *J. Chromatogr.* **776:**205–219.

Zhiping, W., Malmberg, P., Larsson, B., Larsson, K., Larrson, L., and Saraf, A., 1996, Exposure to bacteria in swine-house dust and acute inflammatory reactions in humans, *Am. J. Respir. Crit. Care Med.* **154:**1261–1266.

13

Degradation of Cellulose and Starch by Anaerobic Bacteria

Kevin L. Anderson

1. INTRODUCTION

Cellulose and starch comprise the majority of the world's carbohydrate. Cellulose is the principal polysaccharide of plant cell walls and is the most abundant organic compound in the biosphere (Raven *et al.,* 1999), with annual production estimated at more than 10^9 tons (McGinnis and Shafizadeh, 1991). Starch is the primary storage polysaccharide of plants, and annual production is estimated in excess of 7×10^8 tons (Jenner, 1982).

These polysaccharides also comprise the vast majority of global biomass. Microbial degradation of this biomass is a key component of the global carbon cycle. This bioconversion could also provide industrial energy sources, such as ethanol or methane gas. Interestingly, even though cellulose is biodegradable, cellulosic material comprises some 50% of landfill space (Bayer and Lamed, 1992), indicating the need for even greater levels of microbial degradation.

Despite a plethora of research, attempts to enhance biomass degradation or establish biotechnological applications of degradation have had only limited success. Reasons for this failure vary, but are at least partly due to an insufficient understanding of the biochemistry and physiology of the bacterial degradation process.

Kevin L. Anderson • Department of Biological Sciences, Mississippi State University, Mississippi State, Mississippi 39762.

Glycomicrobiology, edited by Doyle.
Kluwer Academic/Plenum Publishers, New York, 2000.

2. STRUCTURAL ORGANIZATION OF CELLULOSE AND STARCH

2.1. Cellulose

Cellulose is a linear polymer of anhydro-D-glucopyranose units linked by $\beta(1 \rightarrow 4)$-glycosidic bonds. The β-glycosidic linkages cause each glucopyranose unit to rotate 180° from the adjacent glucose unit around the main polymer axis. This axis of rotation forms an alteration in the subunits, making the basic repeating unit cellobiose rather than glucose. However, the polymer chain length is measured in glucose units, and the average length of native cellulose is generally several thousand units (Ward and Seib, 1970), giving a possible molecular weight far in excess of 1 million.

The axial positioning of the hydrogen bonds together with the 180° rotation of the glucose units also results in a polymer with minimal steric hindrance. This, in turn, results in a uniform linear structure that allows different cellulose molecules to pack closely together, giving cellulose a very stable crystalline structure that is difficult to dissolve or hydrolyze. In fact, this close packing enables cellulose fibrils (tightly wound strands of cellulose) to have a strength exceeding an equivalent thickness of steel (Raven *et al.*, 1999).

The chemical structure of cellulose has been known for some time; however, the physical organization and orientation (conformation, crystallinity/packing order, etc.) of cellulose fibrils is still an open question (Reiling and Brickmann, 1995). Studies of these fibrils show that water absorption occurs only in limited regions, and partial acid hydrolysis gives short crystalline rods. Cellulose fibrils contain both crystalline (ordered) and amorphous (disordered) regions, with the crystalline form comprising approximately 70% of native cellulose of cotton and wood (Wood, 1988).

These amorphous regions are surface faces, bends, molecular chain ends, dislocations, and imperfections in the fibrils (Sjöström, 1993). Such regions expose the fibril to enzymatic hydrolysis; generally the greater the degree of swelling (i.e., amorphous content) of the cellulose, the faster the initial rate of hydrolysis (Rowland, 1975). Therefore, amorphous forms of cellulose, such as acid-swollen cellulose or carboxy-methylcellulose (CMC), are often used as substrate for detection of cellulase activity. However, these altered forms cannot always be considered an appropriate substitute for crystalline cellulose. For example, the degree of substitution found in CMC affects the rate of enzymatic action, preventing a suitable correlation between the kinetics of CMC and crystalline cellulose hydrolysis. Hence, CMC is a poor cellulose substitute for the enzymatic analysis of certain cellulases.

2.2. Starch

By contrast, starch molecules are composed of both linear and branched structures. The linear form, amylose, consists of anhydro-D-glucopyranose units with α(1 → 4) glycosidic linkages, which gives starch a helical shape. The mixture of α(1 → 4) and α(1 → 6) glycosidic linkages of amylopectin give it a branched structure, which further prevents a close packing arrangement of starch. Consequently, even though native starch is still relatively insoluble, it is an amorphous molecule that is readily hydrolyzed by many different enzymes or weak acids.

In addition, pullulan is an extracellular product of the fungus, *Aureobasidium pullulans*. This polysaccharide is a linear polymer of repeating maltotriose units linked by α(1 → 6) glucosidic linkages. Hence, the conformational orientation of these glucosidic linkages is different from either amylose or amylopectin. As a result, pullulan has become a popular enzymatic substrate for the study of certain types of starch-degrading enzymes.

3. AMYLOLYTIC ANAEROBIC BACTERIA

3.1. Amylolytic Ecosytems

Starch-degrading enzymes are produced by all major forms of life: animals, plants, and microorganisms. In the microbial world, amylolytic ability is common; there is no specific starch-degrading microbiota. Amylolytic bacteria are virtually ubiquitous in the environment.

Soil contains many bacteria and fungi that actively produce starch-degrading enzymes. Waterlogged soil generates an anaerobic environment, where plant starch is typically degraded by amylolytic clostridia (Alexander, 1977). Since many of these bacteria produce ammonia, which is then oxidized to nitrate by autotrophic bacteria, anaerobic decomposition of plant material contributes to the nitrogen content of the soil.

While humans produce amylolytic enzymes, a considerable amount of starch escapes hydrolysis by pancreatic enzymes in the small intestine (Englyst and Cummings, 1987). These "resistant" starches reach the human colon, where they are potential substrata for the colonic bacteria. In fact, since large portions of starch are not digested or absorbed in the small intestine (Macfarlane and Cummings, 1991), degradation of starch by the anaerobic colonic bacteria may provide an important nutritional role for humans.

As a herbivore, ruminant animals consume large amounts of starch. Since digesta initially enters the rumen first, starch is subjected to microbial degradation prior to entering the small intestine. Hence, anaerobic microorganisms have a significant role in the ruminant's digestion of starch.

Certain starch-degrading enzymes (especially the more thermostable) are extracted from different microbes and utilized in commercial processes. For example, since alcohol-producing yeast is not amylolytic, starch-degrading enzymes are needed to convert the starch to sugars that can be utilized by the yeast. These enzymes are also used for study of the structure of starch materials.

3.2. Starch-Degrading Enzymes

A variety of bacterial enzymes hydrolyze either starch or degradation products of starch. Amylolytic bacteria usually contain several different starch hydrolyzing enzymes, also known as α-glucanases. The characteristics of these enzymes vary widely but are typically divided into three general categories of hydrolytic activity.

Endoamylases randomly hydrolyze the $\alpha(1 \rightarrow 4)$ linkages of amylose, resulting in a variety of maltodextrin endproducts. α-Amylase is the most common bacterial endoamylase and forms end products that have an α-configuration at the number 1 carbon (Vihinen and Mäntsälä, 1989). A few α-amylases also have activity against the $\alpha(1 \rightarrow 6)$ linkage of amylopectin, but the rate of hydrolysis is considerably lower than for $\alpha(1 \rightarrow 4)$ bonds.

α-Amylases have been isolated from many different bacteria and possess a wide variety of chemical properties. Their molecular weights range from 10,000 to 140,000 Da, with an average of 50,000–60,000 Da (Vihinen and Mäntsälä, 1989). The temperature optima for amylase activity usually corresponds to the growth temperature of the source bacterium, and may be greater than 100°C for α-amylases from some thermophiles (Piggott et al., 1984). A variety of pH optima also have been reported for α-amylases, ranging from 2.0 to 10.5 (Vihinen and Mäntsälä, 1989).

As a metalloenzyme, α-amylases contain at least one Ca^{2+} ion (Vallee et al., 1959). Calcium appears to provide both activity and stability, possibly by tightening the binding between domains of the α-amylase (Vihinen and Mäntsälä, 1989). Rare earth metals can substitute for Ca^{2+} in some amylases (Smolka et al., 1971).

Exoamylases hydrolyze the $\alpha(1 \rightarrow 4)$ linkages at the nonreducing end of the starch molecule. The end product formed depends on the type of enzyme, but is generally one predominant dextrin. For example, β-amylases liberate maltose (in the β form), whereas glucoamylases produce only glucose as an end product. Other exoacting enzymes may liberate solely maltotriose or maltotetraose, which are subsequently hydrolyzed by α-glucosidase. In addition, glucoamylases have the ability to cleave the $\alpha(1 \rightarrow 6)$ linkage of amylopectin, whereas β-amylases cannot cleave this linkage.

Debranching enzymes are those capable of cleaving $\alpha(1 \to 6)$ glucosidic linkages. Isoamylases hydrolyze the various branch structures of amylopectin, glycogen, and branched oligosaccharides and dextrins. However, they are unable to cleave the $\alpha(1 \to 6)$ linkage of pullulan. Rather, this linkage is specifically hydrolyzed by pullulanases, yielding maltotriose. Isopullulanases cleave the maltotriose portion of pullulan, liberating isopanose. Some unique α-amylases (also referred to as neopullulanases) cleave the maltotriose portion of pullulan, liberating panose (Kuriki et al., 1988). A fourth group, occasionally found in bacteria, are those enzymes that degrade amylose and form cyclodextrins as end products.

Interestingly, classification based on amino acid sequence rather than activity places all bacterial starch-degrading enzymes into only two families (Henrissat and Bairoch, 1993). α-Amylases and the debranching enzymes are both placed into family 13. β-Amylases comprise a separate family (family 14).

3.3. Starch-Binding and Transport Systems

The large number of extracellular bacterial amylases that have been reported (Vihinen and Mäntsälä, 1989) seem to indicate that secretion of extracellular polysaccharidases is the primary strategy for amylolytic bacteria. These extracellular enzymes degrade the polysaccharide into smaller saccharides (usually mono- or disaccharides), which are then transported into the cell. However, this strategy poses a difficulty for bacteria in highly competitive ecosystems. For example, the microbiota of the bovine rumen consist of numerous microorganisms that cannot utilize starch but readily utilize the smaller saccharide products of polysaccharidase action. Thus, amylolytic ruminal bacteria would be forced to compete for the by-products liberated by their own extracellular enzymes.

Indeed, the ruminal bacterium, Ruminobacter (formerly Bacteroides) amylophilus, was initially reported to produce extracellular amylase activity (Cotta, 1988; McWethy and Hartman, 1977). However, further analysis revealed that all amylase activity in the extracellular fluid could be accounted for by cell lysis (Anderson, 1995). Instead, R. amylophilus appears to employ a strategy similar to that detected in the colonic Bacteroides.

3.3.1. THE BACTEROIDES THETAIOTAOMICRON MODEL

Studies of the colonic bacterium, Bacteroides thetaiotaomicron, found that its starch-degrading enzymes are virtually all cell associated, typically in the periplasm (D'elia and Salyers, 1996; Smith and Salyers, 1989, 1991; Anderson and Salyers, 1989a,b). If these enzymes are periplasmic, contact with the starch molecules apparently requires transporting the polysaccharide across the cell enve-

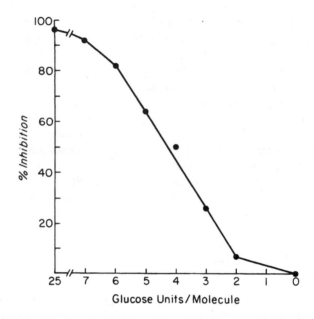

Figure 1. Relative affinity of the starch-binding sites of *Bacteroides thetaiotaomicron* for individual maltodextrins, as indicated by the ability of these maltodextrins to competitively inhibit whole cell binding of radiolabeled starch. Reproduced from Anderson and Salyers (1989a), with permission.

lope. Such transport would likely be facilitated by outer membrane starch-binding proteins, which were first detected by Anderson and Salyers (1989a). They demonstrated that radiolabeled starch is bound by intact cells of *B. thetaiotaomicron*. This binding had the characteristics of a protein-mediated event, and its regulation corresponded to the regulation of the starch-degrading enzymes (Anderson and Salyers, 1989a). Maltose served as an inducer of both amylase activity and starch binding, although higher levels of enzyme activity were detected in starch-grown cells. What is more, the starch-binding sites had a high affinity for maltoheptaose and decreasing affinity for smaller maltodextrins (Fig. 1). Subsequent genetic manipulation revealed that *B. thetaiotaomicron* possesses two binding affinities: one with a high affinity for larger starch oligomers and a second with a high affinity for maltodextrins (Anderson and Salyers, 1989b).

Whole cells of *R. amylophilus* were also found to bind radiolabeled starch in a manner consistent with a protein mediated system (Anderson, 1995). However, unlike *B. thetaiotaomicron,* these starch-binding sites had an equal affinity for amylose and maltodextrins as small as maltotetraose (Fig. 2). The single binding affinity detected on *R. amylophilus* indicates this organism employs only one binding component for utilization of both maltodextrins and starch oligomers.

Mutational analysis of *B. thetaiotaomicron* also provided further evidence

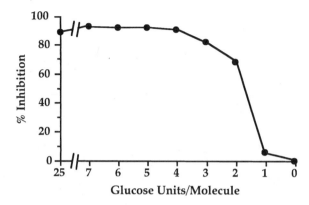

Figure 2. Relative affinity of the starch-binding sites of *Ruminobacter amylophilus* for individual maltodextrins, as indicated by the ability of these maltodextrins to competitively inhibit whole cell binding of radiolabeled starch. Reproduced from Anderson (1995), with permission.

that starch binding is essential for starch utilization. Transposon-generated mutants were screened for their inability to utilize starch as a substrate (Anderson and Salyers, 1989b). All classes of starch-minus mutants obtained were deficient in starch binding. In fact, for two classes of these mutants, lack of starch binding is the only detectable physiological difference from the wild-type strain.

Subsequent analysis by Tancula *et al.* (1992) found that at least one membrane protein is missing from each mutant strain. Using polyclonal antisera, they found that *B. thetaiotaomicron* produces four maltose-inducible proteins: three in the outer membrane and one in the cytoplasmic membrane. One or more of these proteins is missing in three classes of the starch-minus mutants, and all the classes of mutants contain reduced levels of two constitutively produced outer membrane proteins (Tancula *et al.*, 1992). Thus, starch binding and uptake appear to be mediated by a complex of several different polypeptides.

One of these maltose-inducible outer membrane polypeptides (115-kDa) was subsequently determined to be essential for *B. thetaiotaomicron* utilization of maltoheptaose and starch (Reeves *et al.*, 1996). This polypeptide (SusC) is the product of *sus*C, a gene in the starch utilization system operon. Evidence also indicates that Sus C has properties consistent with a binding protein rather than an enzyme (Reeves *et al.*, 1996).

Four additional genes (*sus*D, *sus*E, *sus*F, and *sus*G), located downstream of *sus*C, were also found to be maltose inducible (Reeves *et al.*, 1997). All four genes encode outer membrane polypeptides. Sus D, Sus E, and Sus F are apparently involved in starch binding, with Sus D making the most significant contribution. The inability of *sus*G⁻ mutants to utilize starch indicates that Sus G is essential for utilization of starch by *B. thetaiotaomicron*, but its function is not clear (Reeves *et* *al.,* 1997). This further confirms that this bacterium binds starch by employing an outer membrane protein complex.

However, the role of periplasmic proteins cannot be discounted. While intact cells bind starch, disrupted cells and membrane fractions did not retain starch-binding capability (Anderson and Salyers, 1989a). This may be due to the loss of periplasmic proteins that are loosely associated with the membrane.

In addition to amylase and amylopectinase activity, *B. thetaiotaomicron* also produces pullulanase activity. Smith and Salyers (1989) cloned a pullulanase gene (*pul*I) from *B. thetaiotaomicron,* whose product cleaves the α(1 → 6) linkages of pullulan to liberate maltotriose. However, disruption of *pul*I reduced the total cellular level of pullulanase activity by only 30%, and did not appreciably affect this organism's ability to use pullulan as a growth substrate.

A second pullulan-degrading enzyme (pullulanase II) was subsequently detected in *B. thetaiotaomicron* (Smith and Salyers, 1991). This enzyme cleaves the α(1 → 4) linkages of pullulan to liberate panose and also degrades amylose and has cyclodextrinase activity. This pullulanase–amylase characteristic is similar to the neopullulanases detected in *Thermoactinomyces vulgaris* (Sakano *et al.,* 1982) and *Bacillus stearothermophilus* (Kuriki *et al.,* 1988). However, unlike the neopullulanases of these bacteria, pullulanase II has a distinct preference for α(1 → 4) glucosidic linkages (Smith and Salyers, 1991). Pullulanase II also has a deduced amino acid sequence similar to α-amylases and is most closely related to a cyclodextrinase of *Clostridium thermohydrosulfuricum* (D'elia and Salyers, 1996).

While both pullulanase I and II were obtained from the soluble fraction of *B. thetaiotaomicron,* pullulanase activity also could be detected in membrane fractions (Smith and Salyers, 1991). Disruption of the gene encoding pullulanase II (*sus*A) revealed the presence of membrane-associated pullulanase activity (D'elia and Salyers, 1996), indicating *B. thetaiotaomicron* possesses at least three enzymes with activity against pullulan. This is further confirmed by the ability of a mutant strain (*pul*I⁻, *sus*A⁻) to still utilize pullulan (D'elia and Salyer, 1996).

Interestingly, of the thousands of transposon-generated mutants originally screened, all were deficient in starch binding but only one class was deficient in polysaccharidase activity (Anderson and Salyers, 1989b). Since subsequent research has revealed numerous starch-degrading enzymes in *B. thetaiotaomicron,* the random insertion of a transposon into a gene encoding for one of these enzymes would be likely. Failure to find a polysaccharidase-minus mutant that still binds starch suggests that loss of a single enzyme does not eliminate growth on starch. The apparent redundancy of starch-degrading enzymes may prevent any single enzyme from being essential, such as the two pullulanases (*pul*I and *sus*A). In fact, Anderson and Salyers (1989b) obtained a class of mutants that produce excessive levels of starch-degrading activity, but their growth rate on starch is greatly reduced (D'elia and Salyers, 1996).

Consistent with the mechanisms employed by *B. thetaiotaomicron,* at least three separate amylase activities recently have been detected in *R. amylophilus.* Two of these activities were present in the soluble fraction of the cell and have characteristics consistent with α-amylase (W. E. Alley and K. L. Anderson, unpublished data). The third activity resides in the membrane fraction, and subsequent purification revealed it possesses neopullulanase characteristics (LiAng and Anderson, in manuscript).

3.4. Regulation of Amylolytic Systems

A few studies have dealt with regulation of amylolytic systems of anaerobic bacteria. In general, those species tested have higher amylase activity when cultivated in maltose- or starch-containing medium than glucose medium (Cotta and Whitehead, 1993; Anderson and Salyers, 1989a; Cotta, 1988). The starch-specific membrane proteins of *B. thetaiotaomicron,* which appear to be involved in starch binding and uptake, are similarly regulated (Reeves *et al.,* 1996, 1997; Anderson and Salyers, 1989a). An exception is *R. amylophilus,* which possesses a substrate range limited to maltose, maltodextrins, and starch (Anderson, 1995).

In addition to being starch inducible, amylolytic systems of some anaerobes may be repressed. This type of repression is common in anaerobic and facultative bacteria and enables them to utilize energy substrates in a preferential manner, that is, the ability of *Escherichia coli* to utilize only glucose despite other substrates also present in the medium (Postma *et al.,* 1993). For enteric bacteria this type of repression involves the phosphoenolpyruvate–phosphotransferase system (PEP-PTS) (Postma *et al.,* 1993). Similar repression systems also have been identified in gram-positive bacteria, although some differences from gram-negative organisms have been found (Saier, 1996).

Some evidence indicates that glucose causes repression of amylase synthesis by *Streptococcus bovis* (Cotta and Whitehead, 1993) and *Clostridium acetobutylicum* (Annous and Blaschek, 1990). Cyclic AMP-mediated repression was not detected in *S. bovis* (Cotta and Whitehead, 1993), which is consistent with observations that many gram-positive bacteria employ cAMP-independent repression systems (Saier, 1996). This is further confirmed by the low levels of cAMP generally detected in anaerobic bacteria (Cotta *et al.,* 1994).

However, not all amylolytic bacteria repress their polysaccharolytic systems. Fields *et al.* (1997) detected polysaccharide-inducible membrane proteins produced by *Bacteroides xylanolyticus,* but no glucose-induced repression of these proteins was detected.

Little is also known about the genetic organization of amylolytic systems in anaerobic bacteria. Using the *E. coli* maltose regulon as a model (Schwartz, 1987), it is tempting to assume that starch-degrading genes have a similar organization. Although the chromosomal location of several amylolytic genes of *B. thetaio-*

taomicron has been determined, the organizational nature of these genes has not yet been identified (Reeves *et al.,* 1997; Tancula *et al.,* 1992).

4. CELLULOLYTIC ANAEROBIC BACTERIA

4.1. Cellulolytic Ecosystems

Unlike starch degradation, microorganisms are responsible for virtually all degradation of cellulose (Fenchel and Jorgensen, 1977). Aerobic degradation is typically thought to account for most bioconverison of cellulose, with only a small proportion involving anaerobic microbes. Since lignin degradation was thought to require oxygen, intricate association of plant cellulose with lignin is often considered a limiting factor of anaerobic cellulolysis (Schink, 1988). However, several studies have indicated that anaerobic modification of lignin does occur (Schink, 1988; Brenner *et al.,* 1984). Anaerobic cleavage of several intermonomeric bonds found in lignin is also now known to be possible (Colberg and Young, 1982, 1985; Healy and Young, 1979).

Anaerobic cellulose degradation, in fact, occurs in a variety of ecological niches. The microbial consortium responsible for converting the carbon of cellulose to methane (the methane cycle) consists primarily of anaerobic bacteria (Ljungdahl and Eriksson, 1985). Common anaerobic ecosystems include sediments, soil, composts, sewage, and animal digestive tracts.

4.2. Cellulose-Degrading Enzymes

The efficient enzymatic hydrolysis of cellulose is impaired by its high insolubility (Coughlan, 1992). Lignin and other plant cell wall molecules associated with cellulose also may sterically inhibit the hydrolytic enzymes from binding to cellulose. Enzymatic hydrolysis also can be limited by both the degree of crystallization and polymerization of the cellulose (McGinnis and Shafizadeh, 1991). Consequently, this compact cell wall matrix must be perturbed or opened in some manner before the relatively large enzymes can bind and function catalytically.

All cellulolytic bacteria so far studied possess a battery of cellulase enzymes with different specificities of cellulose chain length, mode of action, and hydrolytic capability toward crystalline regions of cellulose. This diversity appears to be necessary to enable a bacterium to hydrolyze such a physically heterogenic molecule as cellulose. What is more, the structure of cellulose changes during enzymatic hydrolysis, altering the nature of the cellulose, and thereby altering the types of enzymes needed for further hydrolysis.

Proteins capable of hydrolyzing portions of cellulose molecules typically exhibit endoglucanase, exoglucanase (also known as cellobiohydrolase), or β-glucosidase activity (Singh and Hayashi, 1995; Tomme *et al.*, 1995). Endoglucanases preferentially hydrolyze the β(1 → 4) glycosidic bonds of amorphous regions of cellulose fibrils. Exoglucanases apparently attack more systematically, starting at the extremity of the cellulose chain and moving in a processive manner, primarily liberating cellobiose. The least-studied enzyme of cellulose degradation is β-glucosidase or cellobiase, which hydrolyzes cellobiose and to a lesser extent cellodextrins, liberating glucose.

Numerous endoglucanases have been characterized (Henrissat and Bairoch, 1993), partially because they are easily identified by their high specific activity against CMC. Endoglucanases also can hydrolyze cellodextrins, but their activity decreases as the chain length decreases (Singh and Hayashi, 1995). Individual bacterial cells typically produce more than one class of endoglucanase (Warren, 1996; Tomme *et al.*, 1995). However, there is disagreement of molecular weight and chemical composition of the different enzymes detected in any one cellulase system. Thus, there is difficulty assigning a classification and function to various isoenzymic forms of endoglucanases (Béguin, 1990). Adding to the ambiguity, one family of enzymes consist of both exo- and endoglucanase activity (Tomme *et al.*, 1995).

Only a few bacterial exoglucanases have been identified (Liu and Doi, 1998; Stålbrand *et al.*, 1998; Tomme *et al.*, 1995), perhaps due to their low activity on most *in vitro* cellulosic substrates. However, Davies and Henrissat (1995) suggest exoglucanase activity may be a crucial component of the cell's cellulolytic system. The efficiency of cellulose hydrolysis is probably further enhanced by the action of two types of exoglucanases. One hydrolyzes the reducing end of the cellulose molecule and the other hydrolyzes the nonreducing end (Barr *et al.*, 1996; Tomme *et al.*, 1995).

4.3. Cellulase Systems

The chemical composition and extreme insolubility of cellulose make it a difficult substrate for bacteria to degrade and utilize. Bacterial utilization of soluble substrates, such as lactose, typically involve only a few proteins (Wolfe, 1993). By contrast, bacterial cellulolytic systems have been found to consist of numerous proteins, including at least 15 different cellulases (Béguin and Aubert, 1994). Apparently the physical nature of cellulose requires this large number of different proteins, functioning interactively, for effective hydrolysis.

Some cellulolytic bacteria produce cellulase systems that can effectively degrade native cellulose almost completely. These "complete" systems, in bacteria such as *Clostridium thermocellum* and *Ruminococcos albus,* are evidently capable of exploiting the plant biomass as a sole energy source. By comparison, bacteria, such as *Prevotella ruminicola* and *Pseudomonas solanacearum,* possess an "incomplete" system with limited cellulose hydrolytic capability.

The biological significance of these incomplete systems is an open question. The ruminal bacterium, *P. ruminicola,* produces at least one endoglucanase that can hydrolyze CMC and acid-swollen cellulose (Matsushita *et al.,* 1990) but is unable to utilize insoluble forms of cellulose as an energy source. In addition, another ruminal bacterium, *R. albus* SY3, can utilize Avicel but fails to grow on cotton (Wood *et al.,* 1982). However, from an ecological perspective, neither of these bacteria require a complete cellulase system. As part of the ruminal microbiota, they can act in association with other cellulolytic organisms during hydrolysis of the plant biomass. It also is possible that during extensive laboratory cultivation some strains have lost portions of a complete system, retaining an incomplete system.

4.3.1. NONCOMPLEXED SYSTEMS

In general, noncomplexed or "nonaggregated" cellulase systems are found in aerobic bacteria (and aerobic fungi). The common characteristic of these systems is the extracellular secretion of cellulases and related polysaccharidases. As with some amylolytic systems described above, this strategy involves the extracellular hydrolysis of cellulose fibers. The hydrolytic byproducts (e.g., glucose, cellobiose, and cellodextrins) are then transported into the cell and catabolized.

The exact characteristic of the nonaggregated systems of many bacteria, such as *Cellulomonas fimi* and *Thermomonospora fusca,* is uncertain. Both organisms appear to produce cellobiohydrolases similar to those of the fungi, *Trichoderma reesei* (Stålbrand *et al.,* 1998; Irwin *et al.,* 1993). However, unlike fungal systems, the cell fraction of many of these bacteria may contain more cellulase activity than the cell-free fractions (Tomme *et al.,* 1995). What is more, Calza *et al.* (1985) reported that no exoglucanase activity could be detected in the cell-free fraction of *T. fusca.*

The general model for activity of these secreted endo- and exocellulases (primarily based on studies of fungi) suggests a sequential interaction (Tomme *et al.,* 1995). Accordingly, the rate of exoglucanase hydrolysis is limited by the availability of exposed ends of the cellulose fibril. The endoglucanases, acting primarily at "amorphous" regions of the cellulose, provide progressively more sites for exoglucanase activity. However, not all endo- and exoglucanases may act in such a synergistic manner. For example, endoglucanases from *T. fusca* lack clear synergistic interaction with exoglucanases of the fungi, *Trichoderma reesei.* Rather, action of these endoglucanases may depend on how processive the exoglucanase attacks the cellulose (Irwin *et al.,* 1993).

4.3.2. COMPLEXED SYSTEMS

Anaerobic bacteria typically have been found to possess exocellular multienzyme complexes. Electron micrographs originally suggested that cellulolytic bacteria maintain a close proximity with the cellulosic material (Lithium *et al.*, 1978; Akin, 1976). This led to the conclusion that cellulose binding sites or domains exist on the cell surface, enabling the cell to attach to the cellulose molecule. The nature of this domain was first indicated by Ait *et al.* (1979) when they had difficulty separating the enzymes of *C. thermocellum* because the individual subunits rapidly form a complex. The subsequent discovery that the binding domain also contained numerous cellulase enzymes prompted the conclusion that it was a multiprotein complex (Béguin and Lemaire, 1996). The existence of this complex, first termed a cellulosome by Bayer *et al.* (1983), was further substantiated when antibodies agglutinated wild-type cells, while a spontaneously generated mutant, defective in cellulose binding, did not agglutinate (Bayer *et al.*, 1983). In addition, the cellulosome fractions have been isolated by a variety of methods including sepharose, cellulose affinity binding, and ultrafiltration techniques (Bolobova *et al.*, 1994; Morag *et al.*, 1991; Cavedon *et al.*, 1990; Bayer *et al.*, 1983). Antibodies raised against these subunits react with protein complexes on both the exterior of the cells and those attached to cellulose in the supernatant (Bayer *et al.*, 1983; Lamed *et al.*, 1983a,b).

In the last two decades numerous studies have focused on the nature and character of this complex, although most of the work has involved cellulosomes from cellulolytic clostridia. A high degree of homology exists between those systems of *C. thermocellum, C. cellulovorans, C. cellulolyticum,* and *C. papyrosolvens* (Gal *et al.*, 1997). While much of our understanding of cellulosomes is derived from studies of these clostridia, complex or cell-associated cellulase systems also have been detected in other anaerobic bacteria (Tomme *et al.*, 1995).

4.3.2a. Cellulosomes

The exocellular cellulosome complex of *C. thermocellum* is a large protein component that contains approximately 26 polypeptides (Kohring *et al.*, 1990) and sometimes occurs as multicomponent complexes. These complexes have a high affinity for cellulose and an apparent size of 1–2 MDa. Another distinctive characteristic of a cellulosome is its resistance to high concentrations of salt or urea. Dissociation of the cellulosome occurs only in the presence of ethylenediaminetetraacetic acid (EDTA) or distilled water (Choi and Ljungdahl, 1996a), which is accompanied by a distinct reduction of hydrolytic activity (Lamed and Bayer, 1988). Thus, aggregation of cellulosome subunits is apparently required for maximum hydrolytic activity against the crystalline substrate. This concept is supported by cloning experiments showing that endoglucanase activity is enhanced by association with a cellulose binding domain (CBD) (Cirucla et al., 1998; Kataeva et al., 1997; Karita et al., 1996).

Interestingly, spontaneous reaggregation of individual components does not always restore both cellulose binding and cellulose hydrolyzing activity (Beattie et al., 1994). In fact, at most, in vitro reaggregation of all subunits or specific subunits restores only a small portion of the cellulolytic activity (Cirucla et al., 1998; Wu et al., 1988). Apparently only specific associations of the subunits imparts cellulolytic capability, but there is no clear idea of the nature and number of subunits required for maximal activity.

An understanding of the clostridial cellulosome is now sufficient to enable a possible model to be proposed (Fig. 3). It is yet to be determined whether the clostridial model also will apply to cellulosomal systems of nonclostridial bacteria.

The clostridial cellulosome is composed of a scaffoldin polypeptide linked to a single CBD and several cohesin domains (Pagès et al., 1997a). Presumably, CBDs act to associate regions of cellulose with catalytic regions of the enzyme complex contained within the cellulosome. While there are sequence similarities between CBDs from a variety of cellulolytic bacteria, there does not appear to be qualitative similarity of these domains among different bacteria (Pagès et al., 1997b).

Each cohesin domain of C. thermocellum has a Ca^{2+} mediated attachment to a dockerin–catalytic domain (Leibovitz et al., 1997; Choi and Ljungdahl, 1996b), although a similar Ca^{2+} requirement has not been found for cellulosomes of C. cellulolyticum (Gal et al., 1997). The segments between each cohesin domain are glycosylated regions termed linkers (Gerwig et al., 1993). Dockerin domains consist of a highly conserved duplicated segment of about 23 residues (Tokatlidis et al., 1991). Dockerin domains of C. thermocellum interact with one of nine complementary cohesin domains present in the scaffoldin protein (CipA) (Shimon et al., 1997). In addition, Leibovitz et al. (1997) suggest that the other end of CipA is linked to SbdA, a scaffoldin dockerin binding protein.

The glycosylated regions of C. thermocellum's cellulosome contain a large proportion of covalently bound carbohydrates, particularly N-acetylgalactosamine derivatives. Although the function of this chemical architecture is speculative, binding experiments suggest these glycoproteins may be the recognized targets of lectins (Gerwig et al., 1989). Gerwig et al. (1999, 1983), isolated two major oligosaccharides with O-linkage in close association to cellulosomal subunits of C. thermocellum, both containing galactose residues.

Any galactopyranose present in the cellulosome presumably can be detected by an isolectin from Griffonia (BSI-B_4), which has a specific affinity for terminal α-D-galactosyl residues (Doyle, 1994). The binding of this isolectin to the surface of cellulolytic bacteria indicates the presence of these oligosaccharides (Lamed et al., 1987a,b). Furthermore, the lack of BSI-B_4 affinity to the surface of many non-

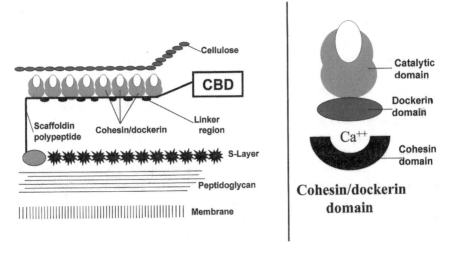

Figure 3. Model of the organization and cell association of the clostridial cellulosome. The scaffoldin polypeptide (left) serves as the platform for the cohesin domains. The location of the cellulose binding domain (CBD) and the number of cohesin domains probably varies among clostridia species. In addition, the nature of the association of the complex with the cell envelope is speculative and may vary among species. The catalytic component of the cohesin–dockerin domain (right) can vary, enabling a single scaffoldin to contain several different types of enzymatic activity.

cellulolytic bacteria indicates such oligosaccharides may be unique for cellulolytic bacteria (Lamed *et al.*, 1987a).

Also, the N-terminus region of the scaffoldin protein from *C. cellulovorans* appears to contain a signal sequence (Gilkes *et al.*, 1991, 1992) that probably functions in the transport of the protein to the exterior regions of the cell. In fact, all known cellulosome polypeptides contain a signal sequence typical of polypeptides secreted through the general secretory pathway (Béguin and Lemaire, 1996). The presence of such signal sequences provides additional evidence of the exocellular nature of the cellulosome complex.

A less understood aspect of the cellulosome is its association with the cell surface. At present, models for cellulosome organization rarely account for interaction with the S-layer or peptidoglycan layer and completely fail to account for possible interactions with the cellular membrane. There is little information of whether cellulosomes form at specific sites along the S-layer or the peptidoglycan layer or how the structure aggregates on the exterior of the cell.

Fujino *et al.* (1993) proposed a model where the scaffoldin polypeptide (CipA) of the *C. thermocellum* cellulosome connects to a polypeptide identified as ORF3p. This polypeptide mediates attachment of the cellulosome to the cell surface by a glycine–proline–threonine–serine-rich peptide linked to a region extending into the S-layer of the cell envelope. S-Layer homologous regions present in SbdA also suggest an interaction of SbdA with the S-layer (Leibovitz et al., 1997). The similarity of the central regions of SdbA with a possible anchoring region of the streptococcal M protein further suggests SbdA mediates anchoring of the cellulosome complex into the cell envelope (Leibovitz and Béguin, 1996).

Various nonproteinaceous components also appear to have an association with cellulosomes. Lipids comprise a portion of the cellulosome complex of C. thermocellum (Bolobova et al., 1994). Bolobova et al. (1994) demonstrated that cellulolytic activity is greatly reduced when extractable lipids are removed from cellulosome fractions, and total removal of all associated lipids required the dissociation of the complex. In fact, the region of the cellulosome most closely associated with cellulose contained a high amount of unsaturated fatty acids (Bolobova et al., 1994). Although little explanation for this has been offered, this work suggests that lipids are intricately involved in the cellulosome's association with crystalline regions of the cellulose fibers.

Membrane protein profiles also suggest a specific involvement of the membrane in cellulose utilization. Electrophoretic profiles of membrane fraction polypeptides from Eubacterium cellulosolvens and C. cellulovorans revealed that these organisms alter their membrane proteins in response to growth on specific substrates, including cellulose (Moon et al., 1996). Since the membrane proteins of these organisms have not been previously studied, no function can yet be assigned to the cellulose-inducible polypeptides. In addition, cellulose-grown cultures of E. cellulosolvens produce a lipoprotein (M_r, 13 kDa) not detected in cultures grown on other carbohydrates (M.-S. Moon and K. L. Anderson, in manuscript).

4.3.2b. Electron Microscopic Analysis of Polycellulosomes

Using transmission electron microscopy (TEM), protuberances have been detected on the surface of C. thermocellum (Mayer et al., 1987; Bayer and Lamed, 1986; Lamed and Bayer, 1986). Such protuberances are presumably composed of numerous cellulosomes (Mayer et al., 1987; Bayer and Lamed, 1986), and are often referred to as a polycellulosome (Béguin and Lemaire, 1996). However, it has yet to be clearly established that these protuberances are composed of cellulase-containing cellulosome complexes. Nor does TEM offer the resolution necessary to distinguish the smaller cellulosomes within the much larger polycellulosome.

Scanning electron microscopy (SEM) also has been used to visualize the ultrastructure of cellulolytic bacteria, and a relationship has been observed of cellulose utilization and presence of protuberances on the cell surface (Lamed et al., 1987a,b, 1991; Miron et al., 1989). SEM analysis was aided by staining samples with cationized ferritin (CF), which binds to certain negatively charged glycoproteins on the bacterial cell surface (Anderson, 1998). The use of this stain is reported to increase the resolution of SEM on surface structures (Lamed *et al.,* 1987b) and is generally regarded as the most effective means of detecting ultrastructures on the exterior surface of cellulolytic bacteria.

There are several problems, though, associated with using CF as a stain for SEM analysis. As a cation, CF will only bind with negatively charged molecules. The cell surface proton level, as affected by chemiosmotic activity of the cell, can affect the cell surface charge. Hence, bacteria with a high proton motive force (Δp) tend to have a net positive surface charge, which can electrostatically repel CF (Anderson, 1998). In fact, Kemper *et al.* (1993) demonstrated that treatment of *Bacillus subtilis* with a proton conductor, such as azide, increases the cellular binding of CF.

What is more, when protons bind to a cellular protein, they may alter that protein's ability to bind cofactors such as Ca^{2+}. Since Ca^{2+} binding seems to be required for the proper functioning and folding of scaffolding proteins in *C. thermocellum,* aggregation may be regulated by the presence of Ca^{2+} (Choi and Ljungdahl, 1996b). In fact, introduction of other cations stimulates reassociation of the complex (Beattie *et al.,* 1994). CF also may provide sufficient cationic charge to stimulate reassociation of the complex or aggregation of other cell surface components (Anderson, 1998). This possibility is supported by a recent study showing that the presence of protuberances on the surface of CF-stained *C. beijerinckii* coincided with growth rate rather than cellulolytic activity (Blair and Anderson, 1998).

4.4. Bacterial Adhesion to Cellulose

Early microscopic observations led researchers to conclude that adhesion of the bacterium was an important factor in plant degradation (Baker and Nasr, 1947). Electron micrographs further confirmed the close proximity of cellulolytic bacteria with cellulosic material (Akin, 1976; Lithium *et al.,* 1978). Intuitively, a close proximity would be expected for bacteria employing a cellulosome system, but it also may be advantageous for bacteria employing a noncomplexed mechanism by making it easier to sequester the products liberated by extracellular enzymes.

Unfortunately, the insoluble nature of cellulose also complicates bacterial adhesion assays. Thus, each protocol has distinct limitations, and a few are particularly laborious. However, some basic trends can be discussed.

Using *C. cellulolyticum* as a model, Gelhaye *et al.* (1992) determined that adhesion to cellulose was site specific. They further speculated that bound bacterial cells are released from cellulose into the liquid phase, possibly because of diminishing amounts of accessible cellulose molecules (Gelhaye *et al.,* 1993). Once released, cells will then adhere at a different site and recolonize. Based on this observation, they proposed that a bacterial culture is in a constant dynamic of adhesion, colonization, release, and readhesion.

Bacterial cells also appear to compete for the same adhesion sites. Adhesion of *C. cellulolyticum* is prevented if *Clostridium* sp. A22 is allowed to adhere to the cellulose first (Gelhaye *et al.,* 1992). *R. albus* is able to outcompete both *Fibrobacter succinogenes* and *Ruminococcus flavefaciens* for adhesion sites and will displace any *R. flavefaciens* that has already adhered (Mosoni *et al.,* 1997).

Morris (1988) also found that EDTA or high salt does not inhibit binding of *R. albus* to cellulose, indicating attachment is not dependent on ionic interaction. Binding by *R. albus* is sensitive to low pH (<5.5) and high temperatures (50°C) (Morris, 1988). Also, CMC and methyl cellulose are inhibitory to cellulose binding (Morris, 1988; Minato and Suto, 1978), possibly by acting as a competitive inhibitor (White *et al.,* 1988).

As discussed earlier, adhesion to substrate appears essential for some amylolytic bacteria; however, it is less certain if substrate adhesion is required for cellulose utilization by anaerobic bacteria. Rasmussen *et al.* (1989) and Morris and Cole (1987) reported that many but not all cellulolytic strains tested bound to crystalline cellulose. In fact, cellulolysis did not always follow adhesion (Morris and Cole, 1987). Gong and Forsberg (1989) isolated adhesion-deficient mutants of *F. succinogenes,* some of which were still capable of growth on crystalline cellulose.

A partial explanation of these results may be provided by a regulatory mutant of *C. thermocellum,* which required an extensive lag period before cellulose adhesion could be detected (Bayer *et al.,* 1996). However, growth of the mutant on cellulose corresponded with the time of adhesion (Bayer *et al.,* 1996). The *F. succinogenes* adhesion mutants, described above, also may be regulatory mutants. Additionally, since the above reports do not include time course evaluations of cellulose adhesion, delayed adhesion cannot be dismissed.

Unlike some amylolytic bacteria, no functional system for genetic manipulation of anaerobic cellulolytic bacteria has yet been developed. Of those anaerobic bacteria that are highly cellulolytic *in vitro,* some progress has been achieved (Anderson *et al.,* 1998; Cocconcelli *et al.,* 1992). However, until such genetic tools are fully exploited, the importance of cellulose binding for cellulolytic activity will likely remain an open question.

4.5. Regulation of Cellulolytic Systems

Bacterial adhesion to cellulose appears to be cellulose inducible. Miron *et al.* (1989) observed that cellulose-grown *F. succinogenes* bound to cellulose at a much greater level than cellobiose-grown cells. They also found a higher level of cellulose binding by *R. albus* and *R. flavefaciens* following adaptation to cellulose as the sole energy substrate. Significant differences between the cellulose-binding levels of cellulose-grown and monosaccharide-grown *C. cellulovorans* also have been detected (Blair and Anderson, 1999).

In many anaerobic bacteria, the formation of a cellulosome is thought to be necessary for the efficient binding and degrading of crystalline cellulose. Assembly of this complex from its subunits may be induced by the presence of cellulose (Matano *et al.,* 1994). The subunits of the cellulosome appear to exhibit an allosteric-type binding with cellulose once cellulose comes in contact with the cell. For cultures of *C. thermocellum* in early exponential growth, the cellulosome apparently binds tightly to crystalline cellulose in the absence of glucose or large amounts of cellobiose. This complex releases from the cell, however, as the culture ages. Once released, the cellulosome remains bound to the cellulose and may retain some catalytic activity (Bayer and Lamed, 1986). The mechanism for this release has not been determined.

In addition, the regulatory control of cellulosome formation, including identification of true inducers, is not well defined. Nochur *et al.* (1993) found a relationship of cellulase activity with the ATP content and proton potential in *C. thermocellum.* Bhat *et al.* (1993) reported that cellobiose is the inducer for cellulosomes of *C. thermocellum.* This study detected several proteins (including cellulase) in the extracellular fraction, but did not determine whether these proteins formed a complex possessing characteristics of a cellulosome complex (i.e., affinity for cellulose, denatured by EDTA, resistant to sodium dodecyl sulfate).

Previous work suggests the possibility that regulation of cellulosome formation involves multiple stages. Matano *et al.* (1994) proposed that cellobiose acts as an inducer of the cellulosome subunits of *C. cellulovorans,* but aggregation into a functional cellulosome requires the presence of crystalline cellulose. Cellulosome aggregation may be similar in *C. thermocellum.* Immunoelectron microscopy reveals that cellobiose-grown cells contain exocellular complexes that reorganize into complex cellulose-binding ultrastructures when cellulose is the substrate (Bayer and Lamed, 1986). However, the specific mechanism(s) responsible for induction of cellulosomal genes or aggregation of the proteins into a complex have yet to be determined. Perhaps aggregation involves a mechanism similar to that used by *Salmonella typhimurium,* which rapidly forms appendages upon contact with the epithelial tissue (Ginocchio *et al.,* 1994).

Since soluble carbohydrates may be present in the microbial ecosystem of cellulolytic bacteria (particularly at the site of plant cell wall degradation), bacteria may prefer these carbohydrates and repress their cellulolytic systems. Béguin and Aubert (1994) suggest that all known cellulolytic microbes repress their cellulolytic systems when a more soluble carbohydrate is available. However, much of their conclusion is derived from work with fungi because regulatory features of most cellulolytic bacteria have not been characterized. The cellulolytic systems of fungi, though, are quite distinct from those of bacteria (Coughlan and Ljungdahl, 1988), and direct comparisons can be misleading.

In fact, it is not clear whether (and which) anaerobic bacteria repress their cellulolytic systems in preference of more soluble substrates. Some studies entail confounding factors that tend to make interpretation of the results difficult. For example, some early studies suggested that soluble carbohydrates will repress cellulolytic activity (Mould et al., 1983; Fusee and Leatherwood, 1972). However, as the soluble carbohydrates in the medium are utilized by the bacterium, acidic metabolites accumulate and lower the medium pH. This lower pH can be inhibitory to cellulolytic activity (Stewart, 1977) and may account for the observed repression effects. El-Shazly et al. (1961) detected a decrease of cellulose degradation when starch was added to the medium. Further analysis, though, revealed that the reduced cellulolytic activity could be due to competition of the amylolytic and cellulolytic bacteria for nitrogen and other nutrients (El-Shazly et al., 1961). Huang and Forsberg (1990) found the cellulolytic activity of F. succinogenes is subject to repression by glucose and cellobiose. On the other hand, Hiltner and Dehority (1983) detected no such repression in several strains of cellulolytic bacteria, including F. succinogenes. Both studies measured the decrease of cellulose concentration in the medium, but used different methods of measurement.

Recently, SEM analysis found that C. cellulovorans produces ultrastructural protuberances only when cells were cultivated in cellulose medium (Blair and Anderson, 1998, 1999). The addition of soluble carbohydrates, though, to cellulose-grown cells causes an almost immediate loss of such protuberances (Blair and Anderson, 1999). The loss of the protuberances was accompanied by the loss of BSI-B$_4$ isolectin binding, and a cellulase-containing protein fraction. Even the addition of CMC causes the same rapid loss of the protuberances. Confounding factors, such as low medium pH or competition for nitrogen sources were not factors.

In addition to these regulatory events, a feature that generally has been overlooked is the potential influence of carbohydrates synthesized by other microorganisms. Recently, Yoo and Day (1996) presented evidence that under certain conditions oligodextrans produced by Leuconostoc mesenteroides and Lipomyces starkeyi could inhibit the growth of Clostridium perfringens and some strains of Salmonella. Since most cellulolytic environments consist of a vast array of microorganisms, the effect one group may have on other groups (including cellulolytics) could be dramatic. In fact, such an effect may not necessarily inhibit growth, but may repress the ability of some microorganisms to utilize certain substrates. This could account for why Butyrivibrio fibrisolvens appears to have cellulolytic capacity in the rumen, but most strains do not retain cellulolytic activity in laboratory medium (Hungate, 1966; Weimer, 1996). Conditions in the rumen may force B. fibrisolvens into a cellulolytic niche that it quickly vacates once those conditions are absent.

5. SUMMARY

Cellulolytic and amylolytic systems of some anaerobic bacteria are now known to be far more complex than simply the production of extracellular enzymes. These systems include the interaction of many components to effectively hydrolyze the polysaccharides. The constituents of these systems appear to form some type of complex: membrane associated for some gram-negative amylolytics and exocellular cellulosomes for some gram-positive cellulolytics.

These complexes house polysaccharide binding proteins, which apparently serve to bring the polysaccharide into proximity to the polysaccharidases. For some bacteria, binding to the starch molecule appears to be an essential step for starch utilization. It is less certain if binding to cellulose molecules is an essential step for cellulolytic bacteria.

In general, both cellulolytic and amylolytic bacterial activity is induced by cultivation in medium containing some form of the respective polysaccharide. However, the overall regulation of the synthesis and aggregation of the cell surface complexes is still uncertain. In addition, while there is evidence that some bacteria repress their polysaccharide utilization systems in preference of utilizing more soluble carbohydrates, such as glucose, it is not certain how universal such repression is in the anaerobic bacterial world.

REFERENCES

Ait, N., Creuzet, N., and Forget, P., 1979, Partial purification of cellulases from *Clostridium thermocellum, J. Gen. Microbiol.* **113**:399–402.

Akin, D. E., 1976, Ultrastructure of rumen bacterial attachment to forage cell walls, *Appl. Env. Microbiol.* **31**:562–568.

Alexander, M., 1977, *Introduction to Soil Microbiology,* 2nd Ed., John Wiley and Sons, New York.

Anderson, K. L., 1995, Biochemical analysis of starch degradation by *Ruminobacter amylophilus* 70, *Appl. Env. Microbiol.* **61**:1488–1491.

Anderson, K. L., 1998, Cationized ferritin as a stain for electron microscopic observation of bacterial ultrastructure, *Biotech. Histochem.* **73**:278–288.

Anderson, K. L., and Salyers, A. A., 1989a, Biochemical evidence that starch breakdown by *Bacteroides thetaiotaomicron* involves outer membrane starch-binding sites and periplasmic starch-degrading enzymes, *J. Bacteriol.* **171**:3192–3198.

Anderson, K. L., and Salyers, A. A., 1989b, Genetic evidence that outer membrane binding of starch is required for starch utilization by *Bacteroides thetaiotaomicron, J. Bacteriol.* **171**:3199–3204.

Anderson, K. L., Megehee, J. A., and Varel, V. H., 1998, Conjugal transfer of transposon Tn1545 into the cellulolytic bacterium *Eubacterium cellulosolvens, Lett. Appl. Microbiol.* **26**:35–37.

Annous, B. A., and Blaschek, H. P., 1990, Regulation and localization of amylolytic enzymes in *Clostridium acetobutylicum* ATCC 824, *Appl. Env. Microbiol.* **56**:2559–2561.

Baker, F., and Nasr, H., 1947, Microscopy in the investigation of starch and cellulose breakdown in the digestive tract, *J. Res. Microsc. Soc.* **67**:27–42.

Barr, B. K., Hsieh, Y.-L., Ganem, B., and Wilson, D. B., 1996, Identification of two functionally different classes of exocellulases, *Biochemistry* **35**:586–592.

Bayer, E. A., and Lamed, R., 1986, Ultrastructure of the cell surface cellulosome of *Clostridium thermocellum* and its interaction with cellulose, *J. Bacteriol.* **167**:828–836.

Bayer, E. A., and Lamed, R., 1992, The cellulose paradox: pollutant par excellence and/or a reclaimable natural resource? *Biodegradation* **3**:171–188.

Bayer, E. A., Kenig, R., and Lamed, R., 1983, Adherence of *Clostridium thermocellum* to cellulose, *J. Bacteriol.* **156**:818–827.

Bayer, E. A., Morag, E., Shoham, Y., Tormo, J., and Lamed, R., 1996, The cellulosome: A cell surface organelle for the adhesion to and degradation of cellulose, in: *Bacterial Adhesion: Molecular and Ecological Diversity* (M. Fletcher, ed.), Wiley-Liss, New York, pp. 155–182.

Beattie, L., Bhat, K. M., and Wood, T. M., 1994, The effect of cations on reassociation of the components of the cellulosome cellulase complex synthesized by the bacterium *Clostridium thermocellum, Appl. Microbiol. Biotechnol.* **40**:740–744.

Béguin, P., 1990, Molecular biology of cellulose degradation, *Annu. Rev. Microbiol.* **44**:219–248.

Béguin, P., and Aubert, J.-P., 1994, The biological degradation of cellulose, *FEMS Microbiol. Rev.* **13**:25–58.

Béguin, P., and Lemaire, M., 1996, The cellulosome: An exocellular multiprotein complex specialized in cellulose degradation, *Crit. Rev. Biochem. Mol. Biol.* **31**:201–236.

Bhat, S., Goodenough, P. W., Owen, E., and Bhat, M. K., 1993, Cellobiose: A true inducer of cellulosome in different strains of *Clostridium thermocellum, FEMS Microbiol. Lett.* **111**:73–78.

Blair, B. G., and Anderson, K. L., 1998, Comparison of staining techniques for scanning electron microscopic detection of ultrastructural protuberances on cellulolytic bacteria, *Biotech. Histochem.* **73**:107–113.

Blair, B. G., and Anderson, K. L., 1999, Regulation of cellulose-inducible structures of *Clostridium cellulovorans, Can. J. Microbiol.* **45**:242–249.

Bolobova, A. V., Zhukov, A. V., and Klosov, A. A., 1994, Lipids and fatty acids in cellulosomes of *Clostridium thermocellum, Appl. Microbiol. Biotechnol.* **42**:128–133.

Brenner, R., Maccubbin, E. A., and Hodson, R. E., 1984, Anaerobic biodegradation of the lignin and polysaccharide components of lignocellulose and synthetic lignin by sediment microflora, *Appl. Env. Microbiol.* **47**:998–1004.

Calza, R. E., Irwin, D. A., and Wilson, D. B., 1985, Purification and characterization of two beta-1,4-endoglucanases from *Thermomonospora fusca, Biochemistry* **24**:7797–7804.

Cavedon, K., Leschine, S. B., and Canale-Parola, E., 1990, Cellulase system of a free-living, mesophilic clostridium (strain C7), *J. Bacteriol.* **172**:4222–4230.

Choi, S. K., and Ljungdahl, L. G., 1996a, Dissociation of the cellulosome of *Clostridium thermocellum* in the presence of ethylenediaminetetraacetic acid occurs with the formation of truncated polypeptides, *Biochemistry* **35**:4897–4905.

Choi, S. K., and Ljungdahl, L. G., 1996b, Structural role of calcium for the organization of the cellulosome of *Clostridium thermocellum*, *Biochemistry* **35**:4906–4910.

Cirucla, A., Gilbert, H. J., Ali, B. R. S., and Hazlewood, G. P., 1998, Synergistic interaction of the cellulosome integrating protein (CipA) from *Clostridium thermocellum* with a cellulosomal endoglucanase, *FEBS Lett.* **422**:221–224.

Cocconcelli, P. S., Ferrari, E., Rossi, F., and Bottazzi, V., 1992, Plasmid transformation of *Ruminococcus albus* by means of high-voltage electroporation, *FEMS Microbiol. Lett.* **94**:203–208.

Colberg, P. J., and Young, L. Y., 1982, Biodegradation of lignin-derived molecules under anaerobic conditions, *Can. J. Microbiol.* **28**:886–889.

Colberg, P. J., and Young, L. Y., 1985, Anaerobic degradation of soluble fractions of [^{14}C-LIGNIN] lignocellulose, *Appl. Env. Microbiol.* **49**:345–349.

Cotta, M. A., 1988, Amylolytic activity of selected species of ruminal bacteria, *Appl. Env. Microbiol.* **54**:772–776.

Cotta, M. A., and Whitehead, T. R., 1993, Regulation and cloning of the gene encoding amylase activity of the ruminal bacterium *Streptococcus bovis*, *Appl. Env. Microbiol.* **59**:189–196.

Cotta, M. A., Wheeler, M. B., and Whitehead, T. R., 1994, Cyclic AMP in ruminal and other anaerobic bacteria, *FEMS Microbiol. Lett.* **124**:355–360.

Coughlan, M. P., 1992, Enzymic hydrolysis of cellulose: an overview, *Bioresource Technol.* **39**:107–115.

Coughlan, M. P., and Ljungdahl, L. G., 1988, Comparative biochemistry of fungal and bacterial cellulolytic enzyme systems, in: *Biochemistry and Genetics of Cellulose Degradation* (J. Aubert, P. Béguin, and J. Millet, eds.), Academic Press, New York, pp. 11–30.

Davies, G., and Henrissat, B., 1995, Structures and mechanisms of glycosyl hydrolases, *Structure* **3**:853–859.

D'elia, J. N., and Salyers, A. A., 1996, Contribution of a neopullulanase, pullulanase, and an α-glucosidase to growth of *Bacteroides thetaiotaomicron* on starch, *J. Bacteriol.* **178**:7173–7179.

Doyle, R. J., 1994, Introduction to lectins and their interactions with microorganisms, in: *Lectin–Microorganism Interactions* (R. J. Doyle and M. Slifkin, eds.), Marcel Dekker, New York, pp. 1–65.

El-Shazly, K., Dehority, B. A., and Johnson, R. R., 1961, Effect of starch on the digestion of cellulose *in vitro* and *in vivo* by rumen microorganisms, *J. Anim. Sci.* **20**:268–271.

Englyst, H. N., and Cummings, J. H., 1987, Resistant starch, a "new" food component: A classification of starch for nutritional purposes, in: *Cereals in a European Context* (I. D. Morton, ed.), Ellis Horwood, Chichester, England, pp. 221–223.

Fenchel, T. M., and Jorgensen, B. B., 1977, Detritus food chains of aquatic ecosystems: The role of bacteria, *Adv. Microb. Ecol.* **1**:1–58.

Fields, M. W., Ryals, P. E., and Anderson, K. L., 1997, Polysaccharide inducible outer membrane proteins of *Bacteroides xylanolyticus* X5–1, *Anaerobe* **3**:43–48.

Fujino, T., Béguin, P., and Aubert, J.-P., 1993, Organization of a *Clostridium thermocellum* gene cluster encoding the cellulosomal scaffolding protein CipA and a protein possibly involved in attachment of the cellulosome to the cell surface, *J. Bacteriol.* **175**:1891–1899.

Fusee, M. C., and Leatherwood, J. M., 1972, Regulation of cellulase from *Ruminococcus, Can. J. Microbiol.* **18**:347–353.

Gal, L., Pagès, S., Gaudin, C., Bélaïch, A., Reverbel-Leroy, C., Tardif, C., and Bélaïch, J.-P., 1997, Characterization of the cellulolytic complex (cellulosome) produced by *Clostridium cellulolyticum, Appl. Env. Microbiol.* **63**:903–909.

Gelhaye, E., Claude, B., Cailliez, C., Burled, S., and Petitdemange, H., 1992, Multilayer adhesion to filter paper of two mesophilic, cellulolytic clostridia, *Curr. Microbiol.* **25**:307–311.

Gelhaye, E., Gehin, A., and Petitdemange, H., 1993, Colonization of crystalline cellulose by *Clostridium cellulolyticum* ATCC 35319, *Appl. Env. Microbiol.* **59**:3154–3156.

Gerwig, G. J., de Waard, P., Vliegnthart, F. G., Morgenstern, E., Lamed, R., and Bayer, E. A., 1989, Novel O-linked carbohydrate chains in the cellulase complex (cellulosome) of *Clostridium thermocellum*, 2-*O*-methyl-*N*-acetylglucosamine as a component of a glycoprotein, *J. Biol. Chem.* **264**:1027–1035.

Gerwig, G. J., Kamerling, J. P., Vliegenthart, J. F. G., Morag, E., Lamed, R., and Bayer, E. A., 1993, The nature of the carbohydrate–peptide linkage region in glycoproteins from the cellulosomes of *Clostridium thermocellum* and *Bacteroides cellulosolvens, J. Biol. Chem.* **288**:26956–26960.

Gilkes, N. R., Henrissat, B., Kilburn, D. G., Miller, R. C., and Warren, R. A. J., 1991, Domains in microbial β-1,4-glycanases: Sequence conservation, function and enzyme families, *Microbiol. Rev.* **55**:303–315.

Gilkes, N. R., Jervis, E., Henrissat, B., Tekant, B., Miller, R. C. J., Warren, R. A. J., and Kilburn, D. G., 1992, The adsorption of a bacterial cellulase and its two isolated domains to crystalline cellulose, *J. Biol. Chem.* **267**:6743–6749.

Ginocchio, C. C., Olmsted, S. B., Wells, C. L., and Galan, J. E., 1994, Contact with epithelial cells induces the formation of surface appendages on *Salmonella typhimurium, Cell* **76**:717–724.

Gong, J., and Forsberg, C. W., 1989, Factors affecting adhesion of *Fibrobactor succinogenes* subsp. *succinogenes* S85 and adherence-defective mutants to cellulose, *Appl. Environ. Microbiol.* **55**:3039–3044.

Healy, J. B., and Young, L. Y., 1979, Anaerobic biodegradation of eleven aromatic compounds to methane, *Appl. Env. Microbiol.* **38**:84–89.

Henrissat, B., and Bairoch, A., 1993, New families in the classification of glycosyl hydrolases based on amino acid sequence similarities, *Biochem. J.* **293**:781–788.

Hiltner, P., and Dehority, B. A., 1983, Effect of soluble carbohydrates on digestion of cellulose by pure cultures of rumen bacteria, *Appl. Env. Microbiol.* **46**:642–648.

Huang, L., and Forsberg, C. E., 1990, Cellulose digestion and cellulase regulation and distribution in *Fibrobacter succinogenes* subsp. *succinogenes* S85, *Appl. Env. Microbiol.* **56**:1221–1228.

Hungate, R. E., 1966, *The Rumen and Its Microbes,* Academic Press, New York.

Irwin, D. C., Specio, M., Walker, L. P., and Wilson, D. B., 1993, Activity studies of eight purified cellulases; Specificity, synergism, and binding domain effects, *Biotechnol. Bioeng.* **42**:1002–1013.

Jenner, C. F., 1982, Storage of starch, in: *Plant Carbohydrates I. Intracellular Carbo-*

hydrates (F. A. Loewus and W. Tanner, eds.), Springer-Verlag, New York, pp. 700–747.

Karita, S., Sakka, K., and Ohmiya, K., 1996, Cellulose-binding domains confer an enhanced activity against insoluble cellulose to *Ruminococcus albus* endoglucanase IV, *J. Ferment. Bioeng.* **81**:553–556.

Kataeva, I., Guglielmi, G., and Béguin, P., 1997, Interaction between *Clostridium thermocellum* endoglucanase CelD and polypeptides derived from the cellulosome-integrating protein ClpA: Stoichiometry and cellulolytic activity of the complexes. *Biochem. J.* **326**:617–624.

Kemper, M. A., Urrutia, M. M., Beveridge, T. J., Koch, A. L., and Doyle, R. J., 1993, Proton motive force may regulate cell wall-associated enzymes of *Bacillus subtilis, J. Bacteriol.* **175**:5690–5696.

Kohring, S., Wiegel, J., and Mayer, F., 1990, Subunit composition and glycosidic activities of the cellulase complex from *Clostridium thermocellum* JW20, *Appl. Env. Microbiol.* **56**:3798–3804.

Kuriki, T., Okada, S., and Imanaka, T., 1988, New type of pullulanase from *Bacillus stearothermophilus* and molecular cloning and expression of the gene in *Bacillus subtilis, J. Bacteriol.* **170**:1554–1559.

Lamed, R., and Bayer, E. A., 1986, Contact and cellulolysis in *Clostridium thermocellum* via extensile surface organelles, *Experientia* **42**:72–73.

Lamed, R., and Bayer, E. A., 1988, The cellulosome of *Clostridium thermocellum, Adv. Appl. Microbiol.* **33**:1–46.

Lamed, R., Setter, E., and Bayer, E. A., 1983a, Characterization of cellulose-binding, cellulase-containing complex in *Clostridium thermocellum, J. Bacteriol.* **156**:828–836.

Lamed, R., Setter, E., Kenig, R., and Bayer, E. A., 1983b, The cellulosome: A discrete cell surface organelle of *Clostridium thermocellum* which exhibits separate antigenic, cellulose-binding and various cellulolytic activities, *Biotechnol. Bioeng. Symp.* **13**:163–181.

Lamed, R., Naimark, J., Morgenstern, E., and Bayer, E. A., 1987a, Specialized cell surface structures in cellulolytic bacteria, *J. Bacteriol.* **169**:3792–3800.

Lamed, R., Naimark, J., Morgenstern, E., and Bayer, E. A., 1987b, Scanning electron microscopic delineation of bacterial surface topology using cationized ferritin, *J. Microbiol. Methods* **7**:233–240.

Lamed, R., Morag, E., Mor-Yosef, O., and Bayer, E. A., 1991, Cellulosome-like entities in *Bacteroides cellulosolvens, Curr. Microbiol.* **22**:27–33.

Leibovitz, E., and Béguin, P., 1996, A new type of cohesion domain that specifically binds the dockerin domain of the *Clostridium thermocellum* cellulosome-integrating protein CipA, *J. Bacteriol.* **178**:3077–3084.

Leibovitz, E., Ohayon, H., Gounon, P., and Béguin, P., 1997, Characterization and subcellular localization of the *Clostridium thermocellum* scaffoldin dockerin binding protein SdbA, *J. Bacteriol.* **179**:2519–2523.

Lithium, M. J., Brooker, B. E., Pettipher, G. L., and Harris, P. J., l978, *Ruminococcus flavefaciens* cell coat and adhesion to cotton cellulose and to cell walls in leaves of perennial ryegrass (Lolium perenne), *Appl. Env. Microbiol.* **35**:156–165.

Liu, C.-C., and Doi, R. H., 1998, Properties of *exeS*, a gene for a major subunit of the *Clostridium cellulovorans cellulosome, Gene* **211**:39–47.

Ljungdahl, L. G., and Eriksson, K.-E., 1985, Ecology of microbial cellulose degradation, *Adv. Microb. Ecol.* **8**:237–299.

Macfarlane, G. T., and Cummings, J. H., 1991, The colonic flora, fermentation and large bowel digestive function, in: *The Large Intestine: Physiology, Pathophysiology and Disease* (S. F. Phillips, J. H. Pemberton, and R. G. Shorter, eds.), Raven Press, New York, pp. 51–92.

Matano, Y., Park, J. S., Goldstein, M. A., and Doi, R. H., 1994, Cellulose promotes extracellular assembly of *Clostridium cellulovorans* cellulosomes, *J. Bacteriol.* **176**:6952–6956.

Matsushita, O., Russell, J. B., and Wilson, D. B., 1990, Cloning and sequencing of a *Bacteroides ruminicola* B$_1$4 endoglucanase gene, *J. Bacteriol.* **172**:3620–3630.

Mayer, F., Coughlan, M. P., Mori, Y., and Ljungdahl, L. G., 1987, Macromolecular organization of the cellulolytic enzyme complex of *Clostridium thermocellum* as revealed by electron microscopy, *Appl. Env. Microbiol.* **53**:2785–2792.

McGinnis, G. D., and Shafizadeh, F., 1991, Cellulose, in: *Wood Structure and Composition* (M. Lewin and I. Goldstein, eds.), Marcel Dekker, New York, pp. 139–181.

McWethy, S. J., and Hartman, P. A., 1977, Purification and some properties of an extracellular alpha-amylase from *Bacteroides amylophilus, J. Bacteriol.* **129**:1537–1544.

Minato, H., and Suto, T., 1978, Techniques for fractionation of bacteria in rumen microbial ecosystems. II. Attachment of bacteria isolated from bovine rumen to cellulose powder *in vitro* and elution of bacteria attached therefrom, *J. Gen. Appl. Microbiol.* **24**:1–16.

Miron, J. Yokoyamo, M. T., and Lamed, R., 1989, Bacterial cell surface structures involved in lucerne cell wall degradation by pure cultures of cellulolytic rumen bacteria, *Appl. Microbiol. Biotechnol.* **32**:218–222.

Moon, M.-S., Blair, B. G., and Anderson, K. L., 1996, Regulation of membrane proteins by the cellulolytic anaerobes *Eubacterium cellulosolvens* and *Clostridium cellulovorans,* in: 96th General Meeting of the American Society for Microbiology, New Orleans, Louisiana, American Society for Microbiology, Washington DC, p. 538.

Morag, E., Halevy, I., and Lamed, R., 1991, Isolation and properties of a major cellobiohydrolase from the cellulosome of *Clostridium thermocellum, J. Bacteriol.* **173**:4155–4162.

Morris, E. J., 1988, Characterization of the adhesion of *Ruminococcus albus* to cellulose, *FEMS Microbiol. Lett.* **51**:113–118.

Morris, E. J., and Cole, O. J., 1987, Relationship between cellulolytic activity and adhesion to cellulose in *Ruminococcus albus, J. Gen. Microbiol.* **133**:1023–1032.

Mosoni, P., Fonty, G., and Gouet, P., 1997, Competition between ruminal cellulolytic bacteria for adhesion to cellulose, *Curr. Microbiol.* **35**:44–47.

Mould, F. L., Orskov, E. R., and Mann, S. O., 1983, Associated effects of mixed feeds. I. Effects of type and level of supplementation and the influence of the rumen pH on cellulolysis *in vivo* and dry matter digestion of various roughages, *Anim. Feed Sci. Technol.* **10**:15–30.

Nochur, S. V., Roberts, M. F., and Demain, A. L., 1993, True cellulase production by *Clostridium thermocellum* grown on different carbon sources, *Biotechnol. Lett.* **15**:641–646.

Pagès, S., Bélaïch, A., Bélaïch, J.-P., Morag, E., Lamed, R., Shoham, Y., and Bayer, E. A., 1997a, Species-specificity of the cohesin-dockerin interaction between *Clostridium thermocellum* and *Clostridium cellulolyticum:* Prediction of specificity determinants of the dockerin domain, *Proteins* **29**:517–527.

Pagès, S., Gal, L., Bélaïch, A., Gaudin, C., Tardif, C., and Bélaïch, J.-P., 1997b, Role of scaffolding protein CipC of *Clostridium cellulolyticum* in cellulose degradation, *J. Bacteriol.* **179**:2810–2816.

Piggott, R. P., Rossiter, A., Ortlepp, S. A., Pembroke, J. T., and Ollington, J. F., 1984, Cloning in *Bacillus subtilis* of an extremely thermostable alpha amylase: Comparison with other cloned heat stable alpha amylases, *Biochem. Biophys. Res. Commun.* **122**:175–183.

Postma, P. W., Lengeler, J. W., and Jacobson, G. R., 1993, Phosphoenolpyruvate:carbohydrate phosphotransferase systems of bacteria, *Microbiol. Rev.* **57**:543–594.

Rasmussen, M. A., White, B. A., and Hespell, R. B., 1989, Improved assay for quantitating adherence of ruminal bacteria to cellulose, *Appl. Env. Microbiol.* **55**:2089–2091.

Raven, P. H., Evert, R. F., and Eichhorn, S. E., 1999, *Biology of Plants,* W. H. Freeman and Co., New York.

Reeves, A. R., D'elia, J. N., Frias, J., and Salyers, A. A., 1996, A *Bacteroides thetaiotaomicron* outer membrane protein that is essential for utilization of maltooligosaccharides and starch, *J. Bacteriol.* **178**:823–830.

Reeves, A. R., Wang, G.-R., and Salyers, A. A., 1997, Characterization of four outer membrane proteins that play a role in utilization of starch by *Bacteroides thetaiotaomicron, J. Bacteriol.* **179**:643–649.

Reiling, S., and Brickmann, J., 1995, Theoretical investigations on the structure and physical properties of cellulose, *Macromol. Theory Simul.* **4**:725–743.

Rowland, S. P., 1975, Selected aspects of the structure and accessibility of cellulose as they relate to hydrolysis in cellulose as a chemical and energy source, in: *Biotechnology and Bioengineering Symposium,* No. 5 (C. Wilke, ed.), Wiley Press, New York, pp. 183–191.

Saier, M. H., 1996, Cyclic AMP-independent catabolite repression in bacteria, *FEMS Microbiol. Lett.* **142**:217–230.

Sakano, Y., Hiraiwa, S.-i., Fukushima, J., and Kobayashi, T., 1982, Enzymatic properties and action patterns of *Thermoactinomyces vulgaris* α-amylase, *Agr. Biol. Chem.* **46**:1121–1130.

Schink, B., 1988, Principles and limits of anaerobic degradation: environmental and technological aspects, in: *Biology of Anaerobic Microorganisms* (A. J. B. Zehnder, ed.), Wiley-Liss, New York, pp. 771–846.

Schwartz, M., 1987, The maltose regulon, in: *Escherichia coli and Salmonella: Cellular and Molecular Biology,* Vol. 2 (F. C. Neidardt, J. L. Ingraham, K. B. Low, B. Magasanik, M. Schaechter, and H. E. Umbarger, eds.), American Society for Microbiology, Washington, DC, pp. 1482–1502.

Shimon, L. J. W., Bayer, E. A., Morag, E., Lamed, R., Yaron, S., Shoham, Y., and Frolow, F., 1997, A cohesion domain from *Clostridium thermocellum:* The crystal structure provides new insights into cellulosome assembly, *Structure* **5**:381–390.

Singh, A., and Hayashi, K., 1995, Microbial cellulases: Protein architecture, molecular properties, and biosynthesis, *Adv. Appl. Microbiol.* **40**:1–44.

Sjöström, E., 1993, *Wood Chemistry. Fundamental and Applications,* 2nd ed., Academic Press, New York.

Smith, K. A., and Salyers, A. A., 1989, Cell-associated pullulanase from *Bacteroides thetaiotaomicron:* Cloning, characterization, and insertional mutagenesis to determine role in pullulan utilization, *J. Bacteriol.* **171:**2116–2123.

Smith, K. A., and Salyers, A. A., 1991, Characterization of a neopullulanase and an α-glucosidase from *Bacteroides thetaiotaomicron* 95–1, *J. Bacteriol.* **173:**2962–2968.

Smolka, G. E., Birnbaum, E. R., and Darnall, D. W., 1971, Rare earth metal ions as substitutes for the calcium ion in *Bacillus subtilis* α-amylase, *Biochemistry* **10:**4556–4561.

Stålbrand, H., Mansfield, S. D., Saddler, J. N., Kilburn, D. G., Warren, R. A. J., and Gilker, N. R., 1998, Analysis of molecular size distributions of cellulose recombinant *Cellulomonas fimi* β-1,4-glucanases, *Appl. Env. Microbiol.* **64:**2374–2379.

Stewart, C. S., 1977, Factors affecting the cellulolytic activity of rumen contents, *Appl. Environ. Microbiol.* **33:**497–502.

Tancula, E., Feldhaus, M. J., Bedzyk, L. A., and Salyers, A. A., 1992, Location and characterization of genes involved in binding of starch to the surface of *Bacteroides thetaiotaomicron, J. Bacteriol.* **174:**5609–5616.

Tokatlidis, K. S., Salamitou, S., Béguin, P., Dhurjati, P., and Aubert, J.-P., 1991, Interaction of the duplicated segment carried by *Clostridium thermocellum* cellulases with cellulosome components, *FEBS Lett.* **291:**185–188.

Tomme, P., Warren, R. A. J., and Gilkes, N. R., 1995, Cellulose hydrolysis by bacteria and fungi, *Adv. Microbiol. Physiol.* **37:**1–81.

Vallee, B. L., Stein, E. A., Summerwell, W. N., and Fischer, E. H., 1959, Metal content of α-amylase of various origins, *J. Biol. Chem.* **234:**2901–2905.

Vihinen, M., and Mäntsälä, P., 1989, Microbial amylolytic enzymes, *Crit. Rev. Biochem. Mol. Biol.* **24:**329–418.

Ward, K., and Seib, P. A., 1970, Cellulose, lichenan, and chitin, in: *The Carbohydrates,* 2nd ed. (W. Pigman, D. Horton, and A. Herp, eds.), Academic Press, New York, pp. 413–445.

Warren, R. A. J., 1996, Microbial hydrolysis of polysaccharides. *Annu. Rev. Microbiol.* **50:**183–212.

Weimer, P. J., 1996, Why can't ruminal bacteria digest cellulose faster? *J. Dairy Sci.* **79:**1496–1502.

White, B. A., Rasmussen, M. A., and Gardner, R. M., 1988, Methylcellulose inhibition of exo-β-1,4-glucanase A from *Ruminococcus flavefaciens* FD1, *Appl. Env. Microbiol.* **54:**1634–1636.

Wolfe, S. L., 1993, Regulation in prokaryotes, in: *Molecular and Cellular Biology* (S. L. Wolfe, ed.), Wadsworth Publishing Co., Belmont, CA, pp. 727–735.

Wood, T. M., 1988, Preparation of crystalline, amorphous and dyed cellulase substrates, *Methods Enzymol.* **160:**19–25.

Wood, T. M., Wilson, C. A., and Stewart, C. S., 1982, Preparation of the cellulase from the cellulolytic anaerobic rumen bacterium *Ruminococcus albus* and its release from the bacterial cell wall, *Biochem. J.* **205:**129–137.

Wu, J. H. D., Orme-Johnson, W. H., and Demain, A. L., 1988, Two components of an extracellular protein aggregate of *Clostridium thermocellum* together degrade crystalline cellulose, *Biochemistry* **27:**1703–1709.

Yoo, S. K., and Day, D. F., 1996, Effects of oligodextran on selected pathogenic bacteria, in: *Joint Meeting of the South Central Branch of the American Society for Microbiology and Mid-South Biochemists,* Tulane University, New Orleans, Louisiana, p. 43.

14

The Cellulosome

An Exocellular Organelle for Degrading Plant Cell Wall Polysaccharides

Edward A. Bayer, Yuval Shoham, and Raphael Lamed

1. INTRODUCTION

The plant cell wall consists of an intricate mixture of polysaccharides, the major components of which are cellulose, hemicellulose (e.g., xylans, pectins, lichenans, etc.), and lignin (Carpita and Gibeaut, 1993). These characteristically robust polymers provide the plant with a stable structural framework and protect the plant cell from the hazards of its surroundings. Despite its recalcitrant nature, the polysaccharides of the plant cell wall offer an exceptional source of carbon and energy, and a variety of microorganisms has evolved that are capable of degrading plant cell wall polysaccharides. These microbes occupy a broad range of habitats: Some are free living and rid the environment of such polysaccharides by converting them to the simple sugars that they assimilate; others are linked closely with cellulolytic animals, residing in the digestive tracts of ruminants and other grazers or in the

Edward A. Bayer • Department of Biological Chemistry, The Weizmann Institute of Science, Rehovot 76100 Israel. *Yuval Shoham* • Department of Food Engineering and Biotechnology, Technion—Israel Institute of Technology, Haifa 32000 Israel. *Raphael Lamed* • Department of Molecular Microbiology and Biotechnology, George S. Wise Faculty of Life Sciences, Tel Aviv University, Ramat Aviv 69978 Israel.

Glycomicrobiology, edited by Doyle.
Kluwer Academic/Plenum Publishers, New York, 2000.

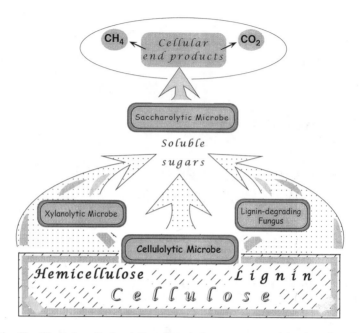

Figure 1. Simplified schematic description of a typical ecosystem comprising degrading plant matter. Cellulolytic, xylanolytic, and ligninolytic microbes combine to decompose the major polysaccharide components to soluble sugars. "Satellite" microorganisms assimilate the excess sugars and other cellular end products, which are ultimately converted to methane and carbon dioxide.

guts of wood-degrading termites and worms (Haigler and Weimer, 1991). In any given ecosystem, the polysaccharide-degrading microbes are not alone, but rely on the complementary contribution of other bacterial and/or fungal species (Bayer and Lamed, 1992; Bayer *et al.*, 1994). The polymer-degrading strains play a primary and crucial role in the ecosystem by converting the plant cell wall polysaccharides to the respective simple sugars and other degradation products (Fig. 1). They are assisted by satellite microbes, which cleanse the microenvironment from the breakdown products, producing in the final analysis methane and carbon dioxide.

In a given polysaccharide-degrading microorganism, the enzymes that catalyze their degradation may occur either in the free state and/or in discrete complexes with other similar types of enzymes. The latter are known as cellulosomes. Cellulosomes are exocellular macromolecular machines, designed for efficient degradation of cellulose and associated plant cell wall polysaccharides (Bayer *et al.*, 1998c). The cellulosome complex is composed of a collection of subunits, each of which comprises a set of interacting functional modules. Thus, one type of cellulosomal module, the cellulose-binding domain (CBD), is selective for binding to the substrate. Another family of modules, the catalytic domains, is specialized for the hydrolysis of the cellulose chains. Yet another complementary pair of domains—the cohesins and dockerins—serves to integrate the enzymatic subunits into the complex and the complex, in turn, into the cell surface. This "Lego"-like arrangement of the modular subunits generates an intricate multicomponent complex, the enzymes of which are bound *en bloc* to the insoluble substrate and act synergistically toward its complete digestion.

Since its initial description in the literature, the cellulosome concept has been subject to numerous reviews (Béguin and Lemaire, 1996; Doi *et al.,* 1994; Felix and Ljungdahl, 1993; Karita *et al.,* 1997; Lamed and Bayer, 1988a,b, 1991, 1993; Lamed *et al.,* 1983b). In this chapter, we will emphasize ecological and evolutionary aspects of the cellulosome and its relationship to its polysaccharide substrates.

2. PLANT CELL WALL POLYSACCHARIDES

Plant cells synthesize a composite matrix of tough polysaccharides on the outer surface of the plasma membranes, called the cell wall (Carpita and Gibeaut, 1993). The cell wall confers a protective covering to the plant cell, providing structure, turgidity, and durability, which renders the cell resistant to the outer elements, including mechanical, chemical, and microbial assault. Different types of plant cell tissues exhibit different ratios of these three major types of cell wall component; the ratio itself, the overall disposition, and the composition and mode of association of the components dictate the general physicochemical properties of the cell wall. On average, the cell wall contains roughly 40% cellulose, 30% hemicellulose, and 20% lignin. The exact composition of polysaccharides in the cell wall of an individual type of plant varies greatly.

Lignin is a heterogeneous, racemic, polydisperse, high-molecular-weight hydrophobic polymer, which consists of nonrepeating aromatic monomers connected via phenoxy linkages (Higuchi, 1990; Lewis and Yamamoto, 1990). Because of its exceptionally recalcitrant chemical structure and its close association with cellulose and hemicellulose, lignin is an important factor in impeding the biodegradation of these plant polysaccharides. The natural microbial utilization of cellulose and hemicellulose depends on prior degradation, penetration, and/or removal of the lignin barrier. The degradation of lignin is limited to filamentous prokaryotes and fungi under aerobic, oxidative conditions. These microbes synthesize a complicated set of lignin-degrading enzymes, which include lignin oxidase, lignin peroxidase, and laccase-type phenol oxidases (Dosoretz and Grethlein, 1991). The characteristics of microbial lignin-degrading systems are outside the scope of this chapter, and the interested reader is referred to previous reviews on this subject (Hatakka, 1994; Kerem and Hadar, 1998; Umezawa and Higuchi, 1991).

2.1. Cellulose

Cellulose is the major constituent of plant matter, and thus represents the most abundant organic polymer on Earth. Its decomposition by microbes constitutes a major part of the carbon cycle. Cellulose is a remarkably stable polymer, consisting of a linear polymer of β-1,4-linked glucose units. Chemically, the repeating unit is simply glucose, but structurally the repeating unit is the disaccharide cellobiose, that is, 4-*O*-(β-D-glucopyranosyl)-D-glucopyranose, since each glucose residue is rotated 180° relative to its neighbor (Fig. 2). The degree of polymerization of the resultant cellulose chains ranges from about 100 to more than 10,000; they are packed tightly in parallel fashion into microfibrils by extensive inter- and intrachain hydrogen-bonding interactions, which account for the rigid structural stability of cellulose. The microfibrils exhibit variable amounts of crystalline and amorphous components, again depending on the degree of polymerization, the extent of hydrogen bonding, and ultimately on the source of the cellulose. The mi-

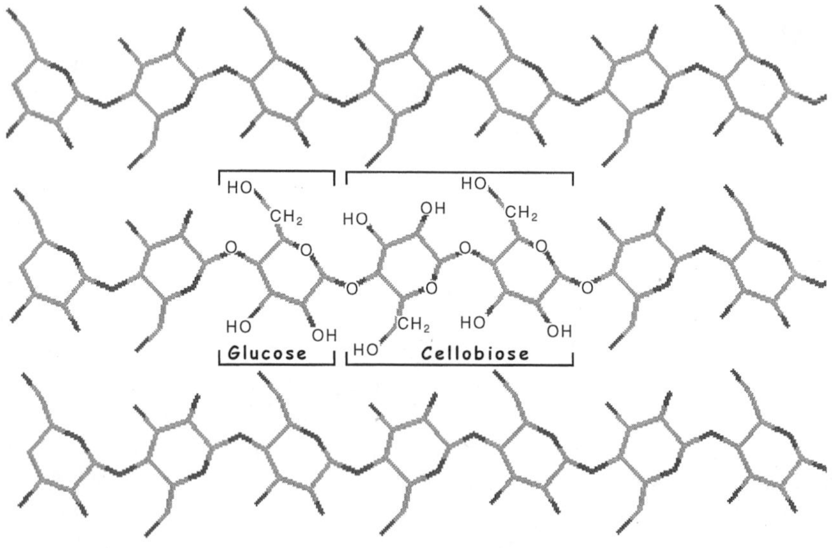

Figure 2. Structure of cellulose. Three parallel chains which form the 0,1,0 face are shown and a glucose moiety and repeating cellobiose unit are indicated. The model was built by Dr. José Tormo, based on early crystallographic data. The diagram was drawn using RasMol 2.6.

crofibrils themselves are further assembled into plant cell walls, the tunic of some sea animals, pellicles from bacterial origin, and so forth. Highly crystalline forms of cellulose include cotton, bacterial cellulose (from *Acetobacter xylinum*), and from the algae *Valonia ventricosa,* which exhibit crystallinity levels of about 45%, 75%, and 95%, respectively. The following reviews are available for more information on the structure of cellulose (Atalla and VanderHart, 1984; Chanzy, 1990; O'Sullivan, 1997).

2.2. Hemicellulose

Hemicelluloses are relatively low-molecular-weight, branched heteropolysaccharides that are associated with both cellulose and lignin, which together build the plant cell wall material (Puls and Schuseil, 1993; Timell, 1967). The main backbone of hemicellulose is usually made of one or two sugars, which determines their classification. For example, the main backbone of xylan is composed of 1,4-linked-β-D-xylopyranose units. Similarly, the backbone of galactoglucomannans is made of linear 1,4-linked β-D-glucopyranose and β-D-mannopyranose units with α-1,6-linked galactose residues. Other common hemicelluloses include arabinogalactan, lichenins (mixed 1,3–1,4-linked β-D-glucans), and glucomannan. Most hemicellulases are based on a 1,4-β-linkage and the main backbone is branched, whereas the individual sugars may be acetylated or methylated.

One of the predominant hemicelluloses is xylan. The structural repeating unit of xylan is β-1,4-xylobiose, like β-1,4-cellobiose is that of cellulose. In fact, the structures of xylobiose and cellobiose are very close, since xylose and glucose are homomorphic sugars, that is, the atoms that compose the pyranose rings exhibit the same configurations. The only difference is at position 5 where a hydrogen atom in xylose is replaced by CH_2OH in glucose. But the analogy stops here. Whereas the linear cellulose chains are absolutely unsubstituted, the linear xylan backbone is highly substituted with a variety of saccharide and nonsaccharide components (Fig. 3).

Hemicelluloses from hardwoods have average degrees of polymerization of 150–200 and constitute about 20 to 30% of the total weight. The xylan backbone of hardwoods contains various side chains, including α-1,3 linked arabinofuranosyl, 4-*O*-methylglucuronic acid linked to the xylose backbone units via α-1,2-glycosidic linkages, and acetyl moieties that esterify the xylose units at the *C*-2 and/or *C*-3 position. In nonacetylated softwood xylans, the substituents are 4-*O*-methyl-D-glucuronosyl and L-arabinofuranosyl residues attached to the mainchain by α-1,3 glycosidic linkages (Puls and Schuseil, 1993). In cereals, such as oat and barley, the lichenins are predominant cell wall components.

In the plant cell wall, xylan is closely associated with other wall components. The 4-*O*-methyl-α-D-glucuronic acid residues can be ester-linked to the hydroxyl

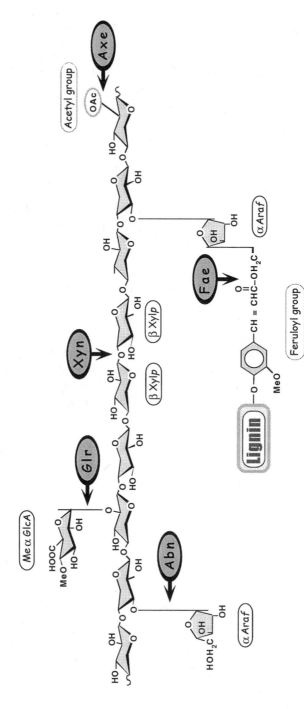

Figure 3. Composition of a typical xylan component of hemicellulose. The xylobiose unit is indicated, as are major substituents: MeαGlcA, Methylglucuronic acid; αAra*f*, arabinofuranosyl; OAc, acetyl group. A presumed lignin attachment site to a feruloyl substituent of xylan also is illustrated. Sites of cleavage by selected hemicellulases and carbohydrate esterases also are shown: Xyn, xylanase; Abn, arabinofuranosidase; Glr, glucuronidase; Axe, acetyl xylan esterase; Fae, ferulic acid esterase.

groups of lignin, providing cross-links between the cell walls and lignin (Das *et al.,* 1984). Similarly, feruloyl substituents serve as cross-linking sites to either lignin or other xylan molecules. Thus, the chemical complexity of xylan is in direct contrast to the chemical simplicity of cellulose. Likewise, the structural diversity of the xylans is in contrast to the structural integrity of the cellulose microfibril. These qualities confer structural consequences to the xylan polymer; rather than a crystallinelike substance, the hemicellulose component adopts a gel-like consistency, providing an amorphous matrix in which the rigid crystalline cellulose microfibrils are embedded.

3. ENZYMES THAT DEGRADE PLANT CELL WALL POLYSACCHARIDES

The chemical and structural complexity of plant cell wall polysaccharides is matched by the diversity and complexity of the enzymes that degrade them. The cellulases and hemicellulases are family members of the broad group of glycosyl hydrolases, which catalyze the hydrolysis of oligosaccharides and polysaccharides in general (Gilbert and Hazlewood, 1993; Kuhad *et al.,* 1997; Viikari and Teeri, 1997; Warren, 1996).

Previously, the type of substrate and manner in which a given enzyme interacted with its substrate were decisive in the classification of the glycosidases, as established first by the Enzyme Commission (EC) and later by the Nomenclature Committee of the International Union of Biochemistry. Enzymes were usually named and grouped according to the reactions they catalyzed. Thus, cellulases, xylanases, mannanases, and chitinases were grouped a priori in different categories. Moreover, enzymes that cleave polysaccharide substrates in the middle of the chain ("endo"-acting enzymes) versus those that clip at the chain ends ("exo"-acting enzymes) also were placed in different groups. For example, in the case of cellulases, the endoglucanases were grouped in EC 3.2.1.4, whereas the exoglucanases (i.e., cellobiohydrolases) were classified as EC 3.2.1.91. The recent trend, however, is to classify the different glycosyl hydrolases into groups based on common structural fold and mechanistic themes (Davies and Henrissat, 1995; Henrissat, 1991; Henrissat and Davies, 1997).

Interestingly, the distinction between endo- and exo-acting enzymes is also reflected by the architecture of the respective class of active site (Fig. 4). The endoglucanases, for example, are commonly characterized by a groove or cleft into which any part of a cellulose chain can fit. On the other hand, the exoglucanases bear tunnel-like active sites, which can only accept a substrate chain via its terminus. One can almost visualize how an endo-acting enzyme has the potential to receive, cleave, and release at various sites along the internal portion of a cellulose chain. The exo-acting enzyme would seem to thread the cellulose chain through

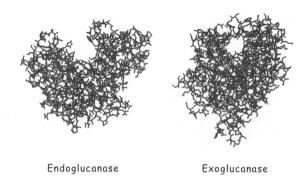

Endoglucanase   Exoglucanase

Figure 4. Structures of a typical endoglucanase and exoglucanase. In each case, the structure is viewed from a perspective that demonstrates the comparative architecture of the respective active site. The endoglucanase (endoglucanase E2 from *Thermomonospora fusca,* PDB code 1TML) is characterized by a deep cleft to accommodate the cellulose chain at any point along its length, whereas the active site of the exoglucanase (cellobiohydrolase CBHI from *Trichoderma reesei,* PDB code 1CEL) forms a tunnel, through which one of the termini of a cellulose chain can be threaded. The structures were drawn using RasMol 2.6.

the tunnel, wherein successive units (e.g., cellobiose) would be cleaved in a sequential manner. The sequential hydrolysis of a cellulose chain is a relatively new notion of growing importance, which has earned the term "processivity" (Davies and Henrissat, 1995), and processive enzymes are considered to be key components that contribute to the overall efficiency of a given cellulase system.

Though instructive, there is growing dissatisfaction with the endo/exo terminology. As our understanding of the nature of catalysis by these enzymes progresses, it has become clear that some enzymes are capable of both endo- and exo-action. Moreover, some glycosyl hydrolase families include both endo- and exo-enzymes, again indicating that the mode of cleavage can be independent of sequence homology and structural fold. In this context, relatively minor changes in the lengths of relevant loops in the general proximity of the active site may dictate the endo- or exo-mode of action without significant differences in the overall fold.

Due to subtle but diverse chemical and structural aspects of the substrates involved, plant cell-wall-degrading enzymes do not follow the same rules as common enzyme standards, such as simple proteases, DNase, RNase, and lysozyme. In fact, the cellulases and hemicellulases are usually very large enzymes, whose molecular masses often exceed those of proteases by factors of 2 to 5 and more. Their polypeptide chains partition into a series of functional modules and linker segments (frequently glycosylated), which together determine their overall activity characteristics and interaction with their substrates and/or with other components of the cellulolytic and hemicellulolytic system.

3.1. Cellulases

The cellulases include the large number of endo- and exoglucanases that hydrolyze β-1,4-glucosidic bonds within the chains that comprise the cellulose polymer (Béguin and Aubert, 1994; Haigler and Weimer, 1991; Tomme *et al.,* 1995a). Thus, in principle, the degradation of cellulose requires the cleavage of a single type of bond. Nevertheless, in practice we find that cellulolytic microorganisms produce a variety of complementary cellulases of different specificities from many different families.

It may seem somewhat surprising that the combined effect of so many different enzymes are required to degrade such a chemically simplistic substrate. This complexity reflects the difficulties an enzyme system encounters upon degrading such a highly crystalline substrate as cellulose. As described in the previous section, cellulases that degrade the cellulose chain can be either "endo-acting" or "exo-acting." Moreover, the degradation of crystalline cellulose should be viewed three-dimensionally and *in situ,* where the cellulose chains are packed within the microcrystal, thus generating the remarkably stable physical properties of the crystalline substrate. The enzymes have to bind to the cellulose surface and localize and isolate suitable chains destined for degradation. It would seem logical that amorphous regions or defects in the crystalline portions of the substrate would be favorable sites for initiation of the process. The structural as opposed to chemical heterogeneity of the substrate dictates the synergistic action of a complex set of complementary enzymes toward its complete digestion.

Various models have been suggested to account for the observed synergy between and among two or more different types of cellulases. For example, an endo-acting enzyme can produce new chain ends in the internal portion of a polysaccharide backbone, and the two newly exposed chains would then be available for action of exo-acting enzymes. In addition, two different types of exoglucanases may exhibit different specificities by acting on a cellulose chain from opposite ends (i.e., the reducing vs. the nonreducing end of the polymer). Likewise, an endoglucanase may be selective for only one of the two sterically distinct glucosidic bonds on the cellulosic surface. In addition, some cellulases may display high levels of activity at the beginning of the degradative process, that is, on the highly crystalline material, whereas others would be selective for newly exposed, partially degraded chains, otherwise embedded within the crystal. Still others would show very high levels of activity after the degradative process has advanced, and cellulose chains that have been freed of the crystalline setting then would be hydrolyzed quite rapidly. A collection of various enzymes, which exhibit complementary specificities and modes of action, would account for the observed synergistic action of the complete cellulase "system" in digesting the cellulosic substrate.

In addition to endo- and exoglucanases, the overall group of cellulases includes the β-glucosidases (EC 3.2.1.21), which hydrolyze terminal, nonreducing β-ᴅ-glucose residues from cello-oligodextrins. In particular, this type of enzyme cleaves cellobiose—the major end product of cellulase digestion—to generate two molecules of glucose. Some β-glucosidases are specific for cellobiose, whereas others show broad specificity for other β-ᴅ-glycosides, for example, xylobiose. Often, the β-glucosidases are associated with the microbial cell surface and hydrolyze cellobiose to glucose before, during, or after the transport process.

3.2. Hemicellulases

In contrast to cellulose degradation, the degradation of the hemicelluloses imposes a somewhat different challenge, since this group of polysaccharides includes widely different types of sugars or nonsugar constituents with different types of bonds. Thus, the complete degradation of hemicellulose requires the action of different types of enzymes. These enzymes, the hemicellulases, can differ in the chemical bond they cleave, or as in the case of the cellulases they may cleave a similar type of bond but with different substrate or product specificity (Biely, 1985; Coughlan and Hazlewood, 1993; Eriksson *et al.,* 1990).

Hemicellulases can be divided into two main types, those that cleave the main chain backbone, that is, xylanases or mannanases, and those that degrade side chain substituents or short end products, such as arabinofuranosidase, glucuronidase, acetyl esterases, and xylosidase. Like the cellulases, hemicellulases can be of the endo- or exo-type. A schematic view of the types of bonds that would be hydrolyzed by different types of hemicellulases is presented in Fig. 3.

Hemicellulolytic microorganisms secrete into the environment at least one type of enzyme that is capable of hydrolyzing a main chain backbone of one of the hemicelluloses. However, many species (fungi in particular) will secrete a complete arsenal of complementary enzymes that will act both on the backbone and on side chain groups, releasing mono- or disaccharides as the end products. In other cases, the only detectable extracellular hemicellulase activity will be a single enzyme that cleaves the main chain backbone, resulting in short, modified, oligosaccharide end products. Thus, fungi, such as *Aspergillus* and *Trichoderma,* secrete an extensive variety and large quantities of extracellular hemicellulases, which bring about the complete degradation of the hemicellulose polymer to soluble mono- and disaccharides. At the other extreme are *Bacillus* strains that secrete a single extracellular xylanase that partially degrades the xylan. The resulting oligosaccharide products enter the cell and are further degraded by intracellular hemicellulases. These two completely different strategies for hemicellulose degradation reflect the different natural habitats the microorganisms occupy.

3.2.1. XYLAN-DEGRADING ENZYMES

Of the hemicellulases, 1,4-β-xylanases are by far the most characterized and studied enzymes. Over 150 sequences of xylanases have been published and currently are available in protein data banks. Endoxylanases (1,4-β-D-xylan xylanhydrolase; EC 3.2.1.8) hydrolyze the 1,4-β-D-xylopyranosyl linkage of xylans, such as D-glucurono-D-xylans and L-arabino-D-xylan. These single-subunit enzymes from both fungi and bacteria exhibit a broad range of physiochemical properties, whereby two main classes have been described: alkaline proteins of low molecular weight ($<$30,000 Da) and acidic proteins of high molecular weight. This general classification scheme correlates with their assignment into glycosyl hydrolases families 10 and 11 (see Section 3.4.1), whereby the former represents the high-molecular-weight xylanases and the latter coincides with the low-molecular-weight enzymes. The two families also differ in their catalytic properties, such that the family 10 enzymes seem to display a greater versatility toward the substrate than that observed for those of family 11, and thus typically are able to hydrolyze highly substituted xylan more efficiently. The family 10 xylanases exhibit a $(\beta/\alpha)_8$ topology, whereas those from family 11 form a β-jelly roll fold. Both families show a retaining catalytic mechanism of hydrolysis (see Section 3.4.1).

The multiplicity of xylanase forms and the variety of their substrate specificities is staggering. Filtrates of *Aspergillus niger, Trichoderma reesei,* and *Talaromyces emersonii* have been found to contain over 13 different xylanases. The variable specificity of different xylanases includes those enzymes that prefer unsubstituted sequences and those that act at main chain bonds adjacent to the substituted residue.

3.2.2. MANNAN-DEGRADING ENZYMES

Glucomannans and galactoglucomannans are branched heteropolysaccharides found in hardwood and softwood. The degradation of these polymers again involve many hydrolytic enzymes, including endo-1,4-β-mannanase (EC 3.2.1.78), β-mannosidase (EC 3.2.1.25), β-glucosidase (EC 3.2.1.21), and α-galactosidase (EC 3.2.1.22). 1,4-β-D-Mannanases hydrolyze linkages of D-mannans and D-galacto-D-mannans. These enzymes, both of the endo- or exo-types, are produced in various microorganisms, including *B. subtilis, A. niger,* and intestinal and rumen bacteria and commonly occur in families 5 and 26 (Section 3.4.1).

3.2.3. LICHENIN-DEGRADING ENZYMES

Lichenase (1,3–1,4-β-D-glucan 4-glucanohydrolase; EC 3.2.1.73) is a mixed linkage β-glucanase, which cleaves the β-1,4 linkages adjacent to the β-1,3 bonds of the lichenin substrate. According to modern structure-based classification, lichenases can be members of families 8, 16 or 17 (see Section 3.4.1).

3.2.4. β-D-XYLOSIDASES

The 1,4-β-D-xylosidases (1,4-β-D-xylan xylohydrolase; EC 3.2.1.37) hydrolyze xylo-oligosacharides (mainly xylobiose) to xylose. These enzymes are found to be both intracellular or extracellular components and are closely associated with hemicellulolytic activities. Monomeric, dimeric, and tetrameric xylosidases have been found with molecular weights of 26,000 to 360,000 Da. Many of the enzymes exhibit transferase activity and can act on various substrates. For example *A. niger* produces an enzyme, classified as a β-xylosidase, that can hydrolyze β-galactosides, β-glucosides, and α-arabinosides, in addition to β-xylosides.

3.2.5. SIDE CHAIN-DEGRADING ENZYMES

α-D-Glucuronidases (EC 3.2.1.39) are the least-studied and characterized xylan debranching enzymes. These enzymes catalyze the cleavage of the α-1,2 glucosidic bond of 4-*O*-methyl-α-D-glucuronic acid side chain. This bond is known to be very stable to acid hydrolysis and the 4-methyl-α-D-glucuronic acid residue has a stabilizing effect on the neighboring xylosidic bonds of the main chain during hydrolysis with 45% formic acid. Several α-glucuronidase genes recently have been cloned and sequenced from several microorganisms, including *Thermotoga maritima, Trichoderma reesei, Aspergillus tubingensis,* and *Bacillus stearothermophilus.*

α-L-Arabinofuranosidases (α-L-arabinofuranoside arabinofuranohydrolase; EC 3.2.1.55) catalyze the hydrolysis of nonreducing terminal α-L-arabinofuranosidic linkages in arabinoxylan, L-arabinan, and other L-arabinose-containing polysaccharides. These enzymes, as with other enzymes that cleave side chain substitutes, are found either in the cell-associated or extracellular form. The genes of arabinofuranosidases are induced by L-arabinose and xylan.

1,4-β-Mannosidases hydrolyze 1,4-linked β-D-mannosyl groups from the nonreducing end. These enzymes (similar to β-xylosidases) hydrolyze mainly the end products of the mannanases, that is, mannobiose and mannotriose. α-Galactosidase removes α-galactosyl units found in branched substituents on softwood *O*-acetylgalactoglucomannans.

3.3. Carbohydrate Esterases

The side chain substituents of xylan are composed not only of sugars but also of acidic residues, such as acetic, ferulic (4-hydroxy-3-methoxycinnamic), or *p*-

coumaric (4-hydroxycinnamic) acids. Carbohydrate esterases that cleave these residues (see Fig. 3) are found in enzyme preparations from both hemicellulolytic and cellulolytic cultures (Borneman *et al.*, 1993). Such enzymes sometimes represent separate modules, separated by linker segments from other cellulolytic or hemicellulolytic catalytic modules in the same polypeptide chain.

3.4. The Modular Nature of Cellulases and Hemicellulases

Cellulases and hemicellulases are modular enzymes, wherein each module or domain comprises a consecutive portion of the polypeptide chain and forms an independently folding, structurally and functionally distinct unit (Gilkes *et al.*, 1991; Teeri *et al.*, 1992). Each enzyme contains at least one catalytic domain that catalyzes the actual hydrolysis of the glycosidic bond. Other domains assist or modify the primary hydrolytic action of the enzyme, thus modifying the overall properties of the enzyme.

3.4.1. THE CATALYTIC DOMAIN

The definitive component of a given enzyme is the catalytic domain. Former EC-based classification schemes according to substrate specificity are now considered somewhat obsolete, since they fail to take into account the structural features of the enzymes themselves. The catalytic domains of glycosyl hydrolases are presently categorized into families according to amino acid sequence homology (Henrissat, 1991; Henrissat *et al.*, 1989, 1998).

All the enzymes of a given glycosyl hydrolase family display the same topology, and the positions of the catalytic residues are conserved with respect to the common fold. In some cases, two or more divergent families can be grouped into "clans," providing that they exhibit the same fold and their catalytic residues appear in equivalent positions. In recent years, various groups have concentrated on the systematic determination of the X-ray crystal structures of individual members of a given family and to identify their catalytic residues (Bayer *et al.*, 1998a; Davies and Henrissat, 1995). This approach has provided a general overview of the structural themes of the glycosyl hydrolases and their interaction with their intriguing set of substrates (Henrissat and Davies, 1997). Our current knowledge concerning the genealogy among the plant cell wall glycosyl hydrolases is summarized in Table I.

The mechanism of catalytic hydrolysis of cellulose and hemicellulose occurs via general acid catalysis and is accompanied by either an overall retention or an inversion of the configuration of the anomeric carbon (Davies and Henrissat, 1995; McCarter and Withers, 1994; White and Rose, 1997). In both cases, cleavage of

Table I

Classification of Catalytic Domains from Cellulases and Hemicellulases[a]

Fold	Clan	Family	Former cellulase-based classification	Enzyme type (substrate specificity)	Mechanism
Structure(s) determined, clans established					
$(\beta/\alpha)_8$	GH-A	1	—	Mainly β-glucosidases and related glycosyl hydrolases	Retaining
$(\beta/\alpha)_8$	GH-A	5	Family A (5 subfamilies)	Mainly endoglucanases	Retaining
$(\beta/\alpha)_8$	GH-A	10	Family F	Mainly xylanases	Retaining
$(\beta/\alpha)_8$	GH-A	17	—	β-Glucosidases, endo-1,3-β-glucosidases, and lichenases	Retaining
$(\beta/\alpha)_8$	GH-A	26	Family I	Mainly endo-1,4-β-mannosidases	Retaining
$(\beta/\alpha)_8$	GH-A	39	—	β-Xylosidases	Retaining
$(\beta/\alpha)_8$	GH-A	51	—	Endoglucanases and arabinofuranosidases	Retaining
β-jelly roll	GH-B	7	Family C	Endoglucanases and cellobiohydrolases	Retaining
β-jelly roll	GH-B	16	—	Mainly β-glucanases (lichenases and laminarinases)	Retaining
β-jelly roll	GH-C	11	Family G	Mainly xylanases	Retaining
β-jelly roll	GH-C	12	Family H	Endoglucanases	Retaining
Structure(s) determined, clans not established					
Distorted (β/α) barrel	—	6	Family B	Endoglucanases and cellobiohydrolases	Inverting
$(\alpha/\alpha)_6$	—	8	Family D	Mainly endoglucanases	Inverting
$(\alpha/\alpha)_6$	—	9	Family E (2 subfamilies)	Mainly endoglucanases	Inverting
β barrel	—	45	Family K	Endoglucanases	Inverting
$(\alpha/\alpha)_6$	—	48	Family L	Processive endoglucanases and/or cellobiohydrolases	Inverting
Structures not determined, clan established					
—	GH-F	43	—	Various hemicellulases	Inverting
—	GH-F	62	—	α-Arabinofuranosidases	*n.d.*[b]
Structures not determined, clans not established					
—	—	3	—	Mainly β-glucosidases	Retaining
—	—	44	Family J	Endoglucanases	Inverting
—	—	52	—	β-Xylosidases	*n.d.*
—	—	55	—	Exo- and endo-1,3-glucanases	*n.d.*
—	—	61	—	Endoglucanases	*n.d.*
—	—	67	—	α-Glucuronidase	*n.d.*

[a] The preparation and organization of this table was greatly assisted by the website (http://afmb.cnrs-mrs.fr/~pedro/DB/db.html) for carbohydrate-modifying enzymes, created and managed by Bernard Henrissat and Pedro Coutinho (Henrissat and Bairoch, 1996; Henrissat and Davies, 1997). The table is an extension and modification of a previously published version (Bayer *et al.*, 1998a).
[b] *n.d.*, Not determined.

the glycosidic bond is catalyzed primarily by two active site carboxyl groups, one that acts as a proton donor and the other as a nucleophile or base. Retaining enzymes function via a double-displacement mechanism, by which a transient covalent enzyme–substrate intermediate is formed (Fig. 5A). In the latter case, the substrate undergoes ring distortion in the active site of the enzyme to attain a "twisted-boat" conformation. In contrast, inverting enzymes employ a single-step concerted mechanism as shown schematically in Fig. 5B. The major structural difference between the two mechanisms is the distance between the acid catalyst and the base. The proton donor in both mechanisms forms a hydrogen bond with the glycosidic oxygen. In retaining enzymes, the distance between the two catalytic residues is about 5.5 Å, whereas in inverting enzymes the distance is about 10 Å. The additional space in the inverting enzymes provides access of a water molecule which participates directly in the hydrolysis of the glycosidic bond, and the resultant product exhibits a stereochemistry opposite to that of the substrate. A variation of the inverting mechanism also has been proposed, whereby two acidic residues together may function as the nucleophile. In any case, the mechanism of hydrolysis is conserved within a given glycosyl hydrolase family (see Table I).

3.4.2. CELLULOSE-BINDING DOMAINS

In addition to the catalytic domain, free cellulases and hemicellulases usually contain at least one cellulose-binding domain (CBD) as an integral part of the polypeptide chain (Linder and Teeri, 1997; Tomme et al., 1995b). The CBD serves predominantly as a targeting agent to direct and attach the catalytic domain to the insoluble crystalline substrate. Like the catalytic domains, the CBDs are categorized into a series of families according to sequence homology. To date, over a dozen different CBD families have been described.

In some cases, the term CBD may be somewhat misleading, since not all of the CBDs bind to crystalline cellulose. Some families (or subfamilies or family members) bind either preferentially or additionally to other insoluble polysaccharides, for example, xylan or chitin. Others prefer less crystalline substrates (e.g., acid-swollen cellulose), single cellulose chains, and/or soluble oligosaccharides. Still others exhibit alternative accessory function(s), a topic that will be described in more detail. Moreover, the CBDs responsible for the primary binding event may further disrupt hydrogen bonding interactions between adjacent cellulose chains of the microfibril (Din et al., 1994), thereby increasing their accessibility to subsequent attack by the hydrolytic domain.

The structures of CBDs from a number of families and subfamilies have been determined (Table II), and an understanding of their structures has provided interesting information regarding the mode of binding to cellulose. Those that bind to crystalline substrates appear to do so via a similar type of mechanism. One of the surfaces of such CBDs (for example, those from families I, II, and III) is charac-

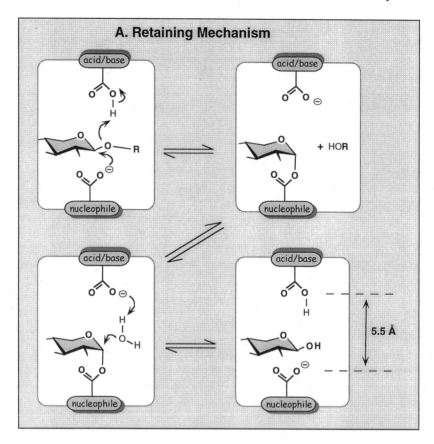

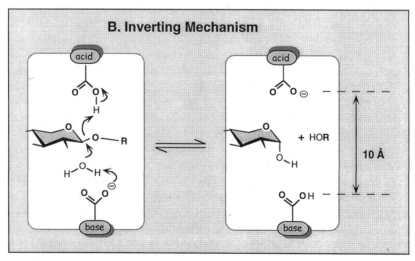

teristically flat and appears to complement the flat surface of crystalline cellulose. Protruding from this surface of the CBD molecule are a series of aromatic amino acid residues that form a planar strip (Mattinen *et al.*, 1997; Tormo *et al.*, 1996). Computer-modeling docking procedures have shown that the aromatic planar strip residues are positioned in such a way that they would stack opposite the glucose rings of a single cellulose chain. This stacking interaction has been proposed to play a major role in the recognition and binding of cellulose. In addition to the planar aromatic strip, a selection of polar amino acid residues on the same surface may be involved in anchoring the CBD to two adjacent cellulose chains. The binding of the CBD to crystalline cellulose thus appears to involve precisely oriented, contrasting hydrophobic and hydrophilic interactions between the reciprocally flat surfaces of the protein and the carbohydrate substrate. Together they provide a selective biological interaction, which contributes to the specificity that a CBD exhibits toward its structure.

In contrast, family IIIc and family IV CBDs preferentially bind to noncrystalline forms of cellulose and clearly have a different function in nature (Johnson *et al.*, 1996b; Sakon *et al.*, 1997; Tomme *et al.*, 1996). For example, the role of family IV CBD may be to recognize, bind to, and deliver an appropriate catalytic module to a cellulose chain that has been loosened or liberated from a more ordered arrangement within the cellulose microfibril. The remarkable role of the family IIIc CBD will be discussed in the following sections.

As with cellulases, xylanases also tend to exhibit a modular structure, being composed of multiple domains joined by linker sequences. In addition to the catalytic domain, these enzymes often contain a CBD and/or a xylan-binding domain (XBD). Unlike the case of various cellulases for which the CBD is usually essential for degradative activity toward insoluble crystalline cellulose, hemicellulases that contain CBDs do not necessarily depend on the CBD for the hydrolysis of xylan. It is likely that these binding domains serve to concentrate and position these soluble enzymes on the insoluble plant cell wall material by utilizing the most abundant cell wall component—cellulose. The immobilized enzyme would then act on the accessible and appropriate hemicellulose components.

Figure 5. The two major catalytic mechanisms of glycosidic bond hydrolysis.(A) The retaining mechanism involves initial protonation of the glycosidic oxygen via the acid–base catalyst with concomitant formation of a glycosyl–enzyme intermediate through the nucleophile. Hydrolysis of the intermediate is then accomplished via attack by a water molecule, resulting in a product that exhibits the same stereochemistry as that of the substrate. (B) The inverting mechanism involves the single-step protonation of the glycosidic oxygen via the acid–base catalyst and concomitant attack of a water molecule, activated by the nucleophile. The resultant product exhibits a stereochemistry opposite to that of the substrate. The type of mechanism is conserved within a given glycosyl hydrolase family and dictated by the active site architecture and atomic distance between the acid–base and nucleophilic residues (aspartic and/or glutamic acids).

Table II
Structures of different CBD families[a]

Structure	Family				
	I	II	IIIa	IIIc	IV
Snapshot[b]					
Size[c]	36	110	155	143	152
Method	NMR	NMR	X-ray crystallography	X-ray crystallography	NMR
Preferred substrate	Microcrystalline cellulose	Microcrystalline cellulose	Microcrystalline cellulose	Single cellulose chain	Amorphous cellulose, soluble oligosaccharides
Reference	Kraulis *et al.* (1989)	Xu *et al.* (1995)	Tormo *et al.* (1996)	Sakon *et al.* (1997)	Johnson *et al.* (1996a)

[a]Modified from Bayer *et al.* (1998a).
[b]Diagrams were drawn using RasMol 2.6. The proposed binding residues are shaded in the snapshot.
[c] Number of amino acid residues.

3.4.3. THE FAMILY-9 ENZYMES: AN EXAMPLE

In this section, we will discuss the subject of enzyme diversity and how a single type of catalytic module can be modified by the class of helper module(s) that flank its C- or N-terminus. The classification of enzymes into families on the basis of sequence and structure homology tells a tale, and as an example we will describe the currently unfolding story of the family 9 cellulases. It should be noted, however, that each glycosyl hydrolase family is a story unto itself. We have much to learn from the variety of members of the different families: their intrinsic similarity and diversity of the associated modules, the consequences of their interactions, and overall comparative properties of the complete multimodular enzyme. We are only at the beginning in the understanding of the complex relationship in the progression from sequence to structure, modular arrangement, activity, and overall function.

In its simplest form, an enzyme would presumably consist of a single catalytic domain, usually with a standard CBD, which would target the enzyme to the crystalline substrate. Indeed, this is the norm for many individual glycosyl hydrolase families. However, in others, for example, the family 9 cellulases, the catalytic domains commonly occur in tandem with a number of accessory modules. Although the story is still rather incomplete, we can discuss the currently available information regarding family 9 and draw several interesting conclusions from the few articles that thus far have been published on this subject.

3.4.3a. Family 9 Theme and Variations

The family 9 catalytic module displays an $(\alpha/\alpha)_6$-barrel fold and inverting catalytic machinery. However, examination of the known sequences of family 9 enzymes reveals relatively few that consist of a solitary catalytic module (Fig. 6A), with the exception of plant cellulases. Microbial family 9 cellulases commonly conform to one of the other themes shown in Fig. 6. In one of these, exemplified by T. fusca cellulase E4, an N-terminal catalytic module is followed immediately downstream by a fused family IIIc CBD (Fig. 6B). This particular CBD imparts special characteristics to the enzyme. A second theme consists of an immunoglobulinlike (Ig) domain immediately upstream to the catalytic domain (Fig. 6C). A variation of the latter theme includes a family IV CBD at the N-terminus of the enzyme, followed by an Ig domain and family 9 catalytic domain (Fig. 6D). In addition to the above-described modular arrangement, each of the free enzyme systems includes a standard CBD that binds strongly to crystalline cellulose (not shown in Fig. 6).

3.4.3b. Family 9 Crystal Structures

Two crystal structures of family 9 cellulases have been determined, representing two subtypes of this particular family of glycosyl hydrolase. These are cel-

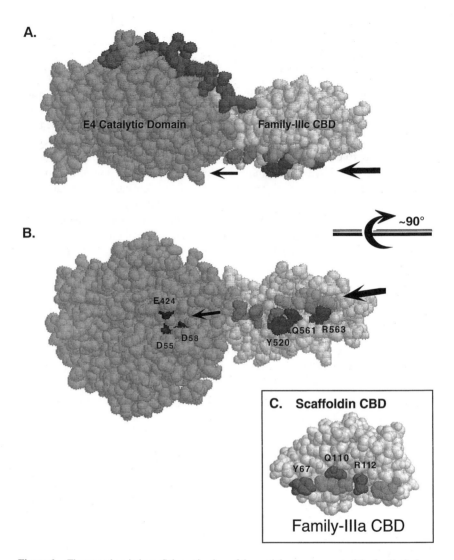

Figure 6. Theme and variations: Schematic view of the modular arrangement of the family 9 glyco-syl hydrolases. (A) The solitary catalytic domain. (B) The catalytic domain and fused family IIIc CBD. (C) Immunoglobulinlike (Ig) domain, fused to the catalytic domain. (D) Successive family IV CBD, Ig, and catalytic domains. The representations of the different modules are based on their known structures and are presented sequentially, left to right, from the *N*- to *C*-terminus. Structures (Spacefill diagrams produced by RasMol 2.6) in A and B are derived from cellulase E4 from *Thermomonospora fusca* (PDB code, 1TF4); those in C and D are from the CelD endoglucanase of *C. thermocellum* (PDB code, 1CLC). The figure used for the family IV CBD in D is derived from the NMR structure of the *N*-terminal CBD of *Cellulomonas fimi* ß-1,4-glucanase CenC (PDB code, 1ULO). The structures in B and C are authentic views of the respective crystallized bidomain protein components. The CBD in D has been placed manually to indicate its *N*-terminal position in the protein sequence, but its spatial position in the quaternary structure and the structure of the linker segment remains unknown.

lulase E4 from *Thermomonospora fusca* (Sakon *et al.,* 1997) and Cel D from *Clostridium thermocellum* (Juy *et al.,* 1992), and their comparison offers exciting implications concerning the modular nature of the cellulases in general. Fortunately, in both cases, one of the neighboring domains cocrystallized with the catalytic module, thus providing primary insight into their combined structures. Previous attempts to cocrystallize multiple domain enzymes have failed, presumably due to the presence of flexible linkers that connect the various domains. In the case of *T. fusca* E4, the catalytic domain and neighboring family IIIc CBD were found to be interconnected by a relatively long but rigid linker sequence, which envelops about half the perimeter of the catalytic domain until it connects to the adjacent CBD (Fig. 7A). The latter module appears to be fused to the catalytic domain via complementary interdomain interactions. On the other hand, in the *C. thermocel-*

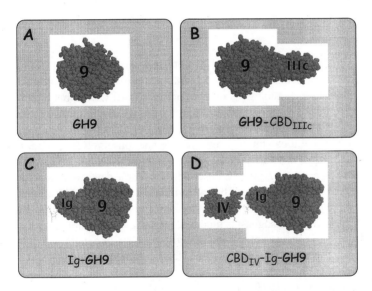

Figure 7. Structural aspects of cellulase E4 from *T. fusca.* (A) "Side view" of the E4 molecule, drawn using RasMol as in Fig. 6. The catalytic module (lightly shaded, at left), the family IIIc CBD (in white, at right), and the intermodular linker (darkly shaded, at top) are shown. The presumed path of a single cellulose chain, from the CBD to the catalytic domain, is shown at the bottom of the structure (arrows). (B) "Bottom view" of the E4 molecule (~90° rotation of A). From this perspective, the proposed catalytic residues, positioned in the active site cleft, are clearly visible. Three residues (darkly shaded) of the family IIIc CBD are homologous in sequence to those of the family IIIa scaffoldin CBD from *C. thermocellum* (the equivalent face is shown in C for comparison). The path of the cellulose chain (arrows) passes through a succession of polar residues (lightly shaded), which conceivably would bind to the incoming cellulose chain and serve to direct it toward the active site acidic residues (shaded) of the catalytic domain. Note that the path of the cellulose chain through the family IIIc CBD binding subsite *en route* to the active site of the catalytic module is not equivalent to the planar strip residues of the other two subfamilies.

lum Cel D, the catalytic domain is adjoined at its *N*-terminus by a seven-stranded Ig-like domain of unknown function. In this case, a short, four-residue linker segment separates the two domains, and their intimate association is maintained largely by an intricate set of complementary interactions among their proximate residues.

The comparison between the E4 and Cel D cellulases indicates that a given type of catalytic module can be structurally and functionally modulated by different types of accessory domains. The resultant crystal structures also taught us that, at least in some enzymes, the intermodular relationship should be viewed as more than simply a "charm bracelet," in that the role of linkers may differ from the assumed role of a flexible extender. In the case of endoglucanase E4, for example, the linker could serve as a clamp that reinforces the family IIIc CBD at its designated position in the molecule.

3.4.3c. Helper Modules

The family IIIc CBD is special. To date, this particular type of CBD has been found in nature associated exclusively with the family 9 catalytic domain. Structurally, the CBD is homologous to the other family III CBDs, but due to characteristic substitutions in many of its surface residues, this particular subtype has been reclassified as a separate subfamily (Bayer *et al.,* 1998b). Notably, many of the recognized cellulose-binding residues are not conserved (Fig. 7B,C), and the question thus arose whether the CBDs in this subfamily would indeed act in a cellulose-binding capacity *per se.* As described above, the determination of the three-dimensional crystal structure of the E4 cellulase exposed the close interrelationship between the family 9 catalytic domain and the family IIIc CBD, which allowed the authors to suggest a functional role for this module. In this context, this type of CBD seems not to bind directly to crystalline cellulose. Instead, the CBD appears to act in concert with the catalytic domain by binding transiently to the incoming cellulose chain, which is then fed into the active site cleft pending hydrolysis (Sakon *et al.,* 1997).

In order to test this hypothesis, permuted forms of the recombinant E4 enzyme were prepared, in which the true crystalline cellulose-binding CBD (from Family II) and/or the family IIIc CBD were deleted (Irwin *et al.,* 1998). The resultant constructs included the intact enzyme, a solitary catalytic domain, the catalytic domain fused to the family IIIc CBD, and the catalytic domain (without the fused CBD) linked to the cellulose-binding, family II CBD. The permutants were examined for their activities on soluble and crystalline forms of cellulose. As expected, removal of the family II CBD prevented efficient binding of the enzyme to crystalline forms of the substrate, and thus inhibited their degradation. On the other hand, the permutant, consisting of the fused catalytic and family IIIc CBD modules, was as active as the intact enzyme on soluble substrates. Removal of the fused CBD, however, posed serious consequences to the action of the enzyme on both soluble and crystalline substrates. Even in the permutant where the true CBD (family II) was present, the enzyme was capable of binding to crystalline cellulose, but in the absence of the family IIIc CBD, hydrolysis of insoluble substrates was severely impaired. The results thus indicated that the family IIIc CBD participates more directly in the catalytic function of the enzyme by promoting the processive character of its action. This general notion was also supported by the study of a similar enzyme from *C. cellulolyticum* (Gal *et al.*, 1997a).

The information derived from the family 9 enzymes suggests that the action of catalytic domains can be modified by accessory modules, which can either supplement or otherwise alter the overall properties of an enzyme (Bayer *et al.*, 1998c). The recurrent appearance in nature of a given type of module adjacent to a specific type of neighboring catalytic domain may indicate a functionally significant theme. In this regard, it is worthwhile noting the proximity of a family VI CBD with the catalytic domain of xylanases. Another recurring theme among cellulolytic enzymes is the frequent association of an N-terminal family IV CBD with an Ig-associated family 9 cellulase (Fig. 6D). Interestingly, this particular theme is characteristic of exoglucanase activity (Zverlov *et al.*, 1998b), whereas an enzyme such as *C. thermocellum* Cel D, which lacks the family IV CBD, has been characterized as a particularly active but typical *endo*glucanase (Béguin *et al.*, 1988). It is tempting to conclude from these structural comparisons that the family IV CBD, which has been demonstrated to exhibit specificity for amorphous cellulose and soluble oligosaccharides, helps to confer exo-acting properties to the family 9 catalytic domain. To prove such a hypothesis, however, rigorous experimental confirmation is required. Additional considerations, for example, the lengths of various loop regions adjacent to the active site, might be more definitive in determining whether an enzyme functions in an endo- or exo-acting mode.

In any event, these observations raise the possibility of a more selective role for certain types of CBD and other modules. Their association with certain types of catalytic domain thus could signify a "helper" role, whereby the helper module could recognize and bind a particular conformation of a cellulose chain (perhaps exposed or generated by the action of another type of CBD or cellulase), and thread the chain into the active site pocket of its neighboring catalytic domain. In doing so, the helper module would provide hydrolytic efficiency and alter the catalytic character of the enzyme.

3.4.4. MULTIFUNCTIONAL ENZYMES

Some cellulases exhibit a more complex architecture in that more than one catalytic domain and/or CBD may be included in the same protein. Examples of such enzymes are the very similar cellulases from *Anaerocellum thermophilum* (Zverlov *et al.*, 1998a) and *Caldocellum saccharolyticum* (Te'o *et al.*, 1995), both of which contain a family 9 and a family 48 catalytic domain. Likewise, family 10 and 11 xylanases may be linked in the same polypeptide chain either to each other, to catalytic domains from Families 5, 16, and 43, or to carbohydrate esterases (Flint *et al.*, 1993; Laurie *et al.*, 1997). Other paired catalytic domains include those from family 44 and either family 5 or 9. Such an arrangement might indicate a close cooperation between two particular catalytic domains, which may lead to synergistic action on the cellulosic substrate, thus portending the advent of cellulosomes on a smaller scale.

4. CELLULOSOMES

Cellulosomes are multienzyme complexes that bind to and catalyze the efficient degradation of cellulosic substrates. The first cellulosome was discovered, described, and defined by us (Bayer *et al.*, 1983; Lamed *et al.*, 1983a,b), using the anaerobic thermophilic bacterium, *Clostridium thermocellum.* In the initial publications in this field, biochemical and immunochemical evidence demonstrated that cellulosomes in this species exist both in cell-associated and in extracellular forms. Ultrastructural staining techniques demonstrated the presence of polycellulosomal protuberancelike organelles on the cell surface. Later, we and others employed a similar approach to search for putative cellulosomes in other cellulolytic organisms (Lamed *et al.*, 1987). Surprisingly, perhaps, antibodies, specific for the *C. thermocellum* cellulosome, agglutinated the cells and cross-reacted with extracellular proteins from a wide range of different species, including *Acetivibrio cellulolyticus, Bacteroides cellulosolvens,* and *Ruminococcus albus.* All the cellulolytic species examined contained protuberancelike organelles on their surfaces (Bayer *et al.*, 1994; Lamed and Bayer, 1988a).

The early biochemical studies on the properties of cellulosomes from different organisms revealed that cellulosomes contain numerous components, many of which were shown to display enzymatic activity. In addition, the cellulosomes contained a characteristic high-molecular-weight component that appeared to lack any enzyme activity. In the original publications, it was shown that this particular component from *C. thermocellum* is highly antigenic and glycosylated (Bayer *et al.*, 1985). The cellulosomal enzymatic subunits from this organism showed a broad range of different cellulolytic and xylanolytic activities (Morag *et al.*, 1991). A current view of the status of the cellulosome from *C. thermocellum* and its interaction with its substrate is shown in Fig. 8.

4.1. Cellulosomal Subunits and Their Modules

The establishment of the cellulosome concept and subsequent sequencing, expression, and biochemical analysis of its individual subunits and their subcom-

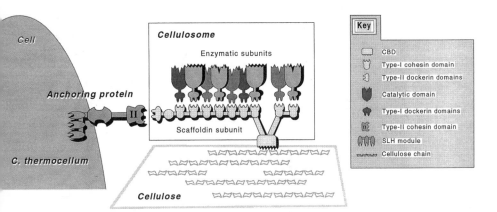

Figure 8. Simplified schematic view of the molecular disposition of the cellulosome and an associated anchoring protein on the cell surface of *C. thermocellum*. The symbols used in the figure for the modules that comprise the different proteins are defined in the key. The progression of cell to anchoring protein to cellulosome to cellulose substrate is illustrated. The SLH module links the parent-anchoring protein to the cell. The cellulosomal scaffoldin subunit performs three separate functions, individually mediated by its resident functional domains: (1) its multiple type I cohesins integrate the cellulosomal enzymes into the complex via their resident type I dockerins, (2) its family IIIa CBD binds to the cellulose surface, and (3) its type II dockerin interacts with the type II cohesin of the exocellular anchoring protein.

ponents revealed several novel types of constituents. The enzyme subunits were found to be united into a complex by means of a hitherto unique class of nonenzymatic, multimodular polypeptide subunit, later termed scaffoldin (Bayer *et al.*, 1994). The scaffoldins contain a definitive type of module, called the cohesin domain. The cellulosomal enzyme subunits, on the other hand, contain a complementary type of module, called the dockerin domain. The interaction between the cohesin and dockerin domains appears to be the definitive molecular mechanism that brings about the integration of the enzyme subunits into the cellulosome complex.

A timely breakthrough in cellulosome research was the cloning and sequencing of two different scaffoldins from two different bacterial species, that is, *C. thermocellum* and *C. cellulovorans* (Gerngross *et al.*, 1993; Shoseyov *et al.*, 1992). The cloning of these two highly related proteins enabled comparative analysis of their sequences. Both were very large, each consisting of about 1850 amino acid residues, comprising multiple types of functional modules. One of these turned out to be a family III CBD, which presumably provided the crystalline cellulose-binding function. In addition, the scaffoldins contained multiple copies of the cohesin domains. It was soon demonstrated that the latter domains interact selectively with the dockerin domains of the cellulosomal enzymes (Salamitou *et al.*, 1994b; Tokatlidis *et al.*, 1991, 1993).

Indeed, the major difference between the free and cellulosomal enzymes is that the free enzymes contain a CBD for guiding the catalytic domain to the substrate, whereas the cellulosomal enzymes lack the type of CBD that binds strongly to crystalline cellulose. Instead, the cellulosomal enzyme subunits carry a dockerin domain, which mediates the incorporation of the entire polypeptide into the cellulosome complex. Otherwise, both the free and cellulosomal enzymes contain very similar types of catalytic domains. The cellulosomal enzymes rely on the CBD of the scaffoldin subunit for binding to crystalline cellulose.

In bacterial cellulosomal subunits, the dockerin domain contains about 70 amino acids and is distinguished by a 22-residue duplicated sequence (Chauvaux *et al.*, 1990), which bears similarity to the well-characterized EF-hand motif of various calcium-binding proteins (e.g., calmodulin and troponin C). A 12-residue segment that is highly conserved with the calcium-binding loop of the EF-hand motif is within this repeated sequence. In fact, all the important calcium-binding residues of the dockerin domain are strictly conserved, which would indicate that the calcium-binding function is an important characteristic of the dockerin domain. This assumption eventually proved to be true.

4.2. Three-Dimensional Structures of Cellulosome Modules

Recombinant forms of several catalytic domains from cellulosomal enzymes furnished appealing subjects for three-dimensional crystallography studies, from which the overall fold and catalytic machinery of representative glycosyl hydrolase families could be determined (Juy *et al.*, 1992; Parsiegla *et al.*, 1998; Souchon *et al.*, 1996). These studies confirmed earlier reports (Kraulis *et al.*, 1989; Rouvinen *et al.*, 1990) that the presence of accessory domains, such as CBDs and dockerins, interfere with the crystallogenesis of cellulases, presumably due to the presence of flexible linker residues that separate the various resident domains of the intact molecule. Hence, the recombinant forms of cellulases and hemicellulases used in crystallography studies usually comprise solitary, isolated modules, unless accessory domains are strongly associated with the catalytic domain, as described in Section 3.4.3b.

More recently, attempts also have been made to determine the structures of individual noncatalytic modules, which define the binding of the cellulosome to its substrate and the integration of its subunits into the complex. Thus far, crystal structures for a family IIIa CBD (Tormo *et al.*, 1996) and two cohesins (Shimon *et al.*, 1997; Tavares *et al.*, 1997) from the scaffoldin of *C. thermocellum* have been successfully determined. Although their sequences and molecular properties are very different, both the CBD and cohesin domain show a surprising similarity in their overall fold. The observed "jelly roll fold" thus may provide a suitable macromolecular framework that would promote efficient protein–protein interactions for both heterologous (e.g., intersubunit, ligand-binding, substrate-binding) and homologous (intrasubunit) interactions among the various cellulosomal domains. Other noncatalytic cellulosomal modules from both the scaffoldin and enzyme subunits may also exhibit a similar type of all-β fold.

A structural model for the dockerins is still lacking. However, sequence homology between the repeated elements of the dockerin domains and the EF-hand motif of various calcium-binding proteins (e.g., calmodulin and troponin C) suggests a similar type of structure, particularly with respect to the calcium-binding loop (Chauvaux *et al.*, 1990). Correlation analysis among the dockerins of distinct specificities has allowed the identification of putative recognition determinants in the dockerin sequence (Pagès *et al.*, 1997). The application of such bioinformatics-based procedures offers great promise both for the prediction of functionally important residues and for the rational selection of mutations in site-directed mutagenesis studies. Such studies are currently being performed for the cellulosomal CBD, cohesins, and dockerins from *C. thermocellum* (Bayer *et al.*, 1998a,b).

4.3. The Cell-Bound Cellulosome of *C. thermocellum*

The scaffoldin of *C. thermocellum* also contains a special type of dockerin domain that is lacking in the scaffoldin of *C. cellulovorans*. This dockerin failed to bind to the cohesins from the same scaffoldin subunit, but instead interacted with a different type of cohesin—termed type II cohesins—identified on the basis of sequence homology (Salamitou *et al.*, 1994b). The type II cohesins were found to be component parts of a group of cell surface "anchoring" proteins on *C. thermocellum* (Leibovitz and Béguin, 1996; Leibovitz *et al.*, 1997; Lemaire *et al.*, 1995; Salamitou *et al.*, 1994a). These anchoring proteins also contained an S-layer homology (SLH) module, previously shown to be associated with the cell surface of gram-positive bacteria (Lupas *et al.*, 1994). The conclusion thus was formulated that the SLH module interacts in some way with the cell surface, the type II cohesins selectively bind the type II dockerins, and the cellulosome (i.e., the scaffoldin subunit together with its adornment of enzyme subunits) is thereby incorporated into the cell surface of *C. thermocellum*.

The scaffoldins from the other clostridial species thus far described all lack type II *dockerin* domains, the inference being that cells of *C. cellulovorans, C. cellulolyticum,* or *C. josui,* for example, apparently would not bear anchoring proteins that contain type II cohesins. It thus follows that either their cellulosomes are not surface bound or, if indeed they are surface components, then their anchoring thereto is accomplished via an alternative molecular mechanism. Until recently, *C. thermocellum* has been the only cellular system for which the presence of type II cohesins, type II dockerins, cell surface anchoring proteins, and the mechanism of cellulosome anchoring to the surface has been described. The other established cellulolytic strains appear to lack such an anchoring apparatus.

The current status of the arrangement of the cellulosome-related proteins and their resident domains in *C. thermocellum* is shown schematically in Fig. 8.

5. ECOLOGY, PHYSIOLOGY, AND EVOLUTION OF CELLULOSOMES

Why cellulosomes? Indeed, in most aerobic systems of both bacteria and fungi, large quantities of cellulases are produced in the free, soluble, extracellular form. The collection of different cellulases apparently act in competition with each other both in attacking the solid cellulosic substrate and for subsequent hydrolysis of degradation intermediates and soluble oligosaccharides. In contrast, cellulosomes usually exist in cell surface form with the enzymes and other functional components together in the same complex.

What advantage does an exocellular cellulosome have for the organism? In *C. thermocellum,* the cellulosomes are initially positioned at the cell–substrate interface. The production of cellulases is conserved, since their random secretion and diffusion into the medium is prevented. Their action is more controlled and efficient, as would be expected of an anaerobic system.

The crystalline cellulose-binding function of the scaffoldin CBD from *C. thermocellum* serves a multiplicity of roles (Bayer *et al.,* 1996). Since the cellulosome is cell-bound, its scaffoldin CBD not only delivers its set of enzymes to the cellulose surface, but it drags along the entire cell as well. The adhesion of the cell to its substrate thus is mediated by the cellulosomal CBD.

Why are the enzymes together in the same complex? It is still not entirely clear whether the major organizational role of the cellulosome is to simply deliver the enzymes to the cellulosic substrate and to bring into proximity the complementary enzymes, which would then work synergistically toward its efficient decomposition (Lamed *et al.,* 1983b). The organization of the enzymes into a defined complex might also enable the protection of product intermediates and facilitate their transfer to other cellulase components for continued hydrolysis. Another potential advantage in such a system is a more effective process of feedback regulation, whereby the microbe would be subject to a refined balance between product levels, enzyme activity, or even cell adhesion and detachment. For example, accumulation of the end product, cellobiose, would inhibit further solubilization of the crystalline substrate until soluble degradation products are adsorbed by the cells. As a consequence, the cells maintain a foothold on the solid substratum and subsequent competitive advantage over other microbes that inhabit the same ecosystem.

5.1. Modulation and Dynamics of Cellulosomal Components

As mentioned in Section 4, there seems to be a link between cell surface protuberances in various cellulolytic organisms and the presence of exocellular cellulosomes (Lamed *et al.,* 1987). However, the nature of this correlation on the molecular level still is not clear. Specifically, it remains to be determined whether cellulosomes are invariably associated with the cell surface or whether in some species the cellulosomes customarily occur as extracellular components.

In *C. thermocellum,* the cellulosomes of exponentially growing cells are tenaciously attached to the surface protuberances via the anchoring proteins. However, when the cells enter the stationary phase of growth, the cellulosomes tend to detach from the cell surface (Bayer *et al.,* 1985; Lamed and Bayer, 1988a). Detachment could occur anywhere along the progression from the cell surface through the anchoring protein, scaffoldin, CBD, and cellulose (see Fig. 8). The actual detachment could potentially could take place at numerous points, including the connection between the SLH domain and the cell surface, the bond between the type-II cohesin and dockerin domains, or at the CBD-cellulose interface. In addition to possible disruption of these biological interactions, cleavage of sensitive peptide bonds within the scaffoldin or anchoring protein also may be an alternative natural mechanism for detachment from the cellulose surface (Lamed *et al.,* in press). In any case, ultrastructural evidence indicates that in the later stages of growth, the cells are detached from the remaining substrate, the surface of which is covered with cell-free cellulosomes.

The production of cellulosomes is constitutive to *C. thermocellum,* but the nature of the substrate seems to dictate the composition of the cellulosomal components (Bayer *et al.,* 1985). In this species, growth is limited to cellulose and its degradation products. Although the cellulosome from *C. thermocellum* hydrolyzes hemicellulases very efficiently, it essentially fails to utilize any of its sugar constituents (Morag *et al.,* 1990). The capacity to digest hemicelluloses is considered to be important for the bacterium to expose the cellulose component of plant matter (Bayer and Lamed, 1992; Bayer *et al.,* 1994). In doing so, it also provides various satellite microorganisms with their preferred substrates.

Growth of *C. thermocellum* on cellobiose tends to alter the content of selected cellulosomal subunits, such that the relative amount of some is reduced, whereas that of others is enhanced (Bayer *et al.,* 1985). The interplay of such cellulosomal constituents is thought to bear physiological consequences. For example, the relative amount of the Cel S subunit, that is, the major cellobiohydrolase in this organism, is markedly reduced when *C. thermocellum* cells are grown on cellobiose. This would suggest that under conditions of high concentration of cellobiose, this particular "processive" enzyme, which is critical to the hydrolysis of crystalline substrates, is not required.

5.2. Interspecies Cellulosomes

Issues pertaining to the ecology, physiology, and regulation of cellulosome production by cellulolytic microorganisms are still quite vague. Within a given microbe, the molecular composition of the cellulase system in general and the cellulosome in particular seems to be a function of several factors. These include evolutionary factors and the range of genes inherited by the microbe, its consequent role in the particular ecosystem, its relationship to other members of its ecosystem, and the nature and range of its substrate(s) at hand in the environment.

Clearly it is too early to provide clear-cut evidence regarding the evolution of cellulosomes. In many species, the presence of cellulosomes only recently has been verified. Nevertheless, the plethora of sequences available in the databases and the new tools for their analysis enable us to speculate as to their possible interspecies relationship.

Recently, numerous cellulosome-related "signature" sequences have been described in several different cellulolytic microorganisms (Table III). The presence of sequences consistent with dockerins and cohesins are currently considered to be indicative of cellulosomes, and these discoveries support the original biochemical evidence (Lamed *et al.*, 1987) that led to the notion that cellulosomes are widely distributed among cellulolytic microorganisms. Most of the new publications have reported dockerin-containing enzymes, although a few new scaffoldins (i.e., containing type I cohesins) also have been described. The list of microorganisms in Table III reveals that cellulosomes are not limited to anaerobic clostridia, but include anaerobic fungi and even an aerobic bacterium.

New data (Bayer *et al.*, 1999; Ding *et al.*, 1998a,b) concerning type II cohesins and/or dockerins in two new strains (i.e., *Acetivibrio cellulolyticus* and *Bacteroides cellulosolvens*) indicate that the presence of anchoring proteins is not limited to *C. thermocellum,* and such discoveries should provide new insight into cellulosome diversity in nature.

5.2.1. INTERSPECIES SCAFFOLDINS

The sequences of four complete cellulosomal scaffoldin genes thus far have been published (Gerngross *et al.*, 1993; Kakiuchi *et al.*, 1998; Pagès *et al.*, 1999; Shoseyov and Doi, 1990). Their polypeptide portions range in molecular size from about 120,000 Da in *C. josui* to 197,000 in *C. thermocellum.* The latter scaffoldin is heavily glycosylated (Gerwig *et al.*, 1991), with oligosaccharides *O*-linked to threonines on linker segments (Gerwig *et al.*, 1993). A very similar type of oligosaccharide, *O*-linked to threonines or serines, has been detected in cellulosomelike entities from *B. cellulosolvens* (Gerwig *et al.*, 1992). On the other hand, the presence of oligosaccharides in the other known scaffoldins has yet to be es-

Table III
Evidence for Cellulosomes in Cellulolytic Microorganisms[a]

Organism	Cellulosome signature sequence(s)		References
	Protein	Domain[b]	
Anaerobic bacteria			
Clostridium	Scaffoldin	Coh-I + CBD +	Béguin and Lemaire (1996);
thermocellum	Surface-anchoring	Doc-II	Gerngross *et al.* (1993);
	proteins	Coh-II	Lamed *et al.* (1983b)
	Enzymes	Doc-I	
Clostridium	Scaffoldin	Coh-I + CBD	Doi *et al.* (1998); Shoseyov
cellulovorans	Enzymes	Doc-I	*et al.* (1992)
Clostridium	Scaffoldin	Coh-I + CBD	Belaich *et al.* (1997)
cellulolyticum	Enzymes	Doc-I	
Clostridium	Scaffoldin	Coh-I + CBD	Karita *et al.* (1997)
josui	Enzymes	Doc-I	
Clostridium	Scaffoldin	Coh-I	S. Leschine, personal
papyrosolvens			communication
Bacteroides	Scaffoldin *or*	Coh-II + CBD	Ding *et al.* (1998b)
cellulosolvens	Surface-anchoring protein		
Acetivibrio	Scaffoldin *and*	Coh-I + CBD +	Ding *et al.* (1998a)
cellulolyticus	Surface-anchoring	Doc-II	
	protein	Coh-II	
Ruminococcus	Enzymes	Doc-I	Kirby *et al.* (1997)
flavefaciens			
Ruminococcus	Enzymes	Doc-I	Karita *et al.* (1997); Ohmiya
albus			*et al.* (1997); Tamaru *et al.* (1997)
Aerobic bacteria			
Vibrio sp.	Enzyme	Fungal-type dockerin	Tamaru *et al.* (1997)
Anaerobic fungi			
Neocallimastix	Enzymes	Fungal dockerins	Fanutti *et al.* (1995)
patriciarum			
Piromyces	Enzymes	Fungal dockerins	Fanutti *et al.* (1995)
Orpinomyces	Enzymes	Fungal dockerins	Li *et al.* (1997)

[a]Modified from Bayer *et al.* (1998a).
[b]Coh-I, Coh-II, Doc-I, Doc-II—type-I and -II cohesins, type-I and -II dockerins, respectively.

tablished. A glance at their sequences indicate short linker regions, compared to those of *C. thermocellum,* with reduced numbers of potential oligosaccharide-bearing hydroxyamino acids.

More recently a novel scaffoldin gene has been identified and sequenced from *Acetivibrio cellulolyticus* (Bayer *et al.,* 1999; Ding *et al.,* 1998a,b). Like the oth-

ClassIScaffoldins

ClassIIScaffoldins

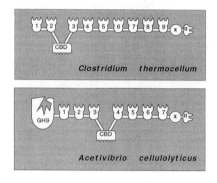

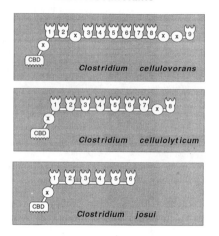

Figure 9. Classification of scaffoldins from different cellulosome species. Class I scaffoldins feature an internal CBD and a *C*-terminal type II dockerin domain. Class II scaffoldins exhibit an *N*-terminal CBD and lack a dockerin domain. Both classes of scaffoldin contain multiple copies of cohesin domains.

er scaffoldins, the corresponding protein contains repeated cohesin units and a family III CBD. Similar to the *C. thermocellum* scaffoldin but unlike the others, the *A. cellulolyticus* scaffoldin contains a *C*-terminal type II dockerin domain. In contrast to all of the known scaffoldins that are clearly noncatalytic in nature, the newly sequenced protein contains a family 9 catalytic module at its *N*-terminus, implying that this particular scaffoldin can function as an enzyme. The flanking regions of the catalytic module are void of any helper modules. The presence of this catalytic module as an integral part of the *A. cellulolyticus* scaffoldin implies that this particular enzyme component must be critical to its degradative function.

On the basis of the limited number of sequences known to date, the scaffoldins thus far can be cataloged in two classes: class I scaffoldins, which contain an internal CBD and a *C*-terminal type II dockerin, and class II scaffoldins, which contain an *N*-terminal CBD but lack a type II dockerin (Fig. 9). It is unclear how the family 9 catalytic module of *A. cellulolyticus* fits into this scheme, that is, whether it is the first example of a common scaffoldin theme or whether it is a unique occurrence in nature. Systematic sequencing of new scaffoldins from other cellulolytic bacteria and fungi undoubtedly will shed light on this aspect of cellulosome structure.

5.2.2. COMPARISON OF INTERSPECIES CELLULOSOMAL COMPONENTS

A decade ago, the range of cellulases and hemicellulases within a given species was assessed mainly by biochemical techniques. In some cases, individual enzymes were isolated and their properties assessed using desired insoluble or soluble substrates. Another approach involved electrophoretic separation of cell-derived or cell-free extracts, and analysis of desired activities using zymograms. There are advantages and disadvantages with each strategy, and the employment of combined complementary approaches is always advisable. More recently, molecular biology techniques have been used to reveal cellulase and hemicellulase genes, which often can be characterized on the basis of sequence homology with related, known genes (Béguin, 1990; Hazlewood and Gilbert, 1993). If further information is required on the structure or action of a given enzyme, the gene then can be expressed in an appropriate host organism and the properties of the product can be characterized.

5.2.2a. Cellulosome Genomics

To date, the bulk of our knowledge concerning cellulosome components has been derived from three different clostridial species. The first cellulosome was described for the anaerobic, thermophilic bacterium, *C. thermocellum,* and in many respects the cellulosome of this bacterium represents a reference for all other works. The other two well-documented cellulosomes are from the mesophilic strains, *C. cellulolyticum* and *C. cellulovorans.*

Throughout the years, 18 complete cellulosomal cellulase and hemicellulase genes of *C. thermocellum* have been sequenced (for a list, see recent review, Bayer *et al.,* 1998c), representing 10 different glycosyl hydrolases families (Fig. 10). These genes are generally scattered over a large portion of the chromosome (Guglielmi and Béguin, 1998). A few small clusters of cellulosomal genes are apparent in the genome, including a scaffoldin-containing cluster that also contains several cell surface anchoring proteins (Fujino *et al.,* 1993). Several noncellulosomal enzymes (not shown in the figure) also have been described from this organism (Morag *et al.,* 1990), such that the cellulase system of *C. thermocellum* displays an exceptional wealth, diversity, and intricacy of enzymatic components, and thus represents the premier cellulose-degrading organism currently known.

The anaerobic mesophile, *C. cellulolyticum,* is a second well-characterized cellulosome-producing bacterium (Belaich *et al.,* 1997). Nine complete cellulase subunits representing four different glycosyl hydrolases families thus far have been described for this strain (Fig. 11). In contrast to *C. thermocellum,* most of the known cellulosomal genes of *C. cellulolyticum* occur mainly in one large cluster (Bagnara-Tardif *et al.,* 1992; Belaich, 1998). A tenth enzyme also is located in the

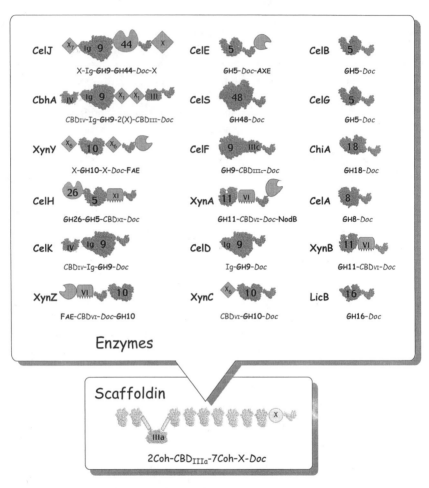

Figure 10. Schematic representation of cellulosomal subunits of *C. thermocellum*. For modules from families with known three-dimensional structures, the RasMol-derived figures of selected family members are used as symbols. The dockerin structure is modeled after a known structure of the EF-hand motif. Numbers denote the different families of modules, as classified according to the CAZyModO Web site (Soutinho and Henrissat, 1999). Abbreviated inscriptions of component modules of the designated subunits appear below the symbolic forms: GH5, family 5, glycosyl hydrolase (GH8, family 8, glycosyl hydrolase, etc.); Coh, cohesin domain; *Doc,* dockerin domain; CBD$_{IV}$, cellulose-binding domain (family IV); FAE, ferulic acid esterase; AXE, acetyl xylan esterase; NodB, enzyme activity similar to AXE, but unrelated in sequence; Ig, immunoglobulinlike domain; X, other modules or linking segments of unknown function.

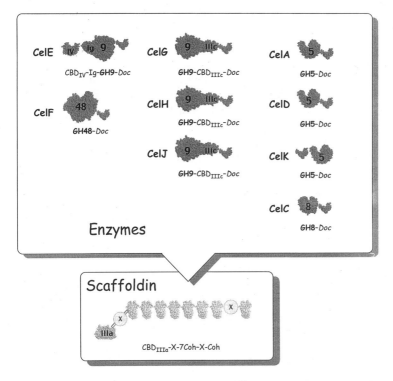

Figure 11. Schematic representation of currently known cellulosomal subunits of *C. cellulolyticum.* The information used for drawing this figure was compiled according to the work of Belaich and colleagues (Belaich *et al.,* 1997; Pagès *et al.,* 1999; Belaich, 1998). Another GH9-CBD$_{IIIc}$ enzyme, CelM, has been partially sequenced and also may be a cellulosomal subunit (J.-P. Belaich and A. Belaich, personal communication). For explanation of symbols and abbreviations, see legend to Fig. 10.

cluster, but its sequence has not been completed and it is not confirmed whether it is cellulosomal or not. This fortunate chromosomal arrangement enabled the investigators to sequence the various genes in a sequential manner and to discover the genes one at a time. The initial gene of the cluster is the scaffoldin gene, followed by cellulase genes from glycosyl hydrolase families 48, 8, and 9. The same sequential pattern of cellulosomal genes in the genome cluster also appears to occur in the two other established cellulosomal systems, that is, from *C. cellulovorans* and *C. josui* (Doi *et al.,* 1998; Kakiuchi *et al.,* 1998; Liu and Doi, 1998). Current information concerning the genome of other cellulosome-producing microbes is still too sketchy to evaluate.

5.2.2b. Cellulosomal Components from *C. thermocellum*

The cellulosomal enzymes from *C. thermocellum* are relatively large proteins, ranging in molecular size from about 40 to 180 kDa (Bayer *et al.*, 1998c; Béguin and Lemaire, 1996; Felix and Ljungdahl, 1993; Lamed and Bayer, 1988b). Examination of Fig. 10 reveals why these enzymes are so big; most of the larger ones contain multiple types of catalytic domains as well as other functional modules as an integral part of a single polypeptide chain [see Table I in Bayer *et al.* (1998c) for list of relevant references].

Many of the *C. thermocellum* cellulosomal enzymes are cellulases, which include both endo- and exo-acting β-glucanases. Some of the important exoglucanases and processive cellulases include Cel S, Cbh A, Cel K, and Cel F. The Cel S subunit is a member of the family 48 glycosyl hydrolases, and this particular family is now recognized as a critical component of bacterial cellulosomes (Morag *et al.*, 1991, 1993; Wang *et al.*, 1993; Wu *et al.*, 1988). The other three processive cellulases are members of the family 9 glycosyl hydrolases. Cel F is an E4-type enzyme [see Fig. 6B, Section 3.4.3 (Navarro *et al.*, 1991)]. The other two are remarkably similar enzymes, which exhibit nearly 95% similarity along their common regions (Kataeva *et al.*, 1998; Zverlov *et al.*, 1998b). The main difference between Cbh A and Cel K is the presence in the former of three extra modules (a family III CBD and two modules of unknown function). The functional significance of these supplementary modules to the activity of Cbh A has not been determined.

The fact that the cellulosome from this organism contains many different types of cellulases is to be expected, of course, if we consider that growth of *C. thermocellum* is restricted to cellulose and its breakdown products, particularly cellobiose. Consequently, it is surprising to discover, in addition to the cellulases, at least five classic xylanases, that is, those belonging to glycosyl hydrolase families 10 and 11. In addition, two of the larger enzymes, Cel H and Cel J, contain hemicellulase components, that is, family 26 and 44 catalytic modules (a mannanase and a xylanase, respectively), together with a standard cellulase module in the same polypeptide chain. It also is interesting to note the presence of carbohydrate esterases together with xylanase or cellulase modules in some of the enzyme subunits, thus conferring the capacity to hydrolyze acetyl or feruloyl groups from hemicellulose substrates. Finally, the *C. thermocellum* cellulosome includes a typical family 16 lichenase and a family 18 chitinase.

Why does this bacterium, which subsists exclusively on cellulosic substrates, need all these hemicellulases? The inclusion of such an impressive array of noncellulolytic enzymes in a strict cellulose-utilizing species would suggest that their major purpose would be to collectively purge the unwanted polysaccharides from the milieu and to expose the preferred substrate: cellulose. The ferulic acid esterases, in concert with the xylanase components of the parent enzymes, could grant the bacterium a relatively simple mechanism by which it could detach the lignin component from the cellulose–hemicellulose composite. The lichenase and chitinase are also intriguing components of the cellulosome. The former would provide the bacterium with added action on cell wall β-glucan components from certain types of plant matter. It is not clear whether the presence of the latter cellulosomal enzyme would reflect chitin-derived substrates from the exoskeletons of insects and/or from fungal cell walls. Whatever the source, the chitin breakdown products, like those of the hemicelluloses, presumably would not be utilized by the bacterium itself but would be passed on to appropriate satellite bacteria for subsequent assimilation.

5.2.2c. Cellulosomal Components from *C. cellulolyticum*

Compared to the set of cellulosomal enzymes in *C. thermocellum,* those in *C. cellulolyticum* appear to be somewhat simplistic. None of the known cellulosomal enzymes yet described for this species contains more than one catalytic module.

All the sequenced enzymes are relatively common cellulases. The largest one, Cel E (estimated at 94 kDa), is equivalent in structure to the family 9 Cel K of *C. thermocellum.* The critical family 48 cellulase is also a major cellulosome component (Reverbel-Leroy *et al.,* 1997). In addition, the gene cluster contains three copies of other family 9 cellulases, all of which contain the thematic fused family IIIc CBD (Belaich, 1998). Another such family 9 enzyme, Cel M, also is included in the cluster (J.-P. Belaich and A. Belaich, personal communication); it has been partially sequenced and also may be a cellulosomal subunit. It is intriguing why this bacterium would produce multiple copies of this particular type of enzyme, and careful biochemical study of all four enzymes should provide evidence to help resolve such issues. The current status of the cellulosome system in this bacterium is rounded off by three family 5 cellulases and a family 8 cellulase.

Biochemical characterization of the *C. cellulolyticum* cellulosome demonstrated a 160-kDa scaffoldin band and up to 16 smaller bands, representing putative enzyme subunits (Gal *et al.,* 1997b). Many of these were clearly identified as known gene products.

The persistent work of these authors on the cellulosomal gene system from this organism provides us with primary information on the diversity of cellulosome composition. It should be noted that the *C. cellulolyticum* system has been recognized only relatively recently. Our knowledge concerning its complement of enzymes is generally a function of the identification and systematic cloning and sequencing of consecutive portions of the gene cluster (Belaich *et al.,* 1997). Only two cellulase genes are currently known outside of the cluster. Further work on the enzyme system of this species may still surprise us with more complicated multimodular enzymes and/or other types of enzymes, such as hemicellulases.

5.2.2d. Cellulosomal Components from *C. cellulovorans*

The scaffoldin subunit of this bacterium holds the distinction of being the first to have been sequenced (Shoseyov *et al.*, 1992). Its sequence was determined even before one of its major functions, that is, the integration of the cellulosomal enzymes into the complex, was recognized. Hence, it was coined Cbp A (cellulose-binding protein A). The full function of the scaffoldin subunit was recognized only later, when the activity of the homologous repeating domains from the scaffoldin of *C. thermocellum* was realized (Gerngross *et al.*, 1993; Tokatlidis *et al.*, 1991). Consequently, subsequent scaffoldins were dubbed "cellulosome-integrating proteins" (Cip).

The characterization of the cellulosomal enzymes from *C. cellulovorans* is much more recent, and their organization on the genome also can be considered an emerging story (Doi *et al.*, 1998; Tamaru *et al.*, 1999). Like *C. cellulolyticum* and *C. josui,* the gene for the scaffoldin subunit is followed downstream by a dockerin-containing family 48 enzyme (Liu and Doi, 1998). The two genes apparently initiate a large cluster similar to that described for *C. cellulolyticum,* although the exact order of the genes in the cluster is as yet unclear. Another major cellulosomal enzyme from family 5 is distant from the scaffoldin gene. Another small gene cluster contains two other putative cellulosomal enzymes: a family 9 endoglucanase and surprisingly a dockerin-containing pectate lyase. One more cluster thus far has been detected, which contains at least four genes, including three endoglucanases and a mannanase. On the basis of these findings, the authors claim that the cellulosome of *C. cellulovorans* contains enzymes that can degrade a variety of plant cell wall polysaccharides, including cellulose, xylan, pectin, and mannan (Tamaru *et al.*, 1999). In addition, at least three noncellulosomal endoglucanases also have been partially or totally sequenced.

5.2.2e. Ecological Significance of Cellulosome Composition

C. thermocellum is known to be a dominant polymer-degrading organism, which universally occupies anaerobic thermophilic environs (Lamed and Bayer, 1991). Very few microorganisms are known to inhabit this niche. Nonetheless, it should be emphasized that in nature the anaerobic thermophilic cellulosic ecosystem is much more prevalent than one would initially consider; hydrolysis of polysaccharide substrates and concomitant microbial growth are both exothermic processes, accompanied by utilization and/or dissipation of oxygen. In any event, the abundance and diversity of the enzyme components of *C. thermocellum,* particularly of its cellulosomal enzymes, would reflect the central and flexible role this bacterium plays in its distinctive ecosystem.

It would appear that the complement of enzymes borne by this organism may serve the other satellite microorganisms in its ecosystem, perhaps as much as it contributes to the exposure and digestion of its own preferred substrate. Degradation of the cell wall polysaccharides by *C. thermocellum* produces vast quantities of soluble sugars, most of which (with the exception of cellobiose) are inconsistent with its substrate-utilizing pattern. By default, the hemicellulolytic and/or saccharolytic strains presumably would receive the pertinent degradation products from the hemicellulose component of plant matter. And the cellobiose? The major end product of cellulose degradation is available for assimilation by *both* the polymer-degrading and saccharolytic strains. However, competition for this soluble sugar is usually not a problem, since there is so much cellobiose formed that it usually causes rapid inhibition of key cellulosomal enzymes, for example, the critical family 48 cellulase (Morag *et al.*, 1991). The action of the saccharolytic strains relieves this inhibitory effect, thus permitting continued degradation of cellulose by the cellulolytic strain.

Although molecular evidence on the cellulosome from the mesophilic bacterium *C. cellulolyticum* is still incomplete, it seems that this strain may be much more restricted than *C. thermocellum* regarding the types of polysaccharides that can be hydrolyzed by its cellulosome. A recent report (Gal *et al.*, 1997b), however, demonstrated cellulosome-associated xylanase activity. In the same work, zymograms showed only endoglucanase activity, whereas no activity could be detected using xylan, crystalline, or amorphous cellulose as target substrates. Thus far, only standard cellulases have been observed in the *C. cellulolyticum* cellulosome; no xylanase or other hemicellulase has yet been sequenced, although a β-xylosidase has been detected (Saxena *et al.*, 1995). Indeed, growth of this bacterium appears to be relatively selective for cellulose and its degradation products, although the growth on other plant cell wall polysaccharides has not been rigorously documented (Giallo *et al.*, 1985; Petitdemange *et al.*, 1984).

In the case of the mesophilic *C. cellulovorans,* the developing picture infers an intricate enzyme system, which may well rival the diversity of the thermophilic *C. thermocellum* cellulosome. The presence of both hemicellulases and cellulases reflects the original evidence (Sleat *et al.*, 1984) that *C. cellulovorans* is capable of assimilating a wide variety of plant cell wall polysaccharides, including cellulose, xylans, pectins, and mannans. The system of this bacterium seems to be designed for the immediate needs of the bacterium for growth and survival in a complex environment (Tamaru *et al.*, 1999).

On the other hand, we have seen above that *C. thermocellum* is restricted to cellulosic substrates for growth. Nevertheless, it does produce a varied complement of cellulosomal and noncellulosomal hemicellulases as well as cellulases. This would argue that simple comparison between the enzyme profile and substrate utilization pattern of a given organism is not necessarily applicable, and perhaps we should interpret such findings within the greater framework of the role a given bacterium plays in its ecosystem. In this context, the cellulosome of *C. ther-*

mocellum may serve as a general polymer-degrading apparatus by providing the satellite microorganisms in the anaerobic thermophilic ecosystem with large amounts of soluble sugars and other cellular end products, including those that the bacterium itself cannot utilize. Hence, such a formidable environment may have led to a concentration of the relevant genes for degradative enzymes in a particularly convenient microbe that is particularly suitable for survival in its ecosystem. In contrast, the cellulosome of *C. cellulolyticum* may play a narrower role in the anaerobic, mesophilic environment, by concentrating mainly on the cellulose component. In this regard, numerous types of mesophilic bacteria, both cellulolytic and hemicellulolytic, are known to inhabit such an ecosystem, and the combined action of their free and cellulosomal enzyme systems would promote efficient degradation of plant cell wall polysaccharides.

The above dissertation is based on the comparative molecular aspects of cellulosome structure and composition from the currently known cellulosomal systems. These are "free-living" strains that occupy physiologically different (mesophilic vs. thermophilic) niches. It is hoped that we will eventually be able to extend such observations to structurally more complex and regulated ecosystems, such as the rumen (Flint, 1994, 1997), in which the cellulase and hemicellulase systems from many different types of microbes combine to digest plant matter in a highly ordered, coordinated, and controlled process. Continued sequencing of new cellulosome-related components, including enzymatic subunits as well as new scaffoldins and genes for anchoring proteins, promise to provide novel information and new surprises. We anticipate the discovery of extensive cellulosomal systems for both rumen bacteria, for example, *Ruminococcus albus* and *Ruminococcus flavefaciens,* and rumen fungi, for example, *Neocallimastix, Orpinomyces,* and *Piromyces.*

5.2.3. PHYLOGENY OF CELLULASE AND CELLULOSOMAL COMPONENTS

Early in the history of the development and establishment of the cellulosome concept, it was noted that the apparent occurrence of cellulosomes in different microorganisms tended to cross ecological, physiological, and evolutionary boundaries (Lamed *et al.,* 1987). Initial biochemical and immunochemical evidence to this effect has been supported by the accumulated molecular biological studies.

Various lines of evidence indicate that the modular enzymes that degrade plant cell wall polysaccharides have evolved from a restricted number of common ancestral sequences. Much of the information in this direction remains as a legacy, inherently encoded in the sequences of the functional domains that comprise the different enzymes. By comparing sequences of the various cellulosomal and noncellulosomal enzymes within and among the different strains, we can gain insight into the evolutionary rationale of the multigene families that comprise the glycosyl hydrolases.

5.2.3a. Horizontal Gene Transfer

It is clear that very similar enzymes that comprise a given glycosyl hydrolase family are prevalent among a variety of different bacteria and fungi, thus indicating that they were not inherited through normal evolutionary processes. The widespread occurrence of such conserved enzymes among phylogenetically different species argues that horizontal transfer of genes has been a major process by which a given microorganism can acquire a desirable enzyme. Once such a transfer event has taken place, the newly acquired gene then would be subjected to environmental pressures of its new surroundings, that is, the genetic and physiological constitution of the cell itself. Following such selective pressure, the sequence of the gene would be adjusted to fit the host cell.

5.2.3b. Gene Duplication

Sequence comparisons also have revealed the presence of very similar genes within a genome that may have very similar or even identical functions. One striking example is the tandem appearance of *cbhA* and *celK* genes in the chromosome of *C. thermocellum*. Other examples are *xynA* and *xynB* also of *C. thermocellum* and *xynA* of the anaerobic fungus *Neocallimastix patriciarum,* which includes two very similar copies of family 11 catalytic modules within the same polypeptide chain. These examples imply a mechanism of gene duplication (Chen *et al.,* 1998; Gilbert *et al.,* 1992), whereby the duplicated gene can serve as a template for secondary modifications that could result in two very similar enzymes with different properties, such as substrate and product specificities. A similar process also could account for the multiplicity of other types of modules (i.e., CBDs, cohesins, or helper modules) within a polypeptide chain.

5.2.3c. Domain Shuffling

Another observation from the genetic composition of the glycosyl hydrolases argues for an alternative type of process that would propagate new or modified types of enzymes. It is clear that many microbial enzyme systems contain individual hydrolases that carry very similar catalytic domains but include different types of accessory modules (Gilkes *et al.,* 1991). An example that demonstrates this phenomenon is the observed species preference of otherwise very similar glycosyl hydrolases for a given family of crystalline cellulose-binding CBD, which is entirely independent of the type of catalytic module borne by the complete enzyme. In this context, the free enzymes of some bacteria, such as *Cellulomonas fimi, Pseudomonas fluorescens,* and *Thermomonospora fusca,* invariably include a family II CBD, irrespective of the type of catalytic domain. In contrast, those of other bacteria, for example, *Bacillus subtilis, Caldocellum saccharolyticum, Erwinia carotovora,* and various clostridia, appear to prefer family III CBDs. Moreover, the position of the CBD in the gene may be different for different genes. For example, the CBD may occur upstream or downstream from the catalytic domain; it may be positioned either internally (sandwiched between two other modules) or at one of the termini of the polypeptide chain. The same pattern is characteristic of several other kinds of modules associated with the plant cell wall hydrolases. This is particularly evident in family 9 cellulases and family 10 xylanases, where the number and types of accessory modules may vary greatly within a given species. Taken together, the information suggests that domain shuffling is an important process by which the properties of such enzymes can be modified.

5.2.3d. Interrelationship of Cellulosome Phylogeny and Ecology

Similar to the observations for free cellulases, the phylogeny of the various cellulosomal components from the cellulosome-producing bacteria does not necessarily reflect the phylogenetic relationship of the bacteria themselves. We have noted (Section *5.2.2a*) the striking similarity of three mesophilic clostridia—*C. cellulovorans, C. cellulolyticum,* and *C. josui*—with respect to the presence of a characteristic cellulosome gene clusters within their genomes and the close relationship among their scaffoldins. It is evident on the basis of 16S rDNA analysis that *C. cellulolyticum* and *C. josui* are indeed separate but very closely related strains (Kakiuchi *et al.,* 1998). On the other hand, the 16S rDNA of *C. cellulovorans* is clearly different and has been classified in another group within the clostridia, representing a phylogenetically separate branch of the clostridial assemblage (Rainey and Stackebrandt, 1993).

Interestingly, the thermophilic *C. thermocellum* is also classified in the same group with *C. cellulolyticum* and *C. josui* (as opposed to *C. cellulovorans).* Nevertheless, the genes of its cellulosome components are widely scattered across the chromosome (Guglielmi and Béguin, 1998), in contrast to the cellulosome clusters of the other three bacteria. We also have described (Section 4.3) that its scaffoldin and association with the cell surface appear to be different than these features of the other three cellulosome systems.

It is also instructive to examine other anaerobic cellulolytic thermophiles, that is, microorganisms that occupy the same niche as does *C. thermocellum. C. stercorarium* is a bacterium that is phylogenetically very close to *C. thermocellum* and occupies essentially the same type of ecosystem. Nevertheless, no cellulosome has been detected in this bacterium. In contrast, its cellulase system appears to be characterized by free cellulases that include a catalytic domain and an authentic crystalline-cellulose-binding CBD. In addition, cellulases and/or hemicellulases from at least two other anaerobic thermophiles, *C. aldocellum saccharolyticum* and *Anaerocellum thermophilum,* have been described (Te'o *et al.,* 1995; Zverlov *et al.,* 1998a). The enzymes from these strains also appear to be noncellulosomal. All these cellulase systems include the distinctive family 48 and family 9/family IIIc cellulases. In the case of *C. saccharolyticum* and *A. thermophilum,* both of the catalytic domains are included in a single, multifunctional polypeptide that also contains multiple copies of cellulose-binding, family III CBDs. In the case of *C. stercorarium,* the enzymes appear to be produced in a simpler form, whereby the respective enzymes are single catalytic domains, targeted by a single family III CBD (Bronnenmeier *et al.,* 1997).

Like the free enzyme systems, the phylogeny of cellulosomal components seems to have been driven by processes that include horizontal gene transfer, gene duplication, and domain shuffling. In cellulolytic–hemicellulolytic ecosystems, the resident microorganisms are usually in close contact, often under difficult conditions and in competition or cooperation with one another toward a common goal: the rapid degradation of recalcitrant polysaccharides and assimilation of their breakdown products.

A possible scenario for the molecular evolution of a cellulase–hemicellulase system in a prospective bacterium could involve the initial transfer of genetic material from one microbe to another in the same ecosystem. The size and type of transferred material could vary, such as a gene or part of gene (e.g., selected functional modules) or even all or part of a gene cluster. The process then could be sustained by gene duplication that would propagate the insertion of repeated modules, for example, the multiple cohesin domains in the scaffoldins, or even smaller units, such as the linker sequences or the duplicated calcium-binding loop of the dockerin domain. Domain shuffling can account for the observed permutations in the arrangement of domains in scaffoldin subunits from different species (Fig. 9). Finally, conventional mutagenesis then would render such products more suitable for the cellular environment or for interaction with other components of the cellulase system.

6. FUTURE CHALLENGES

The available data suggest that there are no set of rules that would enable us at this stage to anticipate the nature of a given cellulase system from a given microorganism. It seems that phylogenetically dissimilar organisms can possess similar types of cellulosomal or noncellulosomal enzyme systems, whereas phylogenetically related organisms that inhabit similar niches may be characterized by different types of enzyme systems. It is clear that in order to shed further light on this apparent enigma, we require more information about more types of enzyme systems. In addition to more sequences and structures, we will need more information—biochemical, physiological, and ecological—in order to sharpen existing notions regarding the enzymatic degradation of plant cell wall polysaccharides or to formulate new ones.

ACKNOWLEDGMENTS

The authors thank Jean-Pierre Belaich, Anne Belaich, Shi-You Ding, and Roy Doi for access to prepublication data. The data concerning Cel H and Cel J were presented in the PhD thesis of Laurent Gal (July 1997, University of Aiz-Marseille); data concerning Cel K and Cel M were kindly provided by Anne and Jean-Pierre Belaich. We are grateful to Linda Shimon (Rehovot, Israel) and Felix Frolow (Tel Aviv University) for continuing collaboration on the structures of cellulosomal components. A contract from the European Commission (Biotechnology Programme, BIO4–97–2303) is sincerely appreciated, as are grants from the Israel Science Foundation (administered by the Israel Academy of Sciences and Humanities, Jerusalem) and the Minerva Foundation (Germany).

REFERENCES

Atalla, R. H., and VanderHart, D. L., 1984, Native cellulose: A composite of two distinct crystalline forms, *Science* **223**:283–285.

Bagnara-Tardif, C., Gaudin, C., Belaich, A., Hoest, P., Citard, T., and Belaich, J.-P., 1992, Sequence analysis of a gene cluster encoding cellulases from *Clostridium cellulolyticum, Gene* **119**:17–28.

Bayer, E. A., and Lamed, R., 1992, The cellulose paradox: Pollutant *par excellence* and/or a reclaimable natural resource? *Biodegradation* **3**:171–188.

Bayer, E. A., Kenig, R., and Lamed, R., 1983, Studies on the adherence of *Clostridium thermocellum* to cellulose, *J. Bacteriol.* **156**:818–827.

Bayer, E. A., Setter, E., and Lamed, R., 1985, Organization and distribution of the cellulosome in *Clostridium thermocellum, J. Bacteriol.* **163**:552–559.

Bayer, E. A., Morag, E., and Lamed, R., 1994, The cellulosome—A treasure-trove for biotechnology, *Trends Biotechnol.* **12**:378–386.

Bayer, E. A., Morag, E., Shoham, Y., Tormo, J., and Lamed, R., 1996, The cellulosome: A cell-surface organelle for the adhesion to and degradation of cellulose, in: *Bacterial Adhesion: Molecular and Ecological Diversity* (M. Fletcher, eds.), Wiley-Liss, New York, pp. 155–182.

Bayer, E. A., Chanzy, H., Lamed, R., and Shoham, Y., 1998a, Cellulose, cellulases and cellulosomes, *Curr. Opin. Struct. Biol.* **8**:548–557.

Bayer, E. A., Morag, E., Lamed, R., Yaron, S., and Shoham, Y., 1998b, Cellulosome structure: Four-pronged attack using biochemistry, molecular biology, crystallography and bioinformatics, in: *Carbohydrases from Trichoderma reesei and Other Microorganisms* (M. Claeyssens, W. Nerinckx, and K. Piens, eds.), The Royal Society of Chemistry, London, pp. 39–67.

Bayer, E. A., Shimon, L. J. W., Lamed, R., and Shoham, Y., 1998c, Cellulosomes: structure and ultrastructure, *J. Struct. Biol.* **124**:221–234.

Bayer, E. A., Ding, S. Y., Shoham, Y., and Lamed, R., 1999, New perspectives in the structure of cellulosome-related domains from different species, in: *Genetics, Biochemistry and Ecology of Lignocellulose Degradation* (K. Ohmiya and K. Hayashi, eds.), Uni Publishers Co, Tokyo, pp. 428–436.

Béguin, P., 1990, Molecular biology of cellulose degradation, *Annu. Rev. Microbiol.* **44**:219–248.

Béguin, P., and Aubert, J.-P., 1994, The biological degradation of cellulose, *FEMS Microbiol. Lett.* **13**:25–58.

Béguin, P., and Lemaire, M., 1996, The cellulosome: An exocellular, multiprotein complex specialized in cellulose degradation, *Crit. Rev. Biochem. Mol. Biol.* **31**:201–236.

Béguin, P., Millet, J., Grépinet, O., Navarro, A., Juy, M., Amit, A., Poljak, R., and Aubert, J.-P., 1988, The *cel* (cellulose degradation) genes of *Clostridium thermocellum,* in: *Biochemistry and Genetics of Cellulose Degradation* (J.-P. Aubert, P. Béguin, and J. Millet, eds.), Academic Press, London, pp. 267–282.

Belaich, J.-P., Belaich, A., Fierobe, H.-P., Gal, L., Gaudin, C., Pagès, S., Reverbel-Leroy, C., Tardif, C., 1998, The cellulolytic system of *Clostridium cellulolyticum,* in: *Genetics, Biochemistry and Ecology of Cellulose Degradation* (K. Ohmiya and K. Hayashi, eds.) Abstracts, Uni Publishers Co., Suzuka, Japan, p.113.

Belaich, J.-P., Tardif, C., Belaich, A., and Gaudin, C., 1997, The cellulolytic system of *Clostridium cellulolyticum, J. Biotechnol.* **57**:3–14.

Biely, P., 1985, Microbial xylanolytic systems, *Trends Biotechnol.* **3**:285–290.

Borneman, W. S., Ljungdahl, L. G., Hartley, R. D., and Akin, D. E., 1993, Feruloyl and *p*-coumaroyl esterases from the anaerobic fungus *Neocallimastix* strain MC-2: Properties and functions in plant cell wall degradation, in: *Hemicellulose and Hemicellulases* (M. P. Coughlan and G. P. Hazlewood, eds.), Portland Press, London, pp. 85–102.

Bronnenmeier, K., Kundt, K., Riedel, K., Schwarz, W., and Staudenbauer, W., 1997, Structure of the *Clostridium stercorarium* gene *cel*Y encoding the exo-1,4-β-glucanase Avicelase II, *Microbiology* **143**:891–898.

Carpita, N. C., and Gibeaut, D. M., 1993, Structural models of primary cell walls in flowering plants: Consistency of molecular structure with the physical properties of the walls during growth, *Plant J.* **3**:1–30.

Chanzy, H., 1990, Aspects of cellulose structure, in: *Cellulose Sources and Exploitation. Industrial Utilization, Biotechnology and Physico-chemical Properties* (J. F. Kennedy, G. O. Philips, and P. A. Williams, eds.), Ellis Horwood, New York, pp. 3–12.

Chauvaux, S., Béguin, P., Aubert, J.-P., Bhat, K. M., Gow, L. A., Wood, T. M., and Bairoch, A., 1990, Calcium-binding affinity and calcium-enhanced activity of *Clostridium thermocellum* endoglucanase D, *Biochem. J.* **265**:261–265.

Chen, H., Li, X., Blum, D., and Ljungdahl, L., 1998, Two genes of the anaerobic fungus *Orpinomyces sp.* strain PC-2 encoding cellulases with endoglucanase activities may have arisen by gene duplication, *FEMS Microbiol. Lett.* **159**:63–68.

Coughlan, M. P., and Hazlewood, G. P., 1993, β-1,4-D-Xylan-degrading enzyme systems: biochemistry, molecular biology and applications, *Biotechnol. Appl. Biochem.* **17:** 259–289.

Coutinho, P. M., and Henrissat, B., 1999, Carbohydrate-Active enZYmes and associated MODular Organization server (CAZyModO Website). [Online]. http://afmb.cnrs-mrs.fr/~pedro/DB/db.html. [10 November 1999, last date accessed],

Das, N. N., Das, S. C., and Mukerjee, A. K., 1984, On the ester linkage between lignin and 4-O-methyl-D-glucurono-D-xylan in jute fiber (*Corchorus capsularis*), *Carbohydr. Res.* **127:**345–348.

Davies, G., and Henrissat, B., 1995, Structures and mechanisms of glycosyl hydrolases, *Structure* **3:**853–859.

Din, N., Damude, H. G., Gilkes, N. R., Miller, R. C. J., Warren, R. A. J., and Kilburn, D. G., 1994, C_1-C_x revisited: Intramolecular synergism in a cellulase, *Proc. Natl. Acad. Sci. USA* **91:**11383–11387.

Ding, S. Y., Steiner, D., Kenig, R., Yaron, S., Morag, E., Shoham, Y., Bayer, E. A., and Lamed, R., 1998a, Domain organization and sequence of the first non-clostridial scaffoldin from *Acetivibrio cellulolyticus* in: *Genetics, Biochemistry and Ecology of Cellulose Degradation* (K. Ohmiya and K. Hayashi, eds.), conference abstract, Uni Publishers Co. Suzuka, Japan, pp. 47–48.

Ding, S. Y., Steiner, D., Kenig, R., Yaron, S., Shoham, Y., Morag, E., Bayer, E. A., and Lamed, R., 1998b, Unique cellulosome-related protein in *Bacteroides cellulosolvens,* in *Genetics, Biochemistry and Ecology of Cellulose Degradation* (K. Ohmiya and K. Hayashi, eds.), conference abstract, Uni Publishers Co. Suzuka, Japan, pp. 49–50.

Doi, R. H., Goldstein, M., Hashida, S., Park, J. S., and Takagi, M., 1994, The *Clostridium cellulovorans* cellulosome, *Crit. Rev. Microbiol.* **20:**87–93.

Doi, R. H., Park, J. S., Liu, C. C., Malburg, L. M., Tamaru, Y., Ichiishi, A., and Ibrahim, A., 1998, Cellulosome and noncellulosomal cellulases of *Clostridium cellulovorans, Extremophiles* **2:**53–60.

Dosoretz, C. G., and Grethlein, H. E., 1991, Physiological aspects of the regulation of extracellular enzymes of *Phanerochaete chrysosporium, Appl. Biochem. Biotechnol.* **28/ 29:**253–265.

Eriksson, K.-E. L., Blanchette, R. A., and Ander, P., 1990, Biodegradation of hemicelluloses, in: *Microbial and Enzymatic Degradation of Wood and Wood Components* Springer-Verlag, Heidelberg, pp. 181–397.

Fanutti, C., Ponyi, T., Black, G. W., Hazlewood, G. P., and Gilbert, H. J., 1995, The conserved noncatalytic 40-residue sequence in cellulases and hemicellulases from anaerobic fungi functions as a protein docking domain, *J. Biol. Chem.* **270:**29314–29322.

Felix, C. R., and Ljungdahl, L. G., 1993, The cellulosome—The exocellular organelle of Clostridium, *Annu. Rev. Microbiol.* **47:**791–819.

Flint, H., 1994, Molecular genetics of obligate anaerobes from the rumen, *FEMS Microbiol. Lett.* **121:**259–267.

Flint, H., 1997, The rumen microbial ecosystem—Some recent developments, *Trends Microbiol.* **5:**483–488.

Flint, H. J., Martin, J., McPherson, C. A., Daniel, A. S., and Zhang, J. X., 1993, A bifunctional enzyme, with separate xylanase and β(1,3–1,4)-glucanase domains, encoded by the xynD gene of *Ruminococcus flavefaciens, J. Bacteriol.* **175:**2943–2951.

Fujino, T., Béguin, P., and Aubert, J.-P., 1993, Organization of a *Clostridium thermocellum* gene cluster encoding the cellulosomal scaffolding protein CipA and a protein possibly involved in attachment of the cellulosome to the cell surface, *J. Bacteriol.* **175**:1891–1899.

Gal, L., Gaudin, C., Belaich, A., Pagès, S., Tardif, C., and Belaich, J.-P., 1997a, CelG from *Clostridium cellulolyticum:* A multidomain endoglucanase acting efficiently on crystalline cellulose, *J. Bacteriol.* **179**:6595–6601.

Gal, L., Pagès, S., Gaudin, C., Belaich, A., Reverbel-Leroy, C., Tardif, C., and Belaich, J.-P., 1997b, Characterization of the cellulolytic complex (cellulosome) produced by *Clostridium cellulolyticum, Appl. Env. Microbiol.* **63**:903–909.

Gerngross, U. T., Romaniec, M. P. M., Kobayashi, T., Huskisson, N. S., and Demain, A. L., 1993, Sequencing of a *Clostridium thermocellum* gene (cipA) encoding the cellulosomal S_L–protein reveals an unusual degree of internal homology, *Mol. Microbiol.* **8**:325–334.

Gerwig, G., Kamerling, J. P., Vliegenthart, J. F. G., Morag (Morgenstern), E., Lamed, R., and Bayer, E. A., 1991, Primary structure of *O*-linked carbohydrate chains in the cellulosome of different *Clostridium thermocellum* strains, *Eur. J. Biochem.* **196**:115–122.

Gerwig, G., Kamerling, J. P., Vliegenthart, J. F. G., Morag (Morgenstern), E., Lamed, R., and Bayer, E. A., 1992, Novel oligosaccharide constituents of the cellulase complex of *Bacteroides cellulosolvens, Eur. J. Biochem.* **205**:799–808.

Gerwig, G., Kamerling, J. P., Vliegenthart, J. F. G., Morag, E., Lamed, R., and Bayer, E. A., 1993, The nature of the carbohydrate-peptide linkage region in glycoproteins from the cellulosomes of *Clostridium thermocellum* and *Bacteroides cellulosolvens, J. Biol. Chem.* **268**:26956–26960.

Giallo, J., Gaudin, C., and Belaich, J.-P., 1985, Metabolism and solubilization of cellulose by *Clostridium cellulolyticum* H10, *Appl. Env. Microbiol.* **49**:1216–1221.

Gilbert, H. J., and Hazlewood, G. P., 1993, Bacterial cellulases and xylanases, *J. Gen. Microbiol.* **139**:187–194.

Gilbert, H. J., Hazlewood, G. P., Laurie, J. I., Orpin, C. G., and Xue, G. P., 1992, Homologous catalytic domains in a rumen fungal xylanase: Evidence for gene duplication and prokaryotic origin, *Mol. Microbiol.* **6**:2065–2072.

Gilkes, N. R., Henrissat, B., Kilburn, D. G., Miller, R. C. J., and Warren, R. A. J., 1991, Domains in microbial β-1,4-glycanases: sequence conservation, function, and enzyme families, *Microbiol. Rev.* **55**:303–315.

Guglielmi, G., and Béguin, P., 1998, Cellulase and hemicellulase genes of *Clostridium thermocellum* from five independent collections contain few overlaps and are widely scattered across the chromosome, *FEMS Microbiol. Lett.* **161**:209–215.

Haigler, C. H., and Weimer, P. J., 1991, Biosynthesis and Biodegradation of Cellulose, Marcel Dekker, New York.

Hatakka, A., 1994, Lignin-modifying enzymes from selected white-rot fungi: Production and role in lignin degradation, *FEMS Microbiol. Rev.* **13**:125–135.

Hazlewood, G. P., and Gilbert, H. J., 1993, Molecular biology of hemicellulases, in: *Hemicellulose and Hemicellulases* (M. P. Coughlan and G. P. Hazlewood, eds.), Portland Press, London, pp. 103–126.

Henrissat, B., 1991, A classification of glycosyl hydrolases based on amino acid sequence similarities, *Biochem. J.* **280**:309–316.

Henrissat, B., and Bairoch, A., 1996, Updating the sequence-based classification of glycosyl hydrolases, *Biochem. J.* **316:**695–696.

Henrissat, B., and Davies, G., 1997, Structural and sequence-based classification of glycoside hydrolases, *Curr. Opin. Struct. Biol.* **7:**637–644.

Henrissat, B., Claeyssens, M., Tomme, P., Lemesle, L., and Mornon, J.-P., 1989, Cellulase families revealed by hydrophobic cluster analysis, *Gene* **81:**83–95.

Henrissat, B., Teeri, T. T., and Warren, R. A. J., 1998, A scheme for designating enzymes that hydrolyse the polysaccharides in the cell walls of plants, *FEBS Lett.* **425:**352–354.

Higuchi, T., 1990, Lignin biochemistry, biosynthesis and biodegradation, *Wood Sci. Technol.* **24:**23–63.

Irwin, D., Shin, D.-H., Zhang, S., Barr, B. K., Sakon, J., Karplus, P. A., and Wilson, D. B., 1998, Roles of the catalytic domain and two cellulose binding domains of *Thermomonospora fusca* E4 in cellulose hydrolysis, *J. Bacteriol.* **180:**1709–1714.

Johnson, P., Joshi, M., Tomme, P., Kilburn, D., and McIntosh, L., 1996a, Structure of the *N*-terminal cellulose-binding domain of *Cellulomonas fimi* Cen C determined by nuclear magnetic resonance spectroscopy, *Biochemistry* **35:**14381–14394.

Johnson, P. E., Tomme, P., Joshi, M. D., and McIntosh, L. P., 1996b, Interaction of soluble cellooligosaccharides with the N-terminal cellulose-binding domain of Cellulomonas fimi CenC. 2. NMR and ultraviolet absorption spectroscopy, *Biochemistry* **35:**13895–13906.

Juy, M., Amit, A. G., Alzari, P. M., Poljak, R. J., Claeyssens, M., Béguin, P., and Aubert, J.-P., 1992, Crystal structure of a thermostable bacterial cellulose-degrading enzyme, *Nature* **357:**39–41.

Kakiuchi, M., Isui, A., Suzuki, K., Fujino, T., Fujino, E., Kimura, T., Karita, S., Sakka, K., and Ohmiya, K., 1998, Cloning and DNA sequencing of the genes encoding *Clostridium josui* scaffolding protein CipA and cellulase CelD and identification of their gene products as major components of the cellulosome, *J. Bacteriol.* **180:**4303–4308.

Karita, S., Sakka, K., and Ohmiya, K., 1997, Cellulosomes, cellulase complexes, of anaerobic microbes: Their structure models and functions, in: *Rumen Microbes and Digestive Physiology in Ruminants,* 14 14, (R. Onodera, H. Itabashi, K. Ushida, H. Yano, and Y. Sasaki, eds.), Japan Scientific Society Press, Tokyo/S. Karger, Basel, pp. 47–57.

Kataeva, I., Li, X.-L., Chen, H., and Ljungdahl, L. G., 1998, CelK—A new cellobiohydrolase from *Clostridium thermocellum* cellulosome: Role of N-terminal cellulose-binding domain, in: *Genetics, Biochemistry and Ecology of Cellulose Degradation* Uni Publishers, Co, Tokyo, pp. 454–460.

Kerem, Z., and Hadar, Y., 1998, Lignin-degrading fungi: Mechanisms and utilization, in: *Agricultural Biotechnology* (A. Altman, ed.), Marcel Dekker, New York, pp. 351–365.

Kirby, J., Martin, J., Daniel, A., and Flint, H., 1997, Dockerin-like sequences in cellulases and xylanases from the rumen cellulolytic bacterium *Ruminococcus flavefaciens,* *FEMS Microbiol. Lett.* **149:**213–219.

Kraulis, P. J., Clore, G. M., Nilges, M., Jones, T. A., Pettersson, G., Knowles, J., and Gronenborn, A. M., 1989, Determination of the three-dimensional solution structure of the C-terminal domain of cellobiohydrolase I from *Trichoderma reesei.* A study using nuclear magnetic resonance and hybrid distance geometry-dynamical simulated anneal-ing, *Biochemistry* **28**:7241–7257.

Kuhad, R. C., Singh, A., and Eriksson, K.-E. L., 1997, Microorganisms and enzymes in-volved in the degradation of plant fiber cell walls, *Adv. Biochem. Eng. Biotechnol.* **57**:45–125.

Lamed, R., and Bayer, E. A., 1988a, The cellulosome concept: Exocellular/extracellular en-zyme reactor centers for efficient binding and cellulolysis, in: *Biochemistry and Ge-netics of Cellulose Degradation* (J.-P. Aubert, P. Beguin, and J. Millet, eds.), Academic Press, London, pp. 101–116.

Lamed, R., and Bayer, E. A., 1988b, The cellulosome of *Clostridium thermocellum, Adv. Appl. Microbiol.* **33**:1–46.

Lamed, R., and Bayer, E. A., 1991, Cellulose degradation by thermophilic anaerobic bac-teria, in: *Biosynthesis and Biodegradation of Cellulose and Cellulose Materials* (C. H. Haigler and P. J. Weimer, eds.), Marcel Dekker, New York, pp. 377–410.

Lamed, R., and Bayer, E. A., 1993, The cellulosome concept—A decade later! in: *Genet-ics, Biochemistry and Ecology of Lignocellulose Degradation* (K. Shimada, S. Hoshi-no, K. Ohmiya, K. Sakka, Y. Kobayashi, and S. Karita, eds.), Uni Publishers Co., Tokyo, Japan, pp. 1–12.

Lamed, R., Setter, E., and Bayer, E. A., 1983a, Characterization of a cellulose-binding, cel-lulase-containing complex in *Clostridium thermocellum, J. Bacteriol.* **156**:828–836.

Lamed, R., Setter, E., Kenig, R., and Bayer, E. A., 1983b, The cellulosome—A discrete cell surface organelle of *Clostridium thermocellum* which exhibits separate antigenic, cel-lulose-binding and various cellulolytic activities, *Biotechnol. Bioeng. Symp.* **13**:163–181.

Lamed, R., Kenig, R., Morag, E., Yaron, S., Shoham, Y., and Bayer, E. A., in press, Non-proteolytic cleavage of aspartyl proline bonds in the cellulosomal scaffoldin subunit from *Clostridium thermocellum, Appl. Biochem. Biotechnol.*

Lamed, R., Naimark, J., Morgenstern, E., and Bayer, E. A., 1987, Specialized cell surface structures in cellulolytic bacteria, *J. Bacteriol.* **169**:3792–3800.

Laurie, J. I., Clarke, J. H., Ciruela, A., Faulds, C. B., Williamson, G., Gilbert, H. J., Rixon, J. E., Millward-Sadler, J., and Hazlewood, G. P., 1997, The NodB domain of a mul-tidomain xylanase from *Cellulomonas fimi* deacylates acetylxylan, *FEMS Microbiol. Lett.* **148**:261–264.

Leibovitz, E., and Béguin, P., 1996, A new type of cohesin domain that specifically binds the dockerin domain of the *Clostridium thermocellum* cellulosome-integrating protein CipA, *J. Bacteriol.* **178**:3077–3084.

Leibovitz, E., Ohayon, H., Gounon, P., and Béguin, P., 1997, Characterization and subcel-lular localization of the *Clostridium thermocellum* scaffoldin dockerin binding protein SdbA, *J. Bacteriol.* **179**:2519–2523.

Lemaire, M., Ohayon, H., Gounon, P., Fujino, T., and Béguin, P., 1995, OlpB, a new outer layer protein of *Clostridium thermocellum,* and binding of its S-layer-like domains to components of the cell envelope, *J. Bacteriol.* **177**:2451–2459.

Lewis, N. G., and Yamamoto, E., 1990, Lignin: occurrence, biogenesis and biodegradation, *Annu. Rev. Plant Physiol. Plant Mol. Biol.* **41**:455–496.

Li, X., Chen, H., and Ljungdahl, L., 1997, Two cellulases, CelA and CelC, from the polycentric anaerobic fungus *Orpinomyces* strain PC-2 contain N-terminal docking domains for a cellulase-hemicellulase complex, *Appl. Env. Microbiol.* **63:**4721– 4728.

Linder, M., and Teeri, T. T., 1997, The roles and function of cellulose-binding domains, *J. Biotechnol.* **57:**15–28.

Liu, C. C., and Doi, R. H., 1998, Properties of exgS, a gene for a major subunit of the *Clostridium cellulovorans* cellulosome, *Gene* **211:**39–47.

Lupas, A., Engelhardt, H., Peters, J., Santarius, U., Volker, S., and Baumeister, W., 1994, Domain structure of the *Acetogenium kivui* surface layer revealed by electron crystallography and sequence analysis, *J. Bacteriol.* **176:**1224–1233.

Mattinen, M.-L., Kontteli, M., Kerovuo, J., Linder, M., Annila, A., Lindeberg, G., Reinikainen, T., and Drakenberg, T., 1997, Three-dimensional structures of three engineered cellulose-binding domains of cellobiohydrolase I from *Trichoderma reesei, Protein Sci.* **6:**294–303.

McCarter, J. D., and Withers, S. G., 1994, Mechanisms of enzymatic glycoside hydrolysis, *Curr. Opin. Struct. Biol.* **4:**885–892.

Morag, E., Bayer, E. A., and Lamed, R., 1990, Relationship of cellulosomal and noncellulosomal xylanases of *Clostridium thermocellum* to cellulose-degrading enzymes, *J. Bacteriol.* **172:**6098–6105.

Morag, E., Halevy, I., Bayer, E. A., and Lamed, R., 1991, Isolation and properties of a major cellobiohydrolase from the cellulosome of *Clostridium thermocellum, J. Bacteriol.* **173:**4155–4162.

Morag, E., Bayer, E. A., Hazlewood, G. P., Gilbert, H. J., and Lamed, R., 1993, Cellulase S$_S$ (CelS) is synonymous with the major cellobiohydrolase (subunit S8) from the cellulosome of *Clostridium thermocellum, Appl. Biochem. Biotechnol.* **43:**147–151.

Navarro, A., Chebrou, M.-C., Béguin, P., and Aubert, J.-P., 1991, Nucleotide sequence of the cellulase gene *celF* of *Clostridium thermocellum, Res. Microbiol.* **142:**927–936.

Ohmiya, K., Sakka, K., Karita, S., and Kimura, T., 1997, Structure of cellulases and their applications, *Biotechnol. Genet. Eng. Rev.* **14:**365–414.

O'Sullivan, A. C., 1997, Cellulose: The structure slowly unravels, *Cellulose* **4:**173–207.

Pagès, S., Belaich, A., Belaich, J.-P., Morag, E., Lamed, R., Shoham, Y., and Bayer, E. A., 1997, Species-specificity of the cohesin-dockerin interaction between *Clostridium thermocellum* and *Clostridium cellulolyticum:* Prediction of specificity determinants of the dockerin domain, *Proteins* **29:**517–527.

Pagès, S., Belaich, A., Fierobe, H.-P., Tardif, C., Gaudin, C., and Belaich, J.-P., 1999, Sequence analysis of scaffolding protein CipC and ORFXp, a new cohesin-containing protein in *Clostridium cellulolyticum:* Comparison of various cohesin domains and subcellular localization of ORFXp, *J. Bacteriol.* **181:**1801–1810.

Parsiegla, G., Juy, M., Reverbel-Leroy, C., Tardif, C., Belaich, J. P., Driguez, H., and Haser, R., 1998, The crystal structure of the processive endocellulase CelF of *Clostridium cellulolyticum* in complex with a thiooligosaccharide inhibitor at 2.0 Å resolution, *EMBO J.* **17:**5551–5562.

Petitdemange, E., Caillet, F., Giallo, J., and Gaudin, C., 1984, *Clostridium cellulolyticum* sp. nov., a cellulolytic mesophilic species from decayed grass, *Int. J. Syst. Bacteriol.* **34:**155–159.

Puls, J., and Schuseil, J., 1993, Chemistry of hemicellulases: Relationship between hemicellulose structure and enzymes required for hydrolysis, in: *Hemicellulose and Hemicellulases* (M. P. Coughlan and G. P. Hazlewood, eds.), Portland Press, London, pp. 1–27.

Rainey, F. A., and Stackebrandt, E., 1993, 16S rDNA analysis reveals phylogenetic diversity among the polysaccharolytic clostridia, *FEMS Microbiol. Lett.* **113**:125–128.

Reverbel-Leroy, C., Pagés, S., Belaich, A., Belaich, J.-P., and Tardif, C., 1997, The processive endocellulase CelF, a major component of the *Clostridium cellulolyticum* cellulosome: Purification and characterization of the recombinant form, *J. Bacteriol.* **179**:46–52.

Rouvinen, J., Bergfors, T., Teeri, T., Knowles, J. K. C., and Jones, T. A., 1990, Three-dimensional structure of cellobiohydrolase II from *Trichoderma reesei, Science* **279**:380–386.

Sakon, J., Irwin, D., Wilson, D. B., and Karplus, P. A., 1997, Structure and mechanism of endo/exocellulase E4 from *Thermomonospora fusca, Nature Struct. Biol.* **4**:810–818.

Salamitou, S., Lemaire, M., Fujino, T., Ohayon, H., Gounon, P., Béguin, P., and Aubert, J.-P., 1994a, Subcellular localization of *Clostridium thermocellum* ORF3p, a protein carrying a receptor for the docking sequence borne by the catalytic components of the cellulosome, *J. Bacteriol.* **176**:2828–2834.

Salamitou, S., Raynaud, O., Lemaire, M., Coughlan, M., Béguin, P., and Aubert, J.-P., 1994b, Recognition specificity of the duplicated segments present in *Clostridium thermocellum* endoglucanase CelD and in the cellulosome-integrating protein CipA, *J. Bacteriol.* **176**:2822–2827.

Saxena, S., Fierobe, H. P., Gaudin, C., Guerlesquin, F., and Belaich, J. P., 1995, Biochemical properties of a β-xylosidase from *Clostridium cellulolyticum, Appl. Environ. Microbiol.* **61**:3509–3512.

Shimon, L. J. W., Bayer, E. A., Morag, E., Lamed, R., Yaron, S., Shoham, Y., and Frolow, F., 1997, The crystal structure at 2.15 Å resolution of a cohesin domain of the cellulosome from *Clostridium thermocellum, Structure* **5**:381–390.

Shoseyov, O., and Doi, R. H., 1990, Essential 170-kDa subunit for degradation of crystalline cellulose by *Clostridium cellulovorans* cellulase, *Proc. Natl. Acad. Sci. USA* **87**:2192–2195.

Shoseyov, O., Takagi, M., Goldstein, M. A., and Doi, R. H., 1992, Primary sequence analysis of *Clostridium cellulovorans* cellulose binding protein A, *Proc. Natl. Acad. Sci. USA* **89**:3483–3487.

Sleat, R., Mah, R. A., and Robinson, R., 1984, Isolation and characterization of an anaerobic, cellulolytic bacterium, *Clostridium cellulovorans,* sp. nov., *Appl. Environ. Microbiol.* **48**:88–93.

Souchon, H., Beguin, P., and Alzari, P. M., 1996, Crystallization of a family 8 cellulase from *Clostridium thermocellum proteins,* **25**:134–136.

Tamaru, Y., Araki, T., Morishita, T., Kimura, T., Sakka, K., and Ohmiya, K., 1997, Cloning, DNA sequencing, and expression of the b-1,4-mannanase gene from a marine bacterium, *Vibrio* sp. strain MA-138, *J. Ferment. Bioeng.* **83**:201–205.

Tamaru, Y., Liu, C.-C., Ichi-ishi, A., Malburg, L., and Doi, R. H., 1999, The *Clostridium cellulovorans* cellulosome and non-cellulosomal cellulases, in: *Genetics, Biochemistry and Ecology of Cellulose Degradation* (Abstracts) (K. Ohmiya and K. Hayashi, eds.), Uni Publishers Co., Tokyo, pp. 488–494.

Tavares, G. A., Béguin, P., and Alzari, P. M., 1997, The crystal structure of a type I cohesin domain at 1.7 Å resolution, *J. Mol. Biol.* **273**:701–713.

Teeri, T. T., Reinikainen, T., Ruohonen, L., Jones, T. A., and Knowles, J. K. C., 1992, Domain function in *Trichoderma reesei* cellulases, *J. Biotechnol.* **24**:169–176.

Te'o, V. S., Saul, D. J., and Bergquist, P. L., 1995, CelA, another gene coding for a multidomain cellulase from the extreme thermophile *Caldocellum saccharolyticum, Appl. Microbiol. Biotechnol.* **43**:291–296.

Timell, T. E., 1967, Recent progress in the chemistry of wood hemicelluloses, *Wood Sci. Technol.* **1**:45–70.

Tokatlidis, K., Salamitou, S., Béguin, P., Dhurjati, P., and Aubert, J.-P., 1991, Interaction of the duplicated segment carried by *Clostridium thermocellum* cellulases with cellulosome components, *FEBS Lett.* **291**:185–188.

Tokatlidis, K., Dhurjati, P., and Béguin, P., 1993, Properties conferred on *Clostridium thermocellum* endoglucanase CelC by grafting the duplicated segment of endoglucanase CelD, *Protein Eng.* **6**:947–952.

Tomme, P., Creagh, A. L., Kilburn, D. G., and Haynes, C. A., 1996, Interaction of soluble cellooligosaccharides with the *N*-terminal cellulose-binding domain of *Cellulomonas fimi* CenC. 1. Binding specificity and calorimetric analysis, *Biochemistry* **35**:13885–13894.

Tomme, P., Warren, R. A. J., and Gilkes, N. R., 1995a, Cellulose hydrolysis by bacteria and fungi, *Adv. Microb. Physiol.* **37**:1–81.

Tomme, P., Warren, R. A. J., Miller, R. C., Kilburn, D. G., and Gilkes, N. R., 1995b, Cellulose-binding domains—Classification and properties, in: *Enzymatic Degradation of Insoluble Polysaccharides* (J. M. Saddler and M. H. Penner, eds.), American Chemical Society, Washington, DC, pp. 142–161.

Tormo, J., Lamed, R., Chirino, A. J., Morag, E., Bayer, E. A., Shoham, Y., and Steitz, T. A., 1996, Crystal structure of a bacterial family III cellulose-binding domain: A general mechanism for attachment to cellulose, *EMBO J.* **15**:5739–5751.

Umezawa, T., and Higuchi, T., 1991, Chemistry of lignin degradation by lignin peroxidases, in: *Enzymes in Biomass Conversion,* ACS Symposium Series 460 (G. F. Leatham and M. E. Himmel, eds.), American Chemical Society, Washington, DC, pp. 236–269.

Viikari, L., and Teeri, T. (eds.), 1997, Biochemistry and genetics of cellulases and hemicellulases and their application, *J. Biotechnol.* **57**:1–228.

Wang, W. K., Kruus, K., and Wu, J. H. D., 1993, Cloning and DNA sequence of the gene coding for *Clostridium thermocellum* cellulase S_S (CelS), a major cellulosome component, *J. Bacteriol.* **175**:1293–1302.

Warren, R. A. J., 1996, Microbial hydrolysis of polysaccharides, *Annu. Rev. Microbiol.* **50**:183–212.

White, A., and Rose, D. R., 1997, Mechanism of catalysis by retaining β-glycosyl hydrolases, *Curr. Opin. Struct. Biol.* **7**:645–651.

Wu, J. H. D., Orme-Johnson, W. H., and Demain, A. L., 1988, Two components of an extracellular protein aggregate of *Clostridium thermocellum* together degrade crystaline cellulose, *Biochemistry* **27**:1703–1709.

Xu, G.-Y., Ong, E., Gilkes, N. R., Kilburn, D. G., Muhandiram, D. R., Harris-Brandts, M., Carver, J. P., Kay, L. E., and Harvey, T. S., 1995, Solution structure of a cellulose-binding domain from *Cellulomonas fimi* by nuclear magnetic resonance spectroscopy, *Biochemistry* **34:**6993–7009.

Zverlov, V. V., Mahr, S., Riedel, K., and Bronnenmeier, K., 1998a, Properties and gene structure of a bifunctional cellulolytic enzyme (CelA) from the extreme thermophile *Anaerocellum thermophilum* with separate glycosyl hydrolase family 9 and 48 catalytic domains, *Microbiology* **144:**457–465.

Zverlov, V. V., Velikodvorskaya, G. V., Schwarz, W.H., Bronnenmeier, K., Kellermann, J., and Staudenbauer, W. L., 1998b, Multidomain structure and cellulosomal localization of the *Clostridium thermocellum* cellobiohydrolase CbhA, *J. Bacteriol.* **180:**3091–3099.

15

The Expression of Polysaccharide Capsules in *Escherichia coli*

A Molecular Genetic Perspective

Ian S. Roberts

1. INTRODUCTION

The production of an extracellular polysaccharide capsule is a common feature of many bacteria (Whitfield and Valvano, 1993). The capsule, which often constitutes the outermost layer of the cell, mediates the interaction between the bacterium and its immediate environment and plays a crucial role in the survival of bacteria in hostile environments. The expression of a polysaccharide capsule may promote the formation of biofilms and stimulate interspecies coaggregation enhancing the colonization of a variety of ecological niches. These include the colonization of industrial pipelines, food preparation machinery, waterpipes, indwelling catheters and prostheses (Costerton *et al.,* 1987). In such instances, the extracellular poly-saccharide may present a permeability barrier to decontaminating agents and an-tibiotics and hinder the effective eradication of the bacteria (Roberts, 1996; Mox-on and Kroll, 1990). In addition, in invasive diseases of man, the expression of a polysaccharide capsule will confer resistance to the non-specific arm of the host's immune system (Moxon and Kroll, 1990).

While the roles of polysaccharide capsules in the above processes are well

Ian S. Roberts • School of Biological Sciences, University of Manchester, Manchester M13 9PT, England.

Glycomicrobiology, edited by Doyle.
Kluwer Academic/Plenum Publishers, New York, 2000.

documented, there is an embarrassing lack of understanding at the molecular level about fundamental aspects of capsule production. The biosynthesis of capsular polysaccharides and their subsequent transport onto the cell surface provide a unique challenge to the bacterium. It must synthesize within the cell a large negatively charged macromolecule consisting of repeating subunits linked in a precise order. This macromolecule must then be transported onto the cell surface where it may or may not be subsequently anchored. Attempting to understand at the molecular level the mechanisms by which polysaccharide capsules are synthesized represents a fascinating biological problem.

In this chapter I will concentrate on the expression of the *serA*-linked group 2 and 3 capsules in *Escherichia coli*. I will describe the genetic organization of group 2 and 3 capsule gene clusters and discuss the mechanisms by which capsule gene diversity at the *serA*-linked locus has been achieved. Subsequently, I describe the regulation of the expression of these capsule gene clusters. Finally, I will highlight our current understanding of the biosynthesis and transport of group 2 polysaccharides and describe our latest results demonstrating the existence of a multiprotein hetero-oligomeric biosynthetic–export complex.

2. *E. COLI* CAPSULES

E. coli can produce in excess of 80 chemically distinct capsular polysaccharides (K-antigens) (Ørskov and Ørskov, 1992). Based on a number of sound biochemical and genetic criteria, the original serological classification was divided into two groups: I and II (Jann and Jann, 1990). In this scheme group I capsules could be distinguished from group II capsules on the basis of the higher molecular weights and lower charge densities of group I capsules, their expression at all growth temperatures and the association of group I capsules with a relatively small range of O-antigens (Jann and Jann, 1990). The genes for group I capsules were located proximal to the *his* operon, while the group II capsule gene clusters were mapped to the *serA* gene (Jann and Jann, 1990). However, the classification into two groups would now appear to be an underestimation for a number of reasons. First, group I capsules could be further defined into two separate capsule types that were referred to as Ia and Ib (Jann and Jann, 1992). Group Ia capsules, typified by the K30-antigen, did not contain amino sugars and resembled the capsular polysaccharides of *Klebsiella* and *Erwinia* species. In addition, strains expressing group Ia capsules were unable to express cell surface colanic acid (Keenleyside *et al.*, 1992). In contrast, group Ib capsules, typified by the K40-antigen, contained amino sugars and were able to coexpress colanic acid (Jayaratne *et al.*, 1993). Second, a separate family of capsule gene clusters, typified by the K10- and K54-antigens, which are distinct to group II capsule gene clusters, were identified at the same *serA* site on the *E. coli* chromosome (Pearce and Roberts, 1995; Russo *et al.*,

1998; Clarke *et al.*, 1999). This group of capsule gene clusters was referred to as group III to avoid any ambiguity with group II capsule gene clusters (Pearce and Roberts, 1995). To attempt to remove any confusion a new classification for *E. coli* capsules has been proposed consisting of four groups (Table I) that is based solely on genetic and biosynthetic criteria (Whitfield and Roberts, 1999). In this system groups 2 and 3 refer to the original II and III, while groups Ia and Ib are now groups 1 and 4, respectively (Whitfield and Roberts, 1999).

3. *E. COLI* GROUP 2 CAPSULES

In contrast to groups 1 and 4 capsular polysaccharides, group 2 capsular polysaccharides are very heterogeneous in composition and can be divided into four subgroups based on their acidic components (Jann and Jann, 1990). In terms of structure and cell surface assembly, group 2 capsular polysaccharides closely resemble the capsular polysaccharides of *Neisseria meningitidis* and *Haemophilus influenzae*. Group 2 capsular polysaccharides are linked via their reducing terminus to α-glycerophosphatidic acid, which is believed to play a role in the formation and stabilization of the capsule structure possibly by anchoring the polysaccharide to the outer membrane via hydrophobic interactions (Jann and Jann, 1990). Interestingly, only 20–50% of capsule preparations contain molecules that are lipid substituted, but this may reflect the lability of the linkage between the α-glycerophosphatidic acid and the reducing sugar. In the case of certain group 2 capsular polysaccharides 2-keto-3-deoxymanno-octonic acid (Kdo) has been shown to be the reducing sugar linked to α-glycerophosphatidic acid, regardless of whether Kdo is present within the repeat structure of the polysaccharide (Finke *et al.*, 1991; Jann and Jann, 1990). In the case of the polysialic acid containing K1 and K92 group 2 capsular polysaccharides, *N*-acetylneuraminic acid is believed to be the reducing sugar (Gotschlich *et al.*, 1981). The reason for this difference is as yet unclear. The conservation of the *kpsU* gene, encoding for a functional CMP-Kdo synthetase, between different group 2 capsule gene clusters (Pazzani *et al.*, 1993) and the elevated levels of CMP-Kdo synthetase at capsule-permissive temperatures (Finke *et al.*, 1991) would suggest a central role for the attachment of Kdo in the expression of group 2 capsules.

4. *E. COLI* GROUP 3 CAPSULES

This group comprises a small group of *E. coli* capsules (Table 1), originally designated I/II, that possess characteristics of both group I and II capsules and could not be readily assigned to either group (Jann and Jann, 1990). In many ways

Table 1
Classification of *E. coli* Capsules

Characteristic	Group			
	1	2	3	4
Former K-antigen group	I A	II	I/II or III	IB (O-antigen capsules)
Coexpressed with O serogroups	Limited range (08, 09, 020, 0101)	Many	Many	Often 08, 09 but sometimes none
Coexpressed with colanic acid	No	Yes	Yes	Yes
Thermostability	Yes	No	No	Yes
Terminal lipid moiety	Lipid A-core in K_{LPS}; unkown for capsular K-antigen	α-glycerophosphate	α-glycerophosphate	lipid A-core in K_{LPS}; unknown for capsular K-antigen
Direction of chain growth	Reducing terminus	Nonreducing terminus	Non-reducing terminus?	Reducing terminus
Polymerization system	Wzy-dependent	Processive	Processive?	Wzy-dependent
Trans-plasma membrane export	Wzx (PST2)	ABC-2 exporter	ABC-2 exporter	Wzx (PST2)
Elevated levels of CMP-Kdo synthetase @ 37°C	No	Yes	No	No
Genetic locus	*cps* near *his* and *rfb*	*kps* near *serA*	*kps* near *serA*	*rfb* near *his*
Thermoregulated (i.e., not expressed below 20°C)	No	Yes	No	No
Model system	Serotype K30	Serotypes K1, K5	Serotypes K10, K54	Serotypes K40, 0111
Similar to	*Klebsiella, Erwinia*	*Neisseria, Haemophilus*	*Neisseria, Haemophilus*	Many genera

these capsular polysaccharides resemble those of group 2, having similar heat lability, composition, and charge density (Jann and Jann, 1990). Typical of group 2 capsules, phospholipid was detected at the reducing end of the K10, K11, K54, and K98 capsular polysaccharides and phospholipid and Kdo at the reducing end of the K10 capsular polysaccharide (Sieberth *et al.*, 1993). The genes for the production of the K10 and K54 capsules also were mapped to the same *serA* locus on the chromosome as the group 2 capsule gene clusters (Ørskov and Nyman, 1974). However, in contrast to group 2 capsules, these capsules are expressed at all growth temperatures, and strains expressing these capsules do not exhibit elevated levels of CMP-Kdo synthetase. The cloning and analysis of the K10 and K54 capsule gene clusters (Clarke *et al.*, 1999; Russo *et al.*, 1998; Pearce and Roberts, 1995) have confirmed that these represent a different distinct group of *E. coli* capsules, and to avoid any ambiguity they have been classified as group 3 capsules (Pearce and Roberts, 1995).

5. THE GENETIC ORGANIZATION AND REGULATION OF *E. COLI* GROUP 2 CAPSULE GENE CLUSTERS

The cloning and analysis of a large number of *E. coli* group 2 capsule gene clusters established that group 2 capsule gene clusters have a conserved modular genetic organization consisting of three regions 1, 2, and 3 (Fig. 1) (Roberts, 1996; Roberts *et al.*, 1986, 1988; Boulnois *et al.*, 1987; Silver *et al.*, 1984). This modular organization, first demonstrated with *E. coli* group 2 capsule gene clusters, would now appear applicable to capsule gene clusters from other bacteria Roberts, 1996). Regions 1 and 3 are conserved in all of the group 2 capsule gene clusters analyzed and encode proteins involved in the transport of group 2 polysaccharides from their site of synthesis on the inner face of the cytoplasmic membrane onto the cell surface. Region 2 is serotype specific and encodes enzymes for the polymerization of the polysaccharide molecule and where necessary for the biosynthesis of the specific monosaccharide components that make up the polysaccharide. The size of the specific region 2 is variable and in part reflects the complexity of the polysaccharide to be synthesised (Boulnois *et al.*, 1992). The region 2 DNA of the K5 and K1 capsule gene clusters has a high (66%) A+T content compared to that of regions 1 (50%) and 3 (57%) (Roberts, 1996). This is typical of genes that encode enzymes for polysaccharide biosynthesis (Roberts, 1995) and would suggest that group 2 capsule diversity has been achieved in part through the acquisition of different region 2 sequences. Amplification by polymerase chain reaction (PCR) of sequences between regions 1 and 2 and between regions 2 and 3 from a number of group 2 capsule gene clusters failed to find any evidence for insertion sequences or site-specific recombination events playing a role in this process (Roberts, 1996). Rather, the acquisition of new region 2 sequences may

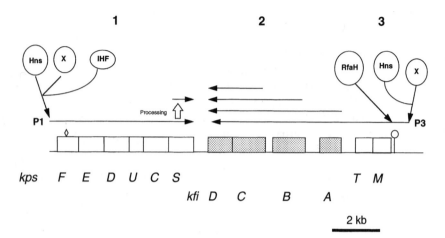

Figure 1. Diagrammatic representation of the *Escherichia coli* K5 capsule gene cluster. The numbers at the top refer to the three functional regions present in *E. coli* group 2 capsule gene clusters with the region 2 gene shaded. P1 and P3 represent the region 1 and region 3 promoters and the arrows denote the major transcripts. The diamond within *kpsF* identifies the intragenic Rho-dependent transcriptional terminator, while the stem loop structure denotes the JUMPstart sequence. The known regulatory proteins are shown with X representing the putative Ara C-like transcriptional activator.

occur through homologous recombination between the flanking regions 1 and 3 of an incoming and resident capsule gene cluster. The observation that 3′ ends of the *kpsS* and *kpsT* genes that flank either side of region 2 (Fig. 1) show the greatest divergence among the conserved region 1 and 3 *kps* genes (Roberts, 1996) would support this hypothesis for acquiring and losing region 2 sequences. The mechanism by which region 2 diversity and therefore the diversity of *E. coli* group 2 capsular polysaccharides has been achieved is still unknown.

Region 1 contains six genes, *kpsFEDUCS*, organized in a single transcriptional unit (Fig. 1). The functions performed by these proteins in the transport of group 2 capsular polysaccharides will be discussed later. A single *E. coli* σ^{70} promoter is located 225 base pairs (bp) 5′ of *kpsF* (Simpson *et al.*, 1996). Analysis of the promoter identified three integration host factor (IHF) binding site consensus sequences. One of these is located 80 bp 5′ to the initiation codon of *kpsF,* while the other two are 60 bp and 110 bp 5′ to the transcription start point. Gel retardation experiments using PCR fragments spanning the region 1 promoter have confirmed that IHF binds to the promoter (Griffiths and Roberts, unpublished results). The observation that mutations in the *himA* and *himD* genes lead to a 20% reduction in expression of the Kps E protein (Simpson *et al.*, 1996) confirm that IHF plays a role in regulating the expression of region 1 at 37°C. Transcription from the region 1 promoter generates an 8.0 kb polycistronic transcript, which is subsequently processed to give a stable 1.3 kb *kpsS*-specific transcript (Fig. 1) (Simpson *et al.,* 1996). The processing of this transcript would appear to be independent of either RNaseIII or RNaseE (Simpson *et al.,* 1996). The processing of mRNA has been implicated in the differential expression of bacterial genes (Bilge *et al.,* 1993; Klug, 1993), and it is possible that the generation of a separate *kpsS*-specific transcript may enable the differential expression of Kps S from the other region 1 proteins. An intragenic Rho-dependent transcriptional terminator is located with the *kpsF* gene. Such intragenic terminators have been implicated in regulating transcription in response to physiological stress (Richardson, 1991). In the case of region 1, under conditions of physiological stress, in which the mRNA message is not being efficiently translated, transcription would cease at the intragenic terminator within *kpsF,* thereby switching off expression of region 1. The observation that mutations in region 1 genes that abolish polysaccharide export out of the cell reduce membrane transferase activity (Bronner *et al.,* 1993) means that the overall effect of terminating transcription within *kpsF* would be to reduce capsule expression under physiological stressful conditions.

Region 3 contains two genes, *kpsM* and *kpsT,* organized in a single transcriptional unit (Fig. 1) (Bliss and Silver, 1996; Roberts, 1996). The promoter has been mapped to 741 bp 5′ to the initiation codon of the *kpsM* gene, and the promoter has a typical *E. coli* σ^{70} −10 consensus sequence but no −35 region (Stevens *et al.,* 1997). No consensus binding sequences for other σ factors were detectable and no IHF binding sites were present in the region 3 promoter (Stevens *et al.,* 1997). However a *cis*-acting regulatory sequence, termed *ops,* which is essential for the action of Rfa H, was identified 33 bp 5′ to the initiation codon of the *kpsM* gene (Stevens *et al.,* 1997). The *ops* sequence of GGCGGTAG is contained within a larger 39 bp regulatory element called JUMPstart (*j*ust *u*pstream of *m*any *p*olysaccharide-associated gene starts) (Hobbs and Reeves 1994). Rfa H regulates a number of gene clusters in *E. coli,* including the *cps, hly, rfa, rfb,* and *tra* operons (Bailey *et al.,* 1997; Whitfield and Roberts, 1999). Rfa H is a homologue of the essential transcription elongation factor Nus G, which is required for Rho-dependent transcription termination and bacteriophage λ N-mediated antitermination. Rfa H is believed to act as a transcriptional elongation factor that permits transcription to proceed over long distances. As such, mutations in *rfa H*- result in increased transcription polarity throughout Rfa H regulated operons without affecting the initiation from the operon promoters (Bailey *et al.,* 1997). It is believed that *ops* sequence in the nascent mRNA molecule recruits Rfa H and possibly other proteins to the transcription complex to promote transcriptional elongation. Recently, it has been proposed that the larger JUMPstart sequence may permit the formation of stem loop structures in the 5′mRNA that mediates the interactions between the mRNA molecule and Rfa H (Marolda and Valvano, 1998). The observations that either a mutation in the *rfaH* gene or deletion of the JUMPstart sequence abolished K5 and K1 capsule production confirmed a role for Rfa

H in regulating group 2 capsule expression in *E. coli* (Stevens *et al.*, 1997). Analysis of the phenotype of an *rfaH* mutant demonstrated that expression of region 2 genes was dramatically reduced and by quantitative reverse transcriptase-PCR (RT-PCR) it was possible to show that this effect was due to a reduction in readthrough transcription across a Rho-dependent terminator in the *kpsT–kfiA* junction (Fig. 1). This is in keeping with Rfa H regulating group 2 capsule expression by permitting transcription originating from the region 3 promoter to proceed through into region 2. The coregulation of a number of cell surface factors by Rfa H is curious and it will be interesting to see how the expression of *rfaH* is regulated and how this relates to the expression of particular cell surface structures under specific environmental conditions.

The genetic organization of region 2 is serotype specific. In the case of the K5 capsule gene cluster, region 2 comprises four genes *kfiABCD* (Petit *et al.*, 1995), while there are six genes in region 2 of the K1 capsule gene cluster (Bliss and Silver, 1996). In both cases, transcription of region 2 is in the same direction of that of region 3, which is important in permitting the regulation of region 2 expression by Rfa H (Stevens *et al.*, 1997; Whitfield and Roberts, 1999). In the K5 capsule gene cluster, promoters have been mapped 5' to *kfiA, kfiB*, and *kfiC* genes. Transcription from the *kfiA* promoter generates a polycistronic transcript of 8.0 kb, while transcription from the *kfiB* or *kfiC* promoter results in transcripts of 6.5 and 3.0 kb, respectively (Petit *et al.*, 1995). This transcriptional organization is surprising, since it generates transcripts with two large untranslated intergenic regions, a gap of 340 bp between the *kfiA* and *kfiB* genes and a gap of 1293 bp between *kfiB* and *kfiC* genes, both of which appear to be untranslated (Petit *et al.*, 1995). The role, if any, of these regions in the mRNA molecule in regulating expression of the region 2 genes is currently unknown. The three region 2 promoters are not temperature regulated, with equivalent transcription at both 37°C and 18°C (Roberts, 1996). However, the region 2 promoters are weak and generate low levels of expression of the region 2 genes, which in the absence of Rfa H-mediated readthrough transcription from the region 3 promoter is insufficient for synthesis of detectable K5 polysaccharide (Stevens *et al.*, 1997). This complex pattern of transcription raises the question of the role of these promoters in the expression of the K5 capsule. One possibility is that these promoters play a role in fine-tuning the expression of the *kfi* genes, or in allowing the bacteria to respond rapidly to temperature changes by maintaining a pool of *kfi*-specific mRNA. Equally, it is possible that these promoters play no role in regulating *kfi* gene expression; rather, they may be remnants following the evolution of the K5 capsule gene cluster and the acquisition of the K5-specific region 2. This process may have occurred either in a single event from another bacterial species in which these promoters were functionally important or in a piecemeal fashion, with each incoming region 2 gene(s) bringing with it its own promoter. Ultimately, provided the transcription of the acquired K5 region 2 was in the same direction as that of region 3, then whatever promoters were also inherited would be irrelevant. If the region 2 promoters play no functional role, it would suggest that acquisition of the K5 region 2 by *E. coli* was a relatively recent event.

Expression of group 2 capsules is temperature regulated with capsule expression at 37°C but not at 18°C. Transcription from the region 1 and 3 promoters is temperature regulated, with no transcription detectable at 18°C (Cieslewicz and Vimr, 1996; Simpson *et al.,* 1996). Temperature regulation is in part controlled by the global regulatory protein Hns, since *hns* mutants show detectable transcription from the region 1 and 3 promoter at 18°C, albeit lower than that seen at 37°C (Rowe and Roberts, unpublished results). This is analogous to the Hns-mediated thermoregulation of the *virB* promoter in *Shigella flexneri* (Dorman and Porter, 1998). In this system, activation of the *virB* promoter has an absolute requirement for the Ara C-like protein Vir F (Dorman and Porter, 1998), and it is clear that Hns regulation involves some form of antagonistic interplay between Hns and an Ara C-like transcriptional activator. Recently, a transcriptional activator for the mediating transcription from the region 1 and 3 promoters has been identified (Rowe, Burton, and Roberts, unpublished results), suggesting that a similar situation may exist in the temperature regulation of transcription from these promoters. At 37°C, the situation is further complicated by the interaction of IHF with the region 1 promoter. IHF tends to act as a facilitator, potentiating the activity of other regulatory proteins (Freundlich *et al.,* 1992), and as such it is likely that IHF interacts with the recently identified Ara C-like transcriptional activator that controls transcription from the region 1 and 3 promoter at 37°C. The lack of any IHF consensus binding sequences in the region 3 promoter (Stevens *et al.,* 1997) confirm that there is no absolute requirement for IHF.

Therefore, in summary, the control of expression of group 2 capsule gene clusters in *E. coli* is complex involving several overlapping regulatory circuits (Fig. 1). The temperature regulation is achieved by temperature-dependent transcription from the region 1 and 3 promoters (Fig. 1). This is mediated by Hns and an Ara C-like transcriptional activator. Superimposed on this system at 37°C is IHF acting at the region 1 promoter, the intragenic terminator within the *kpsF* gene and the processing of the region 1 mRNA to generate a stable *kpsS*-specific transcript. In addition, Rfa H acts to allow transcription from the region 3 promoter to extend into region 2, and thereby result in sufficient expression of region 2 genes for capsular polysaccharide biosynthesis. The net effect of this is that expression of group 2 capsule gene clusters is regulated by two convergent promoters (Fig. 1). However, there are still many unanswered questions concerning the regulation of group 2 capsule gene clusters. In particular, what is the function of the large untranslated regions of 225 bp and 741 bp in the 5′ end of the respective region 1 and 3 mRNA molecules? How are changes in temperature sensed and transduced to induce changes in gene expression and what if any other environmental stimuli may regulate capsule expression?

6. THE GENETIC ORGANIZATION OF *E. COLI* GROUP 3
CAPSULE GENE CLUSTERS

Group 3 capsule gene clusters map near *serA* and are allelic to group 2 capsule gene clusters, being located at the same site on the chromosome. Indeed, the first group 2 (*kps*) genes mapped to the *serA* region of the *E. coli* chromosome by Ørskov and Nyman (1974) were the K10 and K54 capsule gene clusters, which subsequently turned out to be group 3 capsule gene clusters (Pearce and Roberts, 1995). To date, these remain the only group 3 capsule gene clusters that have been analyzed in detail (Pearce and Roberts, 1995; Russo *et al.,* 1998). Group 3 capsule gene clusters have a segmental gene organization reminiscent of group 2 capsule gene clusters. There are two conserved regions 1 and 3 that flank a serotype specific central region 2 (Fig. 2) (Pearce and Roberts, 1995; Clarke *et al.,* 1999). However, within this arrangement the organization of the genes is different. In the group 3 capsule gene clusters region 1 contains four genes encoding homologues of the group 2 region 1 and 3 proteins, Kps D, E, M, and T, while the group 3 region 3 is composed of two genes that encode homologues of the group 2 region 1 proteins Kps C and S (Fig. 2) (Russo *et al.,* 1998; Clarke *et al.,* 1999). The A+T ratio of the group 3 region 1 is 61%, which is significantly higher than the average 51% for the *E. coli* genome (Blattner *et al.,* 1997) and that of region 1 (50%) and region 3 (57%) of the group 2 capsule gene clusters (Clarke *et al.,* 1999). This suggests that region 1 of the group 3 capsule gene clusters may have been acquired by a different route to that of regions 1 and 3 of the group 2 capsule gene clusters. The A+T ratio of the group 3 region 3 is 57%, which is lower than that for region 1, suggesting that regions 1 and 3 of group 3 capsule gene clusters may have been acquired separately in two independent lateral gene transfers.

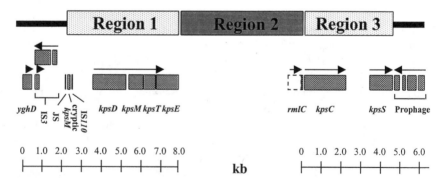

Figure 2. A scale diagram of the *Escherichia coli* K10 capsule gene cluster. The three functional regions are shown and the genes within each region denoted. The direction of transcription is denoted by the arrows. The JUMPstart sequence is shown as JS.

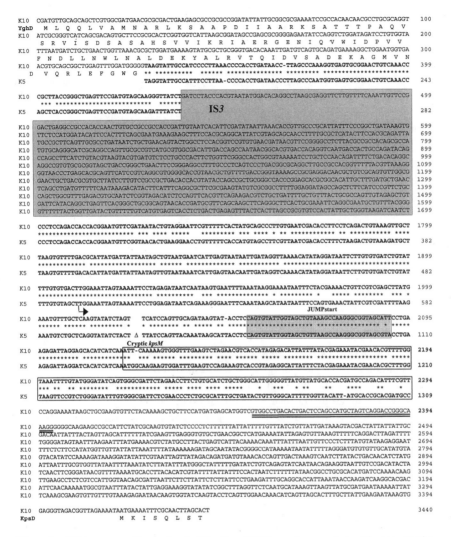

Figure 3. DNA sequence 5′ to region 1 of the *E. coli* K10 capsule gene cluster. The sequence homologous to IS*3* and the JUMPstart sequence are indicated by shaded boxes. The K5 sequence is displayed under the K10 sequence where the two are homologous (boldface). The numbering of the K10 DNA starts within the *yghD* gene. The K5 sequence is numbered from the stop codon of the *yghD* gene. The IS*3* element interrupts regions of homology between K5 and K10. The cryptic *kpsM* gene is indicated by unshaded boxes. The remnant of IS*110* is marked by a solid line. Δ Indicates the region of K10 sequence deleted (531 bp) relative to K5. An arrow marks the transcriptional start site in the K5 sequence.

Analysis of the nucleotide sequence 5′ to the first gene of region 1, $kpsD_{K10}$, revealed the presence of sequences highly homologous to the 5′ end of the $kpsM$ gene and the region 3 promoter from the K5 capsule gene cluster (Fig. 3) (Russo *et al.,* 1998; Clarke *et al.,* 1999). Within this region a JUMPstart motif was detected 1326 bp 5′ to the ATG codon of $kpsD_{K10}$ (Fig. 3), suggesting that group 3 capsule gene clusters also may be regulated by Rfa H. While the transcription start site has not been identified 5′ to region 1 in either the K10 or K54 capsule gene clusters, the high degree of homology between the transcription start site in K5 region 3 promoter and equivalent sequences in the K10 and K54 capsule gene clusters (Fig. 3) would suggest that transcription may be initiated at the same site in both group 2 and 3 capsule gene clusters (Clarke *et al.,* 1999). Although the K10 and K54 region 1 promoter regions are highly homologous to the K5 region 3 promoter, the K10 and K54 sequences contain a deletion that positions the JUMPstart sequence 531 bp closer to the putative transcriptional start site relative to that in K5 (Fig. 3). The significance of this deletion in the regulation of group 3 capsule gene expression is not clear, but it is known that the K10 capsule gene cluster is regulated by Rfa H (Clarke *et al.,* 1999) and the presence of a JUMPstart sequence in K54 (Russo *et al.,* 1998) would imply that this is true for all group 3 capsule gene clusters.

The extent of nucleotide homology between the K5 and K10 and K54 region 1 promoter regions extends approximately 190 bp into the 5′ coding sequence of the K5 $kpsM$ gene (Fig. 3). This DNA sequence does not show significant homology to the functional $kpsM_{K10}$ and $kpsM_{K54}$ genes located 3′ to $kpsD$ (Fig. 2). The identification of a cryptic $kpsM_{K5}$ gene suggests that the groups 2 and 3 promoter regions were acquired from a common ancestor expressing a group 2 capsule. It is possible that the group 3 capsule cluster was derived by the insertion of foreign capsule genes 3′ of the start of the $kpsM$ gene in an exiting group 2 capsule gene cluster. As a consequence, the group 3 capsule determinants are transcribed from the group 2 promoter region, and therefore are regulated by Rfa H. This may imply that Rfa H regulation of capsule expression is an evolutionary advantage to pathogenic *E. coli* strains.

A remnant of IS*110* from *Streptomyces coelicolor* is present 53 bp 3′ to the cryptic $kpsM_{K5}$ gene in K10 and K54 capsule gene clusters (Figs. 2 and 3) and it has been postulated that this insertion sequence (IS) element was involved in the mobilization of the group 3 determinants into the progenitor group 2 strain (Russo *et al.,* 1998). In addition to the IS110 sequence, a region of 99% homology to IS3 (Timmerman and Tu, 1985) was identified 5′ to the K10 JUMPstart sequence (Fig. 3). This insertion element was not identified in the K54 capsule gene cluster and would appear to be specific to the K10 capsule gene cluster. IS elements have been implicated in the duplication of genes in the group I capsule locus of *E. coli* K30 (Drummelsmith *et al.,* 1997) and they have been found near the capsule genes of *Klebsiella pneumoniae* (Wacharotayankun *et al.,* 1993). In addition, remnants of IS600 and IS630 elements may have been involved in lateral transfer of a pathogenicity island into enteropathogenic and enterohemorrhagic *E. coli* (Perna *et al.,* 1998). Conceivably, a block of capsule genes could be mobilized through transposition if they were flanked by IS elements. Although numerous IS3 elements are present on the *E. coli* chromosome (Blattner *et al.,* 1997), a second IS3 within or flanking the K10 capsule gene cluster was not identified (Clarke *et al.,* 1999). However, a second flanking IS3 element could have been lost through subsequent recombination events. Alternatively, capsule genes could be transferred by homologous recombination between IS elements located in *E. coli* and in DNA from another organism.

Analysis of DNA sequence 3' to region 3 (Fig. 2) identified the presence of a prophage related to retronphage (R73 and other CP4-like cryptic prophages found in *E. coli* K-12 (Blattner *et al.,* 1997). Numerous virulence determinants have been associated with lysogenic bacteriophages (Cheetham and Katz, 1995), and CP4-like cryptic prophages have been implicated in the acquisition of the locus of enterocyte effacement (LEE) pathogenicity island of enterohemorrhagic *E. coli* strain, EDL933 (Perna *et al.,* 1998). Therefore, it is possible that bacteriophage transduction may have played a role in the acquisition of the K10 capsule gene cluster and other group 3 capsule gene clusters.

7. THE BIOSYNTHESIS OF *E. COLI* GROUP 2 CAPSULES

The biosynthesis of group 2 capsular polysaccharides occurs on the inner face of the cytoplasmic membrane by the sequential addition of activated sugar residues to the nonreducing end of the growing polysaccharide chain. The polymerization is catalyzed by a processive glycosyltransferase enzyme, which in the case of both the K1 and K5 capsules is incapable of initiating the biosynthetic reaction (Roberts, 1996; Steenbergen and Vimr, 1990). The initiation of group 2 polysaccharide biosynthesis, the nature of the initial acceptor, and the role of lipids in this process are still unresolved. In the case of K1 biosynthesis, polyisoprenoid lipid intermediates have been identified (Troy, 1995), while this has not been demonstrated for the K5 capsular polysaccharide (Finke *et al.,* 1991). It is possible of course that the initiation of biosynthesis of group 2 polysaccharides is not conserved mechanistically and that starting reactions and acceptors used will vary from one group 2 polysaccharide to another. Whatever the scenario, at some point the nascent polysaccharide molecule must be ligated at its reducing end to phosphatidyl-Kdo. The timing of this substitution also is open to conjecture. Mutations in *kpsC* and *kpsS* result in the cytoplasmic accumulation of group 2 polysaccharides, which lack phosphatidyl-Kdo (Bronner *et al.,* 1993); this has been interpreted to suggest that the addition of phosphatidyl-Kdo occurs after the initiation of group 2 polysaccharide biosynthesis and that these reactions are catalyzed by the Kps C and S proteins (Roberts, 1996). However, it is possible the lack of phosphatidyl-KDO at the reducing end of cytoplasmic polysaccharide in these mutants reflects the lability of the glycosyl–phosphate linkage between the polysaccharide and the phosphatidyl-KDO. Therefore, until biochemical activities can be unequivocally assigned to the Kps C and S proteins, their role and the timing of this substitution process remains uncertain.

Subsequently, the substituted nascent polysaccharide chain is transported across the cytoplasmic membrane, periplasm, and outer membrane to be anchored on the bacterial cell surface. The transport process is mediated by the Kps C, D, E, F, M, S, T, and U proteins, which transport the specific group 2 polysaccharide independent of the repeat structure of the particular polysaccharide molecule (Whitfield and Roberts, 1999; Roberts, 1996).

8. THE BIOSYNTHESIS OF THE *E. COLI* K5 POLYSACCHARIDE

The expression of the K5 capsule is the best studied to date of the *E. coli* group 2 capsules (Whitfield and Roberts, 1999; Roberts, 1996). The biosynthesis of the K5 polysaccharide requires the Kfi A–D proteins (Petit *et al.*, 1995) and functions now have been assigned to the Kfi A, C, and D proteins (Petit *et al.*, 1995; Griffiths *et al.*, 1998). The Kfi C protein is the processive bifunctional glycosyltransferase that adds alternating GlcA and GlcNAc residues to the nonreducing end of the growing polysaccharide chain (Fig. 4) (Griffiths *et al.*, 1998). As such, the enzyme has both α- and ß-transferase activities, but is unable to initiate *de novo* polysaccharide biosynthesis in the absence of an appropriate oligosaccharide acceptor (Griffiths *et al.*, 1998). Structure–function analysis of the Kfi C protein identified a secondary structure motif termed domain A characteristic of ß-glycosyltransferases together with two highly conserved aspartic acid residues at positions 301 and 352 (Griffiths *et al.*, 1998). Site-directed mutagenesis in combination with *in vitro* transferase assays using oligosaccharide acceptors with defined nonreducing ends confirmed that this region constitutes the active site for ß-GlcA transferase activity and that the conserved aspartic acids are the catalytically important amino acids (Griffiths *et al.*, 1998). The α-GlcNAc transferase activity was assigned to the first 381 amino acids of the 520 amino acid full-length protein (Griffiths *et al.*, 1998), and attempts to further define the α-GlcNAc transferase active site are currently underway.

Mutations in the *kfiA* and *kfiB* genes abolish any detectable K5 polysaccharide and result in low K5 transferase activity when membranes are assayed *in vitro* (I. S. Roberts, unpublished results). Addition of oligosaccharide acceptors to the *in vitro* assay stimulates K5 transferase activity, suggesting that *kfiA* and *kfiB* mu-

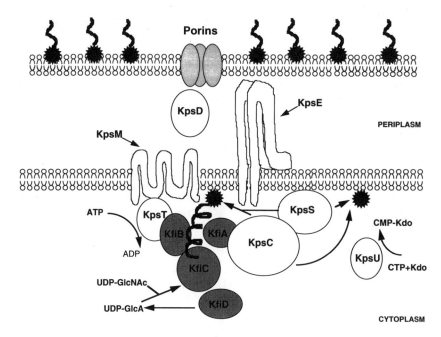

Figure 4. A schematic representation of the K5 biosynthetic–export complex. The phosphatidyl-Kdo is depicted by the filled star shape, while the K5 polysaccharide is shown as the helical structure. The region 2 proteins are shaded. The formation of a membrane adhesion site is not shown, although some of localized membrane perturbation is likely to occur to form a direct continuum between the cytoplasm and the cell surface.

tants are defective in the initiation of K5 biosynthesis (Roberts, 1996). Recently, the Kfi A protein has been purified and shown to be a GlcNAc transferase enzyme (G. Griffiths and I. S. Roberts, unpublished results), thereby raising the possibility that Kfi A is involved in the addition of GlcNAc to a membrane acceptor, which then serves as a substrate for the Kfi C transferase to polymerize the K5 polysaccharide (Fig. 4). The role of the Kfi B protein in this process is as yet unknown. No detectable transferase activity can be assigned to this protein, and the only clue to its likely function is that structural predictions indicate that the Kfi B protein is likely to be a coiled–coil protein (I. S. Roberts unpublished results). As such, the Kfi B protein may play some form of structural role in facilitating the initiation of K5 polysaccharide biosynthesis.

The Kfi D protein has been purified and demonstrated to be a UDP-Glc dehydrogenase, which converts UDP-Glc to UDP-GlcA, one of the two substrates for Kfi C (Petit *et al.*, 1995). The other substrate for Kfi C is UDP-GlcNAc, but there is no need to specifically synthesize this molecule, since there is a pool of

UDP-GlcNAc within *E. coli* due to the incorporation of GlcNAc into peptidoglycan and ECA (Rick and Silver, 1996).

Therefore, the biosynthesis of the *E. coli* K5 polysaccharide can be summarized in the following. The initiation of K5 biosynthesis involves the Kfi A and Kfi B proteins. The Kfi A protein is a GlcNAc transferase that probably adds GlcNAc to an as yet unidentified membrane acceptor, which then acts as a substrate for extension by Kfi C. The Kfi B protein is essential for the initiation reaction to take place, but its role has not been defined. The Kfi D proteins is a UDP-Glc dehydrogenase that generates UDP-GlcA for the polymerization of the K5 polysaccharide. At what point phosphatidyl-KDO is added or whether it is indeed the acceptor for the first reaction is as yet unknown. All these biosynthetic reactions take place on the inner face of the cytoplasmic membrane, and there is now increasing evidence that a multiprotein biosynthetic complex is formed at this site (Rigg *et al.*, 1998).

9. THE TRANSPORT OF *E. COLI* GROUP 2 CAPSULAR POLYSACCHARIDES

The export of capsular polysaccharides in *E. coli* from their site of synthesis on the inner face of the cytoplasmic membrane onto the bacterial cell surface presents a unique challenge to the microorganism. It requires the translocation of a hydrophilic high-molecular-weight negatively charged macromolecule across two lipid bilayers and the intervening periplasmic space. In the case of group 2 capsular polysaccharides a single export pathway, irrespective of the repeat structure of the particular polysaccharide molecule, is used to translocate the polysaccharides from their site of synthesis on the inner face of the cytoplasmic membrane onto the cell surface (Roberts, 1996; Roberts *et al.*, 1988). The Kps C, D, E, F, M, S, T, and U proteins constitute the transport pathway for group 2 polysaccharides (Roberts, 1996). The functions for certain of these proteins have been determined and activities demonstrated *in vitro,* but for the remainder the precise role played is still unclear. As described earlier, the Kps C and S proteins are believed to attach phosphatidyl-Kdo to the reducing end of the nascent polysaccharide chain (Roberts, 1996), albeit that the enzymology of this reaction awaits characterization. The substitution with phosphatidyl-Kdo may permit entry of the polysaccharide molecule into the export pathway (Bronner *et al.,* 1993; Roberts, 1996; Whitfield and Roberts, 1999) and serve as the export "flag" recognized by the translocation machinery. This could explain how chemically different group 2 polysaccharides may be exported by the same pathway. The CMP-Kdo for the attachment of Kdo to phospholipid is generated by Kps U, a group 2-specific CMP-Kdo synthetase enzyme (Pazzani *et al.,* 1993). The subsequent translocation of the substituted group 2 polysaccharide across the cytoplasmic membrane is achieved by the Kps M and T proteins, which constitute an ATP-binding cassette (ABC-2) transporter (Paulsen *et al.,* 1997), in which Kps M is the integral membrane protein and Kps T is the ATPase (Fig. 4) (Smith *et al.,* 1990; Pavelka *et al.,* 1994; Pigeon and Silver, 1994). A model for the action of the Kps MT transporter has been reviewed recently (Bliss and Silver, 1996). Typically, such ABC-2 transporters involved in polysaccharide export require two additional accessory proteins: a cytoplasmic membrane-periplasmic auxiliary protein (MPA) and outer membrane auxiliary protein (OMA) (Paulsen *et al.,* 1997). No OMA protein is encoded by group 2 capsule gene clusters, and this role may be filled by other outer membrane proteins. The MPA family has been further subdivided based on the presence (MPA1) or absence (MPA2) of an ATP-binding domain (Paulsen *et al.,* 1997). Kps E is a member of the MPA2 family and is anchored to the cytoplasmic membrane via an N-terminal *trans*-membrane domain with a large periplasmic domain of 300 amino acids (Whitfield and Roberts, 1999; Roberts, 1996; Rosenow *et al.,* 1995). The C-terminus of the Kps E protein is associated with the outer face of the cytoplasmic membrane via an amphipathic α-helix (T. Hammarton and I. S. Roberts, unpublished results). It is possible that Kps E may act in an analogous fashion to the membrane fusion proteins (MFPs) that are present in analogous ABC transport systems for protein secretion (Dinh *et al.,* 1994). In these systems it has been suggested that MFPs are anchored to the cytoplasmic membrane via their N-termini and interact with the outer membrane via their C-termini, thereby spanning the periplasm and linking the cytoplasmic and outer membranes (Dinh *et al.,* 1994). In the case of Kps E, it could interact with the outer membrane via its periplasmic domain, and thereby generate adhesion sites between the cytoplasmic and outer membrane that are associated with the biosynthetic–export complex (Fig. 4). Chemical cross-linking indicates that Kps E exists as a dimer and that dimerization is mediated through the formation of coiled–coil structure (T. Hammarton and I. S. Roberts, unpublished results).

The Kps D protein is localized in the periplasm and does show some limited homology to the OMA family (Paulsen *et al.,* 1997). However, considering the periplasmic location of Kps D, the significance of this observation is unclear. Mutations in Kps D result in the accumulation of periplasmic polysaccharide that is localized at sites of membrane adhesion (Bliss and Silver, 1996; Roberts, 1996). This suggests that Kps D is involved in the terminal stages of the transport pathway. It has been suggested that Kps D may recruit porins to the export pathway and that porins may provide the route for the egression of group 2 polysaccharides onto the cell surface. There is genetic evidence to support a role for porins in the transport process (Whitfield and Valvano, 1993), and the ability of porins to act in this way might explain why no outer membrane protein is encoded in the *E. coli* group 2 capsule gene cluster, in contrast to capsule gene clusters from other gram-negative bacteria (Roberts, 1996). To date, however, there is no biochemical evidence to demonstrate a role for porins, nor is it clear how the relatively small channels formed by porins (1–2 nm) would facilitate the exit of group 2 polysaccharides. Chemical cross-linking has not demonstrated any protein–protein interaction between the Kps E and D proteins (T. Hammarton, C. Arrecubieta, and I. S. Roberts, unpublished results), and it may be that these proteins interact via the exported polysaccharide molecule. Indeed, recent studies using affinity chromatography of ABC transport systems for protein secretion have demonstrated that substrate binding is required for the assembly of the export complex (Letoffe *et al.,* 1996). Whether the substitution of group 2 polysaccharides with phosphatidyl-Kdo provides the binding domain onto which these proteins associate with the polysaccharide during its export are unclear. The function of Kps F in the transport process is unclear, but it may play a role in the correct assembly of the biosynthetic–export complex (Cieslewicz and Vimr, 1997).

10. THE BIOSYNTHETIC–EXPORT COMPLEX
   FOR *E. COLI* GROUP 2 CAPSULES

While it is convenient to dissect capsule expression into polysaccharide biosynthesis and polysaccharide export, the reality is that these two processes are intimately linked. The observation that mutations that disrupt polysaccharide export are pleiotropic and also affect polysaccharide biosynthesis confirms the linkage between these events (Roberts, 1996; Bronner *et al.,* 1993). Indeed, there is now increasing evidence from studies on expression of the K5 capsule that biosynthesis of group 2 capsules involves a hetero-oligomeric membrane-bound protein complex on the cytoplasmic membrane (Rigg *et al.,* 1998). This complex consists of the proteins (Kfi A–D) required for the polymerization of the K5 polysaccharide, together with the Kps C, M, S, and T proteins that are involved in polysaccharide transport across the cytoplasmic membrane (Fig. 4) (Rigg *et al.,* 1998). The analysis of mutants defective for individual Kps proteins indicated that the Kps C, M, S, and T proteins are required to target the biosynthetic machinery (the Kfi A and Kfi C proteins) for K5 polysaccharide biosynthesis to the cytoplasmic membrane, suggesting that these Kps proteins play critical roles in the formation and stabilization of the biosynthetic–export complex on the cytoplasmic membrane (Rigg *et al.,* 1998). In addition, the association of the Kfi C glycosyltransferase with the complex is dependent on the presence of the Kfi A protein, suggesting that there is some form of hierarchy of association in the formation of the biosynthetic–export complex on the cytoplasmic membrane (G. Griffiths and I. S. Roberts, unpublished results). This order of assembly of the complex would be in keeping with the possible role of the Kfi A protein as the initiating glycosyltransferase that provides the acceptor substrate for the Kfi C enzyme to synthesize the K5 polysaccharide. The conservation of the Kps proteins among all *E. coli* strains expressing group 2 capsules (Roberts, 1996) would indicate that this hetero-oligomeric complex is a common feature in the biosynthesis of group 2 capsules and that these Kps proteins may provide the scaffold onto which the specific capsule biosynthetic proteins associate. The formation of such a multiprotein complex on the cytoplasmic membrane would permit the initiation of polysaccharide synthesis, polysaccharide extension, the addition of phosphatidyl-Kdo and polysaccharide export across the cytoplasmic membrane to be coordinated spatially at one site. In addition, complex formation could improve the efficiency of polysaccharide biosynthesis by increasing the effective concentrations of the necessary proteins at the site of polymer synthesis. It is likely that analogous situations involving hetero-oligomeric protein complexes will exist for the biosynthesis of other capsules in *E. coli* (Whitfield and Roberts, 1999).

The observation that the Kps C and S, together with the Kfi A and C proteins, could be released from the cell by osmotic shock (Rigg *et al.,* 1998) provides further insight into the possible architecture of the biosynthetic–export complex on the cytoplasmic membrane. Cytoplasmic proteins that can be released from *E. coli* by osmotic shock are termed group D proteins (Beacham, 1979) and include the cytoplasmic-bound components of the enterobactin synthase complex, thioredoxin, and Dna K (Rigg *et al.,* 1998). It has been suggested that the release of these cytoplasmic proteins by osmotic shock is a consequence of their association with areas of adhesion between the cytoplasmic and outer membrane (Bayer *et al.,* 1987). These membrane adhesion sites have been implicated as channels for the export of group I capsular polysaccharides (Bayer and Thurow, 1977) and filamentous bacteriophage f1 (Bayer, 1991), as well as the import of colicin A (Guihard *et al.,* 1994). The release following osmotic shock of the Kfi A, Kfi C, Kps C, and Kps S proteins might suggest that the polysaccharide biosynthetic–export complex may be formed at sites of adhesion between the cytoplasmic and outer membrane. The observation that the Kps T protein may be transitorily exposed in the periplasm (Bliss and Silver, 1996) would support the notion that there is a direct link between cytoplasmic components of the biosynthetic complex and the periplasm. The finding that Kps E is associated with areas of membrane adhesion (C. Arrecubieta and I. S. Roberts, unpublished results) would suggest that a multiprotein biosynthetic–export complex is formed that is associated with sites of membrane adhesion (Fig. 4). Such a complex straddling the two membranes and periplasm would provide a direct continuum between the site of polysaccharide biosynthesis in the cytoplasm and its ultimate location on the cell surface, and thereby facilitate the movement of the polysaccharide molecule. Clearly, while there is compelling evidence for the existence of such a complex, the architecture of the complex is still a matter of conjecture as are the specific protein–protein interactions that underpin its formation. Molecular dissection of this complex represents a major challenge in understanding the biosynthesis of group 2 capsules in *E. coli.*

11. CONCLUSIONS

The expression of group 2 polysaccharides in *E. coli* offers an experimentally tractable system in which to study the biosynthesis and export of capsular polysaccharides in gram-negative bacteria. The conservation between different gram-negative pathogens of specific steps in polysaccharide transport makes this a particularly appealing notion. While we have begun to make progress in determing the biochemistry of polysaccharide biosynthesis and transport, there are many unanswered questions that offer exciting future challenges. It is likely that understanding capsule biogenesis at the molecular level will allow the design of potential new antimicrobial agents specifically targeted to inhibit capsule biosynthesis, and thereby permit host clearance of the gram-negative pathogen. With the prospect of a return to a preantibiotic era, such agents could prove invaluable in the fight against gram-negative infections. In addition, understanding at the molecular level the mechanisms of capsule biosynthesis should permit the engineering of polysaccharides in *E. coli* of biomedical importance.

The regulation of group 2 capsule gene expression in *E. coli* appears to be complex, with multiple overlapping regulatory circuits. The challenge will be to decipher this hierarchy and attempt to relate the regulation of capsule expression to the biology of *E. coli*. In particular, what roles do capsules play in the colonization and survival of *E. coli* and how does the regulation of capsule expression relate to these processes. While possible roles for the capsule in invasive disease readily can be assigned, such a situation is atypical for *E. coli*. It will be more pertinent to address the key question of the role of the capsule during growth as a commensal in the gut and the role of the capsule in mediating interactions with other members of the microbial consortium and the host mucosal surface.

Acknowledgments

I gratefully acknowledge the hard work and dedication of previous and current members of my laboratory, without whom none of this work would have been achieved. In addition, I gratefully acknowledge the support of the Lister Institute of Preventive Medicine, the Wellcome Trust, together with the Biotechnology and Biological Sciences Research Council, and the Medical Research Council of the UK.

REFERENCES

Bailey, M. J., Hughes, C., and Koronakis, V., 1997, RfaH and the *ops* element, components of a novel system controlling bacterial transcription elongation, *Mol. Microbiol.* **26**:845–851.

Bayer, M. E., 1991, Zones of membrane adhesion in cryofixed envelope of *Escherichia coli, J Struct. Biol.* **107**:268–280.

Bayer, M. E. and Thurow, H., 1977, Polysaccharide capsule of *Escherichia coli:* Microscope study of its size, structure, and sites of synthesis, *J. Bacteriol.* **130**:911–936.

Bayer, M. E., Baye, M. H., Lunn, C. A., and Pigiet, V., 1987, Association of thioredoxin with the inner membrane and adhesion sites in *Escherichia coli, J. Bacteriol.* **169**:2659–2666.

Beacham, I. R., 1979, Periplasmic enzymes in Gram-negative bacteria. *Int. J. Biochem.* **10**:877–883.

Bilge, S. S., Apostol, J. M., Fullner, K. J., and Moseley, S. L., 1993, Transcription organisation of the F1854 fimbrial adhesin determinant of *Escherichia coli, Mol. Microbiol.* **7**:993–1006.

Blattner, F. R., Plunkett III, G., Bloch, C. A., Perna, N. T., Burland, V., Riley, M., Collado-Vides, J., Glasner, J. D., Rode, C. K., Mayhew, G. F., Gregor, J., Davis, N. W., Kirk-patrick, N. W., Goeden, M. A., Rose, J. D., Mau, B., and Shao, Y., 1997, The complete genome sequence of *Escherichia coli* K-12, *Science* **277**:1453–1462.

Bliss, J. M., and Silver, R. P., 1996, Coating the surface: A model for expression of capsular polysialic acid in *Escherichia coli* K1, *Mol. Microbiol.* **21**:221–231.

Boulnois, G. J., Roberts, I. S., Hodge, R., Hardy, K., Jann, K., and Timmis, K. N., 1987, Analysis of the K1 capsule biosynthesis genes of *Escherichia coli:* Definition of three functional regions for capsule production, *Mol. Gen. Genet.* **208**:242–246.

Boulnois, G. J., Drake, R., Pearce, R., and Roberts, I. S., 1992, Genome diversity at the *serA*-linked capsule locus in *Escherichia coli, FEMS Microbiol. Lett.* **100**:121–124.

Bronner, D., Sieberth, V., Pazzani, C., Roberts, I. S., Boulnois, G. J., Jann, B., and Jann, K., 1993, Expression of the capsular K5 polysaccharide of *Escherichia coli:* Biochemical and electron microscopic analyses of mutants with defects in region 1 of the K5 gene cluster, *J. Bacteriol.* **175**:5984–5992.

Cheetham, B. F., and Katz, M. E., 1995, A role for bacteriophages in the evolution and transfer of bacterial virulence determinants, *Mol. Microbiol.* **18**:201–208.

Cieslewicz, M., and Vimr, E., 1996, Thermoregulation of *kpsF*, the first region 1 gene in the *kps* locus for polysialic acid biosynthesis in *Escherichia coli* K1, *J. Bacteriol.* **178**:3212–3220.

Cieslewicz, M., and Vimr, E., 1997, Reduced polysialic acid capsule expression in *Escherichia coli* K1 mutants with chromosomal defects in *kpsF, Mol. Microbiol.* **26**:237–249.

Clarke, B. R., Pearce, R., and Roberts, I. S., 1999, Genetic organisation of the *Escherichia coli* K10 capsule gene cluster: identification and characterisation of two conserved regions in group III capsule gene clusters encoding polysaccharide transport functions, *J. Bacteriol.* **181**:2279–2285.

Costerton, J. W., Chjeng, K.-J., Geesey, G. G., Ladd, T. I., Nickel, J. C., Dasgupta, M., and Marrie, T., 1987, *Annu. Rev. Microbiol.* **41**:435–466.

Dinh, T., Paulsen, I. T., and Saier, M. H., 1994, A family of extracytoplasmic proteins that allow transport of large molecules across the outer membrane of gram-negative bacteria, *J. Bacteriol.* **176**:3825–3831.

Dorman, C. J., and Porter, M. E., 1998, The *Shigella* virulence gene regulatory cascade: A paradigm of bacterial gene control mechanisms, *Mol. Microbiol.* **29**:677–684.

Drummelsmith, J., Amor, P. A., and Whitfield, C., 1997, Polymorphism, duplication and IS-1-mediated rearrangement in the *his-rfb-gnd* reion of *Escherichia coli* strains with group IA capsular-K-antigens, *J. Bacteriol.* **179**:3232–3238.

Finke, A., Bronner, D., Nikolaev, A. V., Jann, B., and Jann, K., 1991, Biosynthesis of the *Escherichia coli* K5 polysaccharide, a representative of group II capsular polysaccharides: Polymerization *in vitro* and characterization of the product, *J. Bacteriol.* **173**:4088–4094.

Freundlich, M., Ramani, N., Mathew, E., Sirko, A., and Tsui, P., 1992, The role of integration host factor in gene expression in *Escherichia coli, Mol. Microbiol.* **6**:2557–2563.

Gotschlich, E. C., Fraser, B. A., Nishimura, O., Robbins, J. B., and Liu, T-Y., 1981, Lipid on capsular polysaccharides ogf gram-negative bacteria, *J. Biol. Chem.* **256**:8915–8921.

Griffiths, G., Cook, N. J., Gottfridson, E., Lind, T., Lidholt, K., and Roberts, I. S., 1998, Characterization of the glycosyltransferase enzyme from the *Escherichia coli* K5 capsule gene cluster and identification and chracterization of the glucuronyl active site, *J. Biol. Chem.* **273**:11752–11757.

Guihard, G., Boulanger, P., Bénédetti, H., Loubés, R., Bernard, M., and Letellier, L., 1994, Colicin A and Tol proteins involved in its translocation are preferentially located in the contact sites between the inner and outer membranes of *Escherichia coli* cells, *J. Biol. Chem.* **269**:5874–5880.

Hobbs, M., and Reeves, P. R., 1994, The JUMPstart sequence: A 39 bp element common to several polysaccharide gene clusters, *Mol. Microbiol.* **12**:855–856.

Jann, B., and Jann, K., 1990, Structure and biosynthesis of the capsular antigens of *Escherichia coli, Curr. Top. Microbiol. Immunol.* **150**:19–42.

Jann, K., and Jann, B., 1992, Capsules of *Escherichia coli,* expression and biological significance, *Can. J. Microbiol.* **38**:705–710.

Jayaratne, P., Keenleyside, W. J., MacLachlan, P. R., Dodgson, C., and Whitfield, C., 1993, Characterization of *rcsB* and *rcsC* from *Escherichia coli* O9:K30:H12 and examination of the role of the *rcs* regulatory system in expression of group I capsular polysaccharides, *J. Bacteriol.* **175**:5384–5394.

Keenleyside, W. J., Jayaratne, P., MacLachlan, P. R., and Whitfield, C., 1992, The *rcsA* gene of *Escherichia coli* O9:K30:H12 is involved in the expression of the serotype-specific group I K (capsular) antigen, *J. Bacteriol.***174**:8–16.

Klug, G., 1993, The role of mRNA degradation in regulated expression of bacterial photosynthesis genes, *Mol. Microbiol.* **9**:1–7.

Letoff, S., Delepelaire, P., and Wandersman, C., 1996, Protein secretion in gram-negative bacteria: Assembly of the three components of the ABC protein mediated exporters is ordered and promoted by substrate binding, *EMBO J* **15**:5804–58011.

Marolda, C. L., and Valvano, M. A., 1998, Promoter region of the *Escherichia coli* O7-specific lipopolysaccharide gene cluster: Structural and functional characterization of an upstream untranslated mRNA sequence, *J. Bacteriol.* **180**:3070–3079.

Moxon, E. R., and Kroll, J. S., 1990, The role of bacterial polysaccharide capsules as virulence factors, *Curr. Top. Microbiol. Immunol.* **150**:65–86.

Ørskov, I., and Nyman, K., 1974, Genetic mapping of the antigenic determinants of two polysaccharide K antigens, K10 and K54, in *Escherichia coli, J. Bacteriol.* **120**:43–51.

Ørskov, F., and Ørskov, I., 1992, *Escherichia coli* serotyping and disease in man and animals, *Can. J. Microbiol.* **38:**699–704.

Paulsen, I. T., Beness, A. M., and Saier, M. H. J., 1997, Computer-based analyses of the protein constituents of transport systems catalysing export of complex carbohydrates in bacteria, *Microbiology* **143:**2685–2699.

Pavelka, M. S., Hayes, S. F., and Silver, R. P., 1994, Characterisation of KpsT, the ATP-binding component of the ABC-transporter involved in the export of capsular polysialic acid in *Escherichia coli* K1, *J. Biol. Chem.* **269:**20149–20158.

Pazzani, C., Rosenow, C., Boulnois, G. J., Bronner, D., Jann, K., and Roberts, I. S., 1993, Molecular analysis of region 1 of the *Escherichia coli* K5 antigen gene cluster: A region encoding proteins involved in cell surface expression of capsular polysaccharide, *J. Bacteriol.* **175:**5978–5983.

Pearce, R., and Roberts, I. S., 1995, Cloning and analysis of gene clusters for production of the *Escherichia coli* K10 and K54 antigens: identification of a new group of *serA*-linked capsule gene clusters, *J. Bacteriol.* **177:**3992–3997.

Perna, N. T., Mayhew, G. F., Pósfal, G., Elliott, S., Donnenberg, M. S., Kaper, J. B., and Blattner, F. R., 1998, Molecular evolution of a pathogenicity island from enterohemorrhagic *Escherichia coli* O157:H7, *Infect. Immun.* **66:**3810–3817.

Petit, C., Rigg, G. P., Pazzani, C., Smith, A., Sieberth, V., Stevens, M., Boulnois, G., Jann, K., and Roberts, I. S., 1995, Region 2 of the *Escherichia coli* K5 capsule gene cluster encoding proteins for the biosynthesis of the K5 polysaccharide, *Mol. Microbiol.* **17:**611–620.

Pigeon, R. P., and Silver, R. P., 1994, Topological and mutational analysis of KpsM, the hydrophobic component of the ABC-transporter involved in the export of polysialic acid in *Escherichia coli* K1, *Mol. Microbiol.* **14:**871–881.

Richardson, J. P., 1991, Preventing the synthesis of unused transcripts by Rho factor, *Cell* **64:**1047–1049.

Rick, P., and Silver, R. P., 1996, Enterobacterial common antigen and capsular polysaccharides, in: *Escherichia coli and Salmonella: Cellular and Molecular Biology* (F. C. Neidhardt, ed.), ASM Press Washington, DC, pp. 104–122.

Rigg, G. P., Barrett, B., and Roberts, I. S., 1998, The localization of KpsC, S and T and KfiA, C and D proteins involved in the biosynthesis of the *Escherichia coli* K5 capsular polysaccharide: Evidence for a membrane-bound complex, *Microbiology* **144:**2904–2914.

Roberts, I. S., 1995, Bacterial polysaccharides in sickness and in health, *Microbiology* **141:**2023–2031.

Roberts, I. S., 1996, The biochemistry and genetics of capsular polysaccharide production in bacteria, *Annu. Rev. Microbiol.* **50:**285–315.

Roberts, I. S., Mountford, R., High, N., Bitter-Suerman, D., Jann, K., Timmis, K. N., and Boulnois, G. J., 1986, Molecular cloning and analysis of the genes for production of the K5, K7, K12, and K92 capsular polysaccharides of *Escherichia coli, J. Bacteriol.* **168:**1228–1233.

Roberts, I. S., Mountford, R., Hodge, R., Jann, K., and Boulnois, G. J., 1988, Common organization of gene clusters for the production of different capsular polysaccharides (K antigens) in *Escherichia coli, J. Bacteriol.* **170:**1305–1310.

Rosenow, C., Esumeh, F., Roberts, I. S., and Jann, K., 1995, Characterisation and localizsation of the KpsE protein of *Escherichia coli* K5, which is involved in polysaccharide export, *J. Bacteriol.* **177**:1137–1143.

Russo, T. A., Wenderoth, S., Carlino, U. B., Merrick, J. M., and A. J., L., 1998, Isolation, genomic organization, and analysis of the group III capsular polysaccharide genes *kpsD, kpsM, kpsT,* and *kpsE* from an extraintestinal isolate of *Escherichia coli* (CP9, O4/K54/H5), *J. Bacteriol.* **180**:338–349.

Sieberth, V., Jann, B., and Jann, K., 1993, Structure of the K10 capsular antigen from *Escherichia coli* O11:K10:H10, a polysaccharide containing 4,6-dideoxy-4-malonylamino-D-glucose, *Carbohydr. Res.* **246**:219–228.

Silver, R. P., Vann, W. F., and Aaronson, W., 1984, Genetic and molecular analyses of *Escherichia coli* K1 antigen genes, *J. Bacteriol.* **157**:568–575.

Simpson, D. A., Hammarton, T. C., and Roberts, I. S., 1996, Transcriptional organization and regulation of expression of region 1 of the *Escherichia coli* K5 capsule gene cluster, *J. Bacteriol.* **178**:6466–6474.

Smith, A. N., Boulnois, G. J., and Roberts, I. S., 1990, Molecular analysis of the *Escherichia coli* K5 kps locus: Identification and characterisation of an inner-membrane capsular polysaccharide transport system, *Mol. Microbiol.* **4**:1863–1869.

Steenbergen, S. M., and Vimr, E. R., 1990, Mechanism of polysialic acid chain elongation in *Escherichia coli* K1, *Mol. Microbiol.* **4**:603–611.

Stevens, M. P., Clarke, B. R., and Roberts, I. S., 1997, Regulation of the *Escherichia coli* K5 capsule gene cluster by transcription antitermination, *Mol. Microbiol.* **24**:1001–1012.

Timmerman, K. P., and Tu, C. P. D., 1985, Complete sequence of IS3, *Nucleic Acids Res.* **13**:2127–2139.

Troy, F. A., 1995, Sialobiology and the polysialic acid glycotype: occurrence, structure, function, synthesis and glycopathology, in: *Biology of the Sialic Acids* (A. Rosenberg, ed.), Plenum Press, New York, pp. 95–144.

Wacharotayankun, R., Arakawa, Y., Ohta, M., Tanaka, K., Akashi, T., Mori, M., and Kato, N., 1993, Enhancement of extracapsular polysaccharide synthesis in *Klebsiella pneumoniae* by RmpA2, which shows homology to NtrC and FixJ, *Infect. Immun.* **61**:3164–3174.

Whitfield, C., and Roberts, I. S., 1999, Structure, assembly and regulation of expression of capsules in *Escherichia coli, Mol. Microbiol.* **31**:1307–1320.

Whitfield, C., and Valvano, M. A., 1993, Biosynthesis and expression of cell surface polysaccharides in gram-negative bacteria, *Adv. Microb. Physiol.* **35**:135–246.

16

Bacterial Entry and Subsequent Mast Cell Expulsion of Intracellular Bacteria Mediated by Cellular Cholesterol–Glycolipid-Enriched Microdomains

Jeoung-Sook Shin, Zhimin Gao, and Soman N. Abraham

1. INTRODUCTION

Upon gaining entry into the host tissue, a major goal of a pathogen is to resist early clearance by the immune cells of the host. An appreciable number of bacteria avoid detection and clearance by hiding within host cells. However, the intracellular environment is also fraught with danger, and in order to survive a pathogen must be able to circumvent or resist the intrinsic antimicrobial actions of the host cell. One of the most potent antimicrobial acts of the host cell is inducing the fusion of the intracellular vesicles encasing the bacteria or phagosomes with cellular lysosomes following bacterial entry. Successful intracellular pathogens such as *Mycobacterium* spp. (Clemens and Horowitz, 1995; Xu *et al.*, 1990), *Legionella pneumophila* (Clemens and Horowitz, 1995), and *Toxoplasma gondii* (Joiner *et al.*, 1990) resist intracellular killing by preventing fusion of phagosomes with lyso-

Jeoung-Sook Shin, Zhimin Gao, and Soman N. Abraham • Departments of Pathology and Microbiology, Duke University Medical Center, Durham, North Carolina 27710.

Glycomicrobiology, edited by Doyle.
Kluwer Academic/Plenum Publishers, New York, 2000.

somes, whereas others such as *Histoplasma capsulatum* (Eissenberg *et al.,* 1993) allow fusion of intracellular compartments with lysosomes but resist the toxic lysosomal contents. Interestingly, *Listeria monocytogenes* exhibit neither of these traits but avoid death by secreting a variety of enzymes that disrupt the phagosomal membrane and enable the bacteria to escape into the cytoplasm (Marquis *et al.,* 1997). It is now known that each intracellular pathogen modulates the composition of its phagosome to suit its intracellular needs. The specificity of each phagosome to its pathogen is illustrated by the observation that when two pathogens are infecting the same cell, they are rarely found in the same compartment (Hass, 1998). In spite of the large number of studies undertaken on intracellular pathogens and on their host cells, very little is known about how these pathogens direct the host cells into phagocytosing them via nonlethal routes and upon establishing an intracellular niche how they are able to resist the host cell's constant attempts to eliminate them.

We recently have found that in opsonin-deficient conditions, *Escherichia coli,* a classically extracellular pathogen, can enter phagocytic cells of the host without concomitant loss of bacterial viability. The critical determinant on the phagocytes directing the bacteria via the nonlethal route was a specific plasma membrane glycoprotein that was recognized and bound by Fim H, an adhesin moiety expressed by the *E. coli.* Unlike many obligate intracellular bacteria, however, these *E. coli* did not appear to be able to resist the host cell's attempt to eliminate them. Hence, most of the intracellular bacteria were subsequently expelled from the phagocyte without apparent loss of viability to either bacteria or host cell. These intriguing observations point to a novel mode of pathogen–host cell interplay, where the bacteria with the help of their adhesive organelles gain access into the host via a nonlethal phagocytic route, but thereafter the host cell is able to counter the invasion and expel the intruding bacteria back into the extracellular medium. We postulate that these interactions occur *in vivo,* and that in opsonin-deficient environments such as in the urinary tract or at other sites in the body of naive or immunocompromised hosts the temporary intracellular refuge provided by the phagocytic cell might contribute to the bacteria's ability to avoid lethal effects of antibiotic treatment. This notion is supported by the relatively high frequency of recurrent *E. coli* infections in immunocompromised individuals, particularly urinary tract infections (Qualman *et al.,* 1984).

2. DISTINCT INTRACELLULAR FATE OF OPSONIZED AND UNOPSONIZED TYPE 1 FIMBRIATED *E. COLI* FOLLOWING MAST CELL PHAGOCYTOSIS

E. coli is by far the major causative agent of urinary tract infections and much of the success of this pathogen has been attributed to its capacity to bind avidly to the walls of the urinary tract (Hagberg *et al.,* 1981). Most uropathogenic *E. coli* express type 1 fimbriae, which are filamentous appendages of adhesion that radiate peritrichously from the surface of the bacterium (Hagberg *et al.*, 1981; Svanborg-Eden *et al.*, 1984). Each type 1 fimbrium is composed of a major subunit and several minor subunits, including Fim H, a 29-kDa protein that mediates specific binding to mannose containing residues (Abraham *et al.*, 1988; Minion *et al.*, 1986; Orndorff and Falkow, 1984). Although the contribution of type 1 fimbriae to bacterial colonization of the urinary tract is evident, its role in subsequent stages of the infectious process has been questioned, especially since these fimbriae also promoted avid bacterial binding to various phagocytic cells of the host. However, this paradox would be resolved and the contribution of type 1 fimbriae to virulence clarified if it could be shown that the fimbriae-mediated interaction with phagocytes was beneficial to the bacteria. So we sought to demonstrate that the response of a phagocyte to type 1 fimbriated *E. coli* was attenuated compared to its response to antibody-coated bacteria.

We undertook our studies in mast cells, which are found in relatively large numbers in the host–environment interface and which until recently were thought to be involved primarily in immunoglobulin E (IgE)-mediated allergic reactions in the body. It is now known that these cells play a crucial role in modulating the innate and specific immune responses of the host to various infectious bacteria not only through phagocytosing and killing pathogens but also by releasing a myriad of proinflammatory mediators, chemotactic factors, and immunoregulatory cytokines (Abraham and Malaviya, 1997). We developed a system to directly compare the fate of intracellular *E. coli* following internalization via a Fim H-mediated, nonopsonic mechanism to the fate of the same organism internalized via an antibody-mediated opsonic mechanism. The model used for nonopsonic binding to mast cells is *E. coli* strain ORN103 expressing the plasmid pSH2, encoding the entire type 1 fimbrial gene cluster (including Fim H). This strain binds efficiently to mast cells in a Fim H-dependent manner. The isogenic Fim H-minus mutant, *E. coli* ORN103 (pUT2002), exhibits no binding, and purely opsonic binding was induced by coating the Fim H-minus strain with specific antibody raised against *E. coli* ORN103 (pUT2002) in mice. Figure 1 summarizes this experimental system in a diagram.

Intracellular viability studies were performed by modification of methods previously used for other bacteria (Berger and Isberg, 1994). Mast cell monolayers on coverslips were infected with bacteria at a multiplicity of infection (MOI) of 10 or less. Following a 15-min binding period, nonadherent bacteria were washed off and the coverslips were immersed in serum-free culture media containing 100 µg/ml gentamycin. By eliminating extracellular bacteria, we were able to follow the viability of intracellular *E. coli*, exclusively. Intracellular *E. coli* was enumerated by plating onto MacConkey agar plates following vigorous solubilization of the infected mast cells with 0.1% Triton X-100 in phosphate-buffered saline. In sharp contrast to opsonized *E. coli,* over 70% of which was killed in the first hour, the unopsonized *E. coli* remained viable (Table I). Moreover, identical

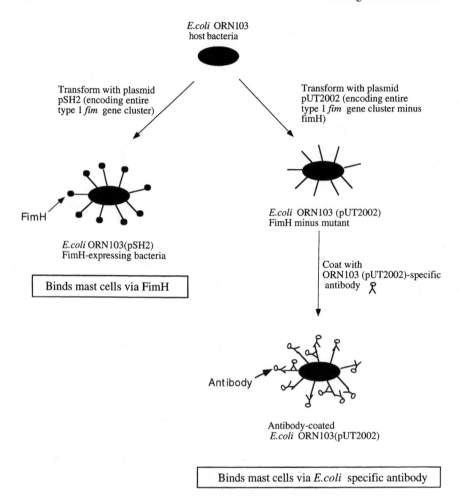

Figure 1. Model system employed to compare intracellular survival of opsonized *E. coli* and unopsonized *E. coli* following mast cell phagocytosis.

results were obtained with another pair of a type 1 fimbriated *E. coli* strain isolated from the urinary tract and its isogenic Fim H minus derivative, which was opsonized with specific antibodies (data not shown). These findings demonstrated that unlike antibody-mediated interactions with phagocytic cells, Fim H-mediated binding did not result in bacterial death.

To explain these different fates of phagocytosed bacteria, we deduced that the Fim H-triggered phagocytosis of bacteria was via a distinct pathway than the route used for opsonized bacteria. A couple of critical observations lent credence to this

Table 1
Intracellular Survival of Opsonized and Unopsonized *E. coli* Following
Mast Cell Phagocytosis

Incubation time (min.)[a]	Intracellular survival (%)[b]	
	Opsonized *E. coli*	Unopsonized *E. coli*
0	100	100
25	61 (± 7)	99 (± 6)
60	31 (± 8)	99 (± 7)

[a]This refers to the time after gentamycin had been rinsed off from the mast cell monolayers.
[b]This is presented as the percentage of the value at 0 min.

notion. First, the phagosome containing type 1 fimbriated *E. coli* was morphologically distinct from that encasing opsonized *E. coli*. Whereas phagosomes containing antibody-coated bacteria were large and spacious, the phagosomes containing unopsonized *E. coli* were markedly more compact (Fig. 2). Second, the pH of phagosomes containing opsonized bacteria was markedly lower and sequestered more toxic oxygen radicals than phagosomes encasing nonopsonized bacteria (Baorto *et al.* 1997), which could explain the viability of nonopsonized *E. coli* within their intracellular compartments. From these observations, it appeared that the specific molecule(s) mediating binding to bacteria was critical in determining the nature of the phagocytic route employed by the mast cell and ultimately the fate of the bacteria. The antibody-opsonized *E. coli* were obviously recognized by Fc-receptors (FcR) on the mast cell, which triggered bacterial uptake via the classical phagosome–lysosome route, resulting in bacterial death. However, the identity of the mast cell moiety(s) initiating phagocytosis of nonopsonized *E. coli* was unknown. We reasoned that we could learn more about this pathway if we could identify the Fim H receptor on the mast cell plasma membrane.

3. IDENTIFICATION OF CD48 AS THE *E. COLI* FIM H RECEPTOR ON MAST CELLS

Our strategy to isolate the Fim H receptor in mast cells is described. Pooled batch cultures of approximately 10^{12} cells from an immortalized mast cell line were prepared and membrane fractions were obtained as described previously (Malaviya *et al.,* 1999). This membrane fraction was passed through a Sepharose-concanavalin-A (ConA) affinity column to isolate candidate mannosylated proteins. We reasoned that since type 1 fimbriae, specifically Fim H, bound to D-mannose, the putative Fim H receptor must contain mannose. Therefore, we used

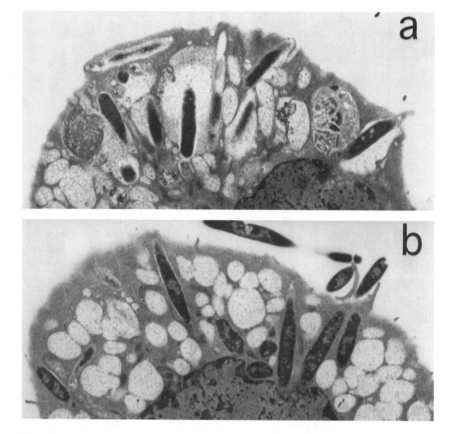

Figure 2. Electron micrographs showing morphologically distinct intracellular compartments of opsonized and unopsonized *E. coli* following phagocytosis. After 1 hr of incubation, (a) most opsonized *E. coli* are encased in relatively spacious compartments (b) whereas most unopsonized *E. coli* are harbored in tight fitting vacuoles. These figures were obtained from Baorto *et al.* (1997).

the mannose-binding ConA lectin to enrich mannose-containing proteins from the mast cell membrane preparations. Next, 100 mM of α-methyl-D-mannoside was employed to elute all proteins bound via their mannose residues. The resulting eluate was subjected to sodium dodecyl sulfate–polyacrylamide gel electrophoresis (SDS-PAGE). Many mannose-containing membrane proteins bound to the ConA column as evidenced by the large number of bands seen after staining of gels with Coomassie Blue (Malaviya *et al.*, 1999). In order to identify the putative Fim H receptor among the mannoside-eluted proteins, we electrophoretically transferred the proteins onto nitrocellulose after SDS-PAGE. The immobilized materials were

Table II
Effect of Phospholipase C (PLC) on the Binding of Fim H-Expressing *E. coli* to Mast Cells[a]

Concentration of PLC (U/ml)	*E. coli* adherence to mast cells (%)[b]
0.00	100
0.10	56 (±15)
0.25	43 (±20)
1.00	19 (±16)
3.00	9 (±13)

[a]Mast cell monolayers were pretreated with varying concentrations of PLC and then exposed to FimH-expressing bacteria for 1 hr. Unbound bacteria were washed away, and the numbers of adherent bacteria were assessed by first treating the mast cell monolayers with 0.1% Triton and then enumerating the number of bacteria (CFU) on agar plates.
[b]This is presented as the percentage of the value from untreated cells.

then exposed to I^{125}-labeled recombinant *E. coli* Fim H that was presented in a complex with its chaperone, Fim C. We found that the Fim H probe specifically bound to a 45-kDa band in the absence, but not in the presence of α-D-mannoside (Malaviya *et al.*, 1999). Furthermore, when the blot was exposed to Fim H-expressing *E. coli* ORN103 (pSH2) and mutant Fim H-deficient *E. coli* ORN103 (pUT2002), only the former bound to the 45-kDa band (Malaviya *et al.*, 1999). The binding reaction of *E. coli* ORN103 (pSH2) was inhibitable by D-mannose. To determine the identity of the 45-kDa band, we purified the protein from the ConA-eluted fraction to homogeneity by fast protein, peptide, and polynucleotide liquid chromatography (FPLC). The sequence of the first 12 amino acid residues in the amino-terminus was determined and found to be 100% homologous to that of CD48, a glycosylphosphatidyl (GPI)-anchored molecule that previously has been reported to be present on mast cells as well as on lymphocytes, macrophages, and endothelial cells (Davis and van der Merwe, 1996; Wong *et al.*, 1990). Several additional experiments were undertaken to confirm that CD48 was the Fim H receptor on mast cells. These showed that (1) pretreatment of mast cells with phospholipase C, an agent that specifically removes GPI-anchored moieties from cell surfaces, markedly reduced binding of Fim H-expressing *E. coli* (Table II); (2) antibodies to CD48, specifically and in a dose-dependent fashion, reduced binding of Fim H-expressing *E. coli* (Table III); and (3) transfection of Chinese hamster ovary (CHO) cells markedly increased binding of type 1 fimbriated bacteria to these cells (Malaviya *et al.*, 1999). The importance of the glycosylation pattern of CD48 was indicated by the findings that Fim H-expressing *E. coli* mediated mannose-inhibitable binding to purified recombinant CD48 and exposing recombinant CD48 to endoglycosidases dramatically reduced the binding of type 1 fimbriated *E. coli* (Malaviya *et al.*, 1999). It is noteworthy that CD48 resembles nonspecific

Table III
Effect of CD48-Specific Antibody on the Binding of Fim H-Expressing
E. coli to Mast Cells[a]

Concentration of antibody (µg/ml)	Number of adherent *E. coli* (%)[b]	
	Anti-CD48-treated	Control antibody[c]treated
0.0	100	100
2.5	95 (±15)	110(±6)
15.0	78(±13)	118(±5)
25.0	46(±7)	118(±4)

[a]The methods employed in this experiment are the same as that in Table II except that antibodies, not PLC, were employed to pretreat mast cells.
[b]This is presented as the percentage of the value in the absence of any antibody.
[c]anti-CD117 antibody was used as the control.

cross-reacting antigen (NCA) and carcinoembryonic antigen (CEA), both of which are previously described receptors for Fim H on granulocytes (Leusch *et al.*, 1991; Sauter *et al.*, 1993), in being a member of the Ig superfamily and being GPI-anchored to the plasma membrane. Figure 3 illustrates some common structural features among these Fim H receptors. These observations cumulatively indicate that CD48 is the definitive receptor on mast cell membranes for the type 1 fimbriated *E. coli*.

4. ASSOCIATION OF CD48 WITH CHOLESTEROL–GLYCOLIPID-ENRICHED MICRODOMAINS AND THE CONTRIBUTION OF THESE STRUCTURES TO BACTERIAL ENTRY

How was CD48 triggering bacterial phagocytosis in the mast cells? Since CD48 does not possess a transmembrane or cytoplasmic domain, this molecule is clearly not able to directly trigger phagocytosis. However, many GPI-anchored proteins including CD48 associate with distinct detergent-insoluble cholesterol–glycolipid-enriched microdomains (CGEM) in the plasma membrane (Brown, 1992). A cluster of proteins that play a prominent role in signal transduction also are associated with these microdomains. These include G proteins (Solomon *et al.*, 1996; Parolini *et al.*, 1996), protein tyrosine kinases such as Lyn and Lck (Brown, 1993; Parolini *et al.*, 1996; Shenoy-Scaria *et al.*, 1993), and protein tyrosine phosphatases such as CD45 (Parolini *et al.*, 1996; Volarevic *et al.*, 1990). Some of these signaling molecules are reported to be intimately associated with various GPI-anchored molecules clustered within the CGEM of the cell. The involvement

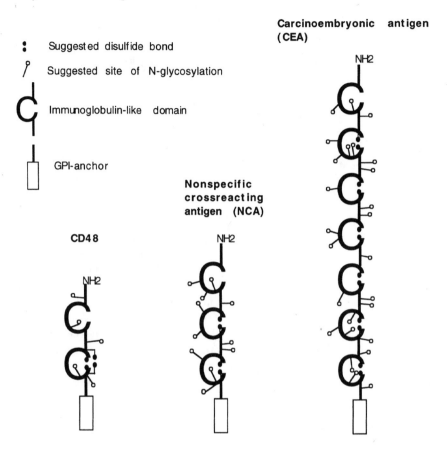

Figure 3. Diagrammatic illustration of structural similarities between Fim H receptors in immune cells. This figure is modified from van der Merwe *et al.* (1996).

of these CGEM-associated signaling proteins in endocytosis was indicated from the observations that when specific ligands bind and aggregate GPI-anchored proteins in the CGEM, the GPI-anchored proteins are readily internalized along with the ligand (Deckert *et al.*, 1996). It is noteworthy that in endothelial and epithelial cells these CGEM contribute to the formation of caveolae, which are distinct flask-shaped plasma membrane invaginations (Parton, 1996; Schnitzer *et al.*, 1995) that play a critical roles in the endocytosis of GPI-anchored proteins (Schnitzer *et al.*, 1996) and cholera toxin (Orlandi and Fishman, 1998). Although the receptor for cholera toxin, GM1, is not GPI-anchored, it is clustered within the CGEM of these cells, and therefore its internalization involves these structures (Ilangumaran *et al.*, 1997).

Table IV

Effect of Filipin on the Entry of FimH-Expressing *E. coli* in Mast Cells[a]

Concentration of filipin (μg/ml)	Number of intracellular *E. coli* (%)[b]
0.0	100
1.0	62 (±15)
1.5	43 (±7)
2.0	31 (18)

[a]Mast cells monolayers were pretreated with varying concentrations of filipin for 20 min, exposed to bacteria, then intracellular viability assays were performed as in Table I.
[b]This is presented as the percentage of the number of intracellular bacteria in the absence of filipin.

One of the characteristics of CGEM-mediated endocytosis is their sensitivity to filipin, a cholesterol-binding agent. Cholesterol maintains the integrity of CGEM in the plasma membrane, and thus the intercalation of sterol-binding agents such as filipin in the plasma membrane abrogates the CGEM's functional activity (Orlandi and Fishman, 1998; Schnitzer *et al.,* 1994). In view of the possibility that CD48 was associating with CGEM and involving these structures in bacterial phagocytosis, we investigated whether uptake of type 1-fimbriated *E. coli* by mast cells could be blocked by filipin. We found that pretreatment of mast cells with increasing amount of filipin reduced bacterial phagocytosis in a dose-dependent manner (Table IV). Thus, the mast cell uptake of type 1-fimbriated *E. coli* triggered by the coupling of FimH with CD48 involves the participation of CGEM, which translocates the bacteria into the mast cell via a pathway that parallels but is distinct from the FcR-mediated classical endosome–lysosome pathway.

5. EXPULSION OF INTRACELLULAR *E. COLI* AND ITS UNDERLYING MECHANISM

Once it has successfully gained entry into the host cell, a pathogen usually grows and persists for a relatively long period of time. We sought to investigate whether this was the case with Fim H-expressing *E. coli* by following its intracellular fate over an extended period of time. Bacteria-infected mast cells were incubated in gentamycin-containing medium and at various time intervals thereafter the number of intracellular bacteria was determined. The viability of infected mast cells also was monitored by dye exclusion assays. To our surprise, it was found that there was a fairly rapid fall in the numbers of intracellular bacteria after a few hours of incubation. It was found that after 16 hr of incubation the intracellular bacterial load had decreased by about 70%. Figure 4 depicts the rate of loss of intracellular bacteria over a 150-hr period in the mast cells. The dramatic reduction in intracellular bacteria could be attributed to either loss of bacterial viability or discharge of bacteria from the mast cell. The second possibility appeared to be the more plausible one, because we were able to demonstrate the presence of viable extracellular bacteria at various time intervals when infected mast cells were incubated in gentamycin-free medium (data not shown). Assays for mast cell viability revealed that loss of bacteria to the extracellular environment was not accompanied by any loss of mast cell viability (data not shown). We also examined cross-sections of infected mast cells by microscopy to see whether the mode of bacterial exiting could be deduced. Unlike the uptake of bacteria that was accompanied by formation of filopodlike structures around the bacteria, (Arock *et al.,* 1998), we found that the expulsion of bacteria was not accompanied by any obvious signs of cell surface activity. The bacterial discharge appeared to result from the fusion of bacterial phagosomes with the plasma membrane (Fig. 5). It would seem that although the mast cell did not appear to have the capability to resist Fim H/CD48-mediated bacterial entry, it was able to subsequently expel the intracellular bacteria. It is noteworthy, however, that not all the intracellular bacteria were expelled even after 150 hr (Fig. 4).

We next sought to investigate the mechanism of bacterial discharge. Since mast cells possess a highly specialized system for exocytosis of its granules, we investigated whether the mast cells were using this system to expel intracellular bacteria. We tested known inhibitors of mast cell degranulation such as cytochalasin D and colchicine on bacteria discharge, but found no appreciable inhibition in the rate of bacteria discharge (data not shown). We also investigated whether bacterial discharge would be accelerated by inducing degranulation in these infected mast cells with mast cell agonists such as phorbol-12-myristate-13-acetate

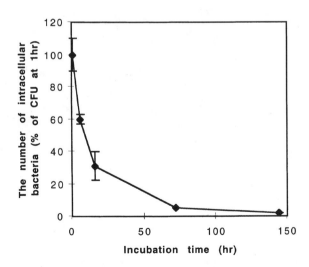

Figure 4.   Loss of intracellular bacteria from infected mast cells over a 150-hr period. See text for methods.

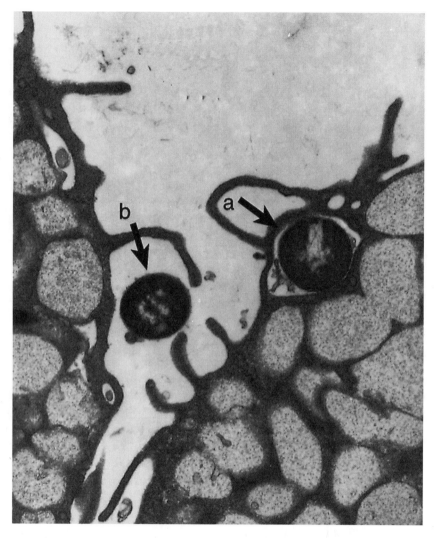

Figure 5.   Electron micrographs showing various stages in the discharge of Fim H-expressing *E. coli* from mast cells. (a) The compartment encasing bacteria seems to fuse with plasma membrane, (b) resulting in bacterial release from the cells.

PMA–ionophore, but these had no significant effect, suggesting that the mode of discharge is through a novel pathway.

Since bacterial entry involved CGEM and since some of the components of these microdomains were likely still to be associated with the intracellular bacteria, we wondered whether CGEM was linked to mast cell discharge of bacteria.

Table V

Effect of β-Methyl-Cyclodextrin and Serum on Mast Cell Discharge of Bacteria[a]

Reagent	Number of residual intracellular bacteria (%)[b]
No treatment	100
β-Methyl-cyclodextrin (1mM)	147 (±1.6)
Fetal bovine serum (10%)	73 (±8.2)

[a]Bacteria-infected mast cells were treated with ß-methyl-cyclodextrin or fetal bovine serum.
[b]After 9 hr treatment with reagents, intracellular bacterial numbers were assessed as previously mentioned.

This notion was strengthened, especially in light of recent findings implicating CGEM in the transport of certain plasma membrane components from Golgi to cell surface (Simons and Ikonen, 1997). For example, in polarized cells apical proteins including several GPI-anchored proteins but not basolateral proteins were transported to the plasma membrane in close association with CGEM (Brown and Rose, 1992; Simons and Ikonen, 1997). Moreover, hemagglutinin of influenza virus was recently shown to be translocated from Golgi to cell surface in association with CGEM in virus-infected cells (Keller and Simons, 1998). It is noteworthy that the translocation of proteins associated with CGEM is highly sensitive to intracellular cholesterol levels, because it is markedly reduced in cholesterol-depleted cells (Hannan and Edidin, 1996; Keller and Simons, 1998).

We sought to investigate whether mast cell discharge of bacteria was mediated by cellular CGEM-trafficking pathway by examining the effect of β-methyl-cyclodextrin, which is known to lower intracellular cholesterol level. We found that in the presence of ß-methyl-cyclodextrin, the amount of bacteria discharged from the mast cells was markedly reduced (Table V). The contribution of cellular CGEM trafficking in the discharge of bacteria was further supported by the finding that the addition of low-density-lipoprotein—enriched serum to the bacteria-infected mast cells significantly increased mast cell expulsion of bacteria (Table V). Thus, mast cell expulsion of bacteria and bacterial entry into mast cells are both mediated by CGEM.

6. CONCLUDING REMARKS

We have shown that type 1 fimbriated *E. coli* can enter mast cells without loss of bacterial viability, indicating an important contribution of the fimbriae to bacterial virulence. The specific mast cell molecule recognized by the Fim H fimbrial adhesin is a mannosylated protein, CD48, which is closely associated with CGEM in the plasma membrane. The CGEM mediated the translocation of the bacteria into the cell via a distinct but attenuated phagocytic pathway. However, the mast cells were subsequently able to counter this bacterial invasion by expelling the bacteria, revealing a novel mode of cellular defense. The rate of expulsion appeared to be directly proportional to the number of intracellular bacteria. Interestingly, the translocation of bacteria back to the plasma membrane appeared to be via the same route employed in the trafficking of CGEM-associated proteins, which is logical considering that the bacterial phagosome primarily comprises CGEM. The bacterial entry into and expulsion from mast cells could reflect a dynamic tussle between the pathogen and the mast cell occurring in opsonin-deficient sites in the body or even at other sites in naive or immunocompromised hosts that have low systemic levels of *E. coli*-specific antibody. This capacity of *E. coli* and probably other ty᠎ ᠎ 1 fimbriae-expressing enterobacteria to hide in phagocytic cells, however te ᠎a-porarily, could contribute to their persistence in spite of a vigorous immune ᠎e-sponse in the host and appropriate antibiotic therapy. Indeed, type 1 fimbriated bacteria have been implicated as important causative agents of recurrent infections in immunocopromised individuals and in the elderly (Kil *et al.,* 1997). Clearly, much remains to be known regarding bacterial entry and expulsion from phagocytic cells, and determining the molecular aspects relating to these events should provide valuable clues on the development of appropriate therapeutic measures.

ACKNOWLEDGMENTS

This work was supported in part from research grants from the NIH (AI 35678 and DK 50814).

REFERENCES

Abraham, S. N., and Malaviya, R., 1997, Mast cells in infection and immunity, *Infect. Immun.* **65**:3501–3508.

Abraham, S. N., Sun, D., Dale, J. B., and Beachey, E. H., 1988, Conservation of the D-mannose-adhesion protein among type 1 fimbriated members of the family *Enterobacteriaceae, Nature* **336**:682–684.

Arock, M., Ross, E., Lai-Kuen, R., Averlant, G., Gao, Z., and Abraham, S. N., 1998, Phagocytic and tumor necrosis factor alpha response of human mast cells following exposure to gram-negative and gram-positive bacteria, *Infect. Immun.* **66**:6030–6034.

Baorto, D. M., Gao, Z., Malaviya, R., Dustin, M. L., van der Merwe, A., Lublin, D. M., and Abraham, S. M., 1997, Survival of FimH-expressing enterobacteria in macrophages relies on glycolipid traffic, *Nature* **389**:636–639.

Berger, K. H., and Isberg, R. R., 1994, Intracellular survival by Legionella, *Methods Cell Biol.* **45**:247–259.

Brown, D. A., 1992, Interaction between GPI-anchored proteins and membrane lipids, *Trends Cell Biol.* **2**:338–343.

Brown, D., 1993, The tyrosine kinase connection: How GPI-anchored proteins activate T cells. *Curr. Opin. Immunol.* **5**:349–354.

Brown, D. A., and Rose, J. K., 1992, Sorting of GPI-anchored proteins to glycolipid-enriched membrane subdomains during transport to the apical cell surface, *Cell* **68:** 533–544.

Clemens, D. L., and Horwitz, M. A., 1995, Characterization of the *Mycobacterium tuberculosis* phagosome and evidence that phagosomal maturation is inhibited, *J. Exp. Med.* **181:**257–270.

Davis, S. J., and van der Merwe, P. A., 1996, Structure and ligand interactions of CD2: Implications for T-cell function, *Immunol. Today* **17:**177–187.

Deckert, M., Ticchioni, M., and Bernard, A., 1996, Endocytosis of GPI-anchored proteins in human lymphocytes: Role of glycolipid-based domains, actin cytoskeleton, and protein kinases, *J. Cell Biol.* **133:**791–799.

Eissenberg, L. G., Goldman, W. E., and Schlesinger, P. H., 1993, *Histoplasma capsulatum* modulates the acidification of phagolysosomes, *J. Exp. Med.* **177:**1605–1611.

Hagberg, L., Jodal, U., Korhonen, T. K., Lidin-Janson, G., Lindberg, U., and Svanborg-Eden, C., 1981, Adhesion, hemagglutination, and virulence of *Escherichia coli* causing urinary tract infections *Infect. Immun.* **31:**564–570.

Hannan, L. A., and Edidin, M., 1996, Traffic, polarity, and detergent solubility of a glycosylphosphatidylinositol-anchored protein after LDL-deprivation of MDCK cells, *J. Cell Biol.* **133:**1265–1276.

Hass, A., 1998, Reprogramming the phagocytic pathway—Intracellular pathogens and their vacuoles, *Mol. Membr. Biol.* **15:**103–121.

Ilangumaran, S., Briol, A., and Hoessli, D. C., 1997, Distinct interactions among GPI-anchored, transmembrane and membrane associated intracellular proteins, and sphingolipids in lymphocyte and endothelial cell plasma membranes, *Biochim. Biophys. Acta* **1328:**227–236.

Joiner, K. A., Fuhrman, S. A., Miettinen, H. M., Kasper, L. H., and Mellman, I., 1990, *Toxoplasma gondii:* Fusion competence of parasitophorous vacuoles in Fc receptor-transfected fibroblasts, *Science* **249:**641–646.

Keller, P., and Simons, K., 1998, Cholesterol is required for surface transport of influenza virus hemagglutinin, *J. Cell Biol.* **140:**1357–1367.

Kil, K. S., Darouiche, R. O., Hull, R. A., Mansouri, M. D., and Musher, D. M., 1997, Identification of a *Klebsiella pneumoniae* strain associated with nosocomial urinary tract infection, *J. Clin. Microbiol.* **35:**2370–2374.

Leusch, H. G., Drzeniek, Z., Hefta S. A., Markos-Pusztai, Z., and Wagener, C., 1991, The putative role of members of the CEA-gene family (CEA, NCA, an BCG) as ligands for the bacterial colonization of different human epithelial tissues, *Zentralbl. Bakteriol.* **275:**118–122.

Malaviya, R., Gao, Z., Tharkavel, K., Menue P. A., Abraham, S. N., 1999, The mast cell tumor necrosis factor α response to fimH-expressing *E. coli* is mediated by glycosylphosphatidylinositol-anchored molecule CD4f, *PNAS,* **96:**8110–8115.

Marquis, H., Goldfine, H., and Portnoy, D. A., 1997, Proteolytic pathways of activation and degradation of a bacterial phospholipase C during intracellular infection by *Listeria monocytogenes, J. Cell Biol.* **137:**1381–1392.

Minion, F. C., Abraham, S. N., Beachey, E. H., and Goguen, J. D., 1986, The genetic determinant of adhesive function in type 1 fimbriae of *Escherichia coli* is distinct from the gene encoding the fimbrial subunit, *J. Bacteriol.* **165:**1033–1036.

Orlandi, P. A., and Fishman, P. H., 1998, Filipin-dependent inhibition of cholera toxin: Evidence for toxin internalization and activation through caveolae-like domains, *J. Cell Biol.* **141**:905–915.

Orndorff, P. E., and Falkow, S., 1984, Organization and expression of genes responsible for type 1 piliation in *Escherichia coli, J. Bacteriol.* **159**:736–744.

Parolini, I., Sargiacomo, M., Lisanti, M. P, and Peschle, C., 1996, Signal transduction and glycophosphatidylinositol-linked proteins (LYN, LCK, CD4, CD45, G proteins and CD55) selectively localize in Triton-insoluble plasma membrane domains of human leukemic cell lines and normal granulocytes, *Blood* **87**:3783–3794.

Parton, R. G., 1996, Caveolae and caveolins, *Curr. Opin. Cell Biol.* **8**:542–548.

Qualman, S. J., Gupta, P. K., and Mendelsohn, G., 1984, Intracellular *Escherichia coli* in urinary malakoplakia: A reservoir of infection and its therapeutic implications., *Am. J. Clin. Pathol.* **91**:35–42.

Sauter, S. L., Rutherfurd, S. M., Wagener, C., Shively, J. E., and Hefta, S. A., 1993, Identification of the specific oligosaccharide sites recognized by type 1 fimbriae from *Escherichia coli* on nonspecific cross-reacting antigen, a D66 cluster granulocyte glycoprotein, *J. Biol. Chem.* **268**:15510–15516.

Schnitzer, J. E., Oh, P., Pinney, E., and Allard, J., 1994, Filipin-sensitive caveolae-mediated transport in endothelium: Reduced transcytosis, scavenger endocytosis, and capillary permeability of select macromolecules, *J. Cell Biol.* **127**:1217–1232.

Schnitzer, J. E., Oh, P., and McIntosh, D. P., 1996, Role of GTP hydrolysis in fission of caveolae directly from plasma membranes, *Science* **274**:239–242.

Shenoy-Scaria, A. M., Timson Gauen, L. K., Kwong, J., Shaw, A. S., and Lublin, D. M., 1993, Palmitylation of an amino-terminal cysteine motif of protein tyrosine kinases p56[lck] and p59[fyn] mediates interaction with glycosyl-phosphatidyl-anchored proteins, *Mol. Cell. Biol.* **13**:6385–6392.

Simons, K., and Ikonen, E., 1997, Functional rafts in cell membranes, *Nature* **387**:569–572.

Solomon, K. R., Rudd, C. E., and Finberg, R. W., 1996, The association between glycosylphosphatidylinositol-anchored proteins and heterotrimeric G protein alpha subunits in lymphocytes, *Proc. Natl. Acad. Sci. USA* **93**:6053–6058.

Svanborg-Eden, C., Bjursten, L. M., Hull, R., Hull, S., Magnusson, K. E., Moldovano, Z., and Leffler, H., 1984, Influence of adhesins on the interaction of *Escherichia coli* with human phagocytes, *Infect. Immun.* **44**:672–680.

van der Merwe, P. A., McNamee, P. N., Davies, E. A., Barclay, A. N., and Davis, S. J., 1995, Topology of the CD2-CD48 cell-adhesion molecule complex: Implications for antigen recognition by T cells, *Curr. Biol.* **5**:74–84.

Volarevic, S., Burns, C. M., Sussman, J. J., and Ashwell, J. D., 1990, Intimate association of Thy-1 and the T-cell antigen receptor with the CD45 tyrosine phosphatase, *Proc. Natl. Acad. Sci. USA* **87**:7085–7089.

Wong, Y. W., Williams, A. F., Kingsmore, S. F, and Seldin, M. F., 1990, Structure, expression, and genetic linkage of the mouse BC M1 (Ox45 or Blast-1) antigen. Evidence for genetic duplication giving rise to the BCM1 region on mouse chromosome 1 and the CD2/LFA3 region on mouse chromosome 3, *J. Exp. Med.* **171**:2115–2130.

Xu, S., Cooper, A., Sturgill-Koszycki, S., van Heyningen, T., Chatterjee, D., Orme, I., Allen, P., and Russell, D. G., 1990, Intracellular trafficking in *Mycobacterium tuberculosis* and *Mycobacterium avium*-infected macrophages, *J. Immunol.* **153**:2568–2578.

17

The Fim H Lectin of *Escherichia coli* Type 1 Fimbriae

An Adaptive Adhesin

David L. Hasty and Evgeni V. Sokurenko

1. INTRODUCTION

1.1. General

A general tenet of bacterial ecology is that the ability to adhere to surfaces is essential for the survival of bacterial species. Indeed, by comparison to the numbers of bacteria that live attached to the surface of an organic or inorganic host, often within a complex biofilm, very few free-floating bacteria can be found in a natural situation outside of the laboratory (Costerton *et al.,* 1995). Over the millennia, many different types of adhesive structures have evolved to accomplish the critical attachment event (DeGraaf and Mooi, 1986; Jann and Jann, 1990; Klemm, 1994; Ofek and Doyle, 1994). One of the best-studied classes of adhesive structures is commonly referred to by the general term *fimbriae* (Duguid *et al.,* 1955), hairlike appendages of the bacterial surface. The primary function of these organelles, so far as is currently known, is in securing attachment. Fimbriae, which

David L. Hasty • Department of Anatomy and Neurobiology, University of Tennessee Memphis, Memphis, Tennessee 38163, and the Veterans' Affairs Medical Center, Memphis, Tennessee 38104. *Evgeni V. Sokurenko* • Department of Anatomy and Neurobiology, University of Tennessee Memphis, Memphis, Tennessee 38163.

Glycomicrobiology, edited by Doyle.
Kluwer Academic/Plenum Publishers, New York, 2000.

are also sometimes called pili (Brinton, 1959, 1965), are particularly common among members of the family *Enterobacteriaceae,* and a great many different structural and functional types have been described (Jann and Jann, 1990; Klemm, 1994; Ofek and Doyle, 1994).

Because the adhesive function of many fimbriae can be inhibited by simple saccharides, they are considered to be lectins. For certain types of fimbriae, the primary structural subunit and the adhesin appear to be one in the same (de Graaf and Gaastra, 1994). For other fimbriae the lectin subunit is presented in very small copy numbers and only (or at least primarily) at the very tips of a polymeric shaft. The primary focus of this chapter is the structure and function of one of the lectins of *Escherichia coli,* the Fim H subunit of type 1 fimbriae. Although the Fim H lectin is only a minor component of the overall type 1 fimbrial superstructure, it plays a prominent role in initiating fimbrial biogenesis (Klemm and Christiansen, 1987) and is almost completely responsible for the binding activity (Sokurenko *et al.,* 1997). Nevertheless, because of the presentation of the fimbrial lectin as a minor component of type 1 fimbriae located primarily on the tip, consideration of Fim H structure and function must be made within the context of its structural scaffold.

1.2. Type 1 Fimbriae

Type 1 fimbriae bearing the Fim H lectin are expressed on the surfaces of virtually all *E. coli* and most other members of the family *Enterobacteriaceae* (e.g., *Klebsiella, Enterobacter,* and *Salmonella* species) (Abraham *et al.,* 1988; Hanson and Brinton, 1988). The ubiquity of type 1 fimbriae alone should signify an extremely important function for enterobacterial populations, but instead this fact has long presented more of a problem than a solution, particularly in epidemiological analyses of the contribution of the various fimbrial types to *E. coli* infections (Johnson, 1991). The fact that type 1 fimbriae are produced by almost all normal and pathogenic isolates led many investigators in the field to the long-held assumption that these adhesive organelles could not be important factors in *E. coli* pathogenesis. In addition, the role(s) of type 1 fimbriae in normal *E. coli* ecology was also long debated. The early idea that type 1 fimbriae must contribute in a direct way to the maintenance of *E. coli* in the primary colonic niche was essentially negated by studies indicating that type 1 fimbriae were not required for intestinal colonization (McCormick *et al.,* 1989, 1993). Interestingly, growth of *E. coli* in cecal mucus greatly stimulates type 1 fimbriae production *in vitro* (Krogfelt *et al.,* 1991).

An inability to produce type 1 fimbriae does not negatively affect colonization of the mammalian large intestine when large numbers of bacteria are instilled (McCormick *et al.,* 1989; Bloch *et al.,* 1992). However, when small numbers of *E. coli* are instilled into the oropharyngeal cavity, lack of type 1 fimbrial expression has a very negative effect on the transient colonization of the oropharyngeal cavity that is apparently required for the organisms to successfully pass the severe barrier provided by the stomach. Whether this is because their numbers increase to a sufficient level, because adhesion induces the expression of required genes, or because the short-term continuous supply enables passage through the stomach at a time when acid production is low is not yet known. Nevertheless, in this indirect yet essential way type 1 fimbriae are important in gaining access to the primary colonic niche (Guerina *et al.*, 1983; Bloch *et al.*, 1992). Numerous studies over the last 20 years have suggested that type 1 fimbriae indeed are important in urinary tract infections (UTIs), particularly cystitis, though this role often has been questioned. Several recent publications have much more clearly documented that type 1 fimbriae indeed are required for the colonization of the urinary bladder (Connell *et al.*, 1996; Langermann *et al.*, 1997; Sokurenko *et al.*, 1998). Thus, a number of studies over the last several years, including recent studies from this laboratory, have made it clear that type 1 fimbriae are important for both commensal and pathogenic organisms.

The precise mechanism whereby type 1 fimbriae could be important adhesive factors in two dramatically different environments—the oropharyngeal cavity and urinary bladder—was an enigma until recent studies from this laboratory (Sokurenko *et al.*, 1992, 1994, 1995, 1997, 1998). As will be reviewed in detail below, we have found that there are naturally occurring mutations in the *fimH* gene affecting the receptor specificity of the Fim H protein, and thereby modulating bacterial tissue tropism. Mutant Fim H adhesins enable *E. coli* to adhere to cells in a new host niche, affecting colonization and playing a role in shifting commensal strains toward a pathogenic phenotype.

2. HISTORICAL PERSPECTIVE

Type 1 fimbriae are so named because they were the first to be described, but they probably also should be designated number 1 because they are the most common type of fimbriae among *E. coli* and indeed most other members of the family *Enterobacteriaceae*. The ability of *E. coli* to agglutinate erythrocytes had been described at the turn of the century (Guyot, 1908), and in the early 1940s it was observed that *E. coli* also agglutinated many other types of cells (Rosenthal, 1943). Interestingly, Rosenthal compared the agglutinating activity of *E. coli* to that of the phytoagglutinins and found them to be similar in several respects, even though at the time the lectin nature of the phytoagglutinins had not yet been discovered. The introduction of the electron microscope enabled the first real look at bacterial surface structures that might be involved in the agglutination reaction. One of the earliest images of fimbriae appeared in an electron microscopic study of flagellation (Houwink and van Iterson, 1950). The "new" organelles were mentioned only in the appendix to the article, and it is a testimony to the superior observational skills of these investigators that, even in what amounts to an afterthought, the authors speculated correctly that these thin, straight filaments (in comparison to flagella) emanating from the surface of *E. coli* were organelles of attachment. It is reasonable to think that the filaments shown in this 1949 study were type 1 fimbriae, because the images certainly bear a remarkable structural resemblance to type 1 fimbriae and because type 1 are the single-most common fimbrial type of *E. coli.*

Collier and colleagues (Collier and de Miranda, 1955) were the first to note the ability of mannose to inhibit the agglutinating properties of *E. coli* in the early 1950s, but it was primarily through the efforts of Duguid's and of Brinton's research groups that the characterization of bacterial agglutination by type 1 fimbriae got an impressively thorough start (Brinton, 1959, 1965; Duguid and Old, 1994; Duguid *et al.,* 1955).

3. *E. COLI* TYPE 1 FIMBRIAE STRUCTURE AND FUNCTION

3.1. *Fim* Genetics

In contrast to other fimbriae, some of which are located on plasmids or present in multiple copies on the chromosome, the cluster of nine genes directly responsible for expression of type 1 fimbriae (Fig. 1) is located on a single 9-kb fragment of DNA that is located between 4530 and 4550 kb in the *E. coli* K-12 chromosome (Orndorff and Falkow, 1984a,b; Klemm *et al.,* 1985; Krallmann-Wenzel *et al.,* 1989; Blattner *et al.,* 1997). Producing several hundred of these relatively large-surface appendages (Fig. 2) consumes a significant fraction of cellular resources, so their expression is under the control of a complex series of regulatory events. One essential characteristic of type 1 fimbrial expression is the

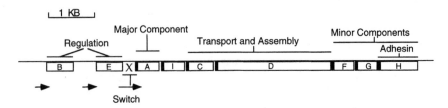

Figure 1.   Genetic organization of the *fim* genes. The location of the genes within the *fim* gene cluster and their roles in regulation or biogenesis are indicated. The location of promoters is indicated by arrows.

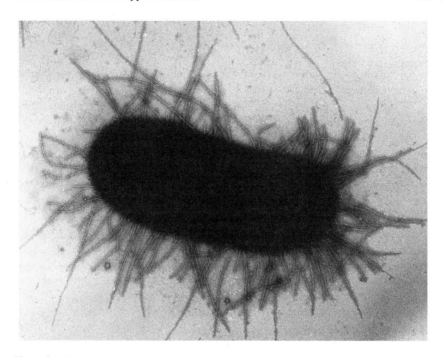

Figure 2.   Electron micrograph of type 1 fimbriated *E. coli* strain illustrating the typical numbers, lengths, and general morphology of the hairlike surface appendages.

ability to vary between "on" and "off" phases. Interestingly, the ability to turn fimbriae off is at least as important as being able to turn them on, such as when the organisms need to evade phagocytes or invade the interstices of a mucous gel (Ofek and Sharon, 1988; May *et al.,* 1993; McCormick *et al.,* 1993). Phase variation is controlled primarily by a phase "switch," a 314-bp invertible segment of DNA immediately upstream of the *fimA* gene encoding the major subunit (Abraham *et al.,* 1985). This element includes a promoter that directs the transcription of *fimA–fimH* when it is in the on orientation. Two regulatory genes at the 5' end of the cluster, *fimB* and *fimE,* encode for proteins that resemble the lambda integrase family of site-specific recombinases (Klemm, 1986; Dorman and Higgins, 1987) and affect the rate of DNA inversion. Fim E promotes inversion to the off orientation preferentially, whereas Fim B promotes inversion in both directions, with minimal preference for off to on inversion (Gally *et al.,* 1996; McClain *et al.,* 1991). Several other accessory proteins bind to the *fim* switch region, resulting in reorientation of the DNA into a conformation more favorable for inversion. These accessory proteins include integration host factor (IHF) (Dorman and Higgins, 1987; Eisenstein *et al.,* 1987; Blomfield *et al.,* 1991, 1997), leucine-responsive regulatory protein (LRP) (Blomfield *et al.,* 1993; Gally *et al.,* 1994), and the histonelike protein (HNS) (Kawula and Orndorff, 1991; Olsen and Klemm, 1994). Recently, it also has been reported that wild-type *E. coli* possess a variety of mutations within or adjacent to the *fim* switch that affect phase variation to different degrees (Leathart and Gally, 1998).

Expression of type 1 fimbriae and several other virulence factors also can be affected in certain strains of *E. coli* by the excision of "pathogenicity islands" (PAIs) (Hacker *et al.,* 1997). The effects of the leucine-specific tRNA locus, *leuX,* on type 1 fimbrial production were first indicated by Newman *et al.* (1994) and PAI II of the uropathogenic *E. coli* strain 536 is inserted at this locus. Because excision of PAI II destroys the *leuX* locus, genes containing the rare leucine codon TTG, which is recognized specifically by the *leuX* product, $tRNA_5^{Leu}$, are not transcribed. Because *fimB* has five such codons (*fimE* has two), the interruption of *leuX* has a dramatic effect on type 1 fimbriae production, as has been shown elegantly for the uropathogenic *E. coli* strain 536 (Ritter *et al.,* 1997). Undoubtedly, there are other phenomena yet to be described that will add to the complexities of fimbrial regulation.

3.2. Fimbrial Biogenesis

E. coli typically produces on the order of several hundred type 1 fimbriae per cell. Fimbrial lengths vary, due to unknown reasons, but typically are 1–2 μm. Using electron microscopy and X-ray diffraction, Brinton (1965) showed in the mid-1960s that type 1 fimbriae consisted of 17-kDa subunits polymerized 3.125 subunits per turn into a right-handed helix having a pitch distance of 2.38 nm, a diameter of 7 nm, and an axial hole of 2 nm. More recent studies of type 1 fimbriae, or P fimbriae, which follow a similar structural plan (Matsui *et al.,* 1973; Bullitt and Makowski, 1995), essentially confirm Brinton's early work. Parenthetically, it is intriguing to think that the axial hole could serve a function other than merely increasing structural stability, and indeed there was the early speculation that the axial hole could be involved in DNA transfer from cell to cell. However, there is little if any evidence at present to support any function other than adhesion.

The function of all but one of the proteins encoded by the *fimA–fimH* region of the *fim* cluster is known (Klemm and Krogfelt, 1994). The proteins are either part of the fimbrial superstructure or those that are involved in fimbrial biogenesis (Fig. 3). Fim A is the primary structural subunit and makes up approximately 98% of the fimbrial mass (Eshdat *et al.,* 1981; Klemm, 1984; Hanson *et al.,* 1988). The function of Fim I is not known at the present time and insertional inactivation of the *fimI* gene has no noticeable effect on fimbriae structure or function (P. Klemm, personal communication). Fim C is a periplasmic chaperone that is es-

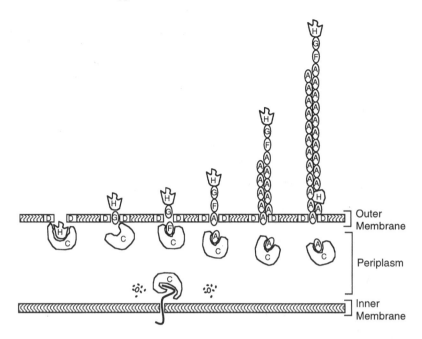

Figure 3. Schematic model for biogenesis of type 1 fimbriae (adapted from Jones *et al.*, 1992). Nascent polypeptides of fimbrial subunits are depicted as being bound by the Fim C chaperone upon translocation through the inner membrane. Polypeptides not protected by the chaperone are thought to be degraded by periplasmic proteases. The Fim H subunit is held by Fim C in a mannose-binding, non-polymerizing form in the periplasm until delivered to the Fim D assembly complex or usher. The Fim H subunit is followed by Fim F and Fim G to form the tip fibrillum. The number of Fim H, Fim G, and Fim F subunits in this illustration is purely hypothetical and not intended to indicate that the actual numbers of the subunits in the tip fibrillum is known. The tip fibrillum is anchored to a helical shaft (see Brinton, 1965) composed primarily by Fim A. Several studies have suggested that Fim H subunits also may be incorporated into the Fim A shaft at long intervals.

sential for proper folding of the nascent polypeptide chains of *fim* gene products, docking at the Fim D outer membrane assembly complex and initiating incorporation of subunits into fimbriae (Jones *et al.*, 1992, 1993; Klemm, 1992). As each fimbrium develops, it extends toward the periphery by adding subunits at the base (Lowe *et al.*, 1987). Fim F, Fim G, and Fim H, whose genes are at the 3′ end of the *fim* cluster, are incorporated into the fimbrial structure in extremely small amounts (Abraham *et al.*, 1987; Hanson *et al.*, 1988, Krogfelt and Klemm, 1988). Maurer and Orndorff (1985) and Minion *et al.* (1986) were the first to show evidence for the possibility that the type 1 fimbrial adhesin was distinct from the primary structural subunit. In the next year, Klemm and Christiansen (1987) as well as Maurer and Orndorff (1987) published *fim* gene sequences and functional analy-

ses and Hanson and Brinton (1988) published chemical analyses of purified fimbriae that confirmed and extended the initial observations. Because Fim F, Fim G,
and Fim H make up a tip fibrillum (Jacob-Dubuisson *et al.,* 1993; Jones *et al.,* 1995)
that is the initial fimbrial structure to polymerize, their presence or absence has significant effects on fimbrial biogenesis. In fact, in the absence of Fim H the bacteria
usually form no fimbriae or a single, unusually long, nonfunctional fimbrium.

3.3. Fim H Receptor Specificity, Pre-1992

We owe much of our understanding of the fine sugar specificity of the FimH
lectin to the studies of Ofek, Sharon, and their colleagues (Ofek *et al.,* 1977; Ofek
and Beachey, 1978; Firon *et al.,* 1983, 1984, 1987; Ofek and Sharon, 1986) and
those of Neeser *et al.* (1986). Prior to 1992, it was assumed that the receptor for
the Fim H adhesin of *E. coli* type 1 fimbriae was a Man_5 unit of *N*-linked glycans
(Fig. 4). The Fim H saccharide-binding pocket was thought to be in the form of a
trisaccharide, and because hydrophobic derivatives of mannosides were much
more potent inhibitors, the ligand-binding pocket was thought to have a closely as-

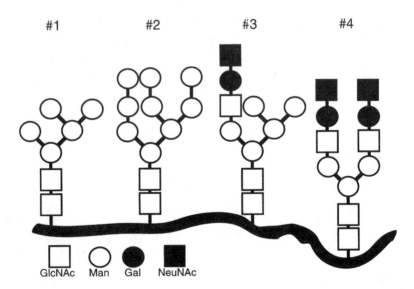

Figure 4.   Schematic diagram of typical N-linked glycans. Saccharides 1 and 2 are examples of high
mannose oligosaccharide chains. Saccharide 3 is an example of a typical hybrid-type glycan unit, one
arm of which bears a trisaccharide. Saccharide 4 is an example of a complex-type glycan unit, both
arms of which are terminally substituted (i.e., no terminal mannose). Prior to 1992, the primary type 1
fimbrial receptor was expected to have the structure of saccharide 1 and 3.

sociated hydrophobic group. Because hybrid type *N*-linked glycans had an exposed trimannose unit, these types of *N*-linked glycans were thought to serve as exclusive receptors. Terminal substitution of the trimannose units with other sugars, even mannose, dramatically reduced the effectiveness of inhibitors. There was no indication that Fim H would bind single-terminal mannose residues. Indeed, monosaccharide receptors are not typical for lectins. Our studies of Fim H structure and function over the last several years have revealed some exciting secrets about this molecule and suggest the strong possibility that the same types of structure–function variation will be seen for Fim H of other enterobacteria.

4. *E. COLI* FIM H PHENOTYPES

4.1. *E. coli* Binding to Human Plasma Fibronectin

Human plasma fibronectin (Fn) was to be a negative control to use in developing an enzyme-linked immunosorbent assay (ELISA) for studying *E. coli* adhesion. Fn has been shown to possess only *O*-linked glycans that do not contain mannose and complex-type *N*-linked glycans with no terminal mannose, which at the time were not expected to react with type 1 fimbriated *E. coli*. It was quite surprising, therefore, to find that Fn was an excellent substratum for mannose-sensitive attachment of the CSH-50 clone *of E. coli* K-12 (Sokurenko *et al.,* 1992). It was much more surprising to find that this *E. coli* strain also was able to bind quite well in a mannose-sensitive manner to Fn domains that possess no glycans at all. Because a recombinant strain bearing the *fim* gene cluster of *E. coli* K-12 strain PC31 [HB101(pPKL4)] produced type 1 fimbriae that did not confer the unusual protein-binding acvitity, we speculated that perhaps this was a spurious finding related in some unknown way to the mutagenesis experiments *E. coli* CSH-50 had been subjected to at Cold Spring Harbor (Miller, 1972). However, a search of the literature indicated that the concept of lectins binding to proteins in a saccharide-sensitive manner was not new (Barondes, 1988) and a survey of *E. coli* isolated from human urine later showed that the protein-binding activity was indeed found in wild-type strains.

4.2. Three Distinct Phenotypes of Fim H

The mannose sensitivity suggested the involvement of the Fim H subunit, but because protein binding had not been described during the almost 40 years of studies of type 1 fimbriated *E. coli,* we remained skeptical of the validity of the observation. To confirm the activity and attempt to determine the fimbrial subunit responsible, *fimH* genes were cloned by polymerase chain reaction (PCR) from CSH-50, PC31, and four human UTI isolates. Plasmid pGB2–24-based plasmids bearing the *fimH* genes (Fig. 5) were transformed into *E. coli* strain AAEC191A, a Δ*fim* derivative of *E. coli* K-12 strain MG1655 (Blomfield *et al.*, 1991), along with plasmid pPKL114 that bears the entire *fim* cluster, but with a translational stop-linker inserted into the 5′ end of the *fimH* gene (Fig. 5). These isogenic strains then expressed type 1 fimbriae that differed only by the Fim H protein that was incorporated. Adhesion of isogenic strains bearing the cloned *fimH* genes to: (1) mannan (a classic type 1 fimbrial receptor from yeast), (2) Fn, (3) periodate-treated Fn (endoglycosidase-treated Fn behaved similarly), and (4) a synthetic peptide of Fn was compared to that of the wild-type strains from which the *fimH* genes had been cloned (Fig. 6). This experiment showed two things. First, the unusual pattern of adhesion was due to the Fim H subunit. Second, it became clear that there was more than one unusual adhesive phenotype (Table I). Two of the wild-type–recombinant pairs of strains exhibited the more expected phenotype and bound only to mannan. This activity was designated the M phenotype. Two other pairs bound to mannan and also bound to Fn, but did not bind to deglycosylated Fn or to synthetic peptide and were designated the MF phenotype. The two last pairs bound to mannan but also bound to Fn, deglycosylated Fn, and synthetic peptide. This activity was designated the MFP phenotype.

We now have reproduced the various Fim H phenotypes by cloning *fimH*

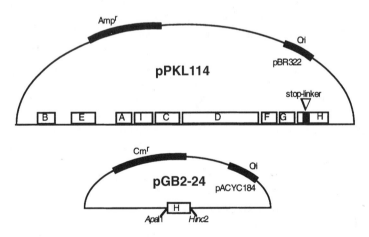

Figure 5. Schematic diagram of the two plasmids used to transform the Δ*fim* K-12 *E. coli* strain AAEC191A. Plasmid pPKL114 contains the entire *fim* gene cluster and a pBR322 replicon but with a stop linker inserted into the *fimH* gene. Because *fimH* is the last gene in the *fim* cluster, no polar effects would be expected. Plasmid pGB2–24 and subsequent derivatives contain *fimH* genes in a pACYC184 replicon and these plasmids complement the *fimH* defect of pPKL114.

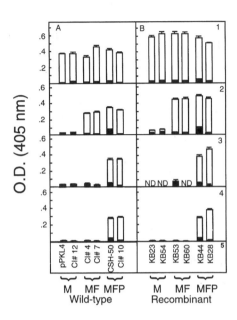

Figure 6. Adhesion of representative (A) wild-type and (B) recombinant M, MF, and MFP class strains to mannan (panel 1), fibronectin (panel 2), periodate-treated fibronectin (panel 3), and a synthetic peptide (panel 4); strain designations are given in panel 5. The recombinant strains are constructed as indicated in Fig. 5 and the text. Open columns indicate bacteria incubated without D-mannose; solid columns indicate bacteria incubated with D-mannose. Values are the means ±SEM (*n* = 4) for each column. ND, not determined. From Sokurenko *et al.* (1994), with permission.

genes into at least four different host strains. The *fimA* genes of these strains have not yet been sequenced, but Fim A is known to be sufficiently variable that it is very unlikely that the subunits are the same in each case. Thus, we believe that differences in the structure of Fim A subunits have little if anything to do with the receptor specificities we have observed. However, because of the nature of the Fim H null strains, they either do or could possess partial or complete *E. coli* K-12 *fimF* and *fimG* sequences. Thus, we cannot yet rule out the possibility that Fim F or Fim G, which are thought to comprise the "stalk" of the tip fibrillum (Jones *et al.*,

Table I
Phenotypic Classes of the Fim H Lectin

Phenotype Class	Receptor Specificity			
	Tri mannoside	Mono mannoside	Non mannosylated saccharide	Non saccharide
M				
M^L	+	−	−	−
M^H	+	+	−	−
MF	+	+	+	−
MFP	+	+	+	+

Residue Number

```
--+++++++++++++++++++    Wild      Recombinant    Adhesive        Source
          111111122      Strain    Strain         Phenotype
21 23567779011116607
1617386038176789361 9
||||||||||||||||||||
MTFVNLGNGSPLGVAIVRHQ    PC31       KB            M[a]            K-12
-----R--------------    CSH50      KB44          MFP             K-12
---A----------------    - -        KB21          M               chimera
---A--------ΔΔΔΔ----    CI#10      KB28          MFP             UTI, Memphis, TN
-N-A----------V-----    CI#3       KB59          M               UTI, Memphis, TN
-N-AH-----L---------    CI#7       KB60          MF              UTI, Memphis, TN
---A------------H--     KS54       KB92          M               UTI, Sweden
---A------------H--     MJ9-3      KB97          M               UTI, Boston, MA

---A---S-N----------    F18        KB91          M               Feces, Kingston, RI
---A---S-N------A---    MJ11-2     KB99          M               UTI, Boston, MA
---A--DS-N------A---    MJ2-2      KB96          M               UTI, Nairobi, Kenya
---A---SEN-------D-     CI#4       KB53          MF              UTI, Memphis, TN
---A---S-N-------C--    CI#12      KB54          M               UTI, Memphis, TN
```

[a] The adhesive subclass of PC31 is based on data using recombinant strain only.

Figure 7. Deduced amino acid sequences of several Fim H variants. The polymorphic sites (sites in which there has been a nonsynonymous mutation in the codon) within the 300 residue Fim H sequence are indicated. The positions are numbered vertically above each polymorphic amino acid, compared to the original Fim H sequence published by Klemm and Christiansen (1987). Positions that do not vary among the Fim H alelles sequenced thus far are not present in the figure. Δ Indicates a deleted residue. Substitutions that affect adhesion phenotype are indicated in boldface type. The sequences are divided into two groups that differ from each other at residues 70 and 78, where Asn/Ser and Ser/Asn substitutions occur. From Sokurenko et al. (1995), with permission.

1995), contribute in important ways to the conformation of the Fim H tip adhesin, and thereby affect the structure of its ligand binding site(s) and the resultant receptor specificity. Nevertheless, we believe that the function of type 1 fimbriae is dictated primarily by the structure of its Fim H lectin subunit. Any additional effects that Fim F or Fim G variants may contribute would obviously add yet another layer of complexity.

The *fimH* genes of several M, MF, and MFP phenotype strains were sequenced and the deduced FimH sequences showed the variation in structure that had been suggested by the phenotypic variation (Fig. 7). However, the Fim H structures within any of the phenotypic classes were not identical, and so it was not possible to determine a consensus motif that would identify any of the phenotypes.

4.3. Quantitative Variation within the M Phenotype Class

As additional wild-type *E. coli* were analyzed in this same manner, we found several additional MF and MFP phenotype isolates, but the majority of the isolates

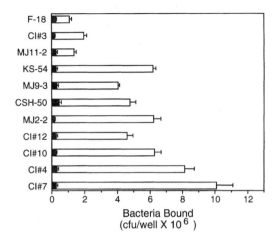

Figure 8. Adhesion of *fimB*-transformed wild-type strains to mannan. Open columns indicate bacteria incubated without α-methyl mannoside; solid columns indicate bacteria incubated with 50 mM α-methyl mannoside. Values are the means ±SEM (*n* = 4). From Sokurenko *et al.* (1995), with permission.

were of the M class. Interestingly, however, within the M class adhesion to mannan varied by up to 15-fold (Fig. 8). The level of adhesion of many of these strains was so low that one wondered why the organism would devote the tremendous resources fimbrial expression requires to generate such an extremely poor adhesin. This observation suggested that a great many *E. coli* strains that have been characterized as non–type 1-fimbriated may simply have been strains that expressed the low-binding Fim H, and therefore were missed in phenotypic assays, especially those involving mannan binding. Recombinant strains varying only by the *fimH* genes showed similar quantitative differences (Fig. 9). No striking differences could be found between the highest binding and lowest binding of the strains in

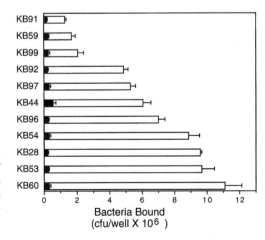

Figure 9. Adhesion of recombinant strains constructed using *fimH* genes cloned from the wild-type strains shown in Fig. 8. Columns and values are as in Fig. 8. From Sokurenko *et al.* (1995), with permission.

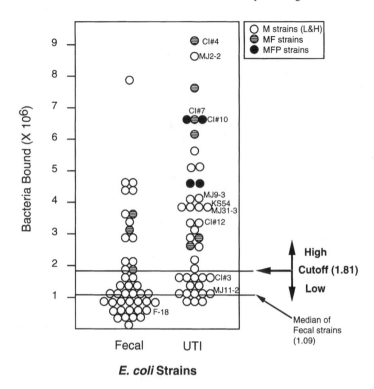

Figure 10.   Adhesion of wild-type isolates from feces and UTI isolates to mannan. All binding was inhibited >80% by α-methyl mannoside. To simplify graphic presentation, data are arranged in groups of 0.25×10^6 bacteria bound per well. Since data are plotted in this way, the actual numbers for circles placed behind the two reference lines fall below the line values, whereas those placed in front of the lines fall above the line values. From Sokurenko *et al.* (1995), with permission.

terms of fimbrial number, fimbrial length, or relative amounts of Fim H protein incorporated into fimbriae. These results suggested that conformational differences in the FimH subunit alone were responsible for the differences in *E. coli* adhesion.

Further efforts to understand the adhesive properties of Fim H variants were focused primarily on the "high"- and "low"-binding strains of the M phenotype, because of their predominance among wild strains. To distinguish between the two, we used the M^L for "low binding to mannan" and M^H for "high binding to mannan." A survey of 42 UTI isolates and 43 isolates from the feces of healthy adults indicated that a majority of the normal fecal isolates exhibited a M^L phenotype, while the UTI isolates were predominantly M^H, binding in threefold higher numbers on average (Fig. 10). The ability of seven isogenic strains constructed as described above to bind to mannan correlated directly with their ability to bind

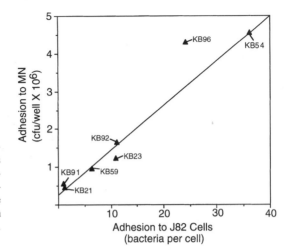

Figure 11. Correlation of the abilities of seven recombinant strains to bind to mannan with their abilities to adhere to the J82 bladder epithelial cell line. Strain numbers are shown. Statistical analysis is given in the text. From Sokurenko *et al.* (1997), with permission.

either to J82 human bladder epithelial cells ($r = 0.97, P > 0.995$) (Fig. 11) or A498 human kidney epithelial cells ($r = 0.93, P > 0.995$). Thus, the quantitative differences observed within the M phenotype also were seen with human cell types. In an attempt to explain the nature of these variations in terms of different receptor specificity or affinity, the adhesion of one M^H strain (KB54) and one M^L strain (KB91) was examined in more detail. Scatchard plot analyses of equilibrium measurements showed that differential binding to mannan could be explained by differences in the numbers of high- and low-affinity sites on the mannan substratum for the two different Fim H lectins (Fig. 12).

Interestingly, these quantitative studies showed that the M^L phenotype strain did exhibit a small number of high-affinity binding sites in mannan, indicating that if the appropriate substratum could be identified, the M^L Fim H could serve as an

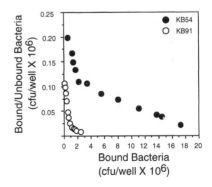

Figure 12. Scatchard plot analyses of the binding of strains KB54 and KB91 to mannan at equilibrium. Data from a single experiment are presented, but the experiment was repeated several times and the results were essentially the same. From Sokurenko *et al.* (1997), with permission.

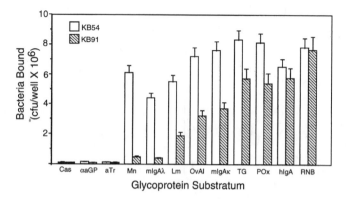

Figure 13. Adhesion of strains KB54 and KB91 to various glycoproteins. Abbreviations: CS, bovine milk casein; TR, human serum *apo*-transferrin; αaGP, human α-acid glycoprotein; MN, yeast mannan; mIgAλ mouse IgAλ; iMC, intestinal mucin; sMC salivary mucin; THP, Tamm-Horsfall protein; LM, human laminin; OVA, chicken egg albumin; m IgAκ mouse IgAκ; POX, horseradish peroxidase; TG, porcine thyroglobulin; hIgA, human IgA; LF, bovine lactoferrin; bRB, bovine RNAse B. Values are means ± standard error of the mean (*n* = 3). From Sokurenko *et al.* (1997), with permission.

effective adhesin. This hypothesis was confirmed when a series of 16 glycoproteins with differing glycosylation patterns was surveyed for their ability to bind type 1 fimbriated *E. coli*. Neither of the M phenotype strains bound to glycoproteins that possessed only complex type *N*-linked glycans. Among the 13 other glycoproteins that possess variable fractions of high mannose and complex type *N*-linked glycans, there was a variable binding of KB54 and KB91 (Fig. 13). Three of the 13 substrata (i.e., human IgA, lactoferrin, and bovine RNase B) bound the

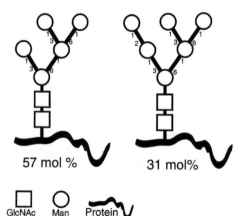

Figure 14. Schematic diagram of bovine RNase B N-linked glycan units. Adapted from Fu *et al.* (1994).

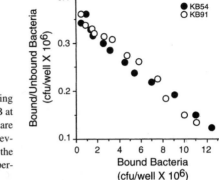

Figure 15. Scatchard plot analyses of binding of strains KB54 and KB91 to bovine RNase B at equilibrium. Data from a single experiment are presented, but the experiment was repeated several times and the results were essentially the same. From Sokurenko *et al.* (1997), with permission.

two strains equally. Of these three, bovine RNase B (bRNase B) was selected for further study because it has only a single *N*-linked glycan per molecule and the structure is relatively uniform, being 57 mole% Man_5-GlcNAc$_2$ units and 31 mole% Man_6-GlcNAc$_2$ units (Fu *et al.*, 1994) (Fig. 14). When adhesion of the M^L and M^H strains to bRNase B was studied in equilibrium binding experiments, the curves in a Scatchard plot were indistinguishable (Fig. 15). These experiments indicated that the M^L Fim H subunit could serve as a very effective adhesin in the presence of an appropriate substratum and it also clearly showed that the two types of FimH subunit utilized different mechanisms of ligand-receptor interaction.

4.4. Quantitative Differences Are due to Variable Ability to Bind to Single Mannosyl Residues

We sought to use simpler mannosides to use in an effort to further examine the ligand–receptor interactions exhibited by these two Fim H subunits. From ear-

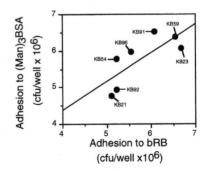

Figure 16. Correlation of the adhesion of seven recombinant strains shown in Fig. 11 to (Man)$_3$BSA with their adhesion to bRB. Strain numbers are shown. From Sokurenko *et al.* (1997), with permission.

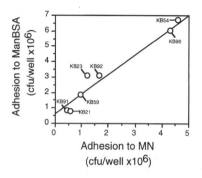

Figure 17. Correlation of the levels of adhesion of the same recombinant strains shown in Fig. 16 to ManBSA with their adhesion to mannan. Strain numbers are shown. From Sokurenko *et al.* (1997), with permission.

lier work, the trimannosyl core of the Man$_5$ units could be predicted to be a major receptor, so we compared the ability of *E. coli* to bind to trimannosyl groups coupled to bovine serum albumin (BSA) with their ability to bind to single mannosyl groups conjugated to BSA. Again, the same seven isogenic strains used above were compared in their ability to bind to Man-BSA and Man$_3$-BSA. All the strains bound relatively well to Man$_3$-BSA and there was a positive correlation of their ability to bind to Man$_3$-BSA and their ability to bind to bRNase B (Fig. 16). Some of the strains bound well to Man-BSA and others bound very poorly. There was a very strong positive correlation between the ability of the strains to bind to Man-BSA and their ability to bind to mannan (Fig. 17). Additional experiments with wild-type strains provided further evidence that type 1 fimbriated *E. coli* bind in relatively similar numbers to Man$_3$-BSA, but there are dramatic differences in their abilities to bind Man-BSA (Fig. 18).

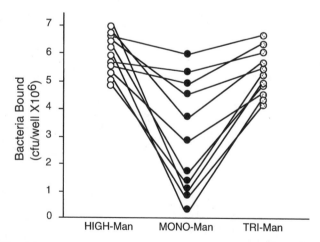

Figure 18. Adhesion of 11 wild-type strains to high-mannose moieties of bovine RNaseB (HIGH-Man), monomannosylated BSA (MONO-Man) and tri-mannosylated BSA (TRI-Man).

5. CONTRIBUTION OF FIM H VARIANTS TO TISSUE TROPISM IN COMMENSAL AND PATHOGENIC NICHES

5.1. Binding to Man-BSA Correlates with Binding to Uroepithelial Cell Lines

To determine whether binding to uroepithelial cells correlated more closely with binding to Man-BSA or Man_3-BSA, a quantitative comparison was performed using the seven isogenic strains. There was a significant correlation ($r = 0.98$, $P > 0.995$) between the level of bacterial adhesion to J82 human bladder epithelial cells and their ability bind to monomannosyl receptors. The same held true when A498 human kidney epithelial cells were used in the adhesion assay. However, there was no correlation between the binding to trimannosyl receptors, Man_3-BSA or bRNase B, and their ability to bind to either of these human cell lines.

5.2. Fim H Variants Differ Quantitatively in Binding to Asymmetric Unit Membranes and Bladder Epithelium *in situ*

Taken together with the results presented earlier indicating that normal human intestinal isolates of *E. coli* adhere to mannan in generally much lower levels than do uropathogenic isolates, we hypothesized that the ability to bind strongly to monomannosyl receptors could be a key factor in the pathogenesis of cystitis. However, although the epithelial cells that were studied had been derived originally from urinary tract tissues, they still are transformed cell lines that may bear little resemblance to transitional epithelium lining the urinary tract. The surface cells of transitional epithelium exhibit a remarkable specialization of their lumenal membrane. The membranes possess rigid plaques connected by more flexible hinge regions such that the apical plasmalemma has an accordionlike appearance in transmission electron micrographs. The plaque areas have been called asymmetric unit membranes (AUMs) due to the increased electron density of the cytoplasmic leaflet. These portions of the membrane have been purified and their integral membrane protein components analyzed in an elegant series of papers by X.-R. Wu *et al.* (1994, 1996). The proteins are called uroplakins (UPs), and four primary protein components have been described: UPIa, UPIb, UPII, and UPIII. Each uroplakin is glycosylated, and although the oligosaccharide components have yet to be thoroughly analyzed, both high mannose type or complex type oligosaccharide chains are present. There is very good evidence from *in vitro* studies that UPs Ia and Ib can serve as receptors for *E. coli* (Wu *et al.*, 1996). The UPs are highly conserved across a variety of species, including bovine and murine tissues. Uroepithelial cell lines such as J82 and A498, however, do not exhibit AUMs

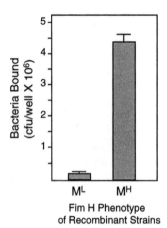

Fim H Phenotype
of Recombinant Strains

Figure 19. Adhesion of M^L and M^H strains to AUMs. From Sokurenko *et al.* (1998), with permission.

and do not appear to express UPs in detectable amounts (X.-R. Wu, personal communication).

To continue testing the hypothesis regarding the tropism of the M^H phenotype for urinary tract epithelium, the ability of *E. coli* strain KB54 and KB91 to bind to AUMs purified from the transitional epithelium of bovine bladder was tested. As had been seen previously when studying the adhesion of these strains to mannan and to uroepithelial cell lines, and maintaining consistency with our general hypothesis, *E. coli* strain KB54 adhered in significantly greater numbers to AUMs than did strain KB91 (Fig. 19). We recently have begun experiments to study the adhesion of KB54 and KB91 to intact mouse bladder. The bladders are excised, cut in half, and stretched slightly as they are pinned into wells of assay plates. Bladders prepared in this way were incubated for 45 min with *E. coli* strains KB54 or KB91 suspended in Earl's balanced salt solution; after washing away the unbound *E. coli,* the bladder segments then were fixed with glutaraldehyde and osmium tetroxide and prepared for scanning electron microscopy (SEM) (Fig. 20). *E. coli* KB91 could be seen to bind to the hexagonal "umbrella cells" in very small numbers (Fig. 21). Although, as anticipated, a great many more of the M^H strain KB54 were bound to the bladder surface, the adhesion occurred in a very striking mosaic pattern in which cells that bound up to 500 *E. coli* KB54 were islands surrounded by cells that bound no bacteria (Fig. 22). The *E. coli*-binding cells appeared to be approximately 10–20% of the total. Overall, however, these results were consistent with the hypothesis that the M^H Fim H phenotype is important in the tropism of certain clones for the urinary tract and also suggest that Fim H variants may serve as a unique tool to study the molecular differentiation of urothelial cell components.

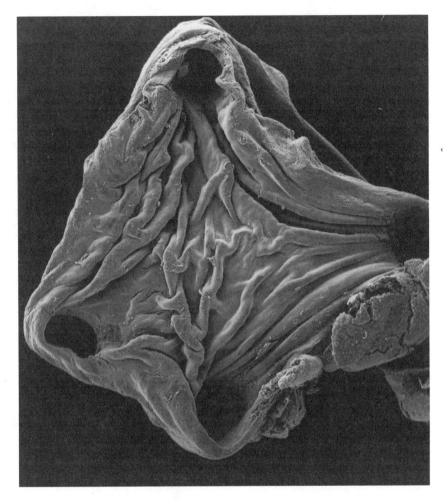

Figure 20. Scanning electron micrograph of half of a mouse bladder pinned into an assay well. The four pinholes holding the bladder in place are obvious.

5.3. Fim H Variants Differ in Colonization of Mouse Urinary Bladder *in vivo*

To test further our hypothesis regarding the contribution of Fim H phenotypes to urinary tract colonization by *E. coli,* we created isogenic strains in a UTI isolate host, *E. coli* strain CI #10, that is genotypically negative for common urovir-

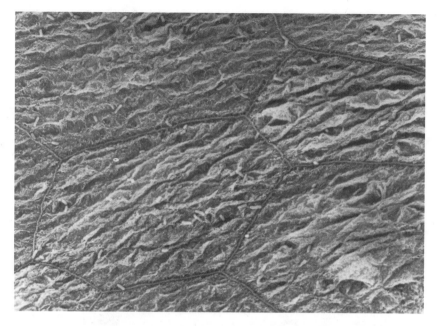

Figure 21.   Scanning electron micrograph of M^L phenotype recombinant *E. coli* strain binding to the surface of mouse bladder epithelial cells. The borders of adjacent epithelial cells can be seen easily due to an obvious ridge that forms at the site of cell–cell junctional complexes. Only a few *E. coli* can be seen bound to the cells.

ulence factors, such as P and P-related fimbriae, S/F1C fimbriae, Dr adhesin, aerobactin, groups II and III capsule, hemolysin, cytotoxic necrotizing factor I, and outer membrane protein T (Sokurenko *et al.,* 1998). The *fimH* gene of CI #10 was insertionally activated using the pCH103 suicide plasmid as described previously (Connell *et al.,* 1996; Shembri *et al.,* 1996), creating CI #10–9. When strain CI #10–9 is complemented with a *fimH* gene that encodes either a M^L or M^H phenotype Fim H (such as those of KB91 or KB54, respectively), a typical M^L or M^H binding pattern is seen *in vitro.* If CI #10–9 is complemented with a *fimH* gene missing a 17-bp segment at the 3' end and encoding an inactive Fim H, no binding is seen *in vitro.* Twenty-four hours following transurethral instillation of equal numbers of these recombinant strains into the bladders of mice, the recombinant strains bearing either M^L or M^H *fimH* genes were recovered from bladders in higher numbers than the strain encoding a nonfunctional Fim H, but the ability of the M^H strain to colonize was at least 15-fold higher than that of the M^L strain (Fig. 23). Thus, in the same genetic background of a uropathognic isolate, the M^H Fim H confers a significant advantage for colonization of the bladder compared with the

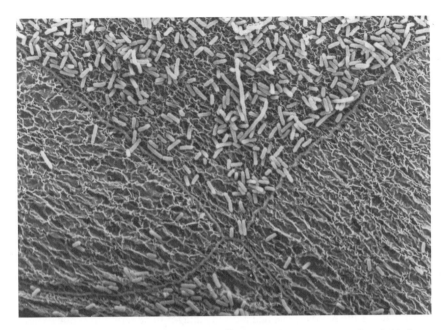

Figure 22. Scanning electron micrograph of a M^H phenotype recombinant *E. coli* strain binding to the surface of mouse bladder epithelial cells. The mosaic pattern of adhesion of this *E. coli* strain is in marked contrast to that of the M^L strain. Cells bearing hundreds of bound *E. coli* are intermingled with cells bearing essentially none.

Figure 23. Colonization of mouse bladders by isogenic *E. coli* expressing nonfunctional Fim H, M^L Fim H, or M^H Fim H subunits. Bars indicate mean colony-forming units per bladder ±SEM. *P* values indicating level of significance between different groups are indicated. From Sokurenko *et al.* (1998), with permission.

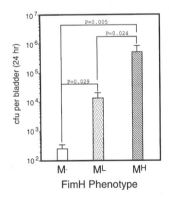

M^L variant. These data provide a rationale for the predominance of the M^H phenotype among UTI isolates and are consistent with our hypothesis regarding the tropism of Fim H phenotypes in cystitis.

5.4. Mutations of Fim H That Increase Urovirulence Are Detrimental for Adhesion to Oropharyngeal Epithelial Cells

The M^H Fim H phenotype would appear at first thought to be a superior adhesin: M^H Fim H confers greater binding than the M^L Fim H phenotype to virtually every model substratum and cell type tested and leads to greatly increased colonization of mouse bladders. However, it is the M^L Fim H phenotype that predominates among intestinal isolates, the primary population of *E. coli*. The results presented thus far failed to explain why the M^H variants have not replaced the M^L phenotype in the general *E. coli* population. Type 1 fimbriae appear to be required not for *E. coli* to colonize the colonic niche, but for transient colonization of the oropharyngeal portal during inter-host transmission (Bloch and Orndorff, 1990; Bloch *et al.*, 1992). Therefore, it was theorized that the M^L variant had a selective advantage (or the M^H variants had a disadvantage) in adhering to oropharyngeal epithelial cells.

Experiments were performed to determine the ability of recombinant strains bearing M^L or M^H Fim H phenotypes to interact with human buccal epithelial cells. Although bacteria bearing M^L or M^H Fim H phenotypes had bound in very different numbers to virtually every other cell or model receptor tested, they interacted with buccal epithelial cells in equivalent numbers (Fig. 24). Because the numbers of bacteria bound to mucosal cells will be a function of the affinity of the adhesin for the epithelial cell receptor and of the interference of inhibitors present

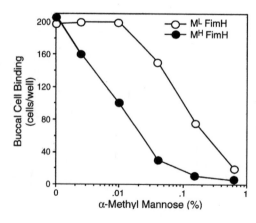

Figure 24. Essentially equivalent binding of *E. coli* to buccal epithelial cells (0% α-methyl mannoside) and inhibition of this interaction by increasing levels of α-methyl-D-mannopyrannoside. From Sokurenko *et al.* (1998), with permission.

Table II
Effect of Soluble Glycoproteins on *E. coli*–Buccal Cell Interactions

	IC_{50}	
Inhibitor	*E. coli* KB91 (M^L Fim H)	*E. coli* KB54 (M^H Fim H)
Yeast mannan	5.0 ± 0.3	0.5 ± 0.2
Bovine intestinal mucin	1.7 ± 0.2	0.6 ± 0.1
Bovine RNase B	0.9 ± 0.2	0.08 ± 0.02
Bovine lactoferrin	2.1 ± 0.2	0.08 ± 0.03

in body fluids bathing the mucosal surfaces, it was hypothesized that any advantage provided by the M^L phenotype in the oropharyngeal cavity must be due to differential effects of inhibitors, since the M^L and M^H variants interacted equally with buccal cells. High levels of α-methyl mannopyranoside completely inhibited the interaction of both variants with buccal cells, but the M^L phenotype was much less sensitive to mannoside and at intermediate levels of the inhibitor it provided for greater levels of attachment (Fig. 24). Free mannose is not usually found in the natural environment, so we tested the effects of several mannose-containing glycoproteins (Table II). In each instance, the M^L phenotype Fim H was the superior adhesin. Since any inhibitors that are likely to affect the binding of *E. coli* with buccal epithelial cells *in situ* should be present in saliva, we also tested the ability of clarified, whole human saliva to inhibit adhesion (Fig. 25). In this case as well, the M^L Fim H phenotype was the superior adhesin. These data provide a strong suggestion that the reduced sensitivity of the M^L-bearing strains to inhibitors confers on *E. coli* a greater ability to bind to buccal epithelial cells; thus, in a physiological environment they would more effectively accomplish the transient colonization of the oropharyngeal cavity that must occur to ensure host-to-host transmission.

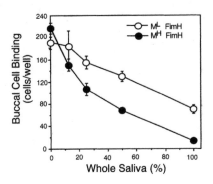

Figure 25. Inhibition of the interaction of *E. coli* and buccal cells by whole, stimulated human saliva. From Sokurenko *et al.* (1998), with permission.

A

fimH allele	Residue number -+++++++++++++++ 1111112 123677791111660 673603816789361 TANGNGSPGVAIVRH	M₁/M₃ ratio	wild strain (source)
1	----S-N--------	0.08	F-18 (fecal)
2	---------------	0.09	KB21 (recomb.)
3	----S-N-----A--	0.15	MJ11-2 (UTI)
4	-V-------------	0.17	K-12 (fecal)
5	N---------V----	0.20	CI3 (UTI)
6	------------H-	0.33	1177 (UTI)
7	---DS-N-----A--	0.63	MJ2-2 (UTI)
8	----S-N------C-	0.72	CI12 (UTI)
9	N-H----L-------	0.77	CI7 (UTI)
10	--------ΔΔΔΔ---	0.91	CI10 (UTI)
11	----SEN-------D	0.93	CI4 (UTI)

B

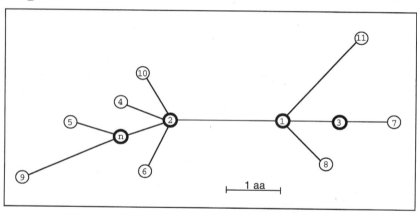

Figure 26. Phylogenetic analysis of Fim H alleles. (A) Amino acid sequences of Fim H variants. The alleles are listed based on an increasing $M_1 : M_3$ binding ratio. The residues listed above the 11 alleles are for the amino acids in original Fim H sequence (17) that vary in the other *fimH* alleles. Only polymorphic residues are shown and the positions are numbered vertically, from −16 to +201. Δ, Deleted residues. (B) Inferred phylogenetic network demonstrating evolutionary relationships of the Fim H alleles shown in Fig. 4A. Each node represents a distinct Fim H allele, numbered as in Fig. 4A. The allele labeled n represents a hypothetical Fim H that differs from allele #2 by the substitution of asp (N) for tyr (T) in the leader sequence (residue, −16) and phenotypically should be equivalent to allele #2. Internal nodes are shown in boldface. The deduced sequences of the 11 Fim H proteins exhibit greater than 99% homology and the network showing their phylogenetic relationships is fully consistent, without any homoplasy. Branch lengths are scaled to the number of amino acids that differ between alleles, as indicated. The deletion of four amino acids in Fim H allele #10 is considered to be a single event, equivalent to one amino acid substitution. From Sokurenko *et al.* (1998), with permission.

6. FIM H VARIANTS AND THE EVOLUTION OF VIRULENCE

6.1. Phylogenetic Analysis of Fim H Alleles

The M^L and M^H pnenotypes are maintained within the population by a form of balancing selection called diversifying selection in which two different genotypes are positively selected in two different environments (Hedrick, 1986). There are several lines of evidence suggesting that M^L is the original phenotype. First, M^L is the predominant phenotype of the population of normal intestinal isolates we have tested. Also, M^L was found to be a superior adhesin for *E. coli*–buccal epithelial cell interaction in the presence of saliva, at least in comparison to the M^H phenotype. Furthermore, phylogenetic analysis of Fim H alleles is also consistent with the concept that the M^L phenotype is the evolutionary original phenotype of Fim H and M^H is the mutant phenotype. We compared the genealogical relationships of 11 unique *fimH* alleles that exhibit various monomannose binding actvities (Fig. 26). All interior nodes of the phylogenetic network are represented by Fim H subunits with a relatively low monomannose-binding capability (M1:M3 ratios of 0.08 to 0.15), whereas their terminal nodes are occupied primarily by distinct M^H phenotype alleles capability (M1:M3 ratios of 0.17 to 0.93). M^H alleles arise from M^L alleles by various nonsynonymous mutations and do not form distinct genetic lineages, indicating that the mutations are random. The position of M^H alleles on outer nodes and the extremely high sequence similarities indicate that these variants were derived recently from M^L alleles. M^H alleles appear to be eliminated eventually from the population, probably because they are more sensitive to inhibitors and therefore are less efficiently spread among hosts. However, since at least a few M^H alleles are found among normal intestinal isolates, it is clear that they are not immediately eliminated from the population, although the half-life of the mutants (e.g., years, tens of years, 100s, etc.) is difficult to assess. The mechanisms whereby a strain minimizes the deleterious effect is not yet clear, but it is likely either that these strains are transmitted to other hosts by a route that bypasses the oropharyngeal cavity, which has yet to be identified, or one or more compensatory mechanisms are acquired. These compensatory mechanisms could involve genes for other adhesins, such as P, S, or Dr, or much larger genetic elements, such as a pathogenicity island (Hacker *et al.,* 1997).

6.2. Pathoadaptive Mutations

The view of bacterial pathogen evolution that currently prevails is that the evolutionary adaptation to the pathological habitat is due to the possession by pathogens of virulence factor genes that are absent from commensal strains (Salyers and Whitt, 1994; Covacci *et al.,* 1997; Falkow, 1997). In particular, the concept that virulence has evolved by quantum leaps through the horizontal transfer of large DNA insertions has become so popular that it has largely eliminated from consideration the possibility that pathogens develop from nonpathogenic, commensal organisms by the modification of existing genes through the effects of selective pressures on randomly mutated commensal traits. While it is clear that the genetic transfer of pathogenicity islands contributes to pathogen evolution, we have introduced the term "pathoadaptive" mutation (Sokurenko *et al.,* 1998) to call attention to the "old" idea that pathogens can evolve by the adaptive mutation of commensal genes. The adaptive mutations of the *fimH* gene is thus far the best and most direct example of a pathoadaptive mutation of which we are aware. We have provided evidence that these mutations increase the fitness of *E. coli* for the urinary tract but very likely have a deleterious effect on interhost transmission *per os.* In fact, it is generally true that mutations that affect the specificity of the original function, optimized over many millions of years, are most likely to be deleterious for the organism in its evolutionarily primary niche (Kimura, 1983). For these "damaged" commensals to survive within the primary niche of a new host where pathoadaptive mutations are maladaptive (i.e., deleterious), compensatory adaptations may be required. It is interesting to speculate that the acquisition of large (e.g., pathogenicity islands) or small (e.g., *pap* operon) genetic elements could actually be driven by an initial pathoadaptive mutation (e.g., the change from M^L to M^H Fim H), either to increase the clone's adaptation to the pathological niche or to compensate for the deleterious effects within the primary niche.

7. SUMMARY AND SPECULATIONS

It is becoming increasingly clear that type 1 fimbriae, the most common adhesin of enterobacteria, are much more complicated and interesting organelles than has been heretofore appreciated. Adaptive mutations of the Fim H lectin, an originally commensal trait, have a dramatic effect on the tropism of *E. coli* and shift the organism to a more virulent phenotype. The adaptive characteristics of Fim H are summarized in Table III. These observations reported in this chapter are important on a number of levels. First, the studies described provide a rational explanation for the contribution of type 1 fimbriae to colonization of both the primary commensal niche and a pathological niche. This has always been an enigma due to the very different physiological environments involved: oropharyngeal cavity, colon, bladder. The studies strongly suggest that any new epidemiological studies evaluating the contribution of type 1 fimbriae to disease must take into consideration the Fim H phenotypes of the organisms isolated from the site of infection. Second, it should be clear that if adaptive mutations occur in the *E. coli fimH* gene, it is likely that similar phenomena occur with other type 1 fimbriated

Table III
Adaptive characteristics of allelic variants of *E. coli* Fim H

	Fim H Phenotype	
	M^L	M^H
Evolutionary relationship	Original	Derived
Cell-binding spectrum	Narrow	Broad
Resistance to inhibitors	Strong	Weak
Fitness in commensal niche	High	Low
Fitness in pathological niche	Low	High

enterobacteria. Third, it is also likely that mutations such as these also occur among the other types of bacterial fimbriae. It is difficult to know precisely how the different classes of the Pap G adhesin fit into this scheme. It has been clearly shown that Pap G classes differ in terms of the receptors they recognize and also documented that these functional differences affect tissue and even species tropism (Strömberg *et al.,* 1990), but the Pap G adhesin classes exhibit less structural homology than the FimH variants, and it is not clear that they are really allelic variants. Fourth, the studies illustrate that bacterial lectins may provide unique tools to study the architecture of host cells, such as the previously undetected mosaicism of urinary bladder umbrella cells.

DEDICATION

We dedicate this chapter to the memory of Dr. Charles C. Brinton, Jr. Dr. Brinton was Professor of Microbiology at the University of Pittsburgh when he died October 21, 1997, after a 3-year battle with amyotropic lateral sclerosis. He was a pioneer in the study of fimbriae . . . excuse me, Charlie, a pioneer in the study of pili. In fact, Charlie coined the term "pili" for filamentous, nonflagellar appendages of enterobacteria. Dr. Brinton was an exceptional man and an exceptional scientist. Despite the debilitating disease he contracted, he continued to work in the laboratory to the last possible moment. No one who knew Charlie will be surprised to hear that. Charlie, we are glad we got to know you and will certainly miss your stimulating insight.

REFERENCES

Abraham, J. M., Freitag, C. S., Clements, J. R., and Eisenstein, B. I, 1985, An invertible element of DNA controls phase variation of type 1 fimbriae of *Escherichia coli, Proc. Natl. Acad. Sci. USA* **82:**5724–5727.

Abraham, S. N., Goguen, J. D., Sun, D., Klemm, P., and Beachey, E. H., 1987, Identification of two ancillary subunits of *Escherichia coli* type 1 fimbriae by using antibodies against synthetic oligopeptides of *fim* gene products, *J. Bacteriol.* **169:**5530–5536.

Abraham, S. N, Sun, D., Dale, J. B., and Beachey, E. H., 1988, Conservation of the D-mannose-adhesion protein among type 1 fimbriated members of the family *Enterobacteriaceae, Nature* **336:**682–684.

Barondes, S. H., 1988, Bifunctional properties of lectins: Lectins redefined, *Trends Biochem. Sci.* **13:**480–482.

Blattner, F. R., Plunkett, G. 3rd, Bloch, C. A., Perna, N. T., Burland, V., Riley, M., Collado-Vides, J., Glasner, J. D., Rode, C. K., Mayhew, G. F., Gregor, J., Davis, N. W., Kirkpatrick, H. A., Goeden, M. A., Rose, D. J., Mau, B., and Shao, Y., 1997, The complete genome sequence of *Escherichia coli* K-12, *Science* **277:**1453–1474.

Bloch, C. A., and Orndorff, P. E., 1990, Impaired colonization by and full invasiveness of *Escherichia coli* K1 bearing a site-directed mutation in the type 1 pilin gene, *Infect. Immun.* **58:**275–278.

Bloch, C. A., Stocker, B. A. D., and Orndorff, P., 1992, A key role for type 1 fimbriae in enterobacterial communicability, *Mol. Microbiol.* **6:**697–701.

Blomfield, I. C., McClain, M. S., Eisenstein, B. I., 1991, Type 1 fimbriae mutants of *Escherichia coli* K12: Characterization of recognized afimbriate strains and construction of new *fim* deletion mutants, *Mol. Microbiol.* **5:**1439–1445.

Blomfield, I. C., Calie, P. J., Eberhardt, K. J., McClain, M. S., and Eisenstein, B. I., 1993, Lrp stimulates phase variation of type 1 fimbriation in *Escherichia coli* K-12, *J. Bacteriol.* **175:**27–36.

Blomfield, I. C., Kulasekara, D. H., and Eisenstein, B. I, 1997, Integration host factor stimulates both FimB- and FimE-mediated site-specific DNA inversion that controls phase variation of type 1 fimbriae expression in *Escherichia coli, Mol. Microbiol.* **23:**705–717.

Brinton, C. C., Jr., 1959, Non-flagellar appendages of bacteria, *Nature* **183:**782–786.

Brinton, C. C., Jr., 1965, The structure, function, synthesis and genetic control of bacterial pili and a molecular model for DNA and RNA transport in gram-negative bacteria, *Trans. NY Acad. Sci.* **27:**1003–1054.

Brinton, C. C., Jr., Buzzell, A., and Lauffer, M. A., 1954, Electrophoresis and phage susceptibility studies on a filament-producing variant of the *E. coli* B bacterium, *Biochim. Biophys. Acta* **15:**533–542.

Bullitt, E., and Makowski, L., 1995, Structural polymorphism of bacterial adhesion pili, *Nature* **373:**164–167.

Collier, W. A., and de Miranda, J. C., 1955, Bacterien–haemagglutination. III. Die hemmung der Coli-haemagglutination durch mannose, *Antonie van Leeuwenhoek* **21:**133–140.

Connell, H., Agace, W., Klemm, P., Schembri, M., Mårild, S., and Svanborg, C., 1996, Type 1 fimbrial expression enhances *Escherichia coli* virulence for the urinary tract, *Proc. Natl. Acad. Sci. USA* **93:**9827–9832.

Costerton, J. W., Lewandowski, Z., Caldwell, D. E., Korber, D. R., and Lappin-Scott, H. M., 1995, Microbial biofilms, *Annu. Rev. Microbiol.* **49:**711–745.

Covacci, A., Falkow, S., Berg, D. E. and Rappuoli, R., 1997, Did the inheritance of a pathognicity island modify the virulence of *Helicobacter pylori*? *Trends Microbiol.* **5:**205–208.

de Graaf, F. K., and Gaastra, W., 1994, Fimbriae of enterotoxigenic *Escherichia coli*, in: *Fimbriae: Adhesion, Genetics, Biogenesis, and Vaccines* (P. Klemm, ed.), CRC Press, Boca Raton, FL, pp. 53–83.

de Graaf, F. K., and Mooi, F. R., 1986,The fimbrial adhesins of *Escherichia coli, Adv. Microb. Physiol.* **28:**65–143.

Dorman, C. J., and Higgins, C. F., 1987, Fimbrial phase variation in *Escherichia coli:* Dependence on integration host factor and homologies with other site-specific recombinases, *J. Bacteriol.* **169:**3840–3843.

Duguid, J. P., and Old, D. C., 1994, Introduction: A historical perspective, in: *Fimbriae: Adhesion, Genetics, Biogenesis, and Vaccines* (P. Klemm, ed.), CRC Press, Boca Raton, FL, pp. 1–7.

Duguid, J. P., Smith, I. W., Dempster, G., and Edmunds, P. N., 1955, Non-flagellar filamentous appendages ("fimbriae") and haemagglutinating activity in *Bacterium coli, J. Pathol. Bact.* **70:**335–348.

Eisenstein, B. I., Sweet, D. S., Vaughn, V., and Friedman, D. I., 1987, Integration host factor is required for the DNA inversion that controls phase variation in *Escherichia coli, Proc. Natl. Acad. Sci. USA* **84:**6506–6510.

Eshdat, Y., Silverblatt, F. J., and Sharon, N., 1981, Dissociation and reassembly of *Escherichia coli* type 1 pili, *J. Bacteriol.* **148:**308–314.

Falkow, S., 1997, What is a pathogen? *ASM News* **63:**359–365.

Firon, N., Ofek, I., and Sharon, N., 1983, Carbohydrate specificity of the surface lectins of *Escherichia coli, Klebsiella pneumoniae,* and *Salmonella typhimurium, Carbohydr. Res.* **120:**235–249.

Firon, N., Ofek, I., and Sharon, N., 1984, Carbohydrate-binding sites of the mannose-specific fimbrial lectins of enterobacteria, *Infect. Immun.* **43:**1088–1090.

Firon, N., Ashkenazi, S., Mirelman, D., Ofek, I., and Sharon, N., 1987, Aromatic alpha-glycosides of mannose are powerful inhibitors of the adherence of type 1 fimbriated *Escherichia coli* to yeast and intestinal epithelial cells, *Infect. Immun.* **55:**472–476.

Fu, D., Chen, L., and O'Neill, R. A., 1994, A detailed structural characterization of ribonuclease B oligosaccharides by 1H NMR spectroscopy and mass spectrometry, *Carbohydr. Res.* **261:**173–186.

Gally, D. L., Rucker, T. J., and Blomfield, I. C., 1994, The leucine-responsive regulatory protein binds to the *fim* switch to control phase variation of type 1 fimbrial expression in *Escherichia coli* K-12, *J. Bacteriol.* **176:**5665–5672.

Gally, D. L., Leathart, J., and Blomfield, I. C., 1996, Interaction of *fimB* and *fimE* with the *fim* switch that controls the phase variation of type 1 fimbriae in *Escherichia coli* K-12, *Mol. Microbiol.* **21:**725–738.

Guerina, N. G., Kessler, T. W., Guerina V. J., Neutra, M. R., Clegg, H. W., Langermann, S., Scannapieco, F. A., and Goldman D. A., 1983, The role of pili and capsule in the pathogenesis of neonatal infection with *Escherichia coli* K1, *J. Infect. Dis.* **148:**395–405.

Guyot, G., 1908, Uber die bakterielle haemagglutination, *Zbl. Bakt. Abt. I. Orig.* **47:**640–653.

Hacker, J., Blum-Oehler, G., Muhldorfer, I., and Tschape, H., 1997, Pathogenicity islands of virulent bacteria: Structure, function and impact on microbial evolution, *Mol. Microbiol.* **23:**1089–1097.

Hanson, M. S., and Brinton C. C., Jr., 1988, Identification and characterization of *E. coli* type-1 pilus tip adhesion protein, *Nature* **332:**265–268.

Hanson, M. S., Hempel, J., and Brinton, C. C., Jr., 1988, Purification of the *Escherichia coli* type 1 pilin and minor pilus proteins and partial characterization of the adhesin protein, *J. Bacteriol.* **170:**3350–3358.

Hedrick, P. W., 1986, Genetic polymorphism in heterogeneous environments—A decade later, *Annu. Rev. Ecol. Syst.* **17:**535–566.

Houwink, A. L., and van Iterson, W., 1950, Electron microscopical observations on bacterial cytology. II. A study on flagellation, *Biochim. Biophys. Acta* **5:**10–44.

Jacob-Dubuisson, F., Heuser, J., Dodson, K., Normark, S., and Hultgren, S. J., 1993, Initiation of assembly and association of the structural elements of a bacterial pilus depend on two specialized tip proteins, *EMBO J.* **12:**837–847.

Jann, K., and Jann, B., eds., 1990, *Bacterial Adhesins,* Springer-Verlag, Berlin.

Johnson, J. R., 1991, Virulence factors in *Escherichia coli* urinary tract infection, *Clin. Microbiol. Rev.* **4:**80–128.

Jones, C. H., Jacob-Dubuisson, F., Dodson, K., Kuehn, M., Slonim, L., Striker, R., and Hultgren, S. J., 1992, Adhesin presentation in bacteria requires molecular chaperones and ushers, *Infect. Immun.* **60:**4445–4451.

Jones, C. H., Pinkner, J. S., Nicholes, A. V., Slonim, L. N., Abraham, S. N., and Hultgren, S. J., 1993, FimC is a periplasmic PapD-like chaperone that directs assembly of type 1 pili in bacteria, *Proc. Natl. Acad. Sci. USA* **90:**8397–8401.

Jones, C. H., Pinkner, J. S., Roth, R., Heuser, J., Nicholes, A. V., Abraham, S. N., and Hultgren, S. J., 1995, FimH adhesin of type 1 pili is assembled into a fibrillar tip structure in the *Enterobacteriaceae, Proc. Natl. Acad. Sci. USA* **92:**2081–2085.

Kawula, T. H., and Orndorff, P. E., 1991, Rapid site-specific DNA inversion in *Escherichia coli* mutants lacking the histonelike protein H-NS, *J. Bacteriol.* **173:**4116–4123.

Kimura, M., 1983, *The Neutral Theory of Molecular Evolution,* Cambridge University Press, Cambridge, United Kingdom.

Klemm, P., 1984, The *fimA* gene encoding the type-1 fimbrial subunit of *Escherichia coli.* Nucleotide sequence and primary structure of the proteins, *Eur. J. Biochem.* **143:**395–399.

Klemm, P., 1986, Two regulatory *fim* genes, *fimB* and *fimE* control the phase variation of type 1 fimbriae in *Escherichia coli, EMBO J.* **5:**1389–1393.

Klemm, P., 1992, FimC, a chaperone-like periplasmic protein of *Escherichia coli* involved in biogenesis of type 1 fimbriae, *Res. Microbiol.* **143:**831–838.

Klemm, P., ed., 1994, *Fimbriae. Adhesion, Genetics, Biogenesis, and Vaccines,* CRC Press, Inc., Boca Raton, FL.

Klemm, P., and Christiansen, G., 1987, Three *fim* genes required for the regulation of length and mediation of adhesion of *Escherichia coli* type 1 fimbriae, *Mol. Gen. Genet.* **208:**439–445.

Klemm, P., and Krogfelt, K. A., 1994, Type 1 fimbriae of *Escherichia coli,* in: *Fimbriae: Adhesion, Genetics, Biogenesis, and Vaccines* (P. Klemm, ed.), CRC Press, Boca Raton, FL, pp. 9–26.

Klemm, P., Jorgensen, B. J., van Die, I., de Ree, H., and Bergmans, H., 1985, The *fim* genes responsible for synthesis of type 1 fimbriae in *Escherichia coli,* cloning and genetic organization, *Mol. Gen. Genet.* **199:**410–414.

Krallmann-Wenzel, U., Ott, M., Hacker, J., and Schmidt, G., 1989, Chromosomal mapping of genes encoding mannose-sensitive (type I) and mannose-resistant F8 (P) fimbriae of *Escherichia coli* O18:K5:H5, *FEMS Microbiol. Lett.* **58:**315–322.

Krogfelt, K. A., and Klemm, P., 1988, Investigation of minor components of *E. coli* type 1 fimbriae: Protein chemical and immunological aspects, *Microb. Pathogen.* **4**:231–238.

Krogfelt, K. A., McCormick, B. A., Burghoff, R. L, Laux, D. C., and Cohen, P. S., 1991, Expression of *Escherichia coli* F-18 type 1 fimbriae in the streptomycin-treated mouse large intestine, *Infect. Immun.* **59**:1567–1568.

Langermann, S., Palaszynski, S., Barnhart, M., Auguste, G., Pinkner, J. S., Burlein, J., Barren, P., Koenig, S., Leath, S., Jones, C. H., and Hultgren, S. J., 1997, Prevention of mucosal *Escherichia coli* infection by FimH-adhesin-based systemic vaccination, *Science* **276**:607–611.

Leathart, J. B. S., and Gally, D. L., 1998, Regulation of type 1 fimbrial expression in uropathogenic *Escherichia coli:* Heterogeneity of expression through sequence changes in the *fim* switch region, *Mol. Microbiol.* **28**:371–381.

Lowe, M. A., Holt, S. C., and Eisenstein, B. I., 1987, Immunoelectron microscopic analysis of elongation of type 1 fimbriae in *Escherichia coli, J. Bacteriol.* **169**:157–163.

Maurer, L., and Orndorff, P. E., 1985, A new locus, *pilE,* required for the binding of type 1 piliated *Escherichia coli* to erythrocytes, *FEMS Microbiol. Lett.* **30**:59–66.

Maurer, L., and Orndorff, P. E., 1987, Identification and characterization of genes determining receptor binding and pilus length of *Escherichia coli* type 1 pili, *J. Bacteriol.* **169**:640–645.

Matsui, Y., Dyer, F. P., and Langridge, R., 1973, X-ray diffraction studies on bacterial pili, *J. Mol. Biol.* **79**:57–64.

May, A. K., Bloch, C. A., Sawyer, R. G., Spengler, M. D., and Pruett, T. L., 1993, Enhanced virulence of *Escherichia coli* bearing a site-targeted mutation in the major structural subunit of type 1 fimbriae, *Infect. Immun.* **61**:1667–1673.

McClain, M. S., Blomfield, I. C., and Eisenstein, B. I., 1991, Roles of *fimB* and *fimE* in site-specific inversion associated with phase variation of type 1 fimbriae in *Escherichia coli, J. Bacteriol.* **173**:5308–5314.

McCormick, B. A., Franklin, D. P., Laux, D. C., and Cohen, P. S., 1989, Type 1 pili are not necessary for colonization of the streptomycin-treated mouse large intestine by type 1-piliated *Escherichia coli* F-18 and *E. coli* K-12, *Infect. Immun.* **57**:3022–3029.

McCormick, B. A., Klemm, P., Krogfelt, K. A., Burghoff, R. L., Pallesen, L., Laux, D. C., and Cohen, P. S., 1993, *Escherichia coli* F-18 phase locked "on" for expression of type 1 fimbriae is a poor colonizer of the streptomycin-treated mouse large intestine, *Microb. Pathog.* **14**:33–43.

Miller, J. H., 1972, *Experiments in Molecular Genetics,* Cold Spring Harbor Laboratory, Cold Spring Harbor, NY.

Minion, F. C., Abraham, S. N., Beachey, E. H., and Goguen, J. D., 1986, The genetic determinant of adhesive function in type 1 fimbriae of *Escherichia coli* is distinct from the gene encoding the fimbrial subunit, *J. Bacteriol.* **165**:1033–1036.

Neeser, J.-R., Koellreutter, B., and Wuersch, P., 1986, Oligomannoside-type glycopeptides inhibiting adhesion of *Escherichia coli* strains mediated by type 1 pili: Preparation of potent inhibitors from plant glycoproteins, *Infect. Immun.* **52**:428–436.

Newman, J. V., Burghoff, R. L., Pallesen, L., Krogfelt, K. A., Kristensen, C. S., Laux, D. C., and Cohen, P. S., 1994, Stimulation of *Escherichia coli* F-18Col− Type-1 fimbriae synthesis by *leuX, FEMS Microbiol. Lett.* **122**:281–287.

Ofek, I., and Beachey, E. H., 1978, Mannose binding and epithelial cell adherence of *Escherichia coli, Infect. Immun.* **22**:247–254.

514

David L. Hasty and Evgeni V. Sokurenko

Ofek, I., and Doyle, R. J., 1994, *Bacterial Adhesion to Cells and Tissues,* Chapman and Hall, London.

Ofek, I., and Sharon, N., 1986, Mannose specific bacterial surface lectins, in: *Microbial Lectins and Agglutinins. Properties and Biological Activity* (D. Mirelman, ed.), John Wiley & Sons, New York, pp. 55–81.

Ofek, I., and Sharon, N., 1988, Lectinophagocytosis: A molecular mechanism of recognition between cell surface sugars and lectins in the phagocytosis of bacteria, *Infect. Immun.* **56:**539–547.

Ofek, I., Mirelman, D., and Sharon, N., 1977, Adherence of *Escherichia coli* to human mucosal cells mediated by mannose receptors, *Nature* **265:**623–625.

Olsen, P. B., and Klemm, P., 1994, Localization of promoters in the *fim* gene cluster and the effect of H-NS on the transcription of *fimB* and *fimE, FEMS Microbiol. Lett.* **116:**95–100.

Orndorff, P. E., and Falkow, S., 1984a, Organization and expression of genes responsible for type 1 piliation in *Escherichia coli, J. Bacteriol.* **159:**736–744.

Orndorff, P. E., and Falkow, S., 1984b, Identification and characterization of a gene product that regulates type 1 piliation in *Escherichia coli, J. Bacteriol.* **160:**61–66.

Ritter, A., Gally, D. L., Olsen, P. B., Dobrindt, U., Friedrich, A., Klemm, P., and Hacker, J., 1997, The Pai-associated leuX specific tRNA$_5^{Leu}$ affects type 1 fimbriation in pathogenic *Escherichia coli* by control of FimB recombinase expression, *Mol. Microbiol.* **25:**871–882.

Rosenthal, L., 1943, Agglutinating properties of *Escherichia coli*. Agglutination of erythrocytes, leucocytes, thrombochtes, speermatozoa, spores of molds, and pollen by strains of *E. coli, J. Bacteriol.* **45:**545–550.

Salyers, A. A., and Whitt, D. D., 1994, *Bacterial Pathogenesis. A Molecular Approach,* ASM Press, Washington, DC.

Shembri, M. A., Pallesen, L., Connell, H., Hasty, D. L., and Klemm, P., 1996, Linker insertion analysis of the FimH adhesin of type 1 fimbriae in an *Escherichia coli fimH*-null background, *FEMS Microbiol. Lett.* **137:**257–263.

Sokurenko, E. V., Courtney, H. S., Abraham, S. N., Klemm, P., and Hasty, D. L., 1992, Functional heterogeneity of type 1 fimbriae of *Escherichia coli, Infect. Immun.* **60:**4709–4719.

Sokurenko, E. V., Courtney, H. S., Ohman, D. E., Klemm, P., and Hasty, D. L., 1994, FimH family of type 1 fimbrial adhesins: Functional heterogeneity due to minor sequence variations among *fimH* genes, *J. Bacteriol.* **176:**748–755.

Sokurenko, E. V., Courtney, H. S., Maslow, J., Siitonen, A., and Hasty, D. L., 1995, Quantitative differences in adhesiveness of type 1 fimbriated *Escherichia coli* due to structural differences in *fimH* genes, *J. Bacteriol.* **177:**3680–3686.

Sokurenko, E. V., Chesnokova, V., Doyle, R. J., and Hasty, D. L., 1997, Diversity of the *Escherichia coli* type 1 fimbrial lectin. Differential binding to mannosides and uroepithelial cells, *J. Biol. Chem.* **272:**17880–17886.

Sokurenko, E. V., Chesnokova, V., Dykhuizen, D. E., Ofek, I., Wu, X.-R., Krogfelt, K. A., Struve, C., Shembri, M. A., and Hasty, D. L., 1998, Pathogenic adaptation of *Escherichia coli* by natural variation of the FimH adhesin, *Proc. Natl. Acad. Sci. USA* **95:**8922–8926.

Strömberg, N., Marklund, B.-I., Lund, B., Ilver, D., Hamers, A., Gaastra, W., Karlsson, K.-

A., and Normark, S., 1990, Host-specificity of uropathogenic *Escherichia coli* depends on differences in binding specificity to Galα1–4Gal-containing isoreceptors, *EMBO J.* **9:**2001–2010.

Wu, X.-R., Lin, L.-H., and Walz, T., 1994, Mammalian uroplakins: A group of highly conserved urothelial differentiation-related membrane proteins, *J. Biol. Chem.* **269:** 13716–13724.

Wu, X.-R., Sun, T.-T., and Medina, J. J., 1996, *In vitro* binding of type 1 fimbriated *Escherichia coli* to uroplakins Ia and Ib: Relation to urinary tract infections, *Proc. Natl. Acad. Sci. USA* **93:**9630–9635.

18

Interactions of Microbial Glycoconjugates with Collectins

Implications for Pulmonary Host Defense

Itzhak Ofek and Erika Crouch

1. INTRODUCTION

1.1. Collectins: A Family of Collagenous C-Type Lectins

The collectins are a family of collagenous carbohydrate binding proteins (collagenous C-type lectins) that interact with complementary sugars in a calcium-dependent manner (Holmskov *et al.,* 1994; Hoppe and Reid, 1994a,b). They include the pulmonary collectins (surfactant proteins A and D, SP-A, and SP-D), serum mannose-binding protein, and the bovine serum lectins, conglutinin, and CL-43. Mannose-binding protein and conglutinin are hepatic proteins that have been implicated previously in various aspects of the systemic response to microbial challenge.

    A variety of observations strongly suggest that the lung collectins, like their systemic counterparts, modulate host–microbial interactions, and thereby participate in the defense of the lung against inhaled microorganisms (Crouch, 1998). Many of these effects involve the recognition of complementary sugars on the sur-

Itzhak Ofek • Department of Human Microbiology, Sackler Faculty of Medicine, Tel Aviv University, Ramat Aviv 69978, Israel.   *Erika Crouch* • Department of Pathology, Washington University School of Medicine, St. Louis, Missouri 63110.

Glycomicrobiology, edited by Doyle.
Kluwer Academic/Plenum Publishers, New York, 2000.

face of the organisms. The collectins also may participate in the recognition or clearance of other complex organic materials with surface glycoconjugates, such as pollens (Malhotra *et al.*, 1993) and dust mite allergens (Wang *et al.*, 1996).

The most compelling evidence for a role of lung collectins in host defense is the observation that otherwise healthy transgenic mice lacking a functional SP-A gene. SP-A (-/-) mice demonstrate increased bacterial proliferation, more intense lung inflammation, and an increased incidence of splenic dissemination following intratracheal inoculation with the group B streptococcus (Korfhagen *et al.*, 1996; Ikegami *et al.*, 1997; LeVine *et al.*, 1997). Other studies suggested defective clearance of *Staphylococcus aureus* and *Pseudomonas aeruginosa*. Significantly, the animals show no obvious associated abnormalities in respiratory function or surfactant lipid metabolism.

This chapter focuses on the structure–function correlations of the two known pulmonary collectins, SP-A and SP-D, as pertains to their interactions with microorganisms. Biochemical properties of potential functional significance and the interactions of these proteins with specific glycoconjugates on the surface of microorganisms will be discussed in the context of their roles in providing innate immunity against pulmonary infections. A comprehensive review on collectin structure, production, and the interactions of these molecules with lipids and host cells may be found in a recent review (Crouch, 1998).

2. GENERAL PROPERTIES OF SP-A AND SP-D

2.1. Collectin Structure

The lung and serum collectins are assembled as oligomers of trimeric subunits (3×43 kDa). Each subunit consists of four major domains: a short cysteine-containing NH_2-terminal domain, a triple helical collagen domain of variable length, a short trimeric coiled–coil domain, and a carboxy-terminal, C-type lectin carbohydrate recognition domain (CRD) (Fig. 1). Interactions between the amino-terminal domains of the collectin subunits are stabilized by interchain disulfide bonds (Crouch *et al.*, 1994a; Haas *et al.*, 1991; Brown-Augsburger *et al.*, 1996).

SP-A ($26–35$ kDa, reduced) is predominantly assembled as 18 mers consisting of 6 trimeric subunits with relatively short collagen domains (Fig. 1). However, smaller oligomeric forms also have been identified in human lung (Hickling *et al.*, 1998). Human SP-A molecules can be assembled as homotrimers or as heterotrimers derived from two genetically different chain types (Voss *et al.*, 1991), but the relative proportions of homo- and heterotrimers accumulating in the lung have not been established.

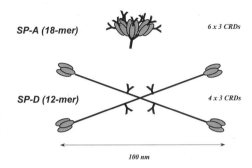

Figure 1. Molecular organization of pulmonary collectins. The predominant molecular forms of SP-A and SP-D are compared assuming maximal spatial separation of the CRDs. The Asn-linked oligosaccharides of SP-D are probably less accessible to lectins expressed by microorganisms than the sugars of SP-A. Reprinted from Crouch (1998).

By contrast, SP-D (43 kDa, reduced) is predominantly assembled as dodecamers (12 mers) consisting of four homotrimeric subunits with relatively long triple helical arms (Crouch *et al.,* 1994a; Holmskov and Jensenius, 1993). Although natural and recombinant rat SP-Ds are almost exclusively assembled as dodecamers, preparations of natural human and bovine SP-D can include a high proportion of trimers (Lu *et al.,* 1992, 1993; Hartshorn *et al.,* 1996a; Holmskov *et al.,* 1995).

2.2. Higher Orders of Collectin Oligomerization

SP-A octadecamers can self-associate to form even larger multimolecular complexes (Hattori *et al.,* 1996a,b). Under some circumstances the aggregated molecules may become cross-linked through the formation of disulfide and nondisulfide covalent bonds (Crawford *et al.,* 1986; Ross *et al.,* 1987; Voss *et al.,* 1992). The functional significance of SP-A multimerization is uncertain. However, differences have been observed between alveolar proteinosis SP-A, which is highly multimerized, and less highly multimerized preparations of natural or recombinant proteins. For example, proteinosis SP-A is more effective than natural or recombinant SP-A in enhancing the adhesion and phagocytosis of mycobacteria by macrophages (Gaynor *et al.,* 1995).

SP-D dodecamers can self-associate at their amino-termini to form highly ordered, stellate multimers with peripheral arrays of trimeric CRDs (Crouch *et al.,* 1994a,b; Hartshorn *et al.,* 1996a). Natural SP-D can contain a high proportion of these multimers with up to eight (or possibly more) SP-D dodecamers. SP-D multimers show higher apparent binding affinity to a variety of ligands and are considerably more potent in mediating microbial aggregation and certain aggregation-dependent interactions with leukocytes (Hartshorn *et al.,* 1996a,b).

2.3. Modulation of Collectin Production and Accumulation *in Vivo*

There is now considerable evidence that the production of these molecules is increased in association with acute injury and epithelial activation (Crouch, 1998; Crouch *et al.*, 1991; Nogee *et al.*, 1989; Kasper *et al.*, 1995; Aderbigbe *et al.*, 1999; Horowitz *et al.*, 1991; Viviano *et al.*, 1995). In addition, regional concentrations of the lung collectins may be influenced by the specific cellular responses to various forms of injury (Aderbigbe *et al.*, 1999; Horowitz *et al.*, 1991). The production and accumulation of the lung collectins are rapidly increased following intratracheal instillation of LPS (McIntosh *et al.*, 1996). Because the mRNAs for both proteins are increased within several hours to a few days following injury, Wright and co-workers suggested that they are pulmonary acute-phase proteins (McIntosh *et al.*, 1996).

2.4. Airspace Distribution of Lung Collectins

Very little SP-A is recovered in solution following high-speed centrifugation of lung washings. Immunoelectron microscopic studies have shown that SP-A is associated with formed lipid-rich components, particularly tubular myelin (Haller *et al.*, 1992; Voorhout *et al.*, 1991), and tubular myelin is nearly absent from the lungs of SP-A(-/-) transgenic mice (Korfhagen *et al.*, 1996). By contrast, the majority of the SP-D remains in the supernatant following high-speed centrifugation (Persson *et al.*, 1989; Kuroki *et al.*, 1992). Recent comparative assays by Honda and co-workers (1996) gave 3.1 ± 0.4 μg/ml for SP-A and 1.3 ± 0.2 μg/ml for lavage from healthy nonsmokers.

3. CARBOHYDRATE BINDING BY LUNG COLLECTINS

3.1. Carbohydrate Binding Domains

Protein and cDNA sequencing studies have shown that the primary sequence of the carboxy-terminal domains of SP-A and SP-D contain characteristic elements of the mannose-type C-type lectin motif (Shimizu *et al.*, 1992; Rust *et al.*, 1991; Drickamer and McCreary, 1987; Sano *et al.*, 1987; Lu *et al.*, 1992). Biochemical and molecular studies have established that these domains are responsible for the carbohydrate-binding activity.

Table 1

Carbohydrate Specificity of Lung Collectins

Collectin	Order of sugar specificity	Method
Human SP-A	ManNAc > Fuc > Mal > Glc > Man	Solid-phase to mannan
	Man, Glc, Gal, Fuc >> GlcNAc,	Affinity chromatography
	Man, Fuc > Glc, Gal >> GlcNac	Direct binding to neoglycoproteins
Rat SP-D	Mal, Inositol > Glc >> Man > Gal,	Solid-phase to maltosyl-BSA
	Lac, Fuc	
	α-Glc-BSA >>> -Glc-BSA	Solid-phase binding assay to
		Glc-BSA neoglycoproteins
Human SP-D	Mal > Fuc, Man > Glc > Gal,	Solid-phase to maltosyl-BSA
	Lac > GlcNAc	

3.2. Carbohydrate Specificity

The carbohydrate selectivities of the two proteins generally are consistent with their subclassification as mannose-type C-type lectins. SP-A and SP-D show calcium-dependent and saccharide-inhibitable interactions with a wide variety of carbohydrate-containing ligands *in vitro* (Tables I & II). The ligands include various neoglycoproteins or saccharide-substituted affinity matrices, purified microbial glycoconjugates, and whole organisms. Although there are some discrepancies in the apparent specificity as determined using the different assay systems,

Table II

Examples of Microbial Glycoconjugates Interacting with Lung Collectins

Microbial ligand	SP-A	SP-D
Gram (-) lipopolysaccharides (LPS)	Lipid A	Core oligosaccharides
Gram (-) capsular polysaccharides		
Klebsiella pneumoniae	Di-mannose or rhamnose units	No binding
Gram (+) lipoteichoic acids (LTA)	?	No binding
Influenza A hemagglutinins (HA)	(HA binds to N-linked sugar of SP-A)	N-linked sugars on HA
Influenza A neuraminidase (NA)	?	N-linked sugars on NA of some strains
Fungal cell wall glycoconjugates	Binds	Binds
Pneumocystis gpA	N-linked sugars	N-linked sugars

important generalizations can be made. Thus, SP-A shows a preference for mannose or ManNAc, whereas SP-D preferentially recognizes the α-anomeric configuration of nonreducing glucopyranosides such as maltose (Persson *et al.,* 1990). Both collectins show comparatively weak interactions with galactose and related sugars. This binding specificity is consistent with known interactions of the lung collectins with known microbial glycoconjugates.

3.3. Influences of Higher-Order Structure on Glycoconjugate Binding

For all the collectins, the major requirements for specific carbohydrate binding include the conserved C-type lectin motif, a conserved tertiary structure stabilized by calcium binding and intrachain disulfide cross-links, and the formation of a trimeric molecule with an appropriate spatial distribution of the constituent CRDs. The three CRDs of a single subunit constitute a single, trimeric, high-affinity ligand binding site (Kishore *et al.,* 1996). High-affinity binding probably requires the simultaneous occupancy of two to three saccharide-binding sites within a single trimeric subunit in apposition to a surface with a comparable spatial distribution of saccharide ligands.

Although the assembly of collectin monomers to form trimeric clusters of C-type CRDs is necessary and sufficient for high-affinity binding, the capacity for bridging interactions between spatially separated ligands depends on an appropriate oligomerization of trimeric subunits. Multivalency also permits even higher-affinity binding interactions. Thus, trimeric CRDs appear to be functionally univalent with regard to their capacity to participate in bridging interactions between large particulate ligands. The apparent dissociation constant for the binding of collectins to highly substituted ligands is typically orders of magnitude higher than is observed with simple test ligands. Whereas the dissociation constant K_d for binding to *E. coli* is approximately 2×10^{-11}M (Kuan *et al.,* 1992), the apparent dissociation constant K_d for SP-D binding to maltosyl$_{30}$ albumin in solution phase-binding assays is approximately 3×10^{-8} M (Persson *et al.,* 1990).

4. MICROBIAL SURFACE GLYCOCONJUGATES RECOGNIZED BY LUNG COLLECTINS

All microbial cells express surface glycoconjugates associated with their cell walls, and some of these glycoconjugates can be recognized by animal lectins (Aarson, 1996). In the following sections we will describe the types of glycoconjugates on microbial surfaces that are currently known to be recognized by lung collectins. In general, the experimental approaches used to obtain the evidence for the involvement of specific microbial glycoconjugates are identical to those usually employed for the study of lectin–carbohydrate interactions. These include collectin-mediated agglutination assays and various competition assays that examine the ability of simple and complex sugars to inhibit binding or agglutination. In some cases, the ability of the isolated glycoconjugates (or their constituent sugars) to directly bind to the collectin and to inhibit interactions with whole organisms has provided more definitive evidence as to the nature and specificity of the collectin–microbe interactions.

Table II lists the types of interactions observed between pathogens and lung collectins. In many cases, only preliminary observations that show agglutinating activity by the collectin are available. In other cases, more complete studies have been carried out, some of which have culminated in determination of the sugar sequence of the glycoconjugate interacting with a specific lung collectin as well as its regulation and assembly on the surface of the microorganisms. Whenever known, the sugar specificity of the interaction is indicated. Examples of collectin–microbial glycoconjugate interactions that have been studied in more detail are discussed below.

4.1. Viral Glycoconjugates

The interactions of lung collectins with influenza A viruses (IAVs) have been extensively characterized and provide a reasonable model system to examine structure–function relationships. IAV attaches to and infects cells by binding through its hemagglutinin (HA) to sialic acid-bearing components on the cell surface, while the neuraminidase is involved in viral production and perhaps inactivation of sialylated host proteins.

The collectins are potent inhibitors of HA-mediated agglutination and also inhibit neuraminidase activity (Malhotra et al., 1994; Hartshorn et al., 1994, 1996b; Caton et al., 1982; Malhotra and Sim, 1995). HA inhibition by SP-D involves the binding of SP-D through its CRD to glycoconjugates expressed near the sialic acid binding site on the hemagglutinin (or neuraminidase) of specific strains of IAV. Higher degrees of valency or multimerization among the various SP-D preparations are associated with increased HA inhibitory activity. At least with some strains of IAV there also is binding to glycoconjugates associated with the viral neuraminidase, and it has been suggested that collectin binding to the neuraminidase can sterically interfere with HA activity (Malhotra et al., 1994; Malhotra and Sim, 1995).

An important aspect of the interaction of collectins with IAV is their ability to cause viral aggregation and to enhance the aggregation-dependent host defense activities of leukocytes (Hartshorn et al., 1996a,b). Among the collectins, SP-D is the most potent at aggregating IAV particles, and multimers of dodecamers are much more potent than dodecamers. Approximately tenfold lower concentrations of SP-D dodecamers are needed to achieve maximal aggregation in light-scattering assays, as compared to SP-A or mannose-binding protein octadecamers (Hartshorn *et al.*, 1996b). SP-D-induced viral aggregates also are much larger than those obtained for SP-A or mannose binding protein.

4.2. Bacterial Glycoconjugates

The lung collectins bind glycoconjugates expressed by a variety of gram-negative bacteria including specific strains of important pulmonary pathogens as *Klebsiella pneumoniae, Pseudomonas aeruginosa, Haemophilus influenzae,* and *Escherichia coli* (Table II). However, their bacterial specificities only partially overlap, and their modes of interaction and the effects of binding on microbial interactions with host defense cells appear distinct.

4.2.1. GRAM-NEGATIVE BACTERIAL LIPOPOLYSACCHARIDES

SP-D specifically binds to core sugars of lipopolysaccharide (LPS) (glucose and/or heptose), which have been identified as major ligands for rat or human SP-D on *E. coli* and *Salmonella minnesota* (Kuan *et al.*, 1992). SP-D also binds to isolated LPS from a variety of other gram-negative bacteria including *K. pneumoniae* and *P. aeruginosa* (Kuan *et al.*, 1992; Lim *et al.*, 1994; Ofek *et al.*, 1997). Dodecamers are potent agglutinins for bacterial strains expressing O-antigen-deficient LPS molecules (e.g., rough strains of *E. coli*), and cause gross aggregation of suspended organisms. SP-D binding to LPS and its effects on bacterial aggregation are blocked by EDTA, competing sugars, LPS and rough mutant forms of LPS, but not by lipid A (Kuan *et al.*, 1992).

Although SP-D does not grossly agglutinate smooth strains of bacteria, it does bind to these strains as evidenced by specific labeling and microaggregation in immunofluorescence assays (Kuan *et al.*, 1992). SP-D also binds weakly to smooth (O-antigen containing) LPS on lectin blots. Because SP-D reacts preferentially with the core region of LPS, which is located near the outer membrane of gram-negative bacteria, a number of surface molecules are likely to interfere with the accessibility of this region to SP-D. As described in Section 4.2.2, capsular material reduces the ability of SP-D to agglutinate *K. pneumoniae*. We infer that the length of the O-antigen similarly influences the accessibility of the core region to the saccharide-binding sites of SP-D. It is well known that O-antigens can mask the accessibility of core determinants to antibody, and it is likely that these structures can sufficiently interfere with collectin binding to the core regions and limit aggregate size.

A   S

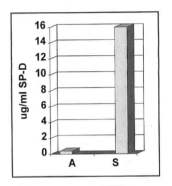

SDS-PAGE/Ag stain
of K50 (cap-) LPS

Minimal amount of SP-D needed
to cause macroscopic aggregation

Figure 2. Modulation of surface expression of LPS- and SP-D-induced agglutination of *Klebsiella pneumoniae* in response to growth conditions. K50 (cap-) strains of *K. pneumoniae,* which show agglutination by SP-D, were grown as aerated (A) or static (S) cultures. Left panel: Outer membrane preparations from organisms grown under these conditions were resolved by SDS-PAGE and visualized by silver staining. Aeration is associated with a predominance of rapidly migrating rough forms, whereas conditions of static culture favor smooth forms with larger O-antigens. Right panel: The minimal concentration of SP-D required for macroscopic aggregation of the organisms is given on the Y-axis. Note that bacteria grown under aerated conditions (which show a predominance of rough LPS) are preferentially agglutinated by SP-D. Reprinted from Crouch (1998).

The molecular size of the O-antigen is determined by the number of its repeating oligosaccharide units, which in turn is influenced by environmental factors such as aeration, pH, and others of the growth conditions (Weiss *et al.,* 1986; McGroarty and Rivera, 1990). We have observed that growth conditions (e.g., aeration) can markedly influence the aggregation of unencapsulated *K. pneumoniae* by SP-D, and that the extent of macroscopic aggregation inversely correlates with the size and complexity of the terminal O-antigen (Fig. 2). Thus, phase variants that express a higher proportion of immature LPS may be preferentially aggregated with SP-D. Immunoelectron microscopic studies have demonstrated preferential localization of binding sites in growth phase cells near the sites of bacterial cell division (Crouch, 1998).

SP-A preferentially binds specifically to the lipid A domain of rough forms of LPS and to purified lipid A (Van Iwaarden *et al.,* 1994; Kalina *et al.,* 1995). The binding to purified rough LPS is calcium-independent and is not inhibited by competing saccharides, but is inhibited or partially reversed by lipid A. Consistent with these findings, human SP-A binds to certain rough but not smooth strains of *E. coli,* with resulting opsonization and enhanced phagocytosis and killing (Pikaar *et al.,* 1995).

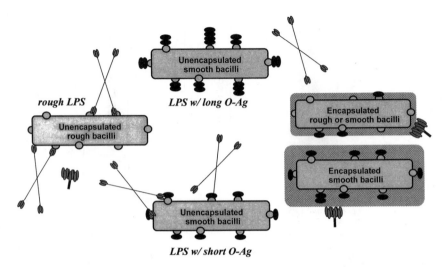

Figure 3. Hypothetical comparison of the interactions of SP-A and SP-D with rough and smooth forms of encapsulated and unencapsulated gram-negative bacteria. SP-D binds to rough or short O-antigen forms of LPS expressed on unencapsulated or weakly encapsulated organisms, a variant phenotype that is likely required for bacterial colonization and invasion. Although encapsulation limits interactions with SP-D, the expression of specific capsular polysaccharides favors SP-A binding (Kabha *et al.,* 1997). In this scheme SP-A and SP-D serve complementary roles in interacting with gram-negative bacteria and modifying their interactions with host–defense cells. Reprinted from Crouch (1998).

4.2.2. GRAM-NEGATIVE CAPSULAR POLYSACCHARIDES

Human SP-A specifically binds to *Klebsiella* serotypes that express Man-α-2/3-Man or LL-Rha-α-2/3-L-Rha sequences in the repeating units of their capsular polysaccharides (Kabha *et al.,* 1997), the same sequences recognized by the macrophage mannose receptor. Significantly, serotypes that do not express these sequences (e.g., K2 serotype) do not react with SP-A or mannose receptor. By contrast, SP-D does not recognize the capsular polysaccharides of *K. pneumoniae,* and the available data suggest that the presence of a well-formed capsule limits interactions of SP-D with underlying LPS molecules (Fig. 3).

4.2.3. GRAM-POSITIVE CELL WALL

Human SP-A shows calcium-dependent binding to clinical isolates of *Staphylococcus aureus* and *Streptococcus pneumoniae* (McNeely and Coonrod, 1993). SP-D also can bind to and agglutinate these organisms (Hartshorn *et al.,* 1998). However, nothing is currently known about the nature of the binding sites on gram-positive bacteria. Various capsular or cell wall glycoconjugates are certainly plausible candidates. In this regard, mannose binding protein (MBP) has been shown to bind to lipoteichoic acids (LTAs), and binding was restricted to LTAs with terminal sugars (Polotsky *et al.*, 1996). Because SP-A can bind to lipid A and the lipid domains of certain surfactant lipids, hydrophobic interactions also could play a role in its binding to gram-positive organisms. For example, LTAs contain fatty acids that could participate in hydrophobic interactions, and other less characterized adhesive hydrophobic components (hydrophobins) are also expressed on some gram-positive bacteria (Ofek and Doyle, 1994).

4.3. Fungal Glycoconjugates

Human SP-A and rat and human SP-D bind to pathogenic unencapsulated but not the capsulated forms of *Cryptococcus neoformans* through a lectin-dependent mechanism (Schelenz *et al.*, 1995). Although the unencapsulated organisms are readily agglutinated by SP-D, there is no significant aggregation by SP-A. More recently, human proteinosis SP-A and SP-D were shown to bind to *Aspergillus fumigatus* conidia in a calcium- and carbohydrate-dependent fashion consistent with binding of the CRD to cell wall glycoconjugates (Madan *et al.*, 1997). In addition, SP-A and SP-D efficiently agglutinated the conidia and enhanced phagocytosis and killing by neutrophils and alveolar macrophages.

Both SP-A and SP-D bind to *Pneumocystis carinii* cysts and trophozoites through interactions with gpA, a mannose and glucose-rich glycoprotein (Limper *et al.*, 1994, 1995; O'Riordan *et al.*, 1995; Williams *et al.*, 1996; Zimmerman *et al.*, 1992). SP-D also may bind to β-glucans associated with the cell wall (Vuk-Pavlovic *et al.*, 1998). Both proteins are associated with *P. carinii in vivo* and are present on the surface of freshly isolated organisms. The proteins bind in a collectin-dependent mechanism that is inhibited by EDTA, competing sugars, or specific antibody. Clusters of organisms in lung washings can be partially disaggregated with EDTA or competing sugars, and "stripped" organisms can be agglutinated by purified SP-D, suggesting that SP-D contributes to the clustering of cysts observed *in vivo*.

4.4. Role in Host Defense

The lung collectins have been shown to interact with a wide range of microorganisms in vitro (Table II), and dissemination of at least one organism is enhanced in SP-A (-/-) mice. The interaction of pulmonary collectins with glycoconjugates on microbial surfaces undoubtedly leads to eradication of the invading pathogen out of the lung. There are two major mechanisms through which lung collectins function in host defense: agglutination and opsonization.

4.4.1. AGGLUTINATION

Because the lung collectins are multimeric and present either 12 or 18 high-affinity saccharide-binding sites, they can readily agglutinate microorganisms expressing the corresponding collectin-specific sugar on their surfaces. Although SP-D is generally more potent as an agglutinin than SP-A under the usual assay conditions and usually leads to the formation of larger aggregates, both collectins can elicit agglutination (Kabha *et al.,* 1997). It is possible that bacterial aggregates are more readily cleared from the mucosal surfaces of the lung through mucocil-iary clearance, which involves the coordinated activity of ciliated epithelial cell lining the airways.

Another mechanism through which aggregation of microorganisms can promote host defense is by enhancing the interaction of microorganisms with phago-cytic cells. For example, the aggregation of influenza A virus by SP-D or SP-A increases the binding of viral particles to neutrophils and enhances neutrophil activation. The specificity of this binding is determined by the viral receptor but not by phagocyte collectin receptors (Hartshorn *et al.,* 1994, 1993a,b, 1996a,b). The concentrations of collectins required to elicit these effects closely correlate with those required for IAV aggregation. It is possible that the altered cellular response to bound virus results from bridging or clustering of the "receptors" by viral aggregates or enhanced viral internalization.

4.4.2. OPSONIZATION

The presence of collectin receptors on macrophages on one hand and the abil-ity of the collectins to bind glycoconjugates to the microbial surface on the other enable these molecules to serve as opsonins (Table III). Although there is evidence that both SP-A and SP-D can function as true opsonins for certain organisms un-der specific assay conditions *in vitro,* enhanced internalization or killing is not an invariant consequence of enhanced binding. Furthermore, many of the published experiments have not excluded direct, nonopsonic effects resulting as a conse-quence of the presence of the collectin in the incubation mixture.

There are two known types of interaction of lung collectins with macro-phages. One mechanism is lectin-independent and involves the binding to specif-ic cell surface receptors. The other involves the lectin-dependent binding of the collectin CRD to macrophage glycoconjugates (Kuan *et al.,* 1994; Manz-Keinke *et al.,* 1991; Wintergerst *et al.,* 1989; Crouch *et al.,* 1995).

The lectin-independent interactions of SP-A with macrophages involve at

Table III

Opsonic Activites of Lung Collectins[a]

| Organism | Enhancement of indicated activity of phagocytes exposed to microorganisms opsonized with | |
	SP-A	SP-D
H. influenzae, type A	Binding[b]	Not tested
H. influenzae, type B	No enhancement[b]	Not tested
K. pneumoniae	Binding and killing (cap + strains)[c]	Binding and killing (cap- strains)[e]
Pneumocystis carinii	Uptake but no killing[d]	No enhancement [d]

[a]The table only includes data for studies in which the microbial–collectin complexes were washed prior to addition to the leukocytes and that show enhancement relative to organisms not complexed with collectins. In most studies, the test protein is added to the bacterial suspension and the organisms are not washed prior to their addition. Thus, the experimental design does not preclude direct activation of the leukocyte by the collectin. Abbreviations: cap+, capsulated bacteria; cap-, unencapsulated bacteria.
[b]McNeely and Coonrod (1993).
[c]Kabha *et al.* (1997).
[d]Williams *et al.* (1996).
[e]Unpublished data.

least two different protein receptors. One corresponds to the C1q receptor (C1qR) (Malhotra *et al.,* 1990, 1992) and the other to an as yet incompletely characterized 210-kDa protein (Chroneos *et al.,* 1996; Kuroki *et al.,* 1988). The expression of macrophage SP-A receptor(s) on host defense cells is subject to complex regulation (Chroneos and Shepherd, 1995; Blau *et al.,* 1994). Interestingly, the expression of macrophage SP-A receptor appears to be inversely related to changes in the expression of mannose receptor (Chroneos and Shepherd, 1995), and SP-A binding can increase mannose receptor expression (Gaynor *et al.,* 1995). By contrast, SP-D does not interact with C1qR (Miyamura *et al.,* 1994). However, SP-D has been shown to be recognized by one or more receptors or binding proteins on alveolar macrophages (Eggleton *et al.,* 1995; Miyamura *et al.,* 1994). It is unclear to what extent the binding moieties may be related to the non-C1qR SP-A receptors. Currently, the best candidate for a macrophage receptor is an SP-D binding protein, designated GP-340 (Holmskov *et al.,* 1997).

4.5. Mechanisms of Host Defense

Following inhalation of microorganisms the agglutinating and opsonic activities of the two collectins may act in concert to provide protection against a wide range of microbial species or intraspecies phase variants. Such a model is consistent with the known interactions of lung collectins with influenza A virus and with *K. pneumoniae.*

Table IV
Antiviral Activities of SP-D

Activity	Sugar-dependent activities of virus incubated with SP-D			
	None	Trimers[a]	Dodecamers	Multimers of dodecamers
Inhibition of				
hemagglutination	−	−	+ +	+ + +
Viral aggregation	−	−	+ +	+ + +
Viral binding to PMN	−	−	+ +	+ + +
Respiratory burst	−	−	+ +	+ + +

[a]Trimeric, single-arm subunits (3 x 43 kDa), wild-type dodecamers, and multimers of SP-D dodecamers are compared.

4.5.1. INFLUENZA A

The interaction of SP-D with the oligosaccharide side chains of the HA and neuraminidase of IAV results in viral aggregation, inhibition of the HA and infectivity, and the enhancement of the ability of the virus to bind to and stimulate the host defense activities of phagocytic cells (Table IV). (Hartshorn et al., 1994, 1996a).

4.5.2. *KLEBSIELLA PNEUMONIAE*

Most *K. pneumoniae* infections occur in compromised hosts. Thus, there is little doubt that innate immunity provided by preexisting host defense mechanisms plays a major role in protecting otherwise healthy individuals against infection by this organism. There also is growing evidence that the interactions of the carbohydrate-binding domains of the lung collectins with cell surface glycoconjugates expressed by *K. pneumoniae* contribute to this process of innate immunity.

The capacity of *K. pneumoniae* to thrive in the environment and in distinct body sites, such as the lung, is related in large part to its intrinsic capacity to regulate the expression of various virulence factors including capsular polysaccharides (Podschun *et al.,* 1993; Tarkkanen *et al.,* 1992). Significantly, the capsular phenotype of *Klebsiella* species can periodically switch from encapsulated to nonencapsulated and vice versa (Sahly and Podschun, 1997). It is interesting that SP-A interacts with some of the encapsulated phenotypes, whereas SP-D preferentially interacts with nonencapsulated organisms (Ofek *et al.,* 1997). The interactions of both SP-A (Kabha *et al.,* 1997) and SP-D (unpublished data) enhance the binding, uptake, and killing of these organisms by macrophages *in vitro*.

4.5.3. HYPOTHESIS

Because the unencapsulated organisms can adhere and colonize mucosal surfaces by expressing adhesins (Ofek and Doyle, 1994), we propose that early during the infectious process the nonencapsulated phenotype predominates in the upper respiratory tract. When this phenotype reaches the lower respiratory tract, it is likely to interact with SP-D. SP-D binding may facilitate the elimination of the invading bacteria through enhanced phagocyte-dependent killing or through enhanced mucociliary clearance of the agglutinated organisms. Although encapsulated phase variants are expected to evade these SP-D mediated defenses, some of these capsular serotypes are recognized by SP-A, permitting clearance via SP-A dependent mechanisms. These same serotypes also are recognized by the macrophage mannose receptor (Athamna *et al.*, 1998). Epidemiological data showing that among serotypes isolated with high frequency from patients with active infection there was a preponderance of bacterial strains with capsular polysaccharides that are not recognized by SP-A and mannose receptor support of this notion (Ofek *et al.*, 1995).

5. CONCLUSIONS

The surfactant-associated proteins, SP-A and SP-D, are members of a family of collagenous host defense lectins. There is increasing evidence that these pulmonary epithelial-derived proteins are important components of the innate immune response to microbial challenge. The lung collectins bind to glycoconjugates expressed by a wide variety of microorganisms *in vitro*. Such binding may cause microbial aggregation with resulting enhancement of mucociliary or leukocyte-mediated clearance. However, SP-A and SP-D also have the capacity to opsonize microorganisms with enhancement of phagocytosis and killing. Complementary or cooperative interactions between SP-A and SP-D and other lectins such as the macrophage mannose receptor could contribute to the efficiency of this defense system. Environmental or growth-phase-dependent modulation of glycoconjugates expressed on the surface of microorganisms could influence the mechanism or effectiveness of lectin-mediated clearance. Collectins may play particularly important roles in settings of inadequate or impaired specific immunity. Studies are needed to examine the possibility that acquired deficiencies of the lung collectins in certain hospitalized patients may render them susceptible to pathogens that otherwise are harmless to healthy individuals.

ACKNOWLEDGMENT

The personal studies cited in the review were supported by NIH grants HL44015 and HL29594.

REFERENCES

Aarson, G. L., 1996, Lectins as defence molecules in vertebrates and invertebrates, *Fish Shellfish Immunol.* **6**:277–289.

Aderibigbe, A. O., Thomas, R. F., Mercer, R. R., Auten, R. L., 1999, Brief exposure to 95% oxygen alters surfactant protein D and mRNA in adult rat alveolar and bronchiolar epithelium, *Am. J. Respir. Cell Mol Biol.* **20**:219–227.

Athamna, A., Ofek, I., Keisari, Y., Markowitz, S., Dutton, G. S., and Sharon, N., 1998, Lectinophagocytosis of encapsulated *Klebsiella* pneumoniae mediated by surface lectin of guinea pig alveolar macrophages and human-monocyte-derived macrophages, *Infect. Immun.* **59**:1673–1682.

Blau, H., Riklis, S., Kravtsov, V., and Kalina, M., 1994, Secretion of cytokines by rat alveolar epithelial cells: Possible regulatory role for SP-A, *Am. J. Physiol.* **266**:(Pt 1):L148–155.

Brown-Augsburger, P., Hartshorn, K., Chang, D., Rust, K., Fliszar, C., Welgus, H. G.,, and Crouch, E. C. 1996, Site-directed mutagenesis of Cys-15 and Cys-20 of pulmonary surfactant protein D. Expression of a trimeric protein with altered anti-viral properties, *J. Biol. Chem.* **271**:13724–13730.

Caton, A., Brownlee, G., Yewdell, J., and Gerhard, W., 1982, The antigenic structure of influenza virus A/PR/8/34 hemagglutinin (H1 subtype), *Cell* **31**:417–427.

Chroneos, Z., and Shepherd, V. L., 1995, Differential regulation of the mannose and SP-A receptors on macrophages, *Am. J. Physiol.* **269**:L721–726.

Chroneos, Z. C., Abdolrasulnia, R., Whitsett, J. A., Rice, W. R., and Shepherd, V. L., 1996, Purification of a cell-surface receptor for surfactant protein A, *J. Biol. Chem.* **271**:16375–16383.

Crawford, S. W., Mecham, R. P., and Sage, H., 1986, Structural characteristics and intermolecular organization of human pulmonary-surfactant-associated proteins, *Biochem. J.* **240**:107–114.

Crouch, E. C., 1998, Collectins and pulmonary host defense, *Am. J. Respir. Cell Mol. Biol.* **19**:177–201.

Crouch, E., Persson, A., Chang, D., and Parghi, D., 1991, Surfactant protein D. Increased accumulation in silica-induced pulmonary lipoproteinosis, *Am. J. Pathol.* **139**:765–776.

Crouch, E., Persson, A., Chang, D., and Heuser, J., 1994a, Molecular structure of pulmonary surfactant protein D (SP-D), *J. Biol. Chem.* **269**:17311–17319.

Crouch, E., Chang, D., Rust, K., Persson, A., and Heuser, J., 1994b, Recombinant pulmonary surfactant protein D. Post-translational modification and molecular assembly, *J. Biol. Chem.* **269**:15808–15813.

Crouch, E. C., Persson, A., Griffin, G. L., Chang, D., and Senior, R. M., 1995, Interactions of pulmonary surfactant protein D (SP-D) with human blood leukocytes, *Am. J. Respir. Cell Mol. Biol.* **12**:410–415.

Drickamer, K., and McCreary, V., 1987, Exon structure of a mannose-binding protein gene reflects its evolutionary relationship to the asialoglycoprotein receptor and nonfibrillar collagens, *J. Biol. Chem.* **262**:2582–2589.

Eggleton, P., Ghebrehiwet, B., Sastry, K. N., Coburn, J. P., Zaner, K. S., Reid, K. B., and Tauber, A. I., 1995, Identification of a gC1q-binding protein (gC1q-R) on the surface of human neutrophils. Subcellular localization and binding properties in comparison with the cC1q-R, *J. Clin. Invest.* **95**:1569–1578.

Gaynor, C. D., McCormack, F. X., Voelker, D. R., McGowan, S. E., and Schlesinger, L. S., 1995, Pulmonary surfactant protein A mediates enhanced phagocytosis of *Mycobacterium tuberculosis* by a direct interaction with human macrophages, *J. Immunol.* **155**:5343–5351.

Haas, C., Voss, T., and Engel, J., 1991, Assembly and disulfide rearrangement of recombinant surfactant protein A *in vitro. Eur. J. Biochem.* **197**:799–803.

Haller, E. M., Shelley, S. A., Montgomery, M. R., and Balis, J. U., 1992, Immunocytochemical localization of lysozyme and surfactant protein A in rat type II cells and extracellular surfactant forms, *J. Histochem. Cytochem.* **40**:1491–1500.

Hartshorn, K. L., Sastry, K., Brown, D., White, M. R., Okarma, T. B., Lee, Y. M., and Tauber, A. I., 1993a, Conglutinin acts as an opsonin for influenza A viruses, *J. Immunol.* **151**:6265–6273.

Hartshorn, K. L., Sastry, K., White, M. R., Anders, E. M., Super, M., Ezekowitz, R. A., and Tauber, A. I., 1993b, Human mannose-binding protein functions as an opsonin for influenza A viruses, *J. Clin. Invest.* **91**:1414–1420.

Hartshorn, K. L., Crouch, E. C., White, M. R., Eggleton, P., Tauber, A. I., Chang, D., and Sastry, K., 1994, Evidence for a protective role of pulmonary surfactant protein D (SP-D) against influenza A viruses, *J. Clin. Invest.* **94**:311–319.

Hartshorn, K., Chang, D., Rust, K., and Crouch, E. C., 1996a, Interactions of recombinant human pulmonary surfactant protein D and SP-D multimers with influenza A, *Am. J. Physiol. (Lung Cell. Mol. Physiol.)* **271**:L753–L762

Hartshorn, K. L., Reid, K. B., White, M. R., Jensenius, J. C., Morris, S. M., Tauber, A. I., and Crouch, E., 1996b, Neutrophil deactivation by influenza A viruses: Mechanisms of protection after viral opsonization with collectins and hemagglutination-inhibiting antibodies, *Blood* **87**:3450–3461.

Hartshorn, K. L., Crouch, E., White, M. R., Colamussi, M. I., Kakkanatt, A., Tauber, B., Shepherd, V., and Sastry, K., 1998, Pulmonary surfactant proteins A and D enhance neutrophil uptake of bacteria, *Am. J. Physiol.* **274**:L958–L969.

Hattori, A., Kuroki, Y., Katoh, T., Takahashi, H., Shen, H. Q., Suzuki, Y., and Akino, T., 1996a, Surfactant protein A accumulating in the alveoli of patients with pulmonary alveolar proteinosis—Oligomeric structure and interaction with lipids, *Am. J. Respir. Cell Mol. Biol.* **14**:608–619.

Hattori, A., Kuroki, Y., Sohma, H., Ogasawara, Y., and Akino, T., 1996b, Human surfactant protein A with two distinct oligomeric structures which exhibit different capacities to interact with alveolar type II cells, *Biochem. J.* **317**:939–944.

Hickling, T. P., Malhotra, R., and Sim, R. B. 1998, Human lung surfactant protein A exists in several different oligomeric states: Oligomer size distribution varies between patient groups, *Mol. Med.* **4**:266–275.

Holmskov, U., and Jensenius, J. C., 1993, Structure and function of collectins: Humoral C-type lectins with collagenous regions, *Behring Inst. Mitt.* **93**:224–235.

Holmskov, U., Malhotra, R., Sim, R. B., and Jensenius, J. C., 1994, Collectins: Collagenous C-type lectins of the innate immune defense system, *Immunol. Today* **15**:67–74.

Holmskov, U., Laursen, S. B., Malhotra, R., Wiedemann, H., Timpl, R., Stuart, G. R., Tornoe, I., Madsen, P. S., Reid, K. B., and Jensenius, J. C., 1995, Comparative study of the structural and functional properties of a bovine plasma C-type lectin, collectin-43, with other collectins, *Biochem. J.* **305**:889–896.

Holmskov, U., Lawson, P., Teisner, B., Tornoe, I., Willis, A. C., Morgan, C., Koch, C., and Reid, K. B. M., 1997, Isolation and characterization of a new member of the scavenger receptor superfamily, glycoprotein-340 (gp-340), as a lung surfactant protein-D binding molecule, *J. Biol. Chem.* **272**:13743–13749.

Honda, Y., Takahashi, H., Kuroki, Y., Akino, T., and Abe, S., 1996, Decreased contents of surfactant proteins A and D in BAL fluids of healthy smokers, *Chest* **109**:1006–1009.

Hoppe, H. J., and Reid, K. B., 1994a, Collectins—Soluble proteins containing collagenous regions and lectin domains—and their roles in innate immunity, *Protein Sci.* **3**:1143–1158.

Hoppe, H. J., and Reid, K. B., 1994b, Trimeric C-type lectin domains in host defence, *Structure* **2**:1129–1133.

Horowitz, S., Watkins, R. H., Auten, R. L., Jr., Mercier, C. E., and Cheng, E. R., 1991, Differential accumulation of surfactant protein A, B, and C mRNAs in two epithelial cell types of hyperoxic lung, *Am. J. Respir. Cell Mol. Biol.* **5**:511–515.

Ikegami, M., Korhagen, T. R., Bruno, M. D., Whitsett, J. A., and Jobe, A. H., 1997, Surfactant metabolism in surfactant protein A-deficient mice, *Am. J. Physiol. (Lung Cell. Mol. Physiol.)* **272**:L479–L485.

Kabha, K., Schmegner, J., Keisari, Y., Parolis, H., Schlepper-Schaefer, J., and Ofek, I., 1997, SP-A enhances phagocytosis of *Klebsiella* by interaction with capsular polysaccharides and alveolar macrophages, *Am. J. Physiol. (Lung Cell. Mol. Physiol.)* **272**:L344–L352.

Kalina, M., Blau, H., Riklis, S., and Kravtsov, V., 1995, Interaction of surfactant protein A with bacterial lipopolysaccharide may affect some biological functions, *Am. J. Physiol.* **268**:L144–151.

Kasper, M., Albrecht, S., Grossmann, H., Grosser, M., Schuh, D., and Muller, M., 1995, Monoclonal antibodies to surfactant protein D: Evaluation of immunoreactivity in normal rat lung and in a radiation-induced fibrosis model, *Exp. Lung Res.* **21**:577–588.

Kishore, U., Wang, J. Y., Hoppe, H. J., and Reid, K. B. M., 1996, The alpha-helical neck region of human surfactant protein D is essential for the binding of the carbohydrate recognition domains to lipopolysaccharides and phospholipids, *Biochem. J.* **318**:505–511.

Korfhagen, T. R., Bruno, M. D., Ross, G. F., Huelsman, K. M., Ikegami, M., Jobe, A. H., Wert, S. E., Stripp, B. R., Morris, R. E., Glasser, S. W., Bachurski, C. J., Iwamoto, H. S., and Whitsett, J. A., 1996, Altered surfactant function and structure in SP-A gene targeted mice, *Proc. Natl. Acad. Sci. U.S.A.* **93**:9594–9599.

Kuan, S. F., Rust, K., and Crouch, E., 1992, Interactions of surfactant protein D with bacterial lipopolysaccharides. Surfactant protein D is an *Escherichia coli*-binding protein in bronchoalveolar lavage, *J. Clin. Invest.* **90**:97–106.

Kuan, S. F., Persson, A., Parghi, D., and Crouch, E., 1994, Lectin-mediated interactions of surfactant protein D with alveolar macrophages, *Am. J. Respir. Cell Mol. Biol.* **10**:430–436.

Kuroki, Y., Gasa, S., Ogasawara, Y., Shiratori, M., Makita, A., and Akino, T., 1992, Binding specificity of lung surfactant protein SP-D for glucosylceramide, *Biochem. Biophys. Res. Commun.* **187**:963–969.

Kuroki, Y., Mason, R. J., and Voelker, D. R., 1988, Alveolar type II cells express a high-affinity receptor for pulmonary surfactant protein A, *Proc. Natl. Acad. Sci. USA* **85:**5566–5570.

LeVine, A. M., Bruno, M., Whitsett, J., and Korfhagen, T. R., 1997, Surfactant protein-A in pulmonary host defense against bacterial pathogens *in vivo, Am. J. Respir. Cell Mol. Biol.* **155:**A214.

Lim, B. L., Wang, J. Y., Holmskov, U., Hoppe, H. J., and Reid, K. B., 1994, Expression of the carbohydrate recognition domain of lung surfactant protein D and demonstration of its binding to lipopolysaccharides of gram-negative bacteria, *Biochem. Biophys. Res. Commun.* **202:**1674–1680.

Limper, A. H., O'Riordan, D. M., Vuk-Pavlovic, Z., and Crouch, E. C., 1994, Accumulation of surfactant protein D in the lung during *Pneumocystis carinii* pneumonia, *J. Euk. Microbiol.* **41:**S98.

Limper, A. H., Crouch, E. C., O'Riordan, D. M., Chang, D., Vuk-Pavlovic, Z., Standing, J. E., Kwon, K. Y., and Adlakha, A., 1995, Surfactant protein D modulates interaction of *Pneumocystis carinii* with alveolar macrophages, *J. Lab. Clin. Med.* **126:**416–422.

Lu, J., Willis, A. C., and Reid, K. B., 1992, Purification, characterization and cDNA cloning of human lung surfactant protein D., *Biochem. J.* **284:**795–802.

Lu, J., Wiedemann, H., Holmskov, U., Thiel, S., Timpl, R., and Reid, K. B., 1993, Structural similarity between lung surfactant protein D and conglutinin. Two distinct, C-type lectins containing collagen-like sequences, *Eur. J. Biochem.* **215:**793–799.

Madan, T., Eggleton, P., Kishore, U., Strong, P., Aggrawal, S. S., Sarma, P. U., and Reid, K. B. M., 1997, Binding of pulmonary surfactant proteins A and D to *Aspergillus fumigatus* conidia enhances phagocytosis and killing by human neutrophils and aleveolar macrophages, *Infect. Immun.* **65:**3171–3179.

Malhotra, R., and Sim, R. B., 1995, Collectins and viral infection, *Trends Microbiol.* **3:**240–244.

Malhotra, R., Thiel, S., Reid, K. B., and Sim, R. B., 1990, Human leukocyte C1q receptor binds other soluble proteins with collagen domains, *J. Exp. Med.* **172:**955–959.

Malhotra, R., Haurum, J., Thiel, S., and Sim, R. B., 1992, Interaction of C1q receptor with lung surfactant protein A, *Eur. J. Immunol.* **22:**1437–1445.

Malhotra, R., Haurum, J., Thiel, S., Jensenius, J. C., and Sim, R. B., 1993, Pollen grains bind to lung alveolar type II cells (A549) via lung surfactant protein A (SP-A), *Biosci. Rep.* **13:**79–90.

Malhotra, R., Haurum, J. S., Thiel, S., and Sim, R. B., 1994, Binding of human collectins (SP-A and MBP) to influenza virus, *Biochem. J.* **304:**455–461.

Manz-Keinke, H., Egenhofer, C., Plattner, H., and Schlepper-Schafer, J., 1991, Specific interaction of lung surfactant protein A (SP-A) with rat alveolar macrophages, *Exp. Cell Res.* **192:**597–603.

McGroarty, E. J., and Rivera, M., 1990, Growth-dependent alterations in production of serotype-specific and common antigen lipopolysaccharides in *Pseudomonas aeruginosa* PAO1. *Infect. Immun.* **58:**1030–1037.

McIntosh, J. C., Swyers, A. H., Fisher, J. H., and Wright, J. R., 1996, Surfactant proteins A and D increase in response to intratracheal lipopolysaccharide, *Am. J. Respir. Cell Mol. Biol.* **15:**509–519.

McNeely, T. B., and Coonrod, J. D., 1993, Comparison of the opsonic activity of human surfactant protein A for *Staphylococcus aureus* and *Streptococcus pneumoniae* with rabbit and human macrophages, *J. Infect. Dis.* **167**:91–97.

Miyamura, K., Leigh, L. E., Lu, J., Hopkin, J., Lopez Bernal, A., and Reid, K. B., 1994, Surfactant protein D binding to alveolar macrophages, *Biochem. J.* **300**:237–242.

Nogee, L. M., Wispe, J. R., Clark, J. C., and Whitsett, J. A., 1989, Increased synthesis and mRNA of surfactant protein A in oxygen-exposed rats, *Am. J. Respir. Cell Mol. Biol.* **1**:119–125.

Ofek, I., and Doyle, J. D., 1994, *Bacterial Adhesion to Cells and Tissues,* Chapman and Hall, New York.

Ofek, I., Goldhar, J., Keisari, Y., and Sharon, N., 1995, Nonopsonic phagocytosis of microorganisms, *Annu. Rev. Microbiol.* **49**:239–276.

Ofek, I., Kabha, K., Keisari, Y., Schlepper-Schaefer, J., Abraham, S. N., McGregor, D., Chang, D., and Crouch, E., 1997, Recognition of *Klebsiella pneumoniae* by pulmonary C-type lectins, *Nova Acta Leopold NF* **75**:43–54.

O'Riordan, D. M., Standing, J. E., Kwon, K.-Y., Crouch, E. C., and Limper, A. H., 1995, Surfactant protein D interacts with *Pneumocystis carinii* and mediates organism adherence to alveolar macrophages, *J. Clin. Invest.* **95**:2699–2710.

Persson, A., Chang, D., Rust, K., Moxley, M., Longmore, W., and Crouch, E., 1989, Purification and biochemical characterization of CP4 (SP-D), a collagenous surfactant-associated protein, *Biochemistry* **28**:6361–6367.

Persson, A., Chang, D., and Crouch, E., 1990, Surfactant protein D is a divalent cation-dependent carbohydrate-binding protein, *J. Biol. Chem.* **265**:5755–5760.

Pikaar, J. C., van Golde, L. M. G., van Strijp, J. A. G., and Van Iwaarden, J. F., 1995, Opsonic activities of surfactant proteins A and D in phagocytosis of gram-negative bacteria by alveolar macrophages, *J. Infect. Dis.* **172**:481–489.

Podschun, R., Sievers, D., Fischer, A., and Ullmann, U., 1993, Serotypes, hemagglutinins, siderophore synthesis, and serum resistance of *Klebsiella* isolates causing human urinary tract infections, *J. Infect. Dis.* **168**:1415–1421.

Polotsky, V. Y., Fischer, W., Ezekowitz, R. A., and Joiner, K. A., 1996, Interactions of human mannose-binding protein with lipoteichoic acids, *Infect. Immun.* **64**:380–383.

Reading, P. C., Morey, L. S., Crouch, E. C., and Anders, E. M., 1997, Collectin-mediated antiviral host defence of the lung: Evidence from influenza virus infection of mice, *J. Virol.* **71**:8204–8212.

Ross, G. F., Ohning, B. L., Tannenbaum, D., and Whitsett, J. A., 1987, Structural relationships of the major glycoproteins from human alveolar proteinosis surfactant, *Biochim. Biophys. Acta* **911**:294–305.

Rust, K., Grosso, L., Zhang, V., Chang, D., Persson, A., Longmore, W., Cai, G. Z., and Crouch, E., 1991, Human surfactant protein D: SP-D contains a C-type lectin carbohydrate recognition domain, *Arch. Biochem. Biophys.* **290**:116–126.

Sahly, H., and Podschun, R., 1997, Clinical, bacteriological, and serological aspects of *Klebsiella* infections and their spondylarthropathic sequelae. *Clin. Diag. Lab. Immunol.* **4**:393–399.

Sano, K., Fisher, J., Mason, R. J., Kuroki, Y., Schilling, J., Benson, B., and Voelker, D., 1987, Isolation and sequence of a cDNA clone for the rat pulmonary surfactant-associated protein (PSP-A), *Biochem. Biophys. Res. Commun.* **144**:367–374.

Schelenz, S., Malhotra, R., Sim, R. B., Holmskov, U., and Bancroft, G. J., 1995, Binding of host collectins to the pathogenic yeast *Cryptococcus neoformans:* Human surfactant protein D acts as an agglutinin for acapsular yeast cells, *Infect. Immun.* **63:**3360–3366.

Shimizu, H., Fisher, J. H., Papst, P., Benson, B., Lau, K., Mason, R. J., and Voelker, D. R., 1992, Primary structure of rat pulmonary surfactant protein D. cDNA and deduced amino acid sequence, *J. Biol. Chem.* **267:**1853–1857.

Tarkkanen, A. M., Allen, B. L., Williams, P. H., Kauppi, M., Haahtela, K., Siitonen, Orskov, I., Orskov, F., Clegg, S., and Korhonen, T. K., 1992, Fimbriation, capsulation, and iron-scavenging systems of *Klebsiella* strains associated with human urinary tract infection, *Infect. Immun.* **60:**1187–1192.

Van Iwaarden, J. F., Pikaar, J. C., Storm, J., Brouwer, E., Verhoef, J., Oosting, R. S., Van Golde, L. M., and van Strijp, J. A., 1994, Binding of surfactant protein A to the lipid A moiety of bacterial lipopolysaccharides, *Biochem. J.* **303:**407–411.

Viviano, C. J., Bakewell, W. E., Dixon, D., Dethloff, L. A., and Hook, G. E., 1995, Altered regulation of surfactant phospholipid and protein A during acute pulmonary inflammation, *Biochim. Biophys. Acta* **1259:**235–244.

Voorhout, W. F., Veenendaal, T., Haagsman, H. P., Verkleij, A. J., van Golde, G. L. M., and Geuze, H. J., 1991, Surfactant protein A is localized at the corners of the pulmonary tubular myelin lattice, *J. Histochem. Cytochem.* **39:**1331–1336.

Voss, T., Melchers, K., Scheirle, G., and Schafer, K. P., 1991, Structural comparison of recombinant pulmonary surfactant protein SP-A derived from two human coding sequences: Implications for the chain composition of natural human SP-A, *Am. J. Respir. Cell Mol. Biol.* **4:**88–94.

Voss, T., Schafer, K. P., Nielsen, P. F., Schafer, A., Maier, C., Hannappel, E., Maassen, J., Landis, B., Klemm, K., and Przybylski, M., 1992, Primary structure differences of human surfactant-associated proteins isolated from normal and proteinosis lung, *Biochim. Biophys. Acta* **1138:**261–267.

Vuk-Pavlovic, Z., Diaz-Montes, T., Standing, J. E., and Limper, A. H., 1998, Surfactant protein-D binds to cell wall ß-glucans, *Am. J. Respir. Cell Mol. Biol.* **157:**A236.

Wang, J. Y., Kishore, U., Lim, B. L., Strong, P., and Reid, K. B. M., 1996, Interaction of human lung surfactant proteins A and D with mite (*Dermatophagoides pteronyssinus*) allergens, *Clin. Exp. Immunol.* **106:**367–373.

Weiss, J., Hutzler, M., and Kao, L., 1986, Environmental modulation of lipopolysaccharide chain length alters the sensitivity of *Escherichia coli* to the neutrophil bactericidal/permeability-increasing protein, *Infect. Immunol.* **51:**594–599.

Williams, M. D., Wright, J. R., March, K. L., and Martin, W. J., 1996, Human surfactant protein A enhances attachment of *Pneumocystis carinii* to rat alveolar macrophages, *Am. J. Respir. Cell Mol. Biol.* **14:**232–238.

Wintergerst, E., Manz-Keinke, H., Plattner, H., and Schlepper-Schafer, J., 1989, The interaction of a lung surfactant protein (SP-A) with macrophages is mannose dependent, *Eur. J. Cell Biol.* **50:**291–298.

Zimmerman, P. E., Voelker, D. R., McCormack, F. X., Paulsrud, J. R., and Martin, W. J., 1992, 120-kD surface glycoprotein of *Pneumocystis carinii* is a ligand for surfactant protein A, *J. Clin. Invest.* **89:**143–149.

Index

539